Minerals and Gemstones

Minerals and Gemstones

300 of the Earth's natural treasures

Dr. David C. Cook
& Dr. Wendy L. Kirk

THUNDER BAY
P · R · E · S · S

San Diego, California

Thunder Bay Press
An imprint of the Advantage Publishers Group
5880 Oberlin Drive, San Diego, CA 92121-4794
www.thunderbaybooks.com

All notations of errors or omissions should be addressed to Thunder Bay Press, Editorial Department, at the above address. All other correspondence (author inquiries, permissions) concerning the content of this book should be addressed to:

Amber Books Ltd
Bradley's Close
74–77 White Lion Street
London N1 9PF
www.amberbooks.co.uk

ISBN-13: 978-1-59223-735-7
ISBN-10: 1-59223-735-5

Library of Congress Cataloging-in-Publication Data available upon request.

Printed in Singapore
1 2 3 4 5 11 10 09 08 07

Project Editor: Sarah Uttridge
Design: Joe Conneally

PICTURE CREDITS
All pictures © DeAgostini Picture Library

CONTENTS

Introduction	6
Native Elements	14
Sulphides	27
Halogenides	70
Oxides and Hydroxides	83
Nitrates, Carbonates and Borates	126
Sulphates, Chromates, Molybdates and Tungstates	150
Phosphates, Arsenates and Vanadates	173
Silicates	201
Other Minerals	310
Glossary	314
Index	316

Introduction

Minerals are mostly naturally occurring, inorganic, crystalline solids found in varying quantities in the Earth and beyond. Gems are usually minerals that are prized for beauty and strength; they have an eye catching colour, lustre or clarity coupled with a durability to make them items of lasting value. These definitions can be somewhat blurred at the edges; this has been reflected in our choice of some of the entries. To help identify minerals we list their properties after each entry.

COLOUR

The most immediately noticeable property, colour is only occasionally useful if the mineral has a distinctive, inherent colour such as yellow sulphur or blue azurite. For minerals whose colour varies with impurity content, it is much less useful. For example pure quartz is colourless, but may be of almost any hue through to black.

Lazurite has a distinct blue colour and sky-blue streak which is helpful in identification.

STREAK

Streak is the colour of a mineral in a finely powdered form. This is usually demonstrated by scratching across unglazed porcelain, crushing a sample or scratching the surface with a knife. The streak tends to remain the same for minerals which appear to be coloured differently in larger masses. It is therefore a more consistent indicator of a mineral. Streak is not useful for most silicate minerals as they are usually white and often too hard to powder easily.

Quartz has a typical crystal form of six-sided prismatic crystals with pointed terminations (as seen in the picture here).

LUSTRE

Lustre describes the nature of light from the surface of a mineral. A *metallic* lustre is shown by opaque minerals such as metals and many sulphides; if imperfect, it is called *submetallic*. A *non-metallic* lustre is a catch-all term for all the rest and is shown by transparent and translucent minerals. It includes:

Adamantine highly reflective like diamond
Vitreous glassy, as in quartz
Resinous like resin, as in amber and opal
Pearly like a pearl, due to alignment of platy minerals under the surface, as in talc and mother-of-pearl
Silky like silk, due to an underlying fibrous structure, as in satin spar, a variety of gypsum
Greasy produced by an irregular surface
Earthy or **dull** matt surface shown by minerals having no lustre

A diamond has few weaknesses and many strengths: it is the hardest mineral known and is the highest on Moh's scale of hardness.

OPACITY OR TRANSPARENCY

It is usual to indicate if a mineral is transparent, translucent or opaque, although this is not a diagnostic property. A mineral may be inherently opaque or be so because it contains many small fragments; translucency (partial transparency) may be a function of the specimen thickness or the number of internal flaws.

HARDNESS

The hardness of a mineral has been defined as its resistance to abrasion or scratching. A practical system for measurement was devised by the Austrian mineralogist Friedrich Mohs in 1812. A set of ten common minerals was chosen of different hardnesses such that each one will scratch the surface of all softer minerals. These were then given the number 1 to 10 in increasing order of hardness. Most literature gives hardness to the nearest half unit, for example as 3.5, as a working approximation.

Mohs' scale of hardness

Tests may be roughly carried out using a fingernail (hardness 2-2.5), a copper coin (hardness 3) or a steel knife (hardness 5.5-6.5); minerals of hardness over 6 will scratch glass.

Number	Mineral	Number	Mineral
1	Talc	6	Orthoclase
2	Gypsum	7	Quartz
3	Calcite	8	Topaz
4	Fluorite	9	Corundum
5	Apatite	10	Diamond

SPECIFIC GRAVITY

This is the weight of a mineral compared to an equal volume of water and can be taken as equivalent to density. Thus the density of water is taken as 1 g/cm^3 (i.e. one cubic centimetre weighs one gram, or one litre weighs one kilogram), and its specific gravity is *defined* as 1 (note: it has no units).

In contrast to the diamond, Talc is the softest mineral, it is the lowest on Moh's scale of hardness.

HABIT AND FORM

The form of the mineral is given first in the properties box and describes the shape adopted by crystals. Some shapes may be well-defined closed geometric forms such as a cube, octahedron or dodecahedron, or describe an open form such as a prism. Some terms are used to describe the appearance of aggregates of crystals. Common terms used to describe single crystals or aggregates include:

Acicular needle-shaped
Bladed flattened like a knife blade
Botryoidal like a bunch of grapes
Dendritic branching like a tree or moss
Fibrous fine strands
Lamellar forming distinctly flat sheets
Mammilated round, mutually interfering masses
Massive crystalline aggregates with no distinct form
Radiating radial arrangement of needles or fibres
Reniform kidney-shaped
Tabular showing broad, flat surfaces

An aggregate of natrolite crystals display an attractive radiating, acicular habit.

An aggregate of small cubic crystals of sal ammoniac, which is not often in a visually appealing habit.

CRYSTAL SYSTEM

Crystals are made up of atoms or molecules arranged in a regular three-dimensional repeated pattern. Each unit which can be seen to repeat in order to build this structure is called a *unit cell*. Crystallography is governed by geometric possibilities (for example a cube can be repeated but a sphere or dodecahedron cannot), rather like the equivalent two-dimensional property of tessellation. Only seven possible patterns are recognised in unit cells. These are referred to as the *crystal system* adopted by that particular mineral. Each crystal system constrains the shape that crystals can adopt. Some crystal shapes can be characteristic of a crystal system such as the cube and the octahedron in the cubic system, but many require specialised knowledge and measurement to be diagnostic. Crystal systems are summarised below.

Crystal System	Unit cell shape
Cubic	all three sides equal in length; all angles 90°
Tetragonal	two sides equal in length; all angles 90°
Orthorhombic	all sides different lengths; all angles 90°
Monoclinic	two angles 90°
Triclinic	no angles at 90°
Hexagonal	prism of regular hexagonal cross-section
Trigonal	prism of regular triangular cross-section

Pseudohexagonal muscovite crystals apparently made up of stacked thin sheets – illustrating the mineral's perfect cleavage in one direction.

CLASSIFICATION

The minerals in this book are divided into groups following a conventional system based on their chemical composition. Classification is firstly on minerals in the same chemical grouping or those having similar properties. The silicates, being the largest group are further subdivided according to structure. All silicates are based on the 'silicate tetrahedron' – a silicon atom bonded to four oxygen atoms which are arranged as if at the corners of a tetrahedron. They are then classified according to how these tetrahedra are joined and arranged.

Classification

1	**Native elements** comprise single elements which occur uncombined in nature.
2	**Sulphides** contain the S^{2-} group and are classed with arsenides, antimonides and tellurides.
3	**Oxides** contain the O^{2-} group.
4	**Carbonates, nitrates and borates**
5	**Sulphates**
6	**Phosphates**
7	**Nesosilicates** contain isolated SiO^{4-} tetrahedra.
8	**Sorosilicates** have two tetrahedra joined as $Si_2O_7^{6-}$ groups.
9	**Cyclosilicates** contain three, four or six tetrahedra joined as a ring.
10	**Inosilicates** have the tetrahedra joined into indefinite chains, usually as single or double chains.
11	**Phyllosilcates** comprise indefinite two-dimensional arrays of tetradedra joined at three corners in hexagonal arrangements.
12	**Tectosilicates** are joined at all four corners into indefinite three-dimensional frameworks containing voids or 'cages'.

Transparent crystals of hyalophane showing some brown iron oxide staining.

Copper

Copper has a characteristic reddish colour darkened by a coating of black copper oxide. Further weathering produces a covering of green copper carbonate. Native copper is quite rare, but the metal is easily obtained from ores. Copper was one of the earliest metals known, and bronze, a hard alloy of copper and tin, has been used since *c.* 3000 BC. Native copper is found in altered copper deposits, cracks in basaltic lavas, and cemented sandstones and conglomerates. The southern shore of Lake Superior (USA) is the best location to find copper, allegedly the source of a piece weighing 381 tonnes (420 tons); others localities include Mansfeld (Germany), Bisbee (Arizona), Tsumeb (Namibia) and Burra Burra (Australia).

Cu

Colour:	light rose-red on fresh surface
Lustre; opacity:	metallic; opaque, translucent green when very thin
Streak:	copper-red
Hardness:	2.5–3
Specific gravity:	8.93
Cleavage; fracture:	absent; hackly, conchoidal
Habit:	rarely hexahedral, tetrahedral, dodecahedral crystals; wiry, arborescent, massive
Crystal system:	cubic

Silver

Silver has been prized, like gold, as one of the 'noble' metals. Jewellery has been made of silver for millennia (c. 4000 BC, the ancient Egyptians were using silver beads). Sterling silver is the term for metal containing at least 92.5 per cent silver. Its high reflectivity makes silver plating an excellent coating for mirrors, utensils and ornaments. Silver's exceptionally high electrical conductivity is utilized in high-quality electronics. Silver occurs in the oxidized zone of hydrothermal sulphide veins associated with other silver-bearing minerals. Good localities to find native silver include Kongsberg (Norway), Freiberg (Germany), Jáchymov (Czech Republic), the Comstock Lode (Nevada, USA) and Cobalt (Canada).

Ag

Colour:	silver-white
Lustre; opacity:	metallic; opaque, highly reflective (95 per cent)
Streak:	silver–white
Hardness:	2.5–3
Specific gravity:	10.5
Cleavage; fracture:	none
Habit:	crystals very rare; wiry or scaly
Crystal system:	cubic

Gold

G old was the first metal known to humans, and has long been highly valued. It is chemically very unreactive and one of the 'noble' metals, used in coinage and jewellery for thousands of years. Gold is graded according to its purity, from pure 24 carat to 9 carat, which contains 37.5 per cent gold. Gold can be made harder and paler by alloying with silver, platinum, zinc or nickel. The largest nuggets include one of 153kg (337lb), found in Chile, and one of 93kg (205lb), found in Hill End (Australia). Gold is found in igneous rocks, often associated with quartz veins, and as placer deposits. The main mining areas are in Wittwatersrand (South Africa), California and Alaska (USA), Australia, South America and Siberia.

Au

Colour:	characteristic yellow
Lustre; opacity:	metallic; opaque
Streak:	yellow
Hardness:	2.5–3
Specific gravity:	19.3
Cleavage; fracture:	none; hackly, malleable
Habit:	crystals octahedral, dodecahedral, hexahedral; grains, nuggets, compact, dendritic
Crystal system:	cubic

Mercury

Mercury, also known as quicksilver, is the only metal that is liquid at room temperature, having a freezing point of –39°C (–38°F). Such mobility is the reason it was named after the mythical messenger of the gods. Mercury occurs as small droplets in cinnabar deposits and in some volcanic rocks. It has a high coefficient of thermal expansion, which makes it useful in thermometers. Mercury forms amalgams with other metals, and these are used in gold extraction, tooth fillings and chemical processes. Locations where mercury is found include Almaden (Spain), Monte Amiata (Italy), Idrija (Slovenia), Moschellandsberg (Germany) and Juan Cavelica (Peru). Almaden is the oldest and largest mercury mine in the world.

Hg

Colour:	tin-white
Lustre; opacity:	metallic; opaque
Streak:	not applicable at room temperature
Hardness:	liquid at room temperature
Specific gravity:	13.6
Cleavage; fracture:	not applicable at room temperature
Habit:	droplets; rhombohedra (at -39°C/-38°F)
Crystal system:	trigonal (at -39°C/-38°F)

Nickel-Iron

Nickel-iron is a term for native iron, as there is invariably some nickel content. Terrestrial nickel-iron is formed in basalts in contact with carbonate rocks. Iron of low nickel content (2–3 per cent), in masses up to 23 tonnes (25 tons), is mined at Ovifak and Disko Island (Greenland). Nickel-iron in the form of meteorites is rare, but of great interest. Characteristic of these meteorites is the 'Widmanstätten pattern' of intergrowths of two different phases, brought out by chemical etching: a dark, low-nickel phase (kamacite) and a lighter one of high nickel content (taenite). The most famous meteorite, with pieces found at Canyon Diablo (Arizona, USA), fell c. 20,000–40,000 years ago, creating a crater 1.6km (1 mile) wide.

Ni–Fe

Colour:	grey to black
Lustre; opacity:	metallic; opaque
Streak:	grey
Hardness:	4–5
Specific gravity:	7.0–7.8
Cleavage; fracture:	perfect in rare iron crystals; hackly
Habit:	never distinct crystals; massive or disseminated grains
Crystal system:	isometric

Platinum

Platinum is a rare and valuable metal, used in laboratory equipment, electrical couplings and jewellery. It is invariably mixed with impurities such as iron, iridium, palladium, rhodium, nickel and/or osmium. Platinum and the related metal palladium are used in chemistry as powerful and versatile catalysts. It was discovered by the Spanish during their conquest of South America in the sixteenth century. Named *platina del Pinto* (Pinto silver) after the Rio Pinto in Colombia, it is also found at Nizhni Tagil and Norilsk (Russia), Ontario (Canada) and Bushveld (South Africa). It occurs in igneous rocks, associated with ilmenite, magnetite and chromite, and as a placer deposit. Like gold, it does not dissolve in any acids except aqua regia.

Pt

Colour:	steel-grey, silvery–white
Lustre; opacity:	metallic; opaque
Streak:	steel-grey, silvery–white
Hardness:	4–4.5
Specific gravity:	14–19 (21.5 when pure)
Cleavage; fracture:	absent; hackly
Habit:	rarely crystals; grains, nuggets, irregular lumps
Crystal system:	cubic
Other:	ductile, malleable

Arsenic

Arsenic is rare as a native element and obtained commercially as a by-product from the smelting of sulphide ores. It is well known as a poison, a property put to positive use in pesticides, preservatives and pharmaceuticals for treating parasitic illnesses. Arsenic has been added to copper to strengthen it since *c.* 2000 BC. The name is derived from the Greek *arsenikos*, for 'masculine' or 'brave'. Native arsenic is found in hydrothermal veins associated with silver, cobalt and nickel ores and in some igneous and metamorphic rocks. It is found in Erzgebirge and the Harz Mountains (Germany), Gikos (Russia), Sterling Hill (New Jersey, USA), Jáchymov and Pribram (Czech Republic) and St Marie-aux-Mines (France).

As

Colour:	light grey tarnishing quickly to dark grey
Lustre; opacity:	metallic; opaque
Streak:	light grey
Hardness:	3.5
Specific gravity:	5.7
Cleavage; fracture:	perfect; uneven, brittle
Habit:	crystals rare, pseudo-cubic; granular, massive, concentrically layered nodules or stalactites
Crystal system:	trigonal

Antimony

Antimony is a metalloid related to arsenic, with some metallic and non-metallic characters. Of its three various forms, the metallic one is more stable and is bright, silvery and hard. It is used in alloys to increase hardness and lower the melting point. Antimony compounds are toxic and have been used as antiparasitic agents in medicine. The word *antimony* is its old Greek name. Native antimony is formed by the alteration of sulphides, such as stibnite, in hydrothermal veins often accompanying silver and arsenic. It is found at St Andreasberg (Germany), Sala (Sweden), Coimbra (Portugal), Pribram (Czech Republic), New Brunswick (Canada), Kern County (California, USA), Sardinia (Italy) and Sarawak (Borneo).

Sb

Colour:	light grey
Lustre; opacity:	metallic; opaque
Streak:	greys
Hardness:	3–3.5
Specific gravity:	6.6–6.7
Cleavage; fracture:	perfect; uneven, very brittle
Habit:	rare crystals as rhombohedra or coarse plates; usually massive and reniform
Crystal system:	trigonal

Bismuth

Native bismuth is rare and is found in hydrothermal veins associated with gold, silver, cobalt, tin, nickel and lead. Bismuth is related to arsenic and antimony. It has a low melting point (271°C/520°F) and is used in low-melting alloys for electrical fuses and fire protection devices. Native bismuth is an ore, although the metal is mostly obtained as a by-product from the smelting of bismuth-rich ores. Many specimens of bismuth on sale are crystallized masses, artificially produced but nevertheless in impressive geometrical squared patterns with an attractive iridescence. Bismuth is found at Kongsberg (Norway), Erzgebirge (Germany), Cornwall (England), Ontario (Canada) and Oruro and Tasna (Bolivia).

Bi

Colour:	lead-grey, pink tarnish
Lustre; opacity:	metallic; opaque
Streak:	silver-white, shiny
Hardness:	2–2.5
Specific gravity:	9.7–9.8
Cleavage; fracture:	perfect; uneven, sectile
Habit:	rare crystals; massive, granular, branching forms
Crystal system:	trigonal

Graphite

Graphite is a form of carbon made of stacked sheets of hexagonally arranged atoms that can quite easily slide past each other, whereas diamond (another carbon form) has a rigid three-dimensional lattice structure. Buckminsterfullerene (C_{60}), a third form, is a recently synthesized molecule found in nature in minute traces. Graphite is used as a lubricant, in crucibles, as a moderator in nuclear reactors and in pencils (as pencil 'lead'). It has a greasy feel and, unusually for a non-metal, conducts electricity. Graphite is found in metamorphosed sediments, basic igneous rocks, pegmatites and quartz veins. Graphite deposits are in Sri Lanka, Madagascar, Russia, South Korea, Mexico, the Czech Republic and Italy.

C

Colour:	black
Lustre; opacity:	dull metallic; opaque
Streak:	black, shiny
Hardness:	1–2
Specific gravity:	2.1–2.2
Cleavage; fracture:	perfect, producing thin flexible sheets
Habit:	rare six-sided flat crystals; scales, foliated or earthy masses, compact lumps
Crystal system:	hexagonal

Diamond

Diamond is the most famous gemstone and the hardest mineral on Earth. Low-quality diamonds are used in industry for cutting and drilling equipment. Bort is a variety of diamond that has a rounded, fibrous, radiate structure, and carbonado is black and microcrystalline. A diamond is formed at very high temperatures and pressures deep within the Earth's mantle (*c.* 80 km/50 miles), and is then brought to the surface through kimberlite pipes. Diamonds are found in ultramafic rocks, especially kimberlite breccias, and as placer deposits. The main mining areas include the Argyle pipe (Western Australia), Kimberley (South Africa), Golconda (India), Diamontina (Minas Gerais, Brazil) and Yakutia (Russia).

C

Colour:	colourless, yellowish to yellow, brown, black, blue, green or red, pink, champagne-tan, cognac-brown
Lustre; opacity:	adamantine, greasy; transparent to opaque
Streak:	none, too hard to abrade
Hardness:	10
Specific gravity:	3.52
Cleavage; fracture:	perfect; conchoidal; uneven
Habit:	octahedral, dodecahedral, cubic crystals
Crystal system:	cubic

Diamond: Gem Varieties

Diamond's unequalled hardness, brilliant lustre and considerable 'fire' make it the most highly prized of gems. Jewellers usually favour the 'brilliant' cut, which gives maximum internal reflection of light and thus minimum loss through the back of the stone, enhancing the natural fire. Diamonds are measured in carats, derived from the weight of a carob seed pod, which is now a standard 0.2g. The Cullinan diamond, originally from South Africa, was of 3106 carats and cut into 104 gemstones, the largest of which was of 531 carats. The Koh-i-Noor diamond (109 carats) is set in the crown of British Queens. The largest deep blue diamond, and some would say the most beautiful, is the Hope diamond (44.5 carats).

C

Colour:	colourless, yellowish to yellow, brown, black, blue, green or red, pink, champagne-tan, cognac-brown
Lustre; opacity:	adamantine, greasy; transparent to opaque
Streak:	none, too hard to abrade
Hardness:	10
Specific gravity:	3.52
Cleavage; fracture:	perfect; conchoidal; uneven
Habit:	octahedral, dodecahedral, cubic crystals
Crystal system:	cubic

Sulphur

Sulphur has a characteristic yellow colour, is very brittle and will often crumble if roughly handled. Also called brimstone, it usually forms in volcanoes as a sublimate around vents and fumaroles. Sulphur can also be found precipitated in hot springs, in shales associated with gypsum and bitumen, and in the cap rock of salt domes. Sulphur is used in gunpowder, matches, sulphuric acid production and is extremely important as the vulcanizing agent for rubber. It has a low melting point (113°C/235°F) and burns to give sulphur dioxide. Sulphur is mined in Louisiana and Texas (USA), Sicily, Japan, Indonesia and the South American Andes. The best examples of crystalline sulphur come from Sicily and the Romagna (Italy).

S

Colour:	yellow, brownish or greenish yellow, orange, white
Lustre; opacity:	resinous, greasy; transparent to translucent
Streak:	none
Hardness:	1.5–2.5
Specific gravity :	2.05–2.08
Cleavage; fracture:	imperfect, fair; irregular, uneven, conchoidal
Habit:	massive, encrusting, powdery and stalactitic
Crystal system:	orthorhombic

Chalcocite

Chalcocite occurs in two distinct forms, forming orthorhombic crystals below 103°C (217°F) and hexagonal crystals above this. Chalcocite is also called chalcosine or copper glance. It is found in hydrothermal veins and in the reduced zones of copper deposits. It is an important copper ore, often produced by the alteration of chalcopyrite, with which it is often associated, and which, in turn, may be altered to malachite, azurite and covellite. Good examples of pseudohexagonal crystals come from the Transvaal (South Africa), Bristol (Connecticut, USA) and Redruth (Cornwall, England). Large deposits are mined in Butte (Montana, USA), Bisbee (Arizona, USA), Chuquicamata (Chile), Tsumeb (Namibia), Peru and Russia.

Cu_2S

Colour:	black or lead grey, often with greenish or bluish tarnish
Lustre; opacity:	metallic; opaque
Streak:	greyish black, sometimes shiny
Hardness:	2.5–3
Specific gravity:	5.5–5.8
Cleavage; fracture:	indistinct; conchoidal, uneven
Habit:	rarely as tabular, pseudohexagonal, striated crystals; mostly dull grey, granular aggregates
Crystal system:	orthorhombic, hexagonal above 103°C (217°F)

Bornite

Bornite, named after the Austrian mineralogist Ignatius von Born (1742–1791), is known as peacock ore because of the beautiful iridescent tarnish shown by many specimens. Too much weathering, however, produces an unattractive black tarnish. Bornite is a dense mineral that settles in magmas and so is found concentrated in igneous rocks, especially mafic rocks. It also occurs in pegmatites, hydrothermal veins and as a secondary mineral in copper deposits. A well-known deposit is found in the copper shales of Mansfeld (Germany). Good crystals come from Cornwall (England), Tsumeb (Namibia) and Butte (Montana, USA). Large quantities are mined in the USA, Mexico, Peru, Chile, Australia and Zambia.

Cu_5FeS_4

Colour:	copper-red or bronze when fresh, tarnishes to an iridescent blue, red and/or purplish surface film
Lustre; opacity:	metallic; opaque _
Streak:	grey-black
Hardness:	3
Specific gravity:	5.1
Cleavage; fracture:	poor/indistinct; conchoidal, indistinct
Habit:	crystalline as cubes, dodecahedra or octahedra; commonly as compact granular masses
Crystal system:	orthorhombic

Acanthite-Argentite

Acanthite occurs in the monoclinic system below 179°C (354°F); above this, cubic argentite is the stable form. As it is usually deposited at high temperatures, acanthite often occurs as cubic pseudomorphs after argentite. Acanthite's metallic lustre quickly blackens and is best seen on a freshly exposed surface. It is named from the Greek for 'arrow'; argentite is named from the Latin *argentum*, meaning 'silver'. It is found in hydrothermal veins with other silver minerals or disseminated in galena deposits. It can also occur in cemented parts of lead and zinc deposits. Acanthite is the main ore for silver and is mined extensively in Mexico, Peru and Honduras. Good crystals come from Kongsberg (Norway) and Freiberg (Germany).

Ag_2S

Colour:	shiny lead grey, black on surface
Lustre; opacity:	metallic; opaque
Streak:	black
Hardness:	2–2.5
Specific gravity:	7.2–7.4
Cleavage; fracture:	poor; conchoidal, uneven
Habit:	distorted, pseudo-cubic crystals in groups; dendritic aggregates, masses and encrustations
Crystal system:	monoclinic (acanthite) – cubic (argentite)

Sphalerite

Sphalerite or zinc blende is an important zinc ore, also providing cadmium, gallium and indium as by-products. Zinc is mostly used as sheets for galvanizing iron, and alloyed with copper to make brass. Sphalerite can be mistaken for galena, hence its name, from the Greek for 'treacherous'. The colour of sphalerite tends to darken as the iron content increases. A content of about 10 per cent imparts a black colour; any higher, and the ore may be called marmatite. Sphalerite forms in pegmatites and hydrothermal veins accompanying galena, acanthite, barite and chalcopyrite. It is found at Alston Moor (England), the Tri-State area (USA), Broken Hill (NSW, Australia), Kapnik (Hungary), Santander (Spain) and Sullivan (Canada).

ZnS

Colour:	usually brown or black, also yellow or reddish, rarely colourless
Lustre; opacity:	resinous, greasy; transparent to translucent to opaque
Streak:	brownish white
Hardness:	3.5–4
Specific gravity:	3.9–4.2
Cleavage; fracture:	perfect; conchoidal, brittle
Habit:	dodecahedral and octahedral crystals common; massive, compact, botryoidal or fibrous
Crystal system:	isometric

Chalcopyrite

Chalcopyrite, or copper pyrites, is the most important copper ore. By-products from copper extraction include silver and gold. It can be distinguished from pyrite by the fact that it does not produce sparks when hit by a hammer and also because it crumbles when cut with a knife due to poor cleavage. It can be distinguished from pyrrhotite because it is non-magnetic, and from gold because it is brittle. Chalcopyrite occurs mainly in hydrothermal veins associated with cassiterite, galena, pyrite, quartz, calcite and/or sphalerite, and in metamorphosed volcanic rocks. It is found in the Copper Belt of Zambia, Rio Tinto (Spain), Katanga (Congo), Cyprus, the Urals (Russia), and Montana, Arizona and Utah (USA).

$CuFeS_2$

Colour:	brass yellow, honey yellow
Lustre; opacity:	metallic; opaque
Streak:	greenish black
Hardness:	3.5
Specific gravity:	4.1–4.3
Cleavage; fracture:	indistinct; conchoidal, uneven
Habit:	pseudo-tetrahedral crystals, uncommon; usually massive and compact, sometimes reniform or mammilated
Crystal system:	tetragonal

Tetrahedrite-Tennanite

Tetrahedrite and tennanite form a continuous series. The replacement of sulphur for tellurium in goldfieldite and of silver for iron in freibergite also occurs. It can act as a geothermometer because the silver content rises as the temperature of formation falls. Most mineral specimens are antimony-rich tetrahedrite. The group have been called fahlerz minerals, and are copper ores that have also been used for silver, mercury and antimony recovery. They occur in hydrothermal veins associated with copper, lead, zinc and silver minerals. Tetrahedrite-tennanite is found in Botés and Kapnik (Romania), Boliden (Sweden), Pribram (Czech Republic), Tsumeb (Namibia) and Butte (Montana, USA).

$(Cu,Fe)_{12}Sb_4S_{13} - (Cu,Fe)_{12}As_4S_{13}$

Colour:	steel grey to brown, occasionally twinned
Lustre; opacity:	metallic; opaque
Streak:	dark grey
Hardness:	3–4.5
Specific gravity:	4.6–5.2
Cleavage; fracture:	none; conchoidal, uneven
Habit:	crystals often modified tetrahedral; massive, granular, compact
Crystal system:	isometric
Other:	melt easily on heating; soluble in nitric acid

Stannite

Stannite, named from the Latin *stannum* for tin, is also known as tin pyrites, bolivianite or bell metal ore. It has been worked as a tin ore in Cornwall and forms part of silver deposits in Bolivia. It is a fairly rare mineral found in hydrothermal veins and pegmatites associated with cassiterite, wolframite, pyrite and arsenopyrite. It may contain other metals such as silver, cadmium and indium, and is called zincian stannite when zinc-rich. Stannite is found at Etna Mine (South Dakota, USA), Seward Peninsula (Alaska, USA), Llallagua (Bolivia), Cinovec (Czech Republic), Zeehan (Tasmania, Australia) and Wheal Rock (Cornwall, England), the last being the type locality for the mineral.

Cu_2FeSnS_4

Colour:	greyish black, steel grey; often has an iridescent olive green or blue surface
Lustre; opacity:	metallic; opaque
Streak:	black
Hardness:	3.5–4
Specific gravity:	4.3–4.5
Cleavage; fracture:	imperfect; conchoidal, uneven
Habit:	crystals rare, tetrahedral or more often pseudooctahedral due to twinning; massive, granular, disseminated
Crystal system:	tetragonal

Wurtzite

Wurtzite is dimorphic with sphalerite, the more common form of zinc sulphide. The zinc can be replaced partially by iron, giving a darker colour. A rarer trigonal polymorph exists, called matraite. Crystals are usually quite small, but can be delightful shapes of somewhat elongated six-sided pyramids on a flat hexagonal base. Wurtzite is found in hydrothermal veins associated with sphalerite, pyrite, chalcopyrite, barite and marcasite. It is named after the French chemist Charles Wurtz (1817–1884). Locations include Thomaston Dam (Connecticut, USA), Butte (Montana, USA), Frisco (Utah, USA), Cornwall (England), Pribram (Czech Republic), Baia Sprie (Romania) and Oruro and Potosi (Bolivia).

ZnS

Colour:	light to dark brown
Lustre; opacity:	adamantine, greasy; translucent to opaque
Streak:	light brown
Hardness:	3.5–4
Specific gravity:	3.98–4.08
Cleavage; fracture:	good; uneven
Habit:	rare crystals usually hemimorphic pyramids, sometimes hexagonal tabular crystals or short prisms; usually concentrically banded crusts, fibrous or columnar
Crystal system:	hexagonal

Greenockite

Greenockite is a cadmium ore named after Lord Greenock, on whose land near Glasgow it was first found. Other cadmium ore minerals are cadmoselite (CdSe) and otavite ($CdCO_3$), and all are quite rare. Commercial production of cadmium, however, is mostly as a by-product in the smelting of zinc sulphide ores. Cadmium is used in rechargeable batteries, alloys and pigments. Greenockite is a secondary mineral formed on the surface of cadmium-rich sphalerites. It is brightly coloured when fresh, but becomes dull on weathering. Tiny crystals are found at Llallagua (Bolivia), Bishopstown (Scotland) and Paterson (New Jersey, USA); encrustations are found at Pribram (Czech Republic), Joplin (Missouri, USA) and Sardinia (Italy).

CdS

Colour:	yellow, orange, red
Lustre; opacity:	adamantine, resinous; translucent
Streak:	orange yellow
Hardness:	3–3.5
Specific gravity:	4.9–5
Cleavage; fracture:	distinct, imperfect
Habit:	prismatic, hexagonal crystals rare, sometimes twinned; usually powdery encrustations or coatings
Crystal system:	hexagonal

Enargite

Enargite is a minor ore of copper and arsenic named from the Greek *enarges*, meaning 'obvious', because of its notably distinct cleavage. It is a dimorph of the mineral luzonite and is often mixed with the analogous antimony mineral stibioluzonite (Cu_3SbS_4). Twinning can give rise to attractive star shapes called 'trillings'. Enargite occurs in hydrothermal veins associated with chalcopyrite, tetrahedrite, bornite, pyrite, barite, quartz and covellite. It is found at Bor (Serbia), Chuquicamata (Chile), Bingham and Tintic (Utah, USA), Butte (Montana, USA), Luzon Island (Phillipines), Morococha, Quiruvilca and Cerro de Pasco (Peru), and Freiberg (Germany). Prisms up to 8cm (3in) have been found at Tsumeb (Namibia).

Cu_3AsS_4

Colour:	grey to black, can have violet or rose-brown internal reflections
Lustre; opacity:	metallic; opaque
Streak:	black
Hardness:	3
Specific gravity:	4.4–4.5
Cleavage; fracture:	perfect; uneven
Habit:	crystals rare tabular, blocky or prismatic often pseudohexagonal and striated ; usually as aggregates or granular masses
Crystal system:	orthorhombic

Galena

Galena is a common sulphide and is the main ore for lead, used and named by the ancient Greeks. Lead is used in batteries, glass, solder and radiation shields. Up to 1 per cent silver is found in galena, which makes it an important silver ore. It is a natural semiconductor and was one of the favourite 'crystals' used as crude diodes in crystal radio sets. Galena occurs in hydrothermal veins associated with chalcopyrite, pyrite, sphalerite barite, calcite, fluorite and quartz. It is mined in the Tri-State area of Missouri, Oklahoma and Kansas (USA), Broken Hill (Australia) and Santa Eulalia (Mexico); crystals are found at Pribram (Czech Republic), Isle of Man (UK) and Sardinia (Italy).

PbS

Colour:	light lead grey, dark lead grey
Lustre; opacity:	metallic; opaque
Streak:	greyish black
Hardness:	2.5
Specific gravity:	7.2–7.6
Cleavage; fracture:	perfect; soft no fracture
Habit:	well-formed crystals; massive, granular
Crystal system:	cubic

Cinnabar

Cinnabar is a toxic, dense red mineral formerly used as a pigment known as vermilion, the use of which began in ancient times. Free mercury can be produced by heating cinnabar above 580°C (1076°F). Mercury compounds are used in fine chemicals and paints, and mercury itself in thermometers and scientific instruments. Cinnabar is found in a variety of rocks often associated with volcanic activity, such as hot springs. It also occurs in hydrothermal veins and as placer deposits. The three most important locations are Almadén (Spain), Monte Amiato (Italy) and Idrija (Slovenia). Other locations include Nikotawa (Russia), Hunan Province (China) and the Altai Mountains of Central Asia.

HgS

Colour:	bright scarlet to brick red
Lustre; opacity:	adamantine; translucent
Streak:	bright red
Hardness:	2–2.5
Specific gravity:	8.1
Cleavage; fracture:	perfect; uneven, splintery
Habit:	rare rhombohedral or thick tabular crystals; earthy films; massive or granular
Crystal system:	trigonal

Pyrrhotite

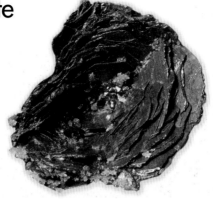

Pyrrhotite is distinguishable from pyrite in being magnetic, hence the alternative name magnetic pyrites. It is somewhat deficient in iron, there being approximately 11 sulphur atoms per 10 iron atoms. The mineral corresponding exactly to the formula FeS, which has been found only in meteorites, is called troilite. Named after the Greek *pyrrotes*, for 'red coloured', pyrrhotite is not a major iron ore, but nickel-rich pyrrhotite deposits are used to extract nickel, cobalt and platinum. It is found in mafic and ultramafic igneous rocks, hydrothermal veins and some high-grade metamorphic rocks. Good crystals come from Trepca (Serbia), Kysbanya (Romania), New York (USA) and Freiberg (Germany).

FeS

Colour:	bronze, bronze red, dark brown
Lustre; opacity:	metallic; opaque
Streak:	grey black
Hardness:	3.5–4
Specific gravity:	4.58–4.65
Cleavage; fracture:	imperfect; uneven
Habit:	prismatic or tabular crystals; massive, granular
Crystal system:	monoclinic (hexagonal when low in sulphur, close to FeS)

Miargyrite

Miargyrite, common among silver ore deposits, is named from the Greek for 'smaller' and 'silver', alluding to its silver content, which is lower than the mineral pyrargyrite. Easily confused with other silver minerals, miargyrite may be distinguished by its unusual deep red streak. Crystals usually grow to only about 1cm (⅜in), but the red internal reflections seen in this mineral make it appealing. It is found in hydrothermal veins associated with pyrargyrite, silver, galena, sphalerite, quartz, calcite and barite. Miargyrite is found at Baia Sprie (Romania), Pribram and Trebsko (Czech Republic), Randsberg (California, USA), Owyhee (Idaho, USA), Copiapo and Tarapaca (Chile), Potosi and Huanchaka (Peru), and Freiberg (Germany).

AgSbS$_2$

Colour:	grey to black with dark red internal reflections
Lustre; opacity:	metallic; opaque
Streak:	cherry red
Hardness:	2.5
Specific gravity:	5.25
Cleavage; fracture:	imperfect/fair; subconchoidal
Habit:	well-formed crystals, often coarse plates, blades, equant or wedge-shaped; granular, massive aggregates, disseminated
Crystal system:	monoclinic

Nickeline

Nickeline, or niccolite, is the first mineral from which nickel was extracted, but is now only a minor ore. The names of both mineral and element come from the German *kupfernickel*, or devil's nickel, because it was impossible to extract copper from the mineral, despite its copper-like appearance. It occurs in hydrothermal veins or is disseminated in basic igneous rocks such as gabbros, and is associated invariably with sulphides of silver, nickel and cobalt. Specimens are often coated with a pale to dark green film of annabergite. Nickeline is found at Eisleben and Freiberg (Germany), Schladming (Austria), Cobalt and Eldorado (Canada), Natsume (Japan), Bou Azzer (Morocco), Anarak (Iran) and La Rioja (Argentina).

NiAs

Colour:	copper-red
Lustre; opacity:	metallic; opaque
Streak:	brownish black
Hardness:	5.5
Specific gravity:	7.78–7.8
Cleavage; fracture:	imperfect; uneven to conchoidal
Habit:	crystals stocky, tabular or pyramidal, rare; columnar; massive; reniform
Crystal system:	hexagonal

Millerite

Millerite is a widespread but uncommon minor ore of nickel, also found in iron-nickel meteorites, albeit in minute quantities. Named after the English mineralogist W.H. Miller (1801–1880), it is also known as hair pyrites after its fine acicular crystals, or capillary pyrite after its occurrence as fine hollow crystals. It is a hydrothermal deposit found in cavities in limestones and dolomites, on barite and as an alteration product of nickel minerals. Millerite is found at Ramsbeck and Kamsdorf (Germany), Kotalahti (Finland), Keokuk (Iowa, USA), St Louis (Missouri, USA), Temagami (Ontario, Canada), Thompson (Manitoba, Canada), Onllwyn (Wales) and Kambalda and Leinster (Australia).

NiS

Colour:	brassy yellow, iridescent on tarnishing
Lustre; opacity:	metallic; opaque
Streak:	greenish black
Hardness:	3–3.5
Specific gravity:	5.2–5.6
Cleavage; fracture:	perfect; uneven
Habit:	acicular crystals, often in tufts or felted masses; rarely granular or massive
Crystal system:	trigonal

Covellite

Covellite is a rare mineral, also called indigo copper, occasionally found as flat, hexagonal crystals up to 10cm (4in) large and attractive rosettes of platy crystals. Covellite crystals are indigo blue and show pleasing iridescent yellow and red flashes. Strongly tarnished samples are coloured purple or black. It was discovered on Vesuvius by the Italian mineralogist N. Covelli (1790–1829). Covellite usually occurs in the oxidized zones of copper deposits and in volcanic sublimates, associated with chalcopyrite, chalcocite, bornite and pyrite. Covellite is found at the Calabona mine (Sardinia) and Vesuvius (Italy), Salzberg (Austria), Butte (Montana, USA), Kennicott (Alaska, USA), Bor (Serbia) and Bou Azzer (Morocco).

CuS

Colour:	indigo blue, strongly iridescent
Lustre; opacity:	metallic; opaque
Streak:	black grey
Hardness:	1.5–2
Specific gravity:	4.6–4.76
Cleavage:	perfect
Habit:	foliated, platy crystal aggregates; rarely hexagonal, flattened crystals; compact masses
Crystal system:	hexagonal

Pyrite

Pyrite, also called iron pyrite or simply pyrites, is also well known as fool's gold. It has often been mistaken for gold due to its yellow metallic lustre. Unlike gold, however, pyrite is hard, brittle and often unmistakably crystalline. Pyrite is used as a source of sulphur, especially for sulphuric acid manufacture. It occurs in igneous rocks, and is found in hydrothermal veins associated with sphalerite, galena, quartz, copper sulphides and gold, the latter two sometimes being commercially extracted from pyrite-rich ores. Pyrite also occurs in some metamorphic rocks and as pseudomorphs infilling the shapes of fossils. Excellent crystals are to be found in many locations. Pyritized fossils are found in Germany, England and Italy.

FeS_2

Colour:	pale brass yellow
Lustre; opacity:	metallic; opaque
Streak:	greenish black
Hardness:	6.5
Specific gravity:	5–5.02
Cleavage; fracture:	poor; uneven to conchoidal
Habit:	striated cubes, octahedra or pyritohedra, sometimes as 'iron cross' twins; compact granular aggregates, nodules, concretions and stalactitic forms, pseudomorphs after fossils
Crystal system:	cubic

Pyrite Varieties

Pyrite is very popular as large cubes and distorted dodecahedra called pyritohedra. Large crystals, especially cubes in metamorphic rocks, can make intriguing specimens with a bright, golden cube embedded incongruously in a contrasting granular or schistose matrix. Discs of sectioned nodules showing a radiating internal structure make attractive ornamental pieces. Pyrite used in jewellery is sometimes called marcasite by jewellers, especially when used as faceted stones set in silver. Pyrite is difficult to cut for jewellery because of its brittleness. It has been used by the ancient Greeks to adorn earrings and pins, and for jewellery in Victorian times. The Incas are said to have used pyrite tablets as mirrors.

FeS_2

Colour:	pale brass yellow
Lustre; opacity:	metallic; opaque
Streak:	greenish black
Hardness:	6.5
Specific gravity:	5–5.02
Cleavage; fracture:	poor; uneven to conchoidal
Habit:	striated cubes, octahedra or pyritohedra, sometimes as 'iron cross' twins; compact granular aggregates, nodules, concretions and stalactitic forms, pseudomorphs after fossils
Crystal system:	cubic

Stibnite

Stibnite, also known as antimonite, is the major ore of antimony. The name is derived from the old Greek word for antimony, *stibi*. Widely distributed but in small deposits, stibnite is mostly found in hydrothermal veins, but can occur in association with hot springs, associated with sulphides of silver, lead and mercury, pyrite, galena and quartz. Stibnite can form striking arrays of roughly aligned or radiating elongated prisms. The largest deposits of stibnite are in Hunan province (China); others locations include Shikoku Island (Japan) and Baia Sprie and Kapnik (Romania). Antimony is used in alloys to harden other metals, especially lead in storage batteries. It is increasingly used in semiconductors.

Sb_2S_3

Colour:	lead grey, bluish lead grey, steel grey, black
Lustre; opacity:	metallic; opaque
Streak:	blackish grey
Hardness:	2
Specific gravity:	4.63
Cleavage; fracture:	perfect; conchoidal
Habit:	granular, prismatic crystals, striated surface or cleavage face
Crystal system:	orthorhombic

Cobaltite

Cobaltite is a major ore of cobalt, the name of which comes from the German *kobold*, meaning 'underground spirit' or 'goblin', in reference to the difficulty of smelting it. Crystal forms of cobaltite are similar to those of pyrite, but the two are easily distinguished by colour. Cobaltite is often weathered to give crusts of pink to bright purple erythrite, called cobalt bloom by miners (evidence of underlying cobalt minerals). It is found in hydrothermal veins or contact-metamorphosed rocks. Excellent crystals come from Tunaberg (Sweden), Skutterud (Norway), Cornwall (England), Española, Cobalt and Sudbury (Ontario, Canada), Broken Hill and Torrington (NSW, Australia), and Bou Azzer (Morocco).

CoAsS

Colour:	reddish silver white, violet steel grey, black
Lustre; opacity:	metallic; opaque
Streak:	greyish black
Hardness:	5.5
Specific gravity:	6.33
Cleavage; fracture:	good; uneven
Habit:	cubes, octahedra or pyritohedra, usually striated; granular or compact masses, disseminated
Crystal system	orthorhombic

Bismuthinite

Bismuthinite is similar in appearance and properties to stibnite, but can be distinguished by its inability to melt in a match flame. Sprays of steel grey prismatic crystals resemble those of stibnite. It is the major ore of bismuth, but the free metal is mostly obtained as a by-product of lead and copper smelting. Substitution of bismuth for lead and copper gives the mineral aikinite ($CuPbBiS_3$), which forms a series with bismuthinite. It occurs in hydrothermal veins associated with tin, silver and cobalt minerals. Bismuthinite is found at Haddam (Connecticut, USA), Llallagua and Tasno (Bolivia), Cerro de Pasco (Peru), Cornwall (England) and Mount Biggenden (Australia).

Bi_2S_3

Colour:	grey, silver white, tin white
Lustre; opacity:	metallic; opaque
Streak:	grey
Hardness:	2
Specific gravity:	6.8–7.2
Cleavage; fracture:	perfect; uneven
Habit:	prismatic, acicular crystals, finely striated; granular and compact aggregates
Crystal system:	orthorhombic

Sylvanite

Sylvanite is one of the few ores of gold, other than native gold itself, and also of silver and tellurium. Gold, normally very unreactive, has a particular affinity for tellurium. Sylvanite is named after Transylvania, where the mineral was first discovered. It is rare and not commercially mined. Tellurium is obtained from the anode slime in copper refining. Although less than 0.1 per cent is usually added to steels, more than half of the tellurium production is used in this way. Sylvanite occurs in hydrothermal veins, associated with calaverite ($AuTe_2$) and petzite (Ag_3AuTe_2). It is found at Baia de Aries (Romania), Kalgoorie (Australia), Bereznyakov and Yaman-Kasy, Ural Mountains (Russia), and Cripple Creek (Colorado, USA).

$AgAuTe_4$

Colour	...wish silver white, white
	...lic; opaque
	...grey
Hardness:	1.5–2
Specific gravity:	7.9–8.3
Cleavage; fracture:	perfect; uneven
Habit:	stubby, prismatic or arborescent crystals; branching encrustations resembling script; granular or bladed masses
Crystal system:	monoclinic

Hauerite

Hauerite is a rare form of manganese disulphide named after the Austrian geologists J. and F. von Hauer. The finest specimens are large octahedral crystals sometimes modified by cubic faces, from the Destricella mine (Raddusa, Sicily). Specimens often comprise hauerite with associated rambergite (MnS). Hauerite occurs in sulphur-rich clay deposits and altered lavas associated with sulphur, realgar, gypsum, aragonite and calcite, and in the ferromanganese deposits in the Pacific Ocean. It is found at Kalinka and Banská Stiavnica (Slovakia), Bohemia (Czech Republic), Jezyorko and Grzybow (Poland), Raddusa (Italy), Yazovsk and Podgornensk (Ural Mountains, Russia) and in salt domes in Texas (USA).

MnS_2

Colour:	brownish grey to brownish black, reddish tints
Lustre; opacity:	metallic; opaque
Streak:	reddish brown
Hardness:	4
Specific gravity:	3.46
Cleavage; fracture:	perfect; uneven
Habit:	octahedral and cubo-octahedral crystals up to 5 cm (2 in), often fractured unevenly; rounded aggregates
Crystal system:	isometric

Ullmannite

Ullmannite is a rare form of nickel antimony sulphide of the cobaltite group, named after the German chemist J. Ullmann (1771–1821). It is closely related to gersdorfite (NiAsS) and will almost invariably contain some arsenic. Ullmannite also forms a series with willyamite, (Co,Ni)SbS. Crystals grow to about 3cm (1¼in) and resemble those seen in pyrite. Ullmannite occurs in hydrothermal veins associated with minerals such as skutterudite, galena, nickeline, pyrrhotite and tetrahedrite. It is found at Siegerland, Harzgerode and Lobenstein (Germany), Waldenstein and Lolling (Austria), Fourstones (Northumberland, England), Durham (England), Broken Hill (NSW, Australia) and Sarrabus (Sardinia, Italy).

NiSbS

Colour:	steel grey, silver white, tin white
Lustre; opacity:	metallic; opaque
Streak:	greyish black
Hardness:	5–5.5
Specific gravity:	6.65
Cleavage; fracture:	good; uneven
Habit:	crystals as cubes, octahedra, dodecahedra and tetrahedra; massive, granular
Crystal system:	cubic

Marcasite

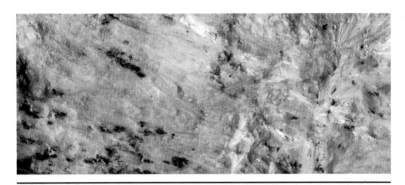

Marcasite is a polymorph of pyrite, and is used, like pyrite, for sulphuric acid production, not as an iron ore. It is called spear pyrites after the arrow- or spear-head shape of twinned crystals; it is also known as white iron pyrite, as it has a paler colour than pyrite. Marcasite was formerly used in jewellery, but is now likely to be pyrite. Marcasite forms in hydrothermal veins, often in lead and zinc ores, is precipitated in sedimentary rocks and can be found replacing fossils. Marcasite is found at Carlsbad and Rammelsberg (Germany), Karlovy Vary (Czech Republic), Derbyshire (England) and the Tri-State mining area (USA). Nodules with a radiating internal structure are found in the chalk of Southeast England.

FeS_2

Colour:	steel grey, silver white
Lustre; opacity:	metallic; opaque
Streak:	greyish black
Hardness:	5.5
Specific gravity:	6.65
Cleavage; fracture:	good; uneven
Habit:	crystals as flat prisms, occasionally as rosettes, often twinned in shapes described as cockscombs and spearheads; massive, granular, crusty aggregates, concretions
Crystal system:	cubic

Arsenopyrite

Arsenopyrite is common in sulphide deposits and is the major ore of arsenic, and is often also rich in silver, gold, cobalt and tin. Arsenic, however, is obtained commercially as a by-product during the refining of sulphide ores. Arsenopyrite has a silvery colour that distinguishes it from pyrite and marcasite, although it can tarnish on exposure. Mispickel and arsenical pyrites are alternative names. It occurs in sulphide deposits, in hydrothermal veins and in some metamorphic rocks. Good crystals are found at Roxbury (Connecticut, USA), Leadville (Colorado, USA) and Valle Anzasca and Val Sugana (Italy). Large deposits are at Boliden (Sweden), Freiberg (Germany), Deloro (Canada), Sulitjelma (Norway) and Cornwall (England).

FeAsS

Colour:	tin white, light steel grey, can have a pink tint
Lustre; opacity:	metallic; opaque
Streak:	black
Hardness:	5
Specific gravity:	6.07
Cleavage; fracture:	distinct; uneven
Habit:	crystals as elongated, striated prisms and cruciform twins; granular masses
Crystal system:	monoclinic

Glaucodot

G laucodot contains up to 25 per cent cobalt, but is not a commercially important ore. It is regarded by some as a member of the arsenopyrite-cobaltite (FeAsS-CoAsS) series, but these crystallize in different systems. The name comes from the Greek for 'blue', after its use in a dark blue glass called smalt. It occurs in hydrothermal veins associated with sulphides and in metamorphosed lavas. Weathering and alteration often produce a bloom of erythrite. Good crystals come from Hakansbö (Sweden) and Huasco (Tarapaca, Chile). Other localities include Cobalt (Ontario, Canada), Oravita (Romania), Skutterud (Norway), Sumpter (Oregon, USA), Franconia (New Hampshire, USA) and Alston (Cumbria, England).

(Cu,Fe)AsS

Colour:	greyish or reddish silver-white
Lustre; opacity:	metallic; opaque
Streak:	black
Hardness:	5
Specific gravity:	5.9–6.01
Cleavage; fracture:	perfect; uneven, fragile
Habit:	prismatic or elongate crystals; granular or radiating fibrous masses
Crystal ystem:	orthorhombic

Skutterudite

Skutterudite, also known as smaltite, is one of the major ores of cobalt. The cobalt is invariably partially replaced by other elements, such as copper, zinc, silver and particularly nickel. When nickel forms the major component, it is referred to as nickel-skutterudite. It forms fine crystals resembling those of pyrite, but with a silvery colour, and is often altered to crimson red erythrite. Skutterudite is formed in hydrothermal veins associated with arsenopyrite, calcite and sulphides of cobalt and nickel. It is named after Skutterud in Norway, where it occurs as fine crystals; it is also found at Cobalt (Canada), Huelva (Spain), Schneeberg (Germany) and Bou Azzer (Morocco). Deposits are mined in Germany, Austria and the Czech Republic.

$(Co,Ni)As_3$

Colour:	white, light steel grey, sometimes with a an iridescent film
Lustre; opacity:	metallic; opaque
Streak:	black
Hardness:	5.5–6
Specific gravity:	6.1–6.9
Cleavage; fracture:	distinct; conchoidal
Habit:	crystals are cubic, octahedral or pyritohedral; compact granular masses
Crystal system:	cubic

Molybdenite

Molybdenite forms two-dimensional sheets, and these, like micas and graphite, give rise to flaky crystals. It is an excellent high-temperature dry lubricant and the main ore of molybdenum. Similar in appearance to graphite, it has a higher density and a more metallic bluish appearance. The name is derived from the Greek *molybdos*, for lead having been mistaken for galena. It mostly occurs in granitic rocks, in pegmatites and pneumatolytic veins, and in contact-metamorphosed rocks. Good crystals are found at Edison (New Jersey, USA), Climax (Colorado, USA), Arendal (Norway), Temikaming (Quebec, Canada) and Kingsgate (NSW, Australia), the latter providing crystals 70 x 5mm (2¾ x ¼in).

MoS_2

Colour:	bluish-grey
Lustre; opacity:	metallic; opaque
Streak:	bright blue-grey
Hardness:	1–1.5
Specific gravity:	4.7
Cleavage; fracture:	perfect
Habit:	tabular hexagonal crystals; usually as bladed, foliated or interwoven masses
Crystal system:	hexagonal

Proustite

Proustite forms beautiful translucent cinnabar-red crystals, but exposure to light and air causes them to darken, becoming semi-opaque with a grey streak. Rarely for a sulphide mineral, it is not metallic or opaque. Resembling pyrargyrite, it was identified only by analysis by the French chemist J. Proust (1755–1826). Proustite is also known as ruby silver ore or light red silver ore. It is found in oxidized zones of hydrothermal veins associated with other silver and sulphide minerals. Proustite crystals of up to 15cm (6in) have been found at Chañarcillo (Chile); other locations are Jáchymov and Pribram (Czech Republic), Chihuahua and Zacatas (Mexico), Annaberg and Wittichen (Germany), and Sarrabus (Sardinia).

Ag_3AsS_3

Colour:	dark red, darkening on exposure to light and air
Lustre; opacity:	adamantine; semi-transparent to translucent
Streak:	brick red, becoming grey in altered specimens
Hardness:	2–2.5
Specific gravity:	5.57
Cleavage; fracture:	good, rhombohedral; conchoida
Habit:	crystals rare, as rhombohedra or scalenohedra, usually striated and distorted; usually massive
Crystal system:	trigonal

Pyrargyrite

Pyrargyrite is related to proustite, with antimony replacing arsenic, both in the hexagonal system. To distinguish it from proustite, pyrargyrite has been called dark red silver ore, but is also referred to as ruby silver. Pyrargyrite crystals can be striking and attractive, but darken fairly quickly on exposure to light and become opaque, posing problems with storage and display. It is formed in hydrothermal veins associated with other silver minerals and by alteration of argentite and silver-rich galena. Good hexagonal crystals occur at St Andreasberg and Freiberg (Germany), and Colquechaca (Bolivia); other localities are Sarrabus (Sardinia), Pribram (Czech Republic), Zacetas (Mexico) and Chañarcillo (Chile).

Ag_3SbS_3

Colour:	black with dark red tints, darkening on exposure to light
Lustre; opacity:	adamantine, submetallic; translucent, becoming opaque on exposure to light
Streak:	purple-red
Hardness:	2.5–3
Specific gravity:	5.85
Cleavage; fracture:	good, rhombohedral; uneven to conchoidal
Habit:	crystals hemimorphic being prisms with varying terminations; granular aggregates, compact masses, disseminated grains
Crystal system:	trigonal

Stephanite

Named after Archduke Victor Stephan (1817–1867), Mining Director for Austria, stephanite is also called black silver ore or brittle silver ore. It is a major ore of silver, alongside acanthite in importance. Good crystals are rare, but can grow to about 6cm (2¼in) and display a brilliant metallic lustre. Twinning often occurs to give pseudohexagonal crystals. Stephanite is found in hydrothermal veins associated with silver, proustite, pyrargyrite, polybasite, tetrahedrite and acanthite. Excellent crystals have been found at St Andreasberg and Freiberg (Germany), and at Zacatecas and Arizpe (Mexico). It is mined at Cobalt in Ontario and Elsa in Yukon (Canada) and is an important ore at the famous Comstock Lode (Nevada, USA).

Ag_5SbS_4

Colour:	iron-black
Lustre; opacity:	metallic; opaque
Streak:	black, shiny
Hardness:	2–2.5
Specific gravity:	6.3
Cleavage; fracture:	poor; uneven
Habit:	crystals rare, prismatic to tabular; granular aggregates, massive, disseminated
Crystal system:	orthorhombic

Polybasite

Polybasite is a minor silver ore and yet another that has been called ruby silver. It forms a series with pearceite, $(Ag,Cu)_{16}As_2S_{11}$, although the antimony-rich polybasite is much more common. Good crystals are rare, but can grow to about 6cm (2¼in) and have a reddish tinge due to internal reflections. Fine crystals are found at Husky Mine (Yukon, Canada) and rosettes of hexagonal plates at Guanajuata (Mexico). Polybasite occurs in hydrothermal veins usually accompanied by quartz, calcite, barite or other silver minerals. Other good localities include St Andreasberg and Freiberg (Germany), and Pribram (Czech Republic). It is mined at Chañarcillo (Chile), Sarrabus (Sardinia) and Colorado (USA).

$(Ag,Cu)_{16}Sb_2S_{11}$

Colour:	iron black, rarely cherry red
Lustre; opacity:	metallic; opaque
Streak:	black
Hardness:	2–3
Specific gravity:	6–6.2
Cleavage; fracture:	poor; uneven
Habit:	bevelled, tabular, pseudohexagonal crystals; bladed, in rosettes, granular compact aggregates, crusts
Crystal system:	monoclinic

Bournonite

Bournonite, also known as endellionite, forms crystals that have a good metallic lustre and interesting shapes. Twinning often produces shapes resembling worn cog wheels, hence the name wheel ore. Such twinned crystals are called 'trillings' and frequently are striated along the 'teeth' of the cog. Some crystals can develop a greyish tarnish, so some care with storage is advised. Bournonite occurs in hydrothermal veins, associated with galena, tetrahedrite, pyrite and especially silver. Excellent crystals are found at Harz (Germany), St Endellion and Lanreath (Cornwall, England), Pribram (Czech Republic), Kapnik-Bnya (Hungary), Sarrabus (Sardinia), Broken Hill (NSW, Australia) and Park City (Utah, USA).

$PbCuSbS_3$

Colour:	dark grey to black
Lustre; opacity:	metallic; opaque
Streak:	steel grey
Hardness:	2.5–3
Specific gravity:	5.7–5.9
Cleavage; fracture:	good; fragile
Habit:	crystals as stubby, tabular, multi-faceted prisms; granular aggregates, disseminated grains
Crystal system:	orthorhombic

Hutchinsonite

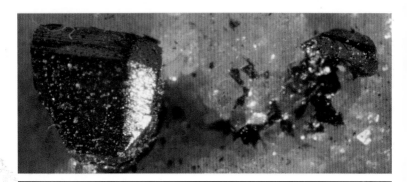

Hutchinsonite is an ore of the metal thallium, others being crooksite (Cu_7TlSe_4), thalcusite ($Cu_8FeTl_2S_4$) and lorandite ($TlAsS_2$). It was named to honour Arthur Hutchinson (1866–1937), Professor of Mineralogy at Cambridge University. Thallium is obtained as a by-product of the smelting of thallium-rich zinc and lead ores (which are likely to contain hutchinsonite). Thallium is very toxic, but is used in some specialist glass manufacture. Hutchinsonite occurs in hydrothermal deposits associated with orpiment, realgar, pyrite, sphalerite and galena. It is found at the Lengenbach Quarry (Binnental, Switzerland), Segen Gottes (Black Forest, Germany), Quiluvrilca (Peru) and Toya-Takarada (Hokkaido, Japan).

$(Tl,Pb)_2As_5S_9$

Colour:	scarlet–vermilion to deep cherry red, strong red internal reflections
Lustre; opacity:	submetallic; subtranslucent to opaque
Streak:	red
Hardness:	1.5–2
Specific gravity:	4.6
Cleavage; fracture:	good; conchoidal, brittle
Habit:	acicular or prismatic crystals; granular masses
Crystal system:	orthorhombic

Boulangerite

Boulangerite is a minor ore of lead named after the French mining engineer C. Boulanger (1810–1849). It can form thin, acicular crystals up to 2cm (¾in) long, resembling fibres. Boulangerite and jamesonite have been called feather ores or plumosite, after the feathery habit of some varieties. Thin fibres of boulangerite are flexible, unlike the brittle jamesonite. Disseminated boulangerite can easily be overlooked, being mistaken for stray hairs. Boulangerite occurs in hydrothermal veins in lead, zinc and antimony deposits. It is found at Pribram (Czech Republic), Molières (France), Bottino (Italy), Noche Buena Mine (Zacatecas, Mexico), Boliden (Sweden), Claustal (Germany) and Stevens County (Washington, USA).

$Pb_5Sb_4S_{11}$

Colour:	grey
Lustre; opacity:	dull, metallic; opaque
Streak:	black
Hardness:	2.5–3
Specific gravity:	5.8–6.2
Cleavage; fracture:	good; brittle but flexible as thin needles
Habit:	acicular crystals; fibrous masses or tufts, disseminated
Crystal system:	monoclinic
Other:	melts easily

Realgar

Realgar, named from the Arabic *rahj al jahr* for 'mine powder', is often found with yellow orpiment (As_2S_3), to which it alters on exposure to light and air. Its colour and transparency contrast with the grey, opaque nature of most other sulphides. Realgar has been used in fireworks (to give a white colour) and paints. Like sulphur, which occurs as rings of eight atoms, it is in the form of alternating sulphur and arsenic atoms As_4S_4. Realgar mostly occurs in veins with orpiment, cinnabar or stibnite, but may be found in limestones, clays, volcanic rocks and hot springs. Crystals are rare, but are found at Nagyag (Romania), Binnental (Switzerland), Matra (Corsica), Manhattan (Nevada, USA) and King County (Washington, USA).

AsS

Colour:	orange-red to red
Lustre; opacity:	resinous; semi-transparent
Streak:	yellow-orange
Hardness:	1.5–2
Specific gravity:	3.5–3.6
Cleavage; fracture:	perfect; conchoidal
Habit:	rare small, stubby, prismatic crystals; compact aggregates; films
Crystal system:	monoclinic

Orpiment

Orpiment is a rare mineral often found in association with realgar. Its name comes from the Latin *auripigmentum*, meaning 'golden pigment'. On standing, orpiment will slowly turn into a powder, a process accelerated by light. Orpiment was traded as a pigment throughout the Roman Empire and used as a medicine in China. Orpiment is deposited around hot springs and volcanic fumaroles, associated with realgar and, sometimes, cinnabar. It also occurs in metamorphosed dolerites and in hydrothermal veins. The major deposits of orpiment are in Zashuran (Iran), Kurdistan (Turkey), Georgia and Manhattan (Nevada, USA). Large crystals have been found at Shimen (Hunan, China) and Quiruvica (La Libertad, Peru).

As_2S_3

Colour:	orange-yellow to yellow
Lustre; opacity:	greasy, pearly on fracture surfaces; translucent to transparent
Streak:	yellow
Hardness:	1.52
Specific gravity:	3.48
Cleavage; fracture:	perfect; flaky
Habit:	crystals rare, as small, flat prisms, occasionally fibrous; crusts or bladed masses, earthy, powdery
Crystal system:	monoclinic

Rammelsbergite

Rammelsbergite, named after the German chemist Karl Rammelsberg (1813–1899), is rare and difficult to distinguish from other sulphides, often being mistaken for gersdorfite (NiAsS). It occurs in hydrothermal veins together with nickel and cobalt minerals. Associated minerals include skutterudite, lollingtonite, safflorite, nickeline, bismuth and silver. Rammelsberg is found at Franklin and Sterling Hill (New Jersey, USA), Mohawk Mine (Michigan, USA), Coniston (Cumbria, England), St Marie-aux-Mines (France), Legnica (Poland), Löllington-Hüttenberg (Austria) and the type locality, Schneeberg (Saxony, Germany). Large crystals come from Bou Azzer (Morocco).

$NiAs_2$

Colour:	tin white with a faint pinkish hue, darker tarnish
Lustre; opacity:	metallic; opaque
Streak:	black
Hardness:	5.5
Specific gravity:	7.1
Cleavage; fracture:	indistinct; uneven
Habit:	small crystals as imperfect prisms; granular, massive, radiating aggregates
Crystal system:	orthorhombic

Jordanite

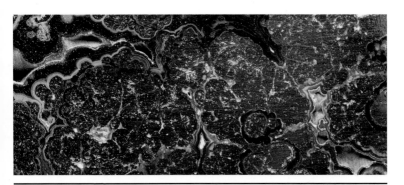

Jordanite is a rare mineral, named after the German mineralogist H. Jordan (1808–1887). It forms a series with antimony-rich geocronite, $Pb_{14}(Sb,As)_6S_{23}$. It occurs in hydrothermal veins and metamorphosed lead-arsenic deposits in dolomite, and may be associated with galena, sphalerite, pyrite and anatase. The most important occurrence is at Lengenbach (Binnental, Switzerland). In the Seravezza marble in Tuscany, Italy, jordanite occurs in cavities, as lustrous, lead-grey hexagonal crystals up to 1cm (½in). Other localities include Săcărîmb (Romania), Wiesloch (Germany), Zuni Mine (Colorado, USA) and Sinking Valley (Pennsylvania, USA). It has also been found in 'black smoker' chimneys on ocean floors.

$Pb_{14}(As,Sb)_6S_{23}$

Colour:	lead-grey, commonly tarnished and iridescent
Lustre; opacity:	metallic; opaque
Streak:	black
Hardness:	3
Specific gravity:	6.4
Cleavage; fracture:	perfect; conchoidal, brittle
Habit:	crystals deeply striated prismatic, and tabular, pseudohexagonal twins; granular, globular and botryoidal aggregates
Crystal system:	monoclinic

Jamesonite

J amesonite, also known as antimony glance or grey antimony, was named for the
Scottish mineralogist Robert Jameson (1774–1854). It is characterized by hairlike fibres, which may be thick and felted, or may occur as individual 'hairs', and is
sometimes also known as feather ore. Jamesonite may be distinguished from the
acicular mineral boulangerite because it has brittle rather than flexible crystals and
from acicular yellow millerite by colour. It occurs in hydrothermal lead-silver-zinc
veins. It may be associated with pyrite, sphalerite, arsenopyrite, siderite, dolomite,
calcite, rhodochrosite or quartz. Notable localities are Cornwall (England), Herja
Mine (Maramures, Romania) and Zacatecas (Mexico).

$Pb_4FeSb_6S_{14}$

Colour:	lead grey, iridescent tarnish
Lustre; opacity:	metallic or silky; opaque
Streak:	grey–black
Hardness:	2.5
Specific gravity:	5.63–5.78
Cleavage; fracture:	perfect basal (perpendicular to length); uneven
Habit:	very elongated acicular or fibrous crystals, occasionally prismatic; usually compact or felted masses, may be radial or plumose
Crystal system:	monoclinic

Cosalite

Cosalite is a rare mineral named after the Cosalá Mine (Sinaloa, Mexico). It occurs in hydrothermal deposits in lead sulphide ores, in contact skarns and in pegmatites. Associated minerals include chalcopyrite, pyrite, sphalerite, skutterudite, bismuth, tremolite, diopside, epidote and quartz. It forms hairlike crystals (found at Carrock Fell Mine, in Cumbria, England) or elongated prisms (as at Limestone Quarry in Saxony, Germany). Other good localities are Kara Oba (Kazakhstan), Braichyroen (Snowdon, Wales), Cobalt (Ontario, Canada) and Hecla Mine, Dundas (Tasmania, Australia). Kudriavite, $(Cd,Pb)Bi_2S_4$, is a closely related mineral found around fumaroles (e.g. Kudriavy Volcano, Kuril Islands, Russia).

$Pb_2Bi_2S_5$

Colour:	lead grey or steel grey to silver-white
Lustre; opacity:	metallic; opaque
Streak:	black
Hardness:	2.5–3
Specific gravity:	6.86–6.99
Cleavage; fracture:	very rare; uneven
Habit:	long prisms, vertically striated, hairlike; radial and granular aggregates
Crystal system:	orthorhombic

Halite

H alite, or rock salt, typically forms through the evaporation of enclosed bodies of sea water. It may form a thick, stratified deposit, or domes where the low-density salt has risen through overlying sediments, as along the Gulf Coast (USA). Halite also forms as a volcanic sublimate or as a cave deposit. It is associated with other evaporites, such as sylvite (KCl), gypsum, anhydrite and dolomite. Rock salt is used as common table salt, and its taste is characteristic. It is an important raw material in the chlor-alkali industry, drilling muds, aluminium purification and many other industries. Well-known deposits include those at Stassfurt (Germany) and in Texas and New Mexico (USA). Good crystals are found at Wieliczka (Galicia, Poland).

NaCl

Colour:	white when pure, greyish, pinkish, bluish, violet, orange
Lustre; opacity:	vitreous, greasy; transparent to translucent
Streak:	white
Hardness:	2.5
Specific gravity:	2.1–2.2
Cleavage; fracture:	perfect cubic; conchoidal
Habit:	crystals cubic or as octahedra, skeletal 'hopper' crystals; granular, massive, rarely stalactitic or fibrous
Crystal system:	cubic

Villiaumite

Villiaumite is a rare mineral named after the French explorer Maxime Villiaume, who collected the mineral from cavities in nepheline syenites at Rouma (Islands of Los, Guinea). It is strongly coloured, with some shades of red being unique in minerals. It is found in nepheline syenites and their pegmatites, in which alkali metals such as sodium, lithium and potassium are concentrated. Associated minerals include nepheline, aegerine, sodalite and zeolites. Villiaumite is soluble in water and displays weak red fluoresecence under shortwave ultraviolet light. Most specimens come from the Kola Peninsula (Russia); good crystals are found at Mont Saint-Hilaire (Quebec, Canada) and Windhoek (Namibia).

NaF

Colour:	carmine red, lavender pink to light orange
Lustre; opacity:	vitreous; transparent
Streak:	white, pinkish
Hardness:	2.5
Specific gravity:	2.79
Cleavage; fracture:	perfect in three directions forming cubes; conchoidal
Habit:	rarely cubic or octahedral crystals; granular aggregates, massive
Crystal system:	cubic

Chlorargyrite

Chlorargyrite, a silver chloride mineral, was named after the Latin *argentum* (silver) and the Greek *chloro* (pale green). Also called cerargyrite or horn silver, it forms a complete solid-solution with bromargyrite (AgBr). It is pale when fresh, but darkens to brownish-purple on exposure to light. Chlorargyrite dissolves in ammonia, but not in nitric acid. It forms as a secondary mineral in the weathered and enriched zones of silver deposits, especially in arid regions, and can form very rich albeit small silver ore deposits. The type locality is Saxony (Germany); it occurs elsewhere at Lake Valley District (New Mexico, USA), Atonopah (Nevada, USA), Broken Hill (NSW, Australia), Freiburg (Saxony, Germany) and Atacama (Chile).

AgCl

Colour:	colourless when fresh, pale grey, yellowish; can darken
Lustre; opacity:	resinous, waxy, adamantine; transparent to translucent
Streak:	white, shining
Hardness:	1.5–2.5
Specific gravity:	5.55 (6.5 in bromargyrite)
Cleavage; fracture:	absent; uneven to subconchoidal, sectile, ductile
Habit:	crystals rare, as cubes, or modified by octahedra; usually massive or in crusts, columnar
Crystal system:	cubic

Iodargyrite

Iodargyrite, or iodyrite, was named after its chemical composition (from the Greek *iodes* meaning 'violet') and the Greek *argyros* for 'silver'. A very rare secondary mineral, it forms in the oxidized parts of silver deposits. The type locality is Albarradón Mine (Zacatecas, Mexico), but the best place for specimens is Broken Hill (NSW, Australia), the world's largest silver-lead-zinc deposit. Iodargyrite is also important for other halides, such as the very rare iodides marshite (copper), miersite (silver and copper), perroudite (a mercury silver halide), and bright yellow crystals of bromargyrite (a silver halide). Other locations include Atacama (Chile), Dzhezkazgan (Kazakhstan) and Nevada (USA).

AgI

Colour:	colourless, becomes yellowish on exposure to light
Lustre; opacity:	greasy to adamantine, pearly on cleavage surfaces; transparent
Streak:	white or yellow, shiny
Hardness:	1–1.5
Specific gravity:	5.7
Cleavage; fracture:	perfect basal; conchoidal, sectile and flexible
Habit:	crystals prismatic, platy, barrel-shaped; scales, powdered
Crystal system:	hexagonal

Fluorite

Fluorite, or fluorspar, is a popular mineral, which can fluoresce as a result of impurities such as yttrium. Some specimens phosphoresce; others thermoluminesce (glow when heated, as at Franklin in New Jersey, USA, and Mont Saint-Hilarie in Quebec, Canada) or triboluminesce (glow when crushed or struck). Fluorite forms in hydrothermal veins as a gangue mineral associated with sulphides (e.g. galena or sphalerite), barite and quartz, and as an accessory in granite and granite pegmatites. It is an important industrial mineral, used as a flux in iron smelting, as a source of fluorine, as special optical lenses, and as a rare gemstone. The main mining areas are Canada, USA, Russia, Mexico and Italy.

CaF_2

Colour:	colourless, white, yellow, green, purple and blue
Lustre; opacity:	vitreous; transparent to translucent
Streak:	white
Hardness:	4
Specific gravity:	3.18
Cleavage; fracture:	perfect octahedral; subconchoidal to uneven
Habit:	crystals common, as cubes, octahedra or rarely dodecahedra; nodular, granular aggregates, earthy masses
Crystal system:	cubic

Fluorite: Gem Varieties

Fluorite shows a wide range of colours. The most famous variety is Blue John, a purple and yellow banded variety from Castleton (Derbyshire, England), which has been used for ornaments and jewellery. The Ancient Greeks and the Ancient Egyptians, and the Chinese have used it for decorative purposes for over 300 years. Rather soft for general use as a gemstone and too well cleaved to be easily cut, it can nonetheless be brightly polished and cut into cabochons, protected by rock crystal (quartz). A deposit of colourful highly silicated fluorite discovered quite recently in Utah has been given a number of fanciful names, such as bertandite and Picasso stone. More properly called opalized fluorite, it makes attractive cabochon gems.

CaF_2

Colour:	colourless, white, yellow, green, purple and blue
Lustre; opacity:	vitreous; transparent to translucent
Streak:	white
Hardness:	4
Specific gravity:	3.18
Cleavage; fracture:	perfect octahedral; subconchoidal to uneven
Habit:	crystals common, as cubes, octahedra or rarely dodecahedra; nodular, granular aggregates, earthy masses
Crystal system:	cubic

Sal-ammoniac

Sal-ammoniac forms in volcanic regions around fumaroles, with solid white crystals forming directly from the bluish ammonium chloride vapour as a sublimate (i.e. there is no liquid phase). It has a pungent, cool and saline taste. Sal-ammoniac forms on Vesuvius (Campania), Etna (Sicily), and other southern Italian volcanoes, Mont Pelée (Martinique), Parícutin (Mexico) and Kilauea (Hawaii). Associated minerals include sulphur, realgar and orpiment. Crystals must be collected quickly, as they will dissolve in the first rain shower. Sal-ammoniac formation is also associated with burning coal seams, as at Duttweiler (Saarland, Germany), and, unusually, with guano on Cicna and Guanape Islands (Peru).

NH_4Cl

Colour:	colourless, white, yellow, reddish or brown
Lustre; opacity:	vitreous; transparent to translucent
Streak:	white
Hardness:	1–2
Specific gravity:	1.53
Cleavage; fracture:	poor; conchoidal to earthy
Habit:	crystals as cubes, octahedra or dodecahedra, also skeletal or dendritic; generally efflorescent or encrusting
Crystal system:	cubic

Cryolite

Cryolite is an uncommon mineral of very restricted distribution. It possesses the unusual property that, if submerged in water, it becomes almost invisible, since it has a refractive index of 1.33, close to that of water. It is soluble in sulphuric acid, producing fumes of hydrofluoric acid. Cryolite occurs only in pegmatites, probably formed as a precipitate from fluoride-rich solutions. The most notable occurrence is in a granitic pegmatite at Ivigtut (Greenland), where it is associated with topaz, siderite, galena, microcline, fluorite and other unusual fluorides. It is also found in a topaz mine at Miask (Urals, Russia). Artificial cryolite has now replaced the natural mineral as a flux in the extraction of aluminium from bauxite.

Na_3AlF_6

Colour:	usually colourless to white, or grey, reddish, brownish
Lustre; opacity:	vitreous to greasy, pearly; transparent
Streak:	white
Hardness:	2.5–3
Specific gravity:	2.95
Cleavage; fracture:	absent, although has three partings; uneven, brittle
Habit:	pseudo-cubic crystals, commonly twinned; granular aggregates often arranged in a parquet-like pattern
Crystal system:	monoclinic

Carnallite

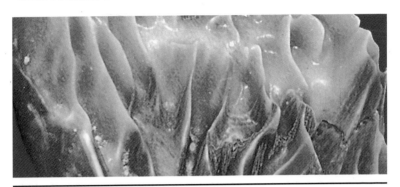

Carnallite is a rare mineral that sometimes fluoresces, and will colour a flame violet due to its potassium content. It has a bitter, salty taste, will readily dissolve in water (emitting a creaking sound), and must be kept in sealed containers. Carnallite is one of the last minerals to precipitate in an evaporating basin, in particular conditions that do not exist today. It occurs with other potassium and magnesium evaporites such as sylvite, kainite and kieserite. It is a valuable source of potash (for fertilizers) and magnesium. The most famous locality is the Stassfurt potash deposit (Germany); it also occurs in Carlsbad (New Mexico, USA), Paradox Basin (Colorado and Utah, USA) and the Perm Basin (Russia).

$KMgCl_3.6H_2O$

Colour:	colourless, milky white, yellow, pink, brown, rarely blue
Lustre; opacity:	vitreous to greasy; transparent to translucent
Streak:	white
Hardness:	2.5
Specific gravity:	1.6 (very light)
Cleavage; fracture:	absent; conchoidal
Habit:	pseudo-hexagonal pyramidal crystals, rare; granular and fibrous aggregates forming layers
Crystal system:	orthorhombic

Atacamite

Atacamite is named from the Atacama Desert in Chile, which is one of the most arid regions in the world. Dark green itself, atacamite is often associated with other coloured minerals such as green malachite, red cuprite, blue-green chrysocolla, and also gypsum. It generally forms through the oxidation of copper minerals in arid, salty conditions, and is found in many places in Chile. It occurs as a fumarolic deposit on Vesuvius (Italy) and Etna (Sicily, Italy), and through the weathering of sulphides formed as black smoker deposits around volcanic vents on the ocean floor. It also forms through the alteration of bronze and copper antiquities, and is found in the slag heaps at Laurion (Attica, Greece).

$Cu_2Cl(OH)_3$

Colour:	bright green, dark emerald green to black-green
Lustre; opacity:	vitreous to adamantine; transparent to translucent
Streak:	apple green
Hardness:	3–3.5
Specific gravity:	3.76
Cleavage; fracture:	good; conchoidal, brittle
Habit:	crystals acicular, striated, or tabular; massive, as fibrous aggregates or granular
Crystal system:	orthorhombic

Cotunnite

Cotunnite forms as a volcanic sublimate, its type locality being Monte Somma-Vesuvius (Italy). It was named in honour of Domenico Cotugno (Cottunius) (1736–1822), an Italian physician and professor of anatomy from Naples (Italy). It was formed during the 1975–76 fissure eruption of Tolbachik volcano (Kamchatka Peninsula, Russia), where it is associated with halite, silver, gold, tenorite, and rare minerals such as burnsite and ponomarevite. It also forms by the alteration of galena in saline environments, and as an alteration product of lead-bearing slag or other archaeological material after immersion in sea water. Hence, like diaboleite, cotunnite occurs in the ancient slag heaps of Laurion (Greece).

$PbCl_2$

Colour:	colourless, pale green, pale yellow, white
Lustre; opacity:	adamantine, silky, pearly; transparent to translucent
Streak:	white
Hardness:	2.5
Specific gravity:	5.8
Cleavage; fracture:	perfect; conchoidal to uneven
Habit:	prismatic or acicular crystals, or skeletal; in aggregates of radiating sprays, massive, granular, crusts, pseudomorphs
Crystal system:	orthorhombic

Boleite

Boleite is an unusual indigo blue colour, and makes a most attractive collector's item, although crystals are rarely cut as gems. It was named after its type locality, Boleó (Baja California, Mexico). Although tetragonal, it is always twinned, and as a result appears as cubes, with corners sometimes cut by octahedral faces. A secondary mineral, it forms from the alteration of sulphide deposits by chlorine-bearing aqueous solutions. It may be associated with atacamite, anglesite, cerussite or gypsum. Noted localities include Broken Hill (NSW, Australia), Mammoth District (Arizona, USA) and the Mendip Hills (Somerset, England). It also forms in Laurion (Greece), where smelter slag has been immersed in the sea.

$Pb_9Cu_8Ag_3Cl_{21}(OH)_{16}.H_2O$

Colour:	light indigo blue, azure-blue, dark blue
Lustre; opacity:	weakly vitreous to pearly on cleavage surfaces; translucent to transparent
Streak:	light green, light blue
Hardness:	3.5
Specific gravity:	4.8–5.1
Cleavage; fracture:	perfect; uneven
Habit:	normally as pseudo-cubic twinned crystals
Crystal system:	tetragonal

Diaboleite

Diaboleite comes from the Greek *dia* ('difference') and boleite (the mineral); it should not be confused with pseudoboleite, $(Pb_5Cu_4Cl_{10}(OH)_8.2H_2O)$, meaning false boleite. Diaboleite is a very rare secondary mineral, its type locality being the Mendip Hills (Somerset, England). The other main locality is in the copper porphyry deposits of Mammoth Mine (Arizona, USA), associated mainly with cerussite, wulfenite, quartz and hemimorphite, and also with boleite, linarite and pseudoboleite. It occurs in the 2000-year-old mineral slags of Laurion (Greece), many of which are now in the sea. This has led to the production of many new minerals, including laurionite, the first slag mineral to be described (in 1887).

$Pb_2CuCl_2(OH)_4$

Colour:	dark blue
Lustre; opacity:	adamantine; transparent to translucent
Streak:	blue
Hardness:	2.5
Specific gravity:	5.48 (extremely heavy for a translucent mineral)
Cleavage; fracture:	perfect; conchoidal
Habit:	very small tabular crystals; also as platy aggregates, grains, or as encrustations
Crystal system:	tetragonal

Zincite

Although zinc oxide itself is colourless, zincite is almost invariably coloured, due to manganese and iron impurities. The more attractive deep red crystals have earned it the name ruby zinc (at Franklin, where it is mined); the red colour is due to the presence of manganese dioxide or hematite. It was one of the first minerals described in an American mineralogy journal. Zincite is found as a primary mineral in metamorphosed ore deposits and as a secondary mineral in weathered zinc ore deposits. It is associated with franklinite, willemite and calcite at Sterling Hill and Franklin (New Jersey, USA): other locations are Tsumeb (Namibia), Kapushi, (Katanga, Congo) and in the ash from Mt St Helens (Washington, USA).

ZnO

Colour:	yellow, orange, red, brown, rarely green or colourless
Lustre; opacity:	subadamantine; transparent, more often translucent
Streak:	yellow–orange
Hardness:	4
Specific gravity:	5.68
Cleavage; fracture:	perfect; irregular
Habit:	hemimorphic crystals, hexagonal prisms terminated differently at either end; granular, massive and foliated in veins
Crystal system:	hexagonal

Cuprite

Cuprite is known as red copper oxide, which distinguishes it from CuO, or black copper oxide. Cuprite oxidizes slowly in air, tarnishing the surface. It is an important copper ore, mainly being found in dull massive bodies. Fine, fibrous crystals are called chalcotricite; ruby copper is a gem-quality crystal found at Santa Rita (New Mexico, USA) and Onganyo (Namibia). It occurs as a secondary mineral in the oxidized zones of copper ore deposits, accompanied typically by malachite, azurite, chalcocite and native copper. Cuprite often occurs with oxides of iron as an earthy red-brown material known as tile ore. It is found at Tsumeb (Namibia), Cornwall (England), Chessy (France) and Bisbee (Arizona, USA).

Cu_2O

Colour:	red to dark red
Lustre; opacity:	submetallic, adamantine, earthy; translucent
Streak:	brownish-red
Hardness:	3.5–4
Specific gravity:	5.8–6.1
Cleavage; fracture:	poor; irregular, conchoidal
Habit:	crystals usually octahedral, sometimes dodecahedral or cubes, rarely needles; mostly massive, granular or earthy
Crystal system:	cubic

Perovskite

Perovskite is mined not for titanium, but for rare earths in it such as niobium and cerium. Named after the Russian mineralogist L. Perovski (1792–1856), perovskite has a structure well known among crystallographers adopted by superconducting ceramics and high-pressure minerals in the Earth's mantle. It occurs in silica-poor igneous rocks and metamophic rocks. Associations include rare earth minerals such as loparite ((Na,Ce)TiO_3), as well as pyrochlore, ilmenite, leucite and titanite. Good crystals have been obtained from Alno (Sweden), Magnet Cove (Arkansas, USA) and the Urals (Russia); other locations include Val Malenco and Vesuvius (Italy), Bagagem (Brazil), Kaiserstuhl (Germany) and Quebec (Canada).

$CaTiO_3$

Colour:	red-brown, grey-black, yellow
Lustre; opacity:	adamantine, submetallic; translucent
Streak:	pale yellow
Hardness:	5.5
Specific gravity:	4.0
Cleavage; fracture:	none; subconchoidal to indistinct
Habit:	crystals pseudocubic, often striated parallel to edges; granular aggregates and reniform masses
Crystal system:	orthorhombic

Tenorite

Tenorite, or black copper oxide, is a rare mineral named after the Italian botanist
Michel Tenore (1781–1861). Earthy specimens containing tenorite are called
melaconite. It is formed in the oxidized zones of copper deposits, associated with
azurite, malachite, chalcocite, cuprite and limonite, and occasionally found in
volcanic sublimates. The bladed crystals are collectable specimens, and botryoidal
tenorite provides a contrasting dull, grey setting as a matrix for bright blue-green
chrysocolla. Tenorite occurs at Val d'Ossola (Italy), Cornwall (England), Leadhills
(Scotland), Rio Tinto (Spain), the Urals and Kamchatka (Russia), Bisbee (Arizona,
USA), Chuquicamata (Chile) and Tsumeb (Namibia).

CuO

Colour:	steel-grey to black
Lustre; opacity:	metallic, dull; opaque
Streak:	black, greenish
Hardness:	3–4
Specific gravity:	5.8–6.4
Cleavage; fracture:	indistinct; conchoidal to irregular
Habit:	crystals usually elongated plates, often striated and serrated edges; scaly and earthy aggregates; encrustations
Crystal system:	monoclinic

Gahnite

Gahnite, known as automolite or zinc-spinel, is named after its discoverer, J. Gahn (1745–1818), the Swedish chemist who discovered manganese. It occurs in granite pegmatites and metamorphic rocks and as a placer deposit. It is associated with galena, sphalerite and magnetite. Gahnite is usually found as small crystals in the type localities Falun (Sweden), Silberberg (Bavaria, Germany) and Tiriolo (Calabria, Italy), but larger crystals have been obtained from mines in Franklin and Sterling Hill (New Jersey, USA). Other deposits are in Connecticut (USA), Minas Gerais (Brazil), Smilovne (Bulgaria) and Victoria Range (New Zealand). A lead-bearing variety called limaite is found at Ponto de Lima (Portugal).

$ZnAl_2O_4$

Colour:	green to bluish green
Lustre; opacity:	vitreous, greasy; translucent, opaque on edges
Streak:	grey
Hardness:	7.5–8
Specific gravity:	4.6
Cleavage; fracture:	indistinct; conchoidal
Habit:	crystals; granular aggregates, grains
Crystal system:	cubic, but twinning is ubiquitous and crystals may appear trigonal

Spinel

The term 'spinel' is also used as a general term for minerals of formula AB_2O_4 such as gahnite, hercynite ($FeAl_2O_4$) and galaxite ($MnAl_2O_4$). Spinel has been mistaken for ruby and sapphire; the so-called Timur Ruby in the British Crown jewels is a ruby spinel. Spinel is found in contact-metamorphosed rocks, in igneous rocks and as a placer deposit. Synthetic spinels have been produced as artificial gemstones since 1910. The word spinel comes from the Latin *spina*, for 'little thorn', after the sharpness of its crystals. The best gem-quality spinels come from gravels in Sri Lanka, Myanmar and Madagascar. Good crystals are found at Vesuvius and Lazio (Italy), Orange County (New York, USA) and Sterling Hill (New Jersey, USA).

$MgAl_2O_4$

Colour:	colourless pure; red, blue, green, black
Lustre; opacity:	vitreous; transparent to translucent
Streak:	white to grey or brown
Hardness:	7.5–8
Specific gravity:	3.5–4.1
Cleavage; fracture:	none, octahedral parting; conchoidals
Habit:	octahedral crystals; granular aggregates; grains
Crystal system:	cubic

Magnetite

Magnetite, or magnetic oxide of iron, is the essential constituent of lodestone, known since ancient times. It is a major iron ore and is mined in vast quantities for the iron and steel industries; slag is further worked to recover vanadium and phosphorus. Swedish magnetite, containing some silicates, is used to make a very hard silicon steel. Magnetite is widespread among many igneous and metamorphic rocks, especially mafic and ultramafic rocks in high temperature mineral veins. It is also found in river and marine sediments and in dune deposits. Good crystals come from Val Malenco and Val de Vizze (Italy), Binnental (Swizerland) and Pfitschtal (Austria). Lodestone is famously found at Magnet Cove (Arkansas, USA).

Fe_3O_4

Colour:	black
Lustre; opacity:	lustre, submetallic, dull; opaque
Streak:	black
Hardness:	5.5–6.5
Specific gravity:	5.2
Cleavage; fracture:	none, octahedral parting; conchoidal
Habit:	crystals octahedral, sometimes dodecahedral; granular or massive
Crystal system:	cubic

Chromite

Chromite is chemically related to magnetite, chromium atoms replacing those of iron; it is also magnetic, but weaker than magnetite. It is the only commercial ore of chromium, which is used in steels, especially stainless steel. Chromium salts are used in electroplating, leather tanning, fireproofing fabrics and in paints. Chromite is used as a refractory for lining furnaces in the ceramic industry. The range in specific gravity is a reflection of its contamination with magnesiochromite ($MgCr_2O_4$), spinel ($MgAl_2O_4$) and related minerals. Chromite occurs in ultrabasic rocks and serpentinites, and as a placer deposit. Large deposits are mined in South Africa, Russia, Albania, Turkey, Zimbabwe and the Philippines.

$FeCr_2O_4$

Colour:	brownish black to black
Lustre; opacity:	metallic to submetallic; opaque
Streak:	dark brown
Hardness:	5.5
Specific gravity:	4.1–5.1
Cleavage; fracture:	none; uneven, conchoidal
Habit:	crystals octahedral, rare; usually massive, granular
Crystal system:	cubic
Other:	weakly magnetic; insoluble in acids

Franklinite

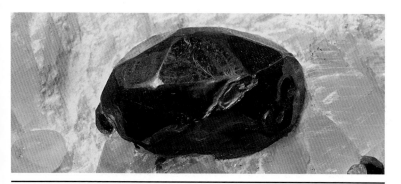

Franklinite is a spinel group mineral named after its type locality of Franklin. White calcite, green willemite and red zincite often accompany franklinite, and specimens displaying combinations of these are very collectable. Franklinite occurs in beds and veins in high temperature metamorphosed dolomites, associated with calcite, zincite, willemite, rhodonite, magnetite and garnet. It is also found in some manganese and iron deposits. The most famous and most mined location is that at Franklin and Sterling Hill (New Jersey, USA), which has produced crystals up to 30cm (12in) large. Other locations are Långban (Sweden), Hranicna (Czech Republic), Atasui (Khazakstan), Sayan (Siberia, Russia) and Western Australia.

$ZnFe_2O_4$

Colour:	black
Lustre; opacity:	metallic, with reddish internal reflections; opaque
Streak:	red–brown
Hardness:	5.5–6.5
Specific gravity:	5.1–5.2
Cleavage; fracture:	imperfect; conchoidal
Habit:	octahedral, sometimes dodecahedral crystals; massive aggregates
Crystal system:	cubic

Minium

Minium, red oxide of lead or simply red lead, has been used as a pigment, especially in anti-rusting paints, although lead's toxicity has lowered demand. Its unusual name comes from the river Minius in northwest Spain. It is formed by the alteration of galena and cerrussite, under extreme oxidizing conditions, and its presence may function as an indicator of the degree of oxidation. Some specimens from Broken Hill (NSW, Australia) are particularly good, but are said to have been the result of a mine fire. Its powdery form can disguise its high density. Minium is found at Badenweiller and Horhausen (Germany), Monteponi (Sardinia, Italy), Bolaños and Zimapan (Mexico), Leadhills (Scotland) and Broken Hill (Australia).

Pb_3O_4

Colour:	light red to red-brown
Lustre; opacity:	dull to greasy; opaque
Streak:	orange-yellow
Hardness:	2–3
Specific gravity:	8.2
Cleavage; fracture:	perfect; earthy
Habit:	rare, scaly crystals; usually powdered or massive aggregates
Crystal system:	tetragonal

Chrysoberyl

Chrysoberyl is hard and durable, making a fine stone that is used in jewellery, though somewhat lacking in 'fire'. A rare, much valued variety is alexandrite, which is green in daylight, but appears reddish in artificial (tungsten filament) light. Cat's eye has inclusions of fine needles of rutile and is most prized when golden yellow-brown. Chrysoberyl is found in pegmatites and mica schists around granite intrusions, and in alluvial and marine deposits. It has been obtained from the Ural Mountains for thousands of years. Alexandrite crystals 1–10cm (½–4in) in length occur in the Urals' Takowaja River. Alluvial deposits in Brazil and Sri Lanka produce cat's eye. Fine transparent yellow-green crystals occur at Espirito Santo (Brazil).

$BeAl_2O_4$

Colour:	colourless, grey, brown, green, yellow
Lustre; opacity:	subadamantine, silky; transparent to translucent
Streak:	white
Hardness:	8.5
Specific gravity:	3.7
Cleavage; fracture:	good; conchoidal
Habit:	prismatic, tabular crystals, often pseudohexagonal twins; grains and pebbles in alluvial deposits
Crystal system:	orthorhombic

Corundum

Extremely hard (equal to silicon carbide and second only to diamond), corundum is a gem mineral (ruby and sapphire) and is used as an abrasive in the form of emery for cutting, grinding and drilling. Grey-coloured masses are often forms that have been altered by hydrothermal solutions to margarite or zoisite. Corundum occurs in igneous rocks, in contact metamorphosed shales and bauxite, in pegmatites and metamorphosed limestones and as a placer deposit. Emery deposits are granular masses, often with magnetite, hematite and spinel. Good localities are Mogok (Myanmar), Sri Lanka, Madagascar, Glebe Hill (South Africa), Ardnamurchan (Scotland), Naxos (Greece) and the central Urals (Russia).

Al_2O_3

Colour:	colourless, grey or brown granular mass; can be red, blue, yellow, green, purple or colourless gems
Lustre; opacity:	adamantine; transparent to semi-opaque
Streak:	white
Hardness:	9
Specific gravity:	4
Cleavage; fracture:	none; uneven to conchoidal
Habit:	rough prisms or barrel shaped crystals bounded by steep pyramids; massive, granular
Crystal system:	trigonal

Corundum: Gem Varieties

Ruby is red, gem-quality corundum widely used in jewellery. The red colour varies with the content of chromium and iron oxides. A very rare pink-orange coloured stone is called papdaradscha. All other corundum gem varieties are called sapphire; the most popular is coloured shades of blue by iron and titanium impurities. Yellow sapphire, formerly known as oriental topaz, occurs alone or banded with blue sapphire to give a green form of sapphire known since medieval times as oriental peridot. Star rubies and sapphires are opalescent and reflect light in the form of three- or six-pointed stars, an effect called asterism. Fine rubies and sapphires come from Myanmar (Burma) Sri Lanka and India.

Al_2O_3

Colour:	colourless, grey or brown granular mass; can be red, blue, yellow, green, purple or colourless gems
Lustre; opacity:	adamantine; transparent to semi-opaque
Streak:	white
Hardness:	9
Specific gravity:	4
Cleavage; fracture:	none; uneven to conchoidal
Habit:	rough prisms or barrel shaped crystals bounded by steep pyramids; massive, granular
Crystal system:	trigonal

Arsenolite

Arsenolite or white arsenic, produced as a by-product of sulphide smelting, is widely used for preparing arsenic compounds, but quite rare in nature. Usually dull in appearance, crystal specimens – e.g. from White Caps (Nevada, USA) and St Etienne (France) – can be quite beautiful under magnification. A dimorphic form of arsenolite is the monoclinic claudite. Arsenolite is toxic and should be handled with care. Arsenolite is formed by alteration of arsenic minerals, including by mine fires. Associated minerals are realgar, orpiment and erythrite. It is found at Annaberg (Germany), Jáchymov (Czech Republic), Cornwall (England), Laurion (Greece), St Marie-aux-Mines (France) and Smolnik (Slovakia).

As_2O_3

Colour:	white, pale blue, pale pink or yellow if contaminated with realgar or orpimiment
Lustre; opacity:	vitreous
Streak:	white
Hardness:	1.5
Specific gravity:	3.7
Cleavage; fracture:	perfect; conchoidal
Habit:	rare octahedral crystals; encrustations, earthy
Crystal system:	cubic

Senarmontite

Senarmontite is dimorphic with the orthorhombic valentinite – also called antimony bloom, with which it is commonly associated. Senarmontite is named after the French mineralogist H. de Sénarmont (1808–1862), whereas valentinite is more intriguingly named after the sixteenth-century alchemist B. Valentinus. Commercially obtained from other antimony-bearing minerals, senarmontite is used in paints, plastics, medicines and especially as a flame-retardant for PVC in aircraft and motor vehicles. It occurs in oxidized antimony-bearing hydrothermal deposits. Large crystals are found at Hamimate Mine (Sensa, Algeria); other locations are Pernek (Slovakia), Mopung Hills (Nevada, USA) and Cornwall (England).

Sb_2O_3

Colour:	colourless to grey
Lustre; opacity:	resinous; translucent to transparent
Streak:	white
Hardness:	2
Specific gravity:	5.3
Cleavage; fracture:	conchoidal; irregular
Habit:	well-formed octahedral crystals; granular, encrusting, massive
Crystal system:	cubic

Ilmenite

Ilmenite is a major ore of titanium. Pure ilmenite is black with a metallic lustre, but its properties can be altered by incorporated giekielite ($MgTiO_3$) and pyrophanite ($MnTiO_3$). Resistant to weathering, it often appears in sands such as at Menaccan Sands (Cornwall, England), where menaccanite was an early reported form of the mineral. The streak is useful to distinguish it from hematite and, unlike magnetite, it is non-magnetic. It is found in igneous rocks, where it settles in magma intrusions with other dense minerals such as hematite and magnetite. Good crystals are found at Kragerö (Norway), Val Devero (Italy), St Gotthard (Switzerland) and Orange County (New York, USA), and in sands at Travancore (India).

$FeTiO_3$

Colour:	iron black
Lustre; opacity:	metallic to submetallic; opaque
Streak:	black to brownish black
Hardness:	5–6
Specific gravity:	4.5–5
Cleavage; fracture:	none, basal partings; conchoidal
Habit:	crystals flat tabular rhombohedra; massive, compact; grains disseminated in igneous rocks
Crystal system:	trigonal

Hematite

Hematite, or red oxide of iron, is named after the Greek for 'blood'. On oxidation of iron-rich fluids, it precipitates, to give rocks a rusty appearance. Red ochre is an earthy clay-rich mixture containing hematite. Rounded opaque bodies, such as kidney ore, have been carved into figures that take a high polish. Rare crystalline forms are petal-shaped aggregates called 'iron roses' and shiny specular hematite. Powdered, it is an abrasive for polishing as jewellers' rouge. It occurs in oxidized igneous rocks and veins, and as a sedimentary cement. Huge quantities are mined at Lake Superior (USA), Quebec (Canada), Brazil and Australia. Beautiful crystals occur at Rio Marino (Elba, Italy), Bahia (Brazil) and Cumbria (England).

Fe_2O_3

Colour:	steel–grey to iron–black, can be iridescent; dull to bright red when massive or earthy
Lustre; opacity:	metallic, dull, earthy; opaque
Streak:	dark red to red-brown
Hardness:	5–6
Specific gravity :	4.9–5.3
Cleavage; fracture:	none; conchoidal
Habit:	crystals tabular or rhombohedral, sometimes with striated or curved faces; massive, laminated or earthy
Crystal system:	trigonal

Quartz

Quartz is the most abundant mineral on the Earth's surface. It forms beautiful crystals as hexagonal prisms, weighing up to 130kg (287lb). Quartz is used in glass-making, ceramics, refractories and abrasives. It produces electricity when strained (piezoelectric) or heated (pyroelectric), used in electrical sensors. Quartz is an essential mineral of many acid igneous and metamorphic rocks, and occurs in most clastic sediments. Some sandstones and their metamorphic equivalent, both called quartzites, are almost entirely quartz. It often occurs veins and fissures; these provide the best crystals. Good specimens occur in the Alps in Switzerland and Austria, at Carrara (Italy), Bourg d'Oisans (France) and the Urals (Russia).

SiO_2

Colour:	colourless or white; tinted many shades
Lustre; opacity:	vitreous; transparent to opaque
Streak:	white
Hardness:	7
Specific gravity:	2.65
Cleavage; fracture:	absent; conchoidal, splintery
Habit:	usually six-sided prismatic crystals terminated by six faces, the prisms faces often horizontally striated; massive, compact, drusy
Crystal system:	trigonal

Quartz: Gem Varieties

Clear, colourless crystals are called rock crystal. The opacity of milky quartz is caused by small, gas or liquid bubbles. Colours are produced by iron hydrates in yellow citrine, ferric oxide in violet amethyst, and titanium or manganese in pink rose quartz. Brown or smoky quartz, found as huge crystals in Brazil, can be produced artificially by irradiating rock crystal. Quartz cat's eye, tiger's eye and hawk's eye have inclusions giving a wavy, striped effect. Rutilated quartz contains rutile needles intersecting at 60° angles. Minute reflective scales in aventurine quartz give a green or brown spangled appearance. Many varieties are used in jewellery and specimens of amethyst are widely collected.

SiO_2

Colour:	colourless or white; tinted many shades
Lustre; opacity:	vitreous; transparent to opaque
Streak:	white
Hardness:	7
Specific gravity :	2.65
Cleavage; fracture:	absent; conchoidal, splintery
Habit:	usually six-sided prismatic crystals terminated by six faces, the prisms faces often horizontally striated; massive, compact, drusy
Crystal system:	trigonal

Cristobalite

Cristobalite is a much rarer form of silicon dioxide than quartz and rarer than tridymite. It has two forms: tetragonal cristobalite, which is stable up to 270°C (518°F), and cubic cristobalite, which is stable above 1470°C (2678°F). Between these temperatures, both forms can exist. Beautiful specimens found at Cerro San Cristobal (Portugal) comprise octahedral cristobalite embedded in transparent but dark, glassy obsidian. Cristobalite occurs in intermediate igneous rocks and as recrystallized opals. Associated minerals are opal, chalcedony and tridymite. Other locations are Monte Dore (France), Glass and Sugarloaf mountains (California, USA), Crater Lake (Oregon, USA), Eiffel (Germany), Sarospatak (Hungary) and Tokatoka (New Zealand).

SiO_2

Colour:	white to yellowish
Lustre; opacity:	vitreous; translucent to transparent
Streak:	white
Hardness:	6.5
Specific gravity:	2.2
Cleavage; fracture:	absent; conchoidal
Habit:	rare crystals; microcrystalline as small balls, fibres (called lussatite), crusts
Crystal system:	tetragonal

Tridymite

Tridymite is stable between 870°C (1598°F) and 1470°C (2678°F) and is a high-temperature polymorph of quartz. Although in a metastable state at normal temperatures, it is widespread and possibly underestimated in abundance due to the difficulty in identification. Tridymite will change into quartz, but the process is very slow. Tridymite is prepared synthetically for use in heat-resistant porcelain and as refractory material in furnaces. Tridymite is found as a sublimate in acid volcanic rocks and in contact metamorphosed sandstones. Tridymite is found in the Eiffel district (Germany), Cerro San Cristobal (Portugal), Pomona (California, USA), Mule Springs (Oregon, USA) and Kamomoto (Japan). It has been found in stony meteorites.

SiO_2

Colour:	colourless, white
Lustre; opacity:	vitreous to pearly; transparent to translucent
Streak:	white
Hardness:	6.5–7
Specific gravity:	2.27
Cleavage; fracture:	absent; conchoidal
Habit:	small pseudo-hexagonal blades; spherical aggregates
Crystal system:	monoclinic

Chalcedony

Chalcedony is made of microscopic quartz crystals and is mostly banded, which imparts much of its appeal as an ornamental stone. Flints and cherts are impure forms, the former being well known as nodules in chalk. It is named from the Ancient Greek town of Kalchedon in Asia Minor. Chalcedony is usually produced by precipitation from aqueous solutions, often associated with hot springs or volcanoes. It can be produced by dehydration of opal. Chalcedony is found as the variety agate in Rio Grande de Sul (Brazil) and Idar-Oberstein (Germany), as carnelian in Brazil, Uruguay and California (USA), as chrysoprase in the Urals (Russia), California and Queensland (Australia), and flint in Southern England.

SiO_2

Colour:	variable, white through shades of brown, red and grey to black
Lustre; opacity:	waxy to vitreous; translucent
Streak:	white
Hardness:	6.5
Specific gravity :	2.6
Cleavage; fracture:	absent; uneven, splintery, conchoidal
Habit:	no crystals; botryoidal or stalactitic, often banded; in fissures and veins, massive or nodular
Crystal system:	none, microcrystalline

Chalcedony: Gem Varieties

Chalcedony, like quartz, has a number of attractive forms coloured by various impurities, but shows different patterns of banding. Agate has curved, often semi-concentric bands. Moss agate is a pale translucent stone with dendritic moss- or tree-like inclusions of iron oxides. The translucent carnelian is orange-red and jasper is a more massive, sometimes striped opaque stone of varying colour, but often red. Chrysoprase is a fine translucent, apple-green; heliotrope, or bloodstone, is dark green and opaque with red spots; and plasma has yellow spots. Onyx has white and black or brown bands, with straight bands rather than curved. Sard is a variety of onyx, and sardonyx has white bands like onyx and red-brown bands like sard.

SiO_2

Colour:	variable, white through shades of brown, red and grey to black
Lustre; opacity:	waxy to vitreous; translucent
Streak:	white
Hardness:	6.5
Specific gravity:	2.6
Cleavage; fracture:	absent; uneven, splintery, conchoidal
Habit:	no crystals; botryoidal or stalactitic, often banded; in fissures and veins, massive or nodular
Crystal system:	none, microcrystalline

Opal

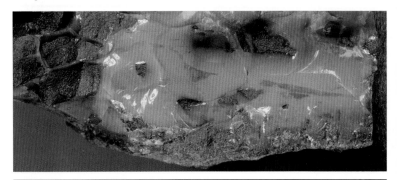

Opal, named from the Sanskrit for 'precious stone', is amorphous hydrated silica. It is made of tiny spheroids, giving internal reflections of light and a play of colours, or 'fire'. Precious opal is beautifully iridescent. Red fire opal and black opal are the most sought-after varieties; the latter has been more expensive than diamond. Opal is produced by the low-temperature weathering and alteration of silica-rich rocks, especially around geysers and hot springs. The skeletons of diatoms, sponges and radiolaria are composed of opal. Opals have been obtained from White Cliffs (NSW, Australia) since the nineteenth century. Mines in the Czech Republic have been worked since Roman times. An artificial form is called Slocum stone.

$SiO_2.nH_2O$

Colour:	white when pure; can be yellow, red, blue, brown or others
Lustre; opacity:	waxy or resinous; transparent to opaque
Streak:	white
Hardness:	5.5–6.5
Specific gravity:	1.8–2.3
Cleavage; fracture:	none; conchoidal, uneven
Habit:	massive, often nodular, stalactitic, cavity-filling, concretions or crusts
Crystal system:	none, amorphous

Pyrolusite

Pyrolusite is an important ore of manganese, a metal used widely in steel-making. It is also used in glass-making for removing coloration due to impurities of iron oxides, hence the name from the Greek words for 'fire' and 'wash'. Pyrolusite occurs with manganite and psilomelane, in a dark earthy ore called wad. Found on the surface of rocks in joints, it forms dendritic markings, often picturesque or fanciful. Pyrolusite is formed by chemical precipitation in lakes or oxidation of manganese ores. Large nodules are found at the bottom of seas and oceans. Mined localities include Platten (Bavaria, Germany), Epleny (Hungary), Cornwall (England), Nikepol (Ukraine), Deccan (India), Minas Gerais (Brazil), Cuba and South Africa.

MnO$_2$

Colour:	grey or grey-black
Lustre; opacity:	metallic; opaque
Streak:	black to bluish-black
Hardness:	6–6.5
Specific gravity:	4.7–5.1
Cleavage; fracture:	poor; uneven
Habit:	rare prismatic crystals; usually concretionary aggregates and earthy masses
Crystal system:	tetragonal

Cassiterite

Cassiterite, named after the Greek for 'tin', is its major ore. Tin is widely used for metal plating in food canning and for alloys, famously in bronze (one of the earliest metals known to the ancients). Twinned crystals known as knee twins are popularly collected. Radiating, fibrous concretions are known as wood tin. Cassiterite is found in hydrothermal veins and pegmatites associated with minerals such as fluorite, scheelite, topaz and tourmaline. Placer deposits are called stream tin. Rarely occurring in workable quantities in the USA, it is found at Erzgebirge, Altenberg and Zinnwald (Germany), Cornwall (England), Brittany (France), Bolivia, Sumatra, Yunnan (China) and Elba and Lake Como (Italy).

SnO_2

Colour:	brown to black
Lustre; opacity:	adamantine; opaque
Streak:	white to yellow
Hardness:	7
Specific gravity:	6.8–7.1
Cleavage; fracture:	imperfect; conchoidal
Habit:	crystals as prisms, bipyramids and needles
Crystal system:	tetragonal

Rutile

Rutile is a form of titanium dioxide often embedded in other minerals as needle-like inclusions of metallic appearance. Such inclusions have been named maiden hair, cat's eye and star forms. Quartz crystals with embedded rutile needles intersecting at angles of 60° are called sagenite or rutilated quartz. Rutile is durable and cuts well, having a high 'fire', but its colour is usually unattractive. Twinning is common, giving elbow-shaped crystals. Pure powdered titanium dioxide is brilliant white and used as a pigment in white paints. Rutile is found in igneous intrusions, pegmatites and metamorphic rocks and placer deposits. Good crystals are found at Binnental (Switzerland) and in Grove Mountain (Georgia, USA).

TiO_2

Colour:	yellow, red, brown, black
Lustre; opacity:	metallic; translucent to opaque
Streak:	yellow browns
Hardness:	6–6.5
Specific gravity :	4.3
Cleavage; fracture:	perfect; uneven, conchoidal
Habit:	crystals usually elongate prisms, often striated; grains, fibrous aggregates and inclusions
Crystal system:	tetragonal

Anatase

Anatase is named from the Greek for 'elongated', after the sharp crystals it often forms. It is a polymorph of titanium dioxide and is formed, like brookite, at lower temperatures than rutile, to which it can be converted by heating. Anatase mostly forms as transparent to opaque 1–3mm (up to ⅛in) isolated crystals embedded in a matrix of accessory minerals. It is easily distinguished from rutile by crystal form and specific gravity. It is found as a secondary mineral in titanium-bearing rocks, in igneous rocks, in metamorphic rocks and as a placer deposit. Associated minerals are brookite, albite, quartz and titanite. Excellent specimens are found at Binnental and Tavetschal (Switzerland) and Val Devero and Vatellina (Italy).

TiO_2

Colour:	yellow to brown, deep blue, black
Lustre; opacity:	adamantine; translucent to opaque
Streak:	white to light brown
Hardness:	5.5–6
Specific gravity:	3.8–3.9
Cleavage; fracture:	perfect; conchoidal
Habit:	usually sharp, bipyramidal crystals, sometimes tabular
Crystal system:	tetrahedral

Brookite

Named after the English mineralogist H. Bruke (1771–1857), brookite is the third polymorph of titanium dioxide. Like anatase, it forms at lower temperatures than rutile. The larger crystals, especially heart-shaped twins, are popular. Brookite is found in veins and fissures in altered igneous and metamorphic rocks, such as granites and gniesses. Associated minerals include anatase, albite, quartz and rutile. It also occurs as a placer deposit. Fine crystals are found at Bourg d'Oisans (Dauphiné, France), Grisons and Uri cantons (Switzerland) and Tyrol and Untersulzbachtal (Austria). The site at Prenteg (Wales) produced some of the finest specimens in the nineteenth century.

TiO$_2$

Colour:	brown to black
Lustre; opacity:	adamantine, submetallic; transparent to translucent
Streak:	yellow, yellow-brown
Hardness:	5.5–6
Specific gravity:	3.9–4.2
Cleavage; fracture:	imperfect; subconchoidal
Habit:	tabular or lamellar crystals, sometimes striated
Crystal system:	orthorhombic

Columbite

Columbite is the major ore of niobium, which is used in high-strength steels and superconducting alloys. The term columbite can also refer to the series of minerals from ferrocolumbite to manganocolumbite, $(Fe,Mn)Nb_2O_6$. Columbite also forms a series with tantalite, $(Fe,Mn)(Ta,Nb)_2O_6$. Niobium, named from Niobe, daughter of Tantalus, was chosen over columbium (after Columbus) as the preferred name in the 1950s, after much debate. Columbite and tantalite are found in granite pegmatites rich in lithium and phosphorus, associated with albite, spodumene, tourmaline, cassiterite, apatite and beryl. It is found at Haddam (Connecticut, USA), Rabenstein (Germany), Ilmen Mountains (Urals, Russia) and Greenland.

$FeNb_2O_6$

Colour:	brown or black
Lustre; opacity:	metallic; nearly opaque, transparent in thin lamellae
Streak:	black to dark red
Hardness:	6
Specific gravity:	5.1
Cleavage; fracture:	good; subconchoidal
Habit:	crystals as complex, stubby prisms, aggregates of thin, tabular crystals; granular, massive
Crystal system:	orthorhombic

Furgusonite

Furgusonite, named after the Scottish politician and landowner Robert Furguson (1767–1840), is an ore of yttrium found widespread but in small amounts. Specimens are usually radioactive and often metamict. Yttrium is used to give red colours on TV screens, in X-ray filters, in superconductors and in alloys. Invariably substituted by a number of the other rare-earth elements, the term furgusonite-Y is used for the yttrium-rich mineral. Fergusonite occurs in rare earth-bearing granite pegmatites and as a placer deposit. The type locality where it was first discovered is Qeqertaussaq Island (Greenland). Major mining areas include Arendal (Norway), Blum mine (Ilmen Mountains, Russia) and Ytterby (Sweden).

$YNbO_4$

Colour:	black, brown, grey, yellow
Lustre; opacity:	submetallic; translucent to opaque
Streak:	brown
Hardness:	5.5–6
Specific gravity:	4.3–5.8
Cleavage; fracture:	indistinct; subconchoidal
Habit:	crystals as prismatic to acicular dipyramids, usually powdery; granular, massive
Crystal system:	tetragonal

Brucite

Brucite, an ore of magnesium, is used to produce magnesia (MgO), magnesium compounds and refractories. Brucite can be quite rich in manganese, weathering to dark brown. Brucite occurs in marbles by alteration of periclase, in hydrothermal veins in metamorphic limestones and chlorite schists, and in serpentinized dunites. Associated minerals include calcite, aragonite, dolomite, magnesite, talc and chrysotile. It is found at Castle Point (Hoboken, New Jersey, USA), Gabbs (Nevada, USA), Asbestos (Quebec, Canada), Vesuvius and Sardinia (Italy), Shetland Islands and Isle of Muck (Scotland), and the Ethyl Mine (Mutorashanga, Zimbabwe). It was first described by Archibald Bruce (1777–1818), an American mineralogist.

$Mg(OH)_2$

Colour:	white, shades of grey, blue and green
Lustre; opacity:	waxy, pearly on cleavage faces; transparent
Streak:	white
Hardness:	2–2.5
Specific gravity:	2.4
Cleavage; fracture:	perfect; uneven, separates into flexible plates
Habit:	tabular crystals in platy or foliate masses, rosettes, sometimes fibrous; granular, massive
Crystal system:	hexagonal

Goethite

Goethite is a major constituent of limonite and ochres, and an important ore mineral. Named after the famous German poet J. Goethe (1749–1832), it is also known as iron hydroxide, needle ironstone or acicular iron ore. Black crystals are attractive but rare. Sprays of acicular crystals are found at Pfíbram. Goethite often forms an attractive matrix in specimens of other minerals such as vanadite. Its perfect cleavage gives it a soft and greasy feel. Goethite is formed by oxidation of iron-rich minerals, usually pyrite, siderite and magnetite. Large amounts are mined at Pribram (Czech Republic) and in Cuba, Alsace-Lorraine (France), Westphalia (Germany), Lake Superior (USA) and Labrador (Canada).

$FeO(OH)$

Colour:	brown to black
Lustre; opacity:	adamantine, submetallic, silky; translucent to opaque
Streak:	yellow-brown
Hardness:	5–5.5
Specific gravity:	4.3
Cleavage; fracture:	perfect; uneven to hackly
Habit:	rare prisms with vertical striations; usually stalactitic, massive or earthy aggregates of radiating fibrous forms
Crystal system:	orthorhombic

Diaspore

Diaspore is most widely found in bauxite, the major ore of aluminium. Other minerals constituting bauxite are closely related, especially boehmite (AlO(OH)) and gibbsite (Al(OH)$_3$). Bauxite is formed by extreme weathering of aluminosilicate rocks in tropical regions. Purer forms of diaspore are rare relative to the abundance of bauxite and good crystals rarer still. Other important associations are with corundum and margarite among emery deposits and with chlorite and chloritoid in metamorphic rocks. The name from the Greek 'to disperse' alludes to its easy disintegration in a flame. Crystals are found in Chester (Massachusetts, USA), Naxos (Greece), Campolongo (Switzerland) and Mramorskoi (Urals, Russia).

AlO(OH)

Colour:	white; greenish, grey or pink as aggregates
Lustre; Opacity:	vitreous or pearly; transparent to translucent
Streak:	white
Hardness:	6.5–7
Specific Gravity :	3.3–3.5
Cleavage; Fracture:	perfect; conchoidal
Habit:	crystals tabular or acicular; foliated or stalactitic aggregates
Crystal System:	orthorhombic

Limonite

Named from the Latin *limus* for 'mud', limonite has been variously called brown ironstone, brown iron and brown haematite. It is a term used to describe a rock made of an ill-defined mixture of mostly amorphous hydrated iron oxides, microcrystalline goethite (FeOOH) and lepidocrocite (FeOOH). Not strictly a mineral, but not quite seen as a rock, limonite is usually included among minerals. It is used as a pigment as yellow ochre and in modelling clay. It is quite hard, but very fragile, easily disintegrating into grains and powders. It is formed by surface oxidation of iron deposits or is left after the dissolution of iron-rich rocks in tropical regions. Limonite is often found as a cubic polymorph after pyrite.

hydrated iron oxides

Colour:	yellow-brown to black
Lustre; opacity:	subvitreous, dull, earthy; opaque
Streak:	yellow brown
Hardness:	5–5.5
Specific gravity:	ca. 4
Cleavage; fracture:	absent; conchoidal, splintery
Habit:	botryoidal, stalactitic, oolitic earthy or porous masses; loose yellow-brown crusts to dark, iridescent bodies when compact
Crystal system:	none, amorphous

Manganite

Manganite, or brown ore of manganese, is a minor ore of manganese. It usually forms rather dull bodies similar to other manganese minerals, but when crystalline it is easier to identify and is quite collectable. Such specimens include lustrous feathery crystals (up to 8mm/⅓in), found in the Kalahari manganese field. Pyrolusite, which can resemble manganite, is softer and has a bluish streak. It is formed by the alteration of other manganese minerals in low-temperature hydrothermal veins, associated with calcite and barite, and in hot springs, associated with psilomelane and pyrolusite. Crystals up to 7–8cm (2¾–3in) large are found in Harz (Germany), Como (Italy) and Negaunee and Marquette (Michigan, USA).

$MnO(OH)$

Colour:	black, opaque
Lustre; opacity:	metallic; opaque, red, translucent in thin plates
Streak:	dark brown
Hardness:	4
Specific gravity:	4.3–4.4
Cleavage; fracture:	perfect; uneven
Habit:	elongate prisms with deep striations lengthwise, usually in bundles; microcrystalline, granular, radially fibrous, oolitic
Crystal system:	monoclinic

Pyrochlore

Pyrochlore is the generic name usually applied to the series from pyrochlore, $(Na,Ca)_2Nb_2O_6(OH,F)$, to microlite, $(Na,Ca)_2Ta_2O_6(OH,F)$. Pyrochlore turns green on heating and is named from the Greek for 'fire' and 'green'; microlite is named from the Greek for 'small stone'. This compositionally diverse mineral is a source of a number of rare earth elements and uranium, and is usually radioactive. It is found in pegmatites and carbonatites, associated with apatite, nepheline, zircon, biotite and forsterite. It is rare and found at Mbeya (Tanzania), Fen (Norway), Kaiserstuhl (Germany), Oka (Canada), Newry (Maine), Haddam (Connecticut, USA), Betanima (Madagascar), Minas Gerais (Brazil) and Varutrask (Sweden).

$(Na,Ca)_2(Nb,Ta,Ti)_2O_6(OH,F,O)$

Colour:	usually brown; greenish, reddish
Lustre; Opacity:	greasy; translucent to opaque
Streak:	yellowish brown
Hardness:	5–5.5
Specific Gravity :	4.3–4.5
Cleavage; Fracture:	distinct, octahedral; conchoidal, uneven
Habit:	crystals usually octahedral, disseminated; granular
Crystal System:	cubic

Thorianite

Thorianite forms a series with uraninite (UO_2) and contains other elements substituting for thorium; a ThO_2 content of about 70 per cent would be typical. It is an ore of thorium, which is used in nuclear fuel elements, refractory materials and incandescent gas mantles. Like all thorium and uranium minerals, thorianite contains radiogenic elements such as lead, which have been formed by radioactive decay. It is found in pegmatites, carbonatites and serpentinites, and as a placer deposit. Thorianite is found at Taolañaro (Madagascar), Transvaal (South Africa), Kola Peninsula (Siberia, Russia) and Easton (Pennsylvania, USA). Thorium is named after the old Scandinavian god of war, Thor.

ThO_2

Colour:	brownish black, greenish, yellow
Lustre; opacity:	metallic; opaque
Streak:	greenish-grey
Hardness:	6.5–7
Specific gravity:	9.7–9.8
Cleavage; fracture:	imperfect; subconchoidal, brittle
Habit:	cubic crystals, often as interpenetrant twins; granular
Crystal system:	cubic

Uraninite

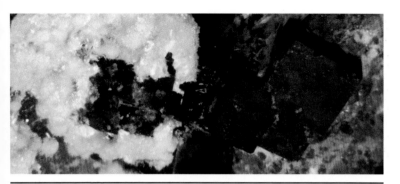

Ores containing uraninite, known as pitchblende, are major ores not only of uranium, but also of radium. Uranium is a major source of energy via nuclear fission. Uraninite is radioactive and found in pegmatites, in sandstones and conglomerates, and in hydrothermal veins in sulphide ores. Pitchblende from Yáchymov (Czech Republic) was used by the Curies in the discovery of polonium and radium in 1898. Radium and uranium are also sources of radioactivity for various uses in industry, science and medicine. Large crystals are found at Wilberforce (Canada). Mining areas include Great Bear Lake (Canada), Shinkolobwe (Zaire), the Colorado Plateau (USA) and Witwatersrand (South Africa).

UO_2

Colour:	brown, grey, black
Lustre; opacity:	greasy, submetallic; opaque
Streak:	black
Hardness:	5.5–6
Specific gravity:	4.3–4.5
Cleavage; fracture:	difficult; conchoidal
Habit:	rare modified cubes; usually dense botryoidal, reniform, massive, colloform; sometimes dendritic
Crystal system:	cubic

Billietite

Billietite is a radioactive mineral found as rare but attractive amber yellow crystals. It is an uncommon alteration product of uraninite, named after the Belgian crystallographer Valère Billiet (1903–1945). Billietite is similar to becquerelite ($Ca(UO_2)_6O_4(OH)_6.6\text{-}8H_2O$), and related to compreignacite ($K_2(UO_2)_6O_4(OH)_6.6\text{-}8H_2O$). Associated minerals are barite, calcite, chalcopyrite, hematite and other rare species such as uranophane, fourmarierite, metabernite, rutherfordine and becquerelite. Billietite is found in Shaba Province (Congo), the La Crouzille, Margnac and Rabéjac mines (France), Menzenschwand (Black Forest, Germany) and the Delta mine (Utah, USA).

$Ba(UO_2)_6O_4(OH)_6.6\text{-}8H_2O$

Colour:	yellow-brown
Lustre; opacity:	adamantine; transparent to translucent
Streak:	yellow
Hardness:	not determined
Specific gravity:	5.3
Cleavage; fracture:	perfect; brittle
Habit:	crystals pseudohexagonal, tabular
Crystal system:	orthorhombic

Plattnerite

Plattnerite is an uncommon dense mineral composed of lead oxide. It mostly occurs as masses, but specimens comprising drusy crusts showing many shiny crystals are fairly common. Specimens of plattnerite with contrasting minerals such as calcite and wulfenite are popular. Plattnerite is named after the German metallurgist K. Plattner (1800–1858). It occurs in oxidized, weathered hydrothermal lead deposits, typically in arid climates, associated with cerussite, smithsonite, hemimorphite, calcite and quartz. It is found in Leadhills and Wanlockhead (Scotland), Anarak and Anjireh (Iran), Tsumeb (Namibia), Idaho, Arizona, Nevada and New Mexico (USA), and Durango and Chihuahua (Mexico).

PbO$_2$

Colour:	black
Lustre; opacity:	adamantine to submetallic; opaque
Streak:	dark brown
Hardness:	5.5
Specific gravity:	9.6
Cleavage; fracture:	good; conchoidal to uneven
Habit:	small, prismatic crystals, sometimes acicular; nodular or botryoidal, fibrous or concentrically zoned, massive
Crystal system:	tetragonal

123

Betafite

Betafite is named after the type locality of Betafo in Madagascar. The hardness, specific gravity and other properties vary widely as composition changes. It is an ore of uranium and consequently radioactive; also used to obtain niobium, tantalum, other rare earth elements and associated thorium. Although black in colour, the surface is invariably altered (metamict), making it appear greenish or yellowish. It is found in pegmatites and adjacent metamorphosed limestones. Crystals up to 100kg (220lb) have been reported. Associated minerals are quartz, feldspars, biotite, zircon and rare earth minerals. Betafite is found at Tangen (Norway), Sludianka (Baikal, Russia), Bancroft (Ontario, Canada) and Val d'Ossola (Italy).

$(Ca,Na,U)_2(Ti,Nb,Ta)_2O_6(OH)$

Colour:	black, with a tint of yellow, brown or green
Lustre; opacity:	waxy, greasy to adamantine; translucent to opaque
Streak:	yellowish white
Hardness:	3–5.5
Specific gravity:	3.7–4.9
Cleavage; fracture:	none; conchoidal to uneven
Habit:	octahedral and dodecahedral crystals, sometimes elongated; granular masses common; crusts
Crystal system:	cubic

Psilomelane

Psilomelane is a term given to a collection of poorly defined hydrated barium-bearing manganese ores, the main component probably being romanechite ($BaMn_5O_{10}.H_2O$). The term wad is used for earthy mixtures of psilomelane, pyrolusite and others. Psilomelane is found in weathed ores and as a replacement deposit in limestones and dolomites. It also forms desert varnish, a dark coating of rocks in dry regions. Romanechite was authenticated at Romanèche (France) and occurs as crystals at Schneeberg (Saxony, Germany) and Oberwolfach (Black Forest, Germany). Psilomelane occurs at Cornwall (England), Virginia, Nevada and New Mexico (USA), Pilbara (Western Australia) and Chihuahua (Mexico).

No fixed formula

Colour:	iron black to steel grey
Lustre; opacity:	dull; opaque
Streak:	brownish black
Hardness:	5–5.5 (6 for romanechite)
Specific gravity:	4.4–4.5
Cleavage; fracture:	absent; conchoidal to uneven
Habit:	rare euhedral crystals (romanechite); massive, fibrous, botryoidal, stalactitic, concretionary, earthy, powdery
Crystal system:	monoclinic (romanechite)

Nitratine

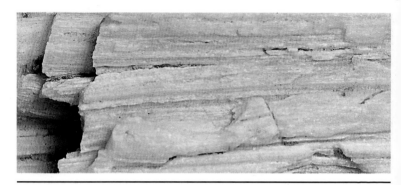

The nitrates as a whole readily dissolve in water, so nitratine (or soda nitre) should be kept in an airtight container with a desiccant. It has a bitter, cooling taste, and colours a flame yellow. It can be distinguished from nitre (salt-petre, KNO3) by the flame test, as nitre turns a flame violet. Nitratine is used in the chemical industry, for fertilizers and in fireworks. It occurs as a bedded deposit in playa lakes in arid regions, associated with halite, gypsum and other evaporites. It may also precipitate from nitrate-bearing ground water. Economically valuable deposits of billions of tons occur in regions adjacent to the type locality of Tarapacá (Chile). It also occurs in arid regions of the Persian Gulf.

NaNO$_3$

Colour:	colourless, white, sometimes greyish, yellowish brown
Lustre; opacity:	vitreous; transparent to translucent
Streak:	white
Hardness:	1.5–2
Specific gravity:	2.2–2.3
Cleavage; fracture:	perfect, rhombohedral; conchoidal, sectile
Habit:	crystals rare; granular or massive encrustations, stalactitic
Crystal system:	trigonal

Calcite

Calcite is a major rock-forming mineral, the principal component of limestone and marble, and an important constituent of igneous carbonatite. It effervesces strongly in cold, dilute hydrochloric acid. Crystals show a wide variety of forms, such as nail-head spar (flat-topped rhombohedra) or dog-tooth spar (sharply pointed). Fine sand-covered crystals come from Fontainebleau (France); calcite with hematite inclusions from Chihuahua (Mexico); and fluorescent calcite from Franklin (New Jersey, USA). An image appears double when viewed through transparent rhombs of Iceland spar, a property known as 'double refraction'. This was used to produce nicol prisms for petrological microscopes.

$CaCO_3$

Colour:	usually colourless or white; may be brown, red, green or black
Lustre; opacity:	vitreous, dull, pearly; transparent to translucent
Streak:	white to greyish
Hardness:	3
Specific gravity:	2.71
Cleavage; fracture:	perfect, rhombohedral; subconchoidal
Habit:	wide variety of crystal forms; stalactitic, granular, massive, fibrous or many other habits
Crystal system:	trigonal

Magnesite

Magnesite, or bitter spar, is much less common than calcite, and will only dissolve with effervescence in hydrochloric acid on warming. It forms when limestone is altered by magnesian solutions, often accompanied by the formation of dolomite, and during the hydrothermal metamorphism of ultramafic rocks to talc schists and serpentinite. An important ore of magnesium, it is heated to produce magnesium oxide (MgO), used in the manufacture of cements and refractory bricks. Powdered magnesia is used in the rubber, paper and pharmaceutical industries. Large deposits are found in Styria (Austria), Manchuria (China), Silesia (Poland), Madras (India) and Euboea (Greece).

$MgCO_3$

Colour:	white or colourless, also greyish or yellowish brown
Lustre; opacity:	vitreous or dull when compact; transparent to translucent
Streak:	white
Hardness:	3.5–4.5
Specific gravity:	3.0–3.2
Cleavage; fracture:	perfect rhombohedral; conchoidal fracture
Habit:	crystals rare and usually rhombohedral; massive, lamellar, granular or fibrous
Crystal system:	trigonal

Siderite

Siderite is a common carbonate, forming series with magnesite, rhodochrosite and smithsonite. It dissolves slowly with effervescence in cold hydrochloric acid, and becomes magnetic on heating. It usually forms in sedimentary deposits, such as in coalfields, as nodules and beds of impure iron carbonate called clay ironstone, and as oolitic ironstone. It is also found in metamorphic iron formations (as at Biwabik, Minnesota, USA). Magnesite-siderite series carbonates occur in carbonatites at Nkombwa (Zambia) and Newania (Rajasthan, India). It also forms in hydrothermal veins associated with barite, fluorite and galena. Large crystals occur at Mont Saint-Hilaire (Quebec, Canada) and Mosojllacta (Colavi, Bolivia).

$FeCO_3$

Colour:	shades of brown, tan, greenish grey
Lustre; opacity:	vitreous, silky, pearly on cleavage; translucent
Streak:	white
Hardness:	3.5–4.5
Specific gravity:	3.96 when pure
Cleavage; fracture:	perfect rhombohedral; uneven to conchoidal
Habit:	crystals rhombohedral, often with curved faces made up of overlapping scales; massive, granular, concretionary, botryoidal
Crystal system:	trigonal

Smithsonite

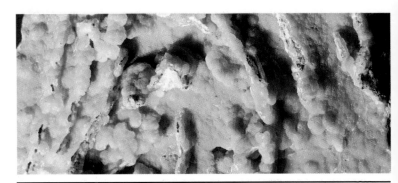

Smithsonite is very attractive and popular, and can fluoresce pale green or blue in ultraviolet light. A secondary mineral, it forms in the weathered zone of zinc deposits and may replace adjacent carbonate rocks. Associated minerals include malachite, azurite, aurichalcite, anglesite and hemimorphite. Pale pink twinned crystals are found in Ojuela Mine (Durango, Mexico); pale green botryoidal smithsonite at Broken Hill (NSW, Australia); and blue-green botryoidal specimens from the Kelly Mine (New Mexico, USA). Cream or yellow globular turkey-fat ore is found in Rush Mine (Arkansas, USA). Other good specimens come from Laurion (Greece), Tsumeb (Namibia) and Chessy (France).

$ZnCO_3$

Colour:	white, grey, yellow, brown, shades of green, pink, purple
Lustre; opacity:	vitreous or pearly; translucent
Streak:	white
Hardness:	4–4.5
Specific gravity:	4.4–4.5
Cleavage; fracture:	perfect rhombohedral; subconchoidal to uneven
Habit:	crystals rare, rhombohedral, with curved faces; massive, botryoidal, reniform, stalactitic, granular or encrustations
Crystal system:	trigonal

Rhodochrosite

Rhodochrosite is a beautiful rose colour, but darkens on exposure through oxidation. A minor ore of manganese, it dissolves with effervescence in hot dilute hydrochloric acid, and can luminesce light pink in long-wave ultraviolet light. Its prime commercial source is the USA. Crystals can be cut as gems, and the banded rock is very attractive when polished; the type known as Ina Rose comes from the oldest mines in Argentina. Rhodochrosite occurs as a metasomatic deposit, as a primary mineral in hydrothermal veins, associated with sulphides and manganese-bearing minerals, and more rarely in pegmatites. It is found in Freiburg (Germany), Las Cabesses (France), Pasto Bueno (Peru) and Colorado (USA).

$MnCO_3$

Colour:	pale to deep rose pink, yellowish grey, brownish or orange
Lustre; opacity:	vitreous to pearly; transparent to translucent
Streak:	white
Hardness:	3.5–4
Specific gravity:	3.5–3.7
Cleavage; fracture:	perfect rhombohedral; uneven
Habit:	crystals rhombohedral, prismatic, scalenohedral or tabular, often curved faces; massive, granular, stalactitic, globular or botryoidal
Crystal system:	trigonal

Dolomite

Dolomite is a major rock-forming mineral. It typically forms rhombohedral crystals, the faces of which may be curved, and can form a saddle-shape. Unlike calcite, it bubbles only weakly in dilute hydrochloric acid. Dolomite is the major constituent of the sedimentary rock called dolomite, most of which has formed from the passage of magnesium-bearing solutions through limestone. It also occurs in marbles, associated with talc, tremolite, diopside or wollastonite, in hydrothermal veins with metallic ores, and in igneous carbonatites. Dolomite has many uses, e.g. in the manufacture of cement, and as a source of magnesium oxide. Very good crystals are found at Eugui (Navarra Province, Spain) and at Touissite (Morocco).

$CaMg(CO_3)_2$

Colour:	colourless, white to cream, pale pink, grey or brown
Lustre; opacity:	vitreous to pearly; transparent to translucent
Streak:	white
Hardness:	3.5–4
Specific gravity:	2.85
Cleavage; fracture:	perfect rhombohedral; subconchoidal
Habit:	crystals rhombohedral, often composed of overlapping scales; massive or granular
Crystal system:	trigonal

Ankerite

Ankerite (brown-spar) is a common mineral that forms a series with dolomite, from which it can be distinguished by colour, which darkens on heating. It effervesces in hydrochloric acid and may fluoresce in long-wave ultraviolet light. Ankerite forms in sedimentary rocks through hydrothermal alteration and low temperature metasomatism; as a gangue mineral in sulphide veins; and in low-grade metamorphic ironstones and sedimentary banded iron formations. It can be associated with dolomite, siderite, quartz, copper sulphides and occasionally gold. The type locality is Styria (Austria). Sharply pointed 'dog-tooth' crystals occur in Chihuahua (Mexico) and crystals up to 5cm (2in) long at Mlynky (Slovakia).

$CaFe(CO_3)_2$

Colour:	white, grey, yellow, yellow-brown; weathers dark brown
Lustre; opacity:	vitreous to pearly; translucent
Streak:	white
Hardness:	3.5–4
Specific gravity:	2.9–3.1
Cleavage; fracture:	perfect rhombohedral; subconchoidal
Habit:	crystals rhombohedral; massive or granular
Crystal system:	trigonal

Witherite

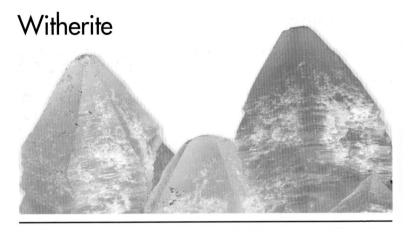

Witherite forms a series with strontianite ($SrCO_3$), from which it is distinguished by a flame test; witherite colours the flame green rather than crimson. It dissolves with effervescence in dilute hydrochloric acid; it sometimes fluoresces light blue, and may phosphoresce. It forms in hydrothermal veins, often by the alteration of barite. Associated minerals include barite, fluorite and galena. It also forms as an anoxic sediment, the barium being sourced through hot springs. An ore of barium, it is used in speciality glasses, in the production of rat poison, and was formerly used for refining sugar. Good crystals are found at Alston Moor (Cumbria) and at Hexham (Northumberland, England).

$BaCO_3$

Colour:	colourless, white, grey, yellow, green or brown
Lustre; opacity:	vitreous or resinous; transparent to translucent
Streak:	white
Hardness:	3–3.5
Specific gravity:	4.29
Cleavage; fracture:	one distinct; uneven
Habit:	twinned crystals of pseudo-hexagonal form, commonly striated; massive, botryoidal, granular, fibrous, columnar
Crystal system:	orthorhombic

Aragonite

Aragonite is less widespread than its polymorph calcite. It also dissolves with effervescence in dilute hydrochloric acid, but lacks the excellent rhombohedral cleavage of calcite. Aragonite crystals are often twinned, giving a hexagonal appearance. It is the main constituent of shells of many recent and fossil organisms. Although precipitating from warm marine waters, it inverts to calcite over time. It occurs as sinter from hot springs, or as dripstone in caves, and is the constituent of pearls in oysters. It occurs in amygdales in basalt, and is the stable polymorph in high-pressure metamorphic rocks. Aragonite is an ornamental stone: onyx is a popular banded variety, and flos-ferri is a coral-like form.

$CaCO_3$

Colour:	colourless to white, or yellowish
Lustre; opacity:	vitreous; transparent to translucent
Streak:	white
Hardness:	3.5–4
Specific gravity:	2.95
Cleavage; fracture:	distinct; subconchoidal
Habit:	crystals prismatic, elongated, often in radiating groups; coral-like form known as flos-ferri; columnar, stalactitic or encrusting
Crystal system:	orthorhombic

Strontianite

Strontianite was named after Strontian in Scotland, and is the mineral from which Sir Humphrey Davy separated strontium in 1807. It is common in carbonatites (e.g. Kola Peninsula, Russia); forms in hydrothermal veins in limestone, marl and chalk; occurs rarely in sulphide-rich veins associated with galena, sphalerite and chalcopyrite; and can form concretions in limestones and marls, occasionally replacing celestite. Fine crystals occur at Strontian (Scotland) and Oberdorf (Styria, Austria). It used to be mined from veins in limestones at Drensteinfurt (Westphalia, Germany). It is a principal ore of strontium, used for special glass for televisions and VDUs; strontium salts are used for the red colour in fireworks.

$SrCO_3$

Colour:	white, colourless, yellow, greenish or brownish
Lustre; opacity:	vitreous to resinous; transparent to translucent
Streak:	white
Hardness:	3.5
Specific gravity:	3.78
Cleavage; fracture:	perfect in one direction; subconchoidal to uneven
Habit:	crystals prismatic, pseudohexagonal twins, or radiating aggregates; massive, fibrous, granular or concretionary
Crystal system:	orthorhombic

Cerussite

Cerussite is a popular mineral due to its high sparkle and its yellow fluorescence. It often forms chevron twins, star-shaped pseudo-hexagonal twins, or complex twins of snowflake appearance. It forms in the oxidation zones of lead veins, associated with galena, sphalerite, pyromorphite, smithsonite, anglesite or goethite. Cerussite is toxic, but was used as a cosmetic by Queen Elizabeth I. It was formerly mined at the Llanfurnach silver-lead mine in Wales and large deposits were worked in southern Kazakhstan. It is mined today at Anguran Mine (Zanjan, Iran). Beautiful 'snowflake' twins occur at Mt Isa Mine (Queensland, Australia) and at Tsumeb (Namibia), where they are known as Jack Straw.

$PbCO_3$

Colour:	colourless, white or greyish
Lustre; opacity:	adamantine, vitreous, resinous; transparent to translucent
Streak:	white
Hardness:	3–3.5
Specific gravity:	6.56 (high for a transparent mineral)
Cleavage; fracture:	two directions, distinct; conchoidal or uneven, very brittle
Habit:	crystals common, often tabular but may be acicular; massive, granular, compact or stalactitic
Crystal system:	orthorhombic

Malachite

Malachite is a highly regarded semi-precious mineral, mined for about 6000 years. Its light and dark green banding make it extremely popular for carvings and jewellery, and it was used both in the Winter Palace in Leningrad and the Taj Mahal, India. Malachite is a common secondary mineral, formed by the weathering of copper deposits. Associated minerals are often very colourful: striking blue azurite and chrysocolla, black mottramite and red limonite. Good crystals are rare, but occur at Chessy (France) and at Onganja mine (Namibia). It has been mined in the Urals (Russia), individual blocks weighing many kilograms, but the most famous ore deposit is in Katanga Province (Democratic Republic of Congo).

$Cu_2CO_3(OH)_2$

Colour:	bright green and commonly banded
Lustre; opacity:	vitreous to silky; translucent to opaque
Streak:	pale green
Hardness:	3.5–4
Specific gravity:	3.9–4
Cleavage; fracture:	perfect; subconchoidal to uneven
Habit:	acicular crystals or tabular pseudomorphs after azurite, twinning common; botryoidal with concretionary structure
Crystal system:	monoclinic

Azurite

Azurite is a striking azure blue mineral, popular with collectors, but rarely faceted. It was used as a pigment during the Middle Ages and Renaissance (blue verditer). It forms as a secondary mineral in the oxidized zone of copper deposits associated with carbonate rocks, occurring with malachite, chrysocolla, cerussite or smithsonite. It forms better crystals than malachite, which sometimes pseudomorphs it through hydration of the azurite. There are more than a hundred different crystal forms; good crystals are found at Chessy (France), hence the synonym chessylite, Sardinia (Italy), Broken Hill (NSW, Australia), Tsumeb (Namibia), Laurion (Greece) and Oujda (Morocco).

$Cu_3(CO_3)_2(OH)_2$

Colour:	intense azure blue, paler when earthy, darker in crystals
Lustre; opacity:	vitreous, adamantine or dull; transparent to opaque
Streak:	pale blue
Hardness:	3.5–4
Specific gravity:	3.78
Cleavage; fracture:	perfect, prismatic; conchoidal
Habit:	crystals complex, often tabular or short prismatic, twinning common; massive, nodular or earthy
Crystal system:	monoclinic

Hydrozincite

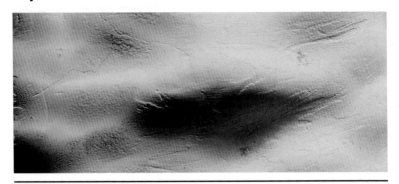

Hydrozincite, or zinc bloom, forms in the oxidation zones of zinc deposits through the alteration of sphalerite, zincite and other zinc minerals. It usually occurs as masses or crusts with internal fibrous structures, frequently with smithsonite. It readily dissolves in hydrochloric acid and often fluoresces strongly. When visited at night and with ultraviolet lights, the Trotter mine dump at Franklin (New Jersey, USA) makes a most unusual sight, with willemite fluorescing bright green, calcite bright red, and hydrozincite blue-white. Hydrozincite has been mined as a zinc ore, such as in the Picos de Europa near Santander (Spain), Nevada (USA) and currently at Mae Sod (Thailand) and Skorpion Mine (Namibia).

$Zn_5(CO_3)_2(OH)_2$

Colour:	usually white to grey, may be yellowish or brownish
Lustre; opacity:	crystals pearly, silky or dull; transparent to translucent
Streak:	white, dull to shining
Hardness:	2–2.5
Specific gravity:	4
Cleavage; fracture:	perfect; uneven
Habit:	crystals small and very rare, sharply pointed; usually massive, earthy, botryoidal, encrusting or stalactitic
Crystal system:	monoclinic

Rosasite

Rosasite is a very attractive rare bluish green mineral discovered in 1908 at Rosas Mine (Sardinia, Italy). A rare secondary mineral, it occurs in the oxidation zone of zinc-copper deposits, typically formed by the action of zinc-bearing solutions on primary copper minerals, but it can also be of post-mine origin. It may be associated with colourful minerals, such as limonite, hydrozincite, malachite, aurichalcite, smithsonite, cerussite or hemimorphite. Single crystals can be found at the Summit Mine (Montana); elsewhere in the USA it is found in California and New Mexico. Rosasite occurs in Kisil Espe (Turkestan), Ojuela Mine (Durango, Mexico), and Tsumeb (Namibia). It is a minor ore of copper.

$(Cu,Zn)_2CO_3(OH)_2$

Colour:	green to bluish green or sky blue
Lustre; opacity:	silky, vitreous or dull; translucent to opaque
Streak:	paler than colour
Hardness:	4.5
Specific gravity:	4.0–4.2
Cleavage; fracture:	in two directions at right angles; fibrous, splintery
Habit:	commonly radiating tufts of fibrous crystals; as crusts, or botyroidal masses or nodules
Crystal system:	monoclinic

Aurichalcite

Aurichalcite is a popular mineral characterized by delicate acicular crystals, and great care must be taken when handling. It was named either after the Greek word *oreichalchos*, which means 'mountain copper', or possibly after *aurichalcum* which means 'yellow copper ore'. Aurichalcite effervesces in cold hydrochloric and nitric acids, and in ammonia. It forms in the oxidized zones of zinc-copper deposits, and is occasionally used as an ore. It can be associated with red limonite and other colourful minerals such as azurite, smithsonite and malachite. Localities include Rosas Mine (Sardinia, Italy), Chessy (Rhône, France), Laurion (Attica, Greece), Tsumeb (Namibia) and notably Ojuela Mine (Mapimi, Durango, Mexico).

$(Zn,Cu)_5(CO_3)_2(OH)_6$

Colour:	pale green to greenish blue and sky blue
Lustre; opacity:	silky or pearly; transparent to translucent
Streak:	pale green or blue
Hardness:	1–2
Specific gravity:	3.6–3.9 (above average for non-metallic minerals)
Cleavage; fracture:	perfect; uneven or fibrous, fragile
Habit:	delicate acicular crystals, often striated, radiating tufts; as feathery encrustations or columnar, laminated, granular
Crystal system:	orthorhombic

Phosgenite

Phosgenite is a rare lead mineral of high lustre, which fluoresces yellow. It dissolves with effervescence in dilute nitric acid. It forms in the oxidation zone of hydrothermal lead ore deposits, especially in the presence of sea water. It has also been found in slag heaps from the ancient lead-silver mines at Laurion (Attica, Greece), which are now in the sea. Phosgenite was used as a white pigment in ancient Egyptian cosmetics, but it was probably synthesized from smelted lead oxides, as there was no access to the natural mineral. Enormous crystals occur at Monteponi and Montevecchio (Sardinia, Italy), in association with cerussite and anglesite; other notable localities are Tsumeb (Namibia), and Dundas (Tasmania).

$Pb_2(CO_3)Cl_2$

Colour:	colourless, white, pale brown or pale yellowish brown
Lustre; opacity:	adamantine; transparent to translucent
Streak:	white
Hardness:	2.5–3
Specific gravity:	6.13
Cleavage; fracture:	two distinct, prismatic; conchoidal, sectile and flexible
Habit:	crystals usually prismatic, may be tabular, weakly striated; also massive or granular
Crystal system:	tetragonal

Trona

Trona is deposited from saline lakes in arid regions, associated with halite, natron, gypsum or thenardite. It has an alkaline taste, is soluble and fluoresces white or blue. A natural source of sodium carbonate (soda ash), it is used to make paper, detergents, soap, glass, food and paper. The world's biggest reserves are in Wyoming (USA) and Lake Magadi (Kenya). In ancient Egypt, salt deposits in the lower Nile delta were exploited for natrum (largely trona), and used in medicine, mummification and to make cement (when mixed with caustic soda from silicate minerals and Nile silt). Notable occurrences include Searles Lake (California), Soda Lake (Nevada, USA) and the Otjiwalundo salt pan (Namibia).

$Na_3(CO_3)(HCO_3).2H_2O$

Colour:	colourless, grey, white, yellowish, brown
Lustre; opacity:	vitreous to dull, earthy; translucent
Streak:	white
Hardness:	2.5–3.0
Specific gravity:	2.17
Cleavage; fracture:	perfect; uneven to subconchoidal
Habit:	crystals prismatic or tabular, fibrous to compact aggregates; earthy crusts or efflorescences
Crystal system:	monoclinic

Gaylussite

Gaylussite, also known as natrocalcite, is a rare mineral named after the French scientist J.L. Gay-Lussac (1778–1850). It dehydrates when exposed to air, slowly crumbling and leaving calcium carbonate. It is slightly soluble in water, and effervesces in hydrochloric acid; it may luminesce creamy white. It is found in clay, shales and evaporites from soda lakes, along with natron, borax or calcite. Large crystals occur in the desert sands of Namibia and at Lake Amboseli (Kenya). It has also formed as an alteration product at Oldoinyo Lengai (Tanzania), the only volcano erupting natrocarbonatite lava. It occurs at Searles Lake (California, USA), Mt Erebus (Antarctica) and the Gobi Desert (Mongolia).

$Na_2Ca(CO_3)_2 \cdot 5H_2O$

Colour:	white, yellowish-white
Lustre; opacity:	vitreous, dull; transparent to translucent
Streak:	grey–white
Hardness:	2.5–3
Specific gravity:	1.99 (well below average)
Cleavage; fracture:	very good; conchoidal
Habit:	crystals prismatic or wedge-shaped; massive and encrusting
Crystal system:	monoclinic

Borax

Borax desiccates on exposure to air, turning to a white powder, tincalconite ($Na_2B_4O_7.5H_2O$). It dissolves in water, has a bittersweet taste and may fluoresce blue-green. One of four main ores of boron, it forms in arid regions on the edges of playa lakes and is associated with other evaporites, especially borates. The source of the boron is probably thermal springs of volcanic origin. Commercially important deposits are exploited in California (USA); huge crystals can be found at Clear Lake and Borax Lake. Borax is used in cleaning products, water softeners, pesticides, timber preservatives and agriculture. It readily converts to boric acid, which is used as , for example, a mild antiseptic and in nuclear power plants.

$Na_2B_4O_7.10H_2O$

Colour:	colourless, white, tinged with grey, blue or green
Lustre; opacity:	vitreous, resinous, dull; transparent to opaque
Streak:	white
Hardness:	2–2.5
Specific gravity:	1.7
Cleavage; fracture:	perfect; conchoidal
Habit:	crystals short, prismatic; massive
Crystal system:	monoclinic

Ulexite

Ulexite forms as an evaporite in boron-rich lake basins in arid regions, associated with other borate minerals. It sometimes forms rounded masses of fine fibrous crystals known as cotton balls, as found in Death Valley (California, USA), or as parallel fibrous aggregates. Light is transmitted up each fibre, thus displaying fibre optics. If a specimen is polished flat at both ends (perpendicular to the fibres) and placed over a print, for example, the image appears at the top of the specimen. This unusual property has made it very popular, and it is sometimes referred to as television stone. The best locality for compact fibres is Boron, Kern County (California, USA). Turkey is the leading producer of boron ore, followed by the USA.

$NaCaB_5O_9.8H_2O$

Colour:	white or colourless
Lustre; opacity:	vitreous or silky
Streak:	white
Hardness:	2.5
Specific gravity:	1.96
Cleavage; fracture:	perfect; uneven
Habit:	crystals very rare; rounded aggregates with fibrous internal structure, tufted masses
Crystal system:	triclinic

Colemanite

Colemanite is one of the main ores of boron (see also borax, kernite and ulexite). It forms in warm and arid regions in alkaline playa lakes deficient in sodium and carbonate; the boron is probably sourced from geothermal springs. One boron mineral may change to another during diagenesis, e.g. colemanite can form from ulexite through contact with calcium-bearing ground water. Associated minerals include other borates, gypsum, celestine and calcite. More than 80 per cent of the world production of borates comes from the USA (including colemanite from Death Valley, California) and Turkey. Boron is used in the chemical industry and in the manufacture of steel, paints, heat-resistant glass and fluxes for welding.

$Ca_2B_6O_{11}.5H_2O$

Colour:	colourless, white, yellowish or grey
Lustre; opacity:	vitreous, adamantine; transparent to translucent
Streak:	white
Hardness:	4–4.5
Specific gravity:	2.44
Cleavage; fracture:	perfect; uneven or conchoidal
Habit:	crystals short, prismatic; compact, granular
Crystal system:	monoclinic

Kernite

Kernite is a rare mineral, the last of the four principal ores of boron. Specimens are difficult to keep because they effloresce (lose water), becoming opaque and dull. Kernite forms in intermittent lakes, supplied with boron from geothermal springs. It can form from borax with a rise in temperature and pressure, thus tending to concentrate at the base of stratified borate deposits. It can be associated with colemanite, ulexite and borax. Crystals up to 8.75m (29ft) long have been reported from Boron (Kern County, California, USA), a commercial source of kernite. Crystals up to 2.5m (8ft) occur at Tincalayu Mine (Salta, Argentina). Other important occurrences include Catalonia (Spain) and the Kirka deposit (Eskisehir, Turkey).

$Na_2B_4O_7.4H_2O$

Colour:	colourless when fresh, turning white
Lustre; opacity:	vitreous, dull, silky; transparent to opaque
Streak:	white
Hardness:	2.5–3
Specific gravity:	1.9
Cleavage; fracture:	perfect; splintery
Habit:	rare short, prismatic crystals; usually cleaved masses with fibrous habit
Crystal system:	monoclinic

Thenardite

Named after the French chemist L.J. Thénard (1777–1857), thenardite is anhydrous sodium sulphate. In damp air, samples will absorb water to give the hydrated form, mirabilite or Glauber salt ($Na_2SO_4.10H_2O$). This requires that the mineral be stored in closed containers. It is used in the manufacture of sodium salts and as a desiccant in the chemical industry. Mostly found in salt lakes in especially hot and dry tropical areas and deserts, thenardite is associated with minerals, including mirabilite, gypsum, glauberite, epsomite and halite. It is common in the deserts of the USA (California, Arizona, Nevada), the Sahara, Chile, Central Asia, Egypt and Sudan. Sometimes it is found in fumaroles, such as in Vesuvius (Italy).

Na_2SO_4

Colour:	colourless, white or grey-white; yellowish or reddish tinges
Lustre; opacity:	vitreous; transparent to translucent
Streak:	white
Hardness:	2.5–3
Specific gravity:	2.7
Cleavage; fracture:	perfect; uneven to hackly
Habit:	crystals, sometimes large, are usually bipyramidal, occasionally prismatic with striated faces; crusts
Crystal system:	orthorhombic

Gypsum

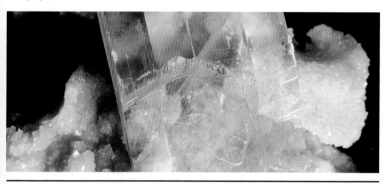

The most common sulphate mineral, gypsum has relatively poor water-solubility and is often the first mineral deposited on evaporation of sea water. Varieties include selenite (colourless and transparent), satin spar (fibrous) and alabaster (massive and granular). Alabaster, usually white or in pastel shades, has been widely used for ornamental carvings for thousands of years. Satin spar is cut into cabochons and polished to give a 'cat's eye' effect. Its low thermal conductivity is utilized to insulate buildings ('drywalling'). Gypsum usually occurs as massive beds among other evaporites. It is often found as a cap rock on salt domes. It is found at Stassfurt (Germany), Paris basin (France) and Nova Scotia (Canada).

$CaSO_4.2H_2O$

Colour:	colourless or white; shades of grey, yellow or pink
Lustre; opacity:	vitreous, pearly on cleavage faces; transparent to translucent
Streak:	white
Hardness:	2
Specific gravity:	2.32
Cleavage; fracture:	perfect; conchoidal, splintery
Habit:	simple tabular crystals, often with curved faces; rosette-shaped aggregates and fibrous or granular masses
Crystal system:	monoclinic

Anhydrite

Anhydrite is formed by evaporation of sea water above 42°C (108°C); gypsum forms below this. It is mostly produced by dehydration of buried gypsum, which causes the rock to shrink, forming crevices and caverns. Anhydrite and gypsum are also the names of rocks comprising large masses of these minerals. Such rocks are used for producing sulphuric acid and ammonium sulphate (for fertilizer), paper fillers and plasters (notably plaster of Paris). Anhydrite often forms thick beds in the lower strata of evaporite deposits and is abundant as the cap to salt domes. It is found at Stassfurt (Germany), the Paris basin (France) and Louisiana and Texas (USA). Unusual purple crystals are found at Bex (Switzerland).

$CaSO_4$

Colour:	colourless to bluish when transparent, white, pink or mauve when massive; may be further discoloured by impurities
Lustre; opacity:	vitreous to pearly; transparent to translucent
Streak:	white
Hardness:	3–3.5
Specific gravity:	2.9–3
Cleavage; fracture:	good; uneven, splintery
Habit:	usually massive or granular; sometimes fibrous or lamellar; rarely crystals
Crystal system:	orthorhombic

Celestine

Celestine, or celestite, named after the Latin for 'celestial', is the main ore of strontium. The most attractive crystals are a beautiful pale sky blue, but colourless celestine is also common. Large crystals (2–3kg/4½–6½lb) occur at Put-in-Bay (Ohio, USA). Strontium gives a strong crimson colour to a flame, useful in fireworks and flares. It is distinguished from barite by its lower density and very different flame test (barium gives green). Celestine occurs in impure limestones associated with sulphur, and in evaporites with gypsum, anhydrite or halite. Good crystals of celestine come from Lake Erie (USA), Bristol (England) and Sicily (Italy). Celestine, as a strontium ore, is mined in England, Russia and Tunisia.

$SrSO_4$

Colour:	colourless or pale blue
Lustre; opacity:	vitreous, pearly on cleavage; transparent to translucent
Streak:	white
Hardness:	3–3.5
Specific gravity:	3.9–4
Cleavage; fracture:	perfect; uneven
Habit:	tabular or prismatic crystals; can be fibrous or granular
Crystal system:	orthorhombic

Anglesite

As with most lead minerals, anglesite is notably dense. Anglesite, named after Anglesey, Wales, is an ore of lead formerly much used as a pigment in white paints, but now largely superseded by titanium oxide. Attractive yellow crystals are popular, but are quite soft and fragile. Anglesite occurs in the oxidized zones of galena deposits, associated with cerussite, mimetite and pyromorphite; and rarely as a sublimate in volcanoes. Good crystals come from Tsumeb (Namibia), Anglesey (Wales), Musen (Germany), Phoenixville (USA) and Monteponi (Sardinia). Sulphur-covered crystals are from Los Lamentos (Chihuahua, Mexico). It is mined in Leadhills (Scotland), Derbyshire (England) and Arizona and Utah (USA).

$PbSO_4$

Colour:	colourless or white; can be tinged brown, grey, blue, green, purple or more usually pale yellow
Lustre; opacity:	adamantine, sometimes resinous; transparent to opaque
Streak:	white
Hardness:	2.5–3
Specific gravity :	6.2–6.4
Cleavage; fracture:	perfect; conchoidal
Habit:	crystals prismatic, tabular or pyramidal; massive, compact or granular
Crystal system:	orthorhombic

Barite

Barite (also baryte, barytes or heavy spar) is notable for its high density, the Greek *baros* meaning 'heavy'. Large crystals (up to 1m/3¼ft) and roseate clusters of grains called desert roses are popular among collectors. Stalagmitic barite may be sectioned to reveal attractive concentric zoning. Large quantities are mined for paint and paper manufacture, as drilling muds in the oil industry and as a source of metallic barium. Barite is found in lead and zinc mines as a gangue mineral (mining waste) with calcite, fluorite and quartz, occurring in veins in lead, zinc, silver, iron and nickel ores. Good desert roses come from Oklahoma and Kansas (USA); large crystals are to be found in Cumbria and Cornwall (England).

$BaSO_4$

Colour:	colourless or white; red, brown, yellow or blue
Lustre; opacity:	vitreous, pearly on cleavage; transparent to translucent
Streak:	white
Hardness:	3–3.5
Specific gravity:	4.3–4.6
Cleavage; fracture:	perfect; uneven
Habit:	crystals tabular, sometimes prismatic; often massive, sometimes fibrous or lamellar clusters, or granular stalagmites
Crystal system:	orthorhombic

Spangolite

A beautiful rare blue-green mineral, spangolite usually occurs as aggregates of small crystals on a matrix; specimens with minerals such as azurite, brochantite and chrysocolla make pleasing displays. Spangolite is named after the engineer and avid mineral collector Norman Spang (1843–1922), who owned the original specimen. Spangolite is widespread, but found only in small quantities. Spangolite occurs in oxidized copper sulphide ores associated with azurite, malachite, adamite and tyrolite. It is found at Blanchard mine (New Mexico, USA), Tintic (Utah, USA), Mujaba Hill (Nevada, USA), St Day (Cornwall, England), Fontana Rossa (Corsica), Broken Hill (NSW, Australia), Laurion (Greece) and Arenas (Sardinia).

$Cu_6AlSO_4(OH)_{12}Cl.3H_2O$

Colour:	green or greeny-blue
Lustre; opacity:	vitreous; transparent
Streak:	pale yellow-green
Hardness:	3
Specific gravity:	3.1
Cleavage; fracture:	perfect; conchoidal
Habit:	short prisms and hexagonal plates
Crystal system:	hexagonal

Brochantite

Brochantite, named after the French geologist A. Brochant de Villiers (1772–1840), is a bright green mineral popular as a micromount. It is very difficult to distinguish from fibrous forms of antlerite and atacamite, although the latter is softer. Brochantite is formed as a secondary mineral in the oxidized zone of copper deposits, usually in dry climates. It is associated with malachite, atacamite, antlerite, limonite, cuprite, chrysocolla and cyanotrichite. Popular localities are Chuquicamata (Chile), Tsumeb (Namibia), Sardinia, Rezbanya (Romania), Ain Barbar (Algeria) and Broken Hill (NSW, Australia). Artificial brochantite is produced, so collectors should be careful with sources.

$Cu_4SO_4(OH)_6$

Colour:	bright green
Lustre; opacity:	vitreous to pearly; transparent to translucent
Streak:	light green
Hardness:	3.5–4
Specific gravity:	3.97
Cleavage; fracture:	perfect; uneven to conchoidal
Habit:	small, stubby or acicular, striated crystals, sprays of small needles; fibrous or granular crusts and nodules; rarely massive
Crystal system:	monoclinic

157

Alunite

A lunite is the main constituent of the rock alum stone, mined for alum production since the fifteenth century. Alum has the formula $KAl(SO_4)2.12H_2O$ and is used for papermaking and treating skins. Alum and alunite have been used to stem bleeding of small cuts. Most deposits were formed when rocks rich in alkali feldspar (pegmatites, syenites, trachytes) were altered by sulphate-rich water, in turn produced from altered pyrite. Sometimes alunite occurs in veins cutting across schists and in fumaroles. It is often found with halloysite and kaolinite. Alunite is mined at La Tolfa (Italy), Nevada (USA), France, Hungary, Greece, Almeria (Spain) and Bullah Delah (NSW, Australia).

$KAl_3(SO_4)_2(OH)_6$

Colour:	white; can be grey, yellowish or reddish
Lustre; opacity:	vitreous; transparent to translucent
Streak:	white
Hardness:	3.5–4
Specific gravity:	2.6–2.9
Cleavage; fracture:	good; conchoidal
Habit:	crystals as rhombohedra or plates; usually granular or earthy masses
Crystal system:	hexagonal

Linarite

L inarite is a rare mineral usually occurring with copper and lead ores. The name is derived from Linares, where it was first recognized in 1839. Linarite can be distinguished from the similar mineral azurite, as the latter fizzes with acids. Linarite is found in the oxidized zones of mixed lead and copper deposits, associated with aurichalcite, cerrussite, malachite and hemimorphite. The best crystals, more than 10cm (4in) in length, come from Mammoth Mine (Tiger, Arizona, USA); others come from Kisamori (Akita, Japan), Linares plateau (Jean, Spain), Tsumeb (Namibia), Red Gill (Cumbria, England), Serra de Capitillas (Argentina), Rosas and San Giovani (Sardinia), Leadhills (Scotland).

$PbCuSO_4(OH)_2$

Colour:	royal blue
Lustre; opacity:	vitreous; transparent to translucent
Streak:	blue
Hardness:	2.5
Specific gravity:	5.3–5.5
Cleavage; fracture:	perfect; conchoidal
Habit:	acicular or tabular crystals, often slender prisms like tourmaline; encrustations, aggregates
Crystal system:	monoclinic

Glauberite

Glauberite is used for the extraction of Glauber salt ($Na_2SO_4.10H_2O$), after which it is named, as a mordant in textile dying and in pharmaceuticals. Glauberite is very water-soluble and can be dissolved out of rocks and infilled with other mineral pseudomorphs. Glauberite may be found in buried sediments of geological age or in modern salt lakes and other evaporites. Associated minerals include halite, thenardite, calcite and gypsum. Occasionally it is a sublimate in fumaroles. Good crystals of glauberite are found in Villarubia (Spain), Lorraine (France) and salt lakes in California and Arizona (USA). Large masses occur around Saltzburg (Austria), Ruthenia (Russia), Texas (USA) and New Mexico (USA).

$Na_2Ca(SO_4)_2$

Colour:	white, tinged yellow or brick-red
Lustre; opacity:	vitreous; transparent to translucent
Streak:	white
Hardness:	2.5–3
Specific gravity :	2.8
Cleavage; fracture:	perfect; conchoidal
Habit:	tabular, prismatic or bipyramidal crystals, sometimes with striated faces and rounded edges; compact masses and crusts
Crystal system:	monoclinic

Halotrichite

Halotrichite is isomorphous with pickeringite (MgAl$_2$(SO$_4$)$_4$.22H$_2$O), which is present in most specimens. Masrite is a variety containing manganese and cobalt; apjohnite, dietrichite, redingtonite, wupatkiite and bilinite variously have cobalt, manganese, zinc, nickel and/or chromium replacing the iron or aluminium. Halotrichite, also known as iron alum or feather alum, is named from the Latin for 'hair salt' because of its form as fine fibrous needles. It is found as a pyrite alteration product in aluminous rocks and in fumaroles. It occurs at Sulfatara (Naples, Italy), Falun (Sweden), Recsk (Hungary), Istria (Croatia), Dubnik (Slovakia), Copiapo (Atacama, Chile) and Alum Mountain (New Mexico, USA).

FeAl$_2$(SO$_4$)$_4$.22H$_2$O

Colour:	colourless, white, yellowish, greenish
Lustre; opacity:	vitreous; transparent to translucent
Streak:	white
Hardness:	1.5
Specific gravity:	1.9
Cleavage; fracture:	imperfect; conchoidal
Habit:	acicular, prismatic crystals; crusts and aggregates
Crystal system:	monoclinic

Cyanotrichite

Cyanotrichite is named from the Greek for 'blue hair', after the appearance of felted masses of crystalline needles. Related to but rarer than halotricite, it is also known as lettsomite or velvet copper. Specimens as attractive blue tufts on a matrix are best viewed magnified, as the crystals are usually small. Cyanotrichite occurs in the oxidized zones of copper deposits, associated with minerals such as azurite, malachite and limonite. It is found at Cap Garonne (France), Laurion (Greece), Nemaqualand (South Africa), Mednorudnyansk (Russia), Bisbee (New Mexico, USA), Grandview (Arizona, USA), Moldova Noua (Romania), Traversella (Piedmont) and Rio Marina (Elba, Italy) and St Day (Cornwall, England).

$Cu_4Al_2SO_4(OH)_{12}.2H_2O$

Colour:	orange-red, hyacinth red
Lustre; opacity:	vitreous; transparent to translucent
Streak:	ochre yellow
Hardness:	2.5
Specific gravity:	2.1
Cleavage; fracture:	perfect; conchoidal, irregular
Habit:	elongated prismatic crystals as radiating fibrous aggregates of botryoidal or reniform shape
Crystal system:	monoclinic

Botryogen

Botryogen is a rare dark orange mineral named from the Greek *botrys* for 'grape' and *genos* for 'yield', after its characteristic habit. A popular variety with small aggregates of botryogen on halotrichite needles is found at Smolník (Slovakia). Botryogen, also known as red iron vitriol, is formed by the action of sulphate-rich fluids on mafic minerals. It often occurs near pyrite deposits in arid regions, associated with epsomite, copiapite and voltaite. It is found at Rammelsberg (Germany), Falun (Sweden), Villé Valley (France), Chuquicamata (Chile), Knoxville (Tennessee, USA), San Juan (Argentina), Queensland (Australia), Madeni-Zakli (Iran) and Paracutin (Mexico).

$MgFe(SO_4)_2OH.7H_2O$

Colour:	orange-red, hyacinth red
Lustre; opacity:	vitreous; transparent to translucent
Streak:	ochre yellow
Hardness:	2.5
Specific gravity:	2.1
Cleavage; fracture:	perfect; conchoidal, irregular
Habit:	elongated prismatic crystals as radiating fibrous aggregates of botryoidal or reniform shape
Crystal system:	monoclinic

Kröhnkite

Named after B. Kröhnke, the first person to analyze it, kröhnkite is a rare blue mineral easily mistaken for others such as chalcanthite. Kröhnkite is abundant in large royal blue crystals in the Chuquicamata mine, which is the largest open-pit copper mine in the world. Twinning in crystals can result in attractive heart-shaped forms. It is formed from sulphate-rich fluids in desert regions and found among copper ores often associated with chalcanthite, antlerite and atacamite. Good localities for finding kröhnkite are Chuquicamata and Quetena mines (Chile), Capo Calamita (Elba, Italy), Wheal Hazard (Cornwall, England), Recsk copper deposit (Matra Mountains, Hungary) and Broken Hill (NSW, Australia).

$Na_2Cu(SO_4)_2 \cdot 2H_2O$

Colour:	blue to light blue
Lustre; opacity:	vitreous; transparent
Streak:	white
Hardness:	2.5–3
Specific gravity:	2.9
Cleavage; fracture:	perfect; conchoidal
Habit:	prismatic crystals; fibrous and granular crusts and aggregates
Crystal system:	monoclinic

Kainite

Kainite is a double salt formed as a mixture of potassium chloride (sylvite) and magnesium sulphate (epsomite). It is named somewhat obscurely after the Greek *kainos*, meaning 'new' or 'contemporary', alluding to its occurrence in rocks of recent origin. Kainite is a common mineral in evaporites, but difficult to identify. Kainite is used in fertilizers and as a source of potassium compounds. It is found among evaporite deposits associated with halite, sylvite and carnallite; and as a volcanic sublimate. Localities where kainite has been mined are Hallstadt (Austria), Eddy County (New Mexico, USA), Kalus (Ukraine) and Sicily (Italy). It is a major mineral found in the large salt deposits of middle and northern Germany.

$KMgSO_4Cl.3H_2O$

Colour:	white, yellow, greyish
Lustre; opacity:	vitreous
Streak:	white
Hardness:	3
Specific gravity:	2.1–2.2
Cleavage; fracture:	good
Habit:	rare thick, tabular crystals; usually granular or fibrous aggregates
Crystal system:	monoclinic

Devilline

Devilline is an attractive blue-green mineral named after the French chemist H. Deville (1818–1881). It is also called devillite or herrengrundite after a locality in Slovakia. A sky-blue zinc-bearing variety is called serpierite or zincian devilline. Devilline is a rare secondary mineral found in copper deposits associated with gypsum, azurite and malachite. It is found in Cornwall (England), Pania Dolina (Slovakia), Tsumeb (Namibia), Montgomery (Pennsylvania, USA) and Vezzani, (Corsica). Serpierite has been found at Kamariza (Laurion, Greece), Ross Island (Ireland) and Broken Hill (NSW, Australia). A light blue manganese–bearing variety campigliaite has been identified at Temerino (Tuscany, Italy).

$CaCu_4(SO_4)_2(OH)_6$

Colour:	blue-green to emerald green
Lustre; opacity:	vitreous to pearly; transparent to translucent
Streak:	light green
Hardness:	2.5
Specific gravity:	3.1
Cleavage; fracture:	perfect; brittle
Habit:	crystals often lamellar in rosettes or needle-like clusters; crusts, botryoidal aggregates
Crystal system:	monoclinic

Epsomite

Commonly known when pure as Epsom or bitter salts, epsomite is a white water-soluble hydrate of magnesium sulphate. It is used as a mordant in hide tanning and textile dying, and dried for use as a desiccant. It is precipitated from hot springs, fumaroles and saline waters. The thermal waters at Epsom are well known for encrustations of epsomite. It is found on the walls of mines near weathered pyrite deposits. Crystals up to 2–3m (6½–10ft) come from Kruger Hills (Washington, USA). Epsomite is light and fragile, and loses water in dry air, becoming dull in appearance. It is mined at Carlsbad (New Mexico, USA), Saxony (Germany), Sedlec (Czech Republic) and Valle Antrona (Italy).

$MgSO_4.7H_2O$

Colour:	white; can be tinged yellow, green, red
Lustre; opacity:	vitreous to silky; transparent to translucent
Streak:	white
Hardness:	2–2.5
Specific gravity:	1.68
Cleavage; fracture:	perfect; conchoidal
Habit:	crystals rare; usually crusts, stalactites and earthy masses
Crystal system:	orthorhombic

Römerite

Römerite is an attractive yellow-brown mineral named after the German geologist Friedrich Römer (1809–1869). Violet-brown aggregates and dark orange crystals are popular, attractive occurrences. It is formed in oxidized zones of iron sulphide deposits, associated with other sulphates, especially halotrichite, copiapite ($Fe_5(SO_4)_6(OH)_2.20H_2O$) and melantesite ($FeSO_4.7H_2O$). It is found in Dresden (Saxony), Rammelsberg (Harz) and Valdsassen (Bavaria, Germany), Rtyne and Pribram (Czech Republic), Madeni Zakh (Iran), Tierra Amarilla and Chuquicamata (Chile), Blyava (Southern Urals) and Kamchatka (Russia).

$Fe_3(SO_4)_4.14H_2O$

Colour:	brown to yellow-brown, occasionally violet-tinged
Lustre; opacity:	vitreous, resinous; translucent
Streak:	yellow-brown
Hardness:	3–3.5
Specific gravity:	2.17
Cleavage; fracture:	perfect, good; uneven
Habit:	tabular or pseudocubic crystals; crusts, granular layers and stalactites
Crystal system:	triclinic

Crocoite

Crocoite, also called red lead ore, is named from the Greek *krokos*, meaning 'saffron' or 'crocus', after its bright orange-red colour. It is the mineral from which chromium was first extracted, but is not used as an ore commercially. Crocoite is one of the few naturally occurring chromates, which are always strongly coloured. Pure crocoite forms the pigment chrome yellow. It is formed where chromic-acid rich hydrothermal solutions have attacked lead deposits. It is associated with cerrusite, limonite, anglesite, phoenicochroite and vauquelinite. Crystals up to 15cm (6in) have been found at Dundas (Tasmania); other localities include Minas Gerais (Brazil), Nontron (France) and Mammoth Cave (Arizona, USA).

$PbCrO_4$

Colour:	yellow, orange, red
Lustre; opacity:	adamantine; translucent
Streak:	orange
Hardness:	2.5–3
Specific gravity:	6
Cleavage; fracture:	perfect; conchoidal
Habit:	prismatic crystals, often striated, often hollow, acicular crystals; massive, granular
Crystal system:	monoclinic

Scheelite

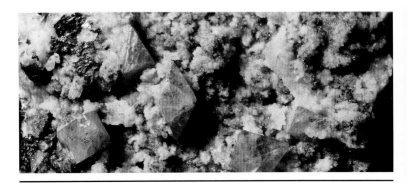

Scheelite is a major ore of tungsten, used in lightbulb filaments, high-strength steels and tungsten carbide cutting tools. Named after the Swedish chemist K. Scheele (1742–1786), it forms crystals with a high lustre and fire, and was used as an early synthetic diamond. Fluorescing pale blue under UV light, scheelite is used as a scintillator. The fluorescence becomes greener as the content of powellite ($CaMoO_4$) increases. Scheelite is found in pegmatites, hydrothermal veins and medium-grade metamorphic rocks. Crystals of a weight up to 500g (1lb 2oz) are reported from Minas Gerais (Brazil) and of a size up to 7.5cm (3in) in Tong Wha (Korea); other localities are Sardinia, Traversella, Val di Fiemme and Valsugana (Italy).

$CaWO_4$

Colour:	grey-white, yellow, brown, reddish, greenish
Lustre; opacity:	vitreous to adamantine; transparent to opaque
Streak:	white
Hardness:	4.5–5
Specific gravity:	5.9–6.1
Cleavage; fracture:	imperfect; conchoidal
Habit:	crystals as bipyramids, pyramids, plates; massive aggregates, granular, crusts, pseudomorphs
Crystal system:	tetragonal

Wolframite

Wolframite is a major ore of tungsten, its name coming from *wolfram*, the German word for tungsten. Its composition lies within the ferberite-heubnerite series ($FeWO_4$–$MnWO_4$). The colour darkens as the iron content increases. Tungsten is widely used, but some demand for its use in hardened metals has been taken by depleted uranium. It occurs in pegmatites, hydrothermal veins, pneumatolytic zones and placer deposits. Associated minerals include scheelite, cassiterite, pyrite and sphalerite. Important deposits are in southern China, Queensland (Australia), La Paz (Bolivia), Arizona (USA) and New Mexico (USA); superb crystals at Hualapón (Pasto Bueno, Peru) and Quartz Creek (Idaho, USA).

$(Fe,Mn)WO_4$

Colour:	yellow to reddish-brown to blackish brown
Lustre; opacity:	submetallic, resinous; opaque
Streak:	brown to black
Hardness:	5–5.5
Specific gravity:	7.1–7.5
Cleavage; fracture:	perfect; uneven
Habit:	vertically striated tabular or lamellar crystals; massive, granular
Crystal system:	monoclinic

Wulfenite

Wulfenite, or yellow lead ore, a minor ore of molybdenum, is named after the Austrian mineralogist Franz Wülfen (1728–1805). Popular specimens comprise clusters of large square red crystals among smaller ones. As with many lead minerals, wulfenite has a high refractive index (2.3–2.4), which enhances its appearance. Wulfenite occurs in the oxidized zones of lead deposits, associated with cerussite, vanadinite, pyromorphite and mimetite. The best localities include: Red Cloud (Arizona, USA), noted for superb red crystals; Rezbanya (Romania) for reddish crystals; Bleiberg (Austria) for yellow crystals; Tsumeb (Namibia) and Phoenixville (Pennsylvania, USA) for colourless crystals.

$PbMoO_4$

Colour:	yellow, orange, red, rarely white
Lustre; opacity:	resinous; transparent to opaque
Streak:	yellowish-white
Hardness:	2–3
Specific gravity:	6.7–6.9
Cleavage; fracture:	imperfect; conchoidal
Habit:	crystals square tabular or stubby pyramids, often as clusters; massive, granular or earthy aggregates
Crystal system:	tetragonal

Purpurite

Purpurite is a very rare mineral, much sought-after by collectors. Its name comes from its characteristic colour when fresh, although it may show surface alteration to a darker brown. It forms through the oxidation and leaching of manganese-iron phosphates, particularly lithiophyllite ($LiMnPO_4$), found in granite pegmatites. It has also been known to form through the reaction of sea water with bat guano, which provides the phosphorus. Associated minerals include lithiophyllite and heterosite. As well as the type locality Faires tin mine (North Carolina, USA), it occurs in the Varuträsk pegmatite (Västerbotten, Sweden), Mangualde (Portugal), La Vilate quarry (Chanteloupe, France) and the Gunong Keriang cave (Malaysia).

$(Mn,Fe)PO_4$

Colour:	reddish purple to deep rose red
Lustre; opacity:	satiny on fresh fractures, or earthy; translucent to opaque
Streak:	pale purple to pale red
Hardness:	4–4.5
Specific gravity:	3.2–3.4
Cleavage; fracture:	good, surfaces may be curved or crinkled; uneven, brittle
Habit:	does not occur as crystals; forms small irregular masses, cleavage fragments may reach 20cm (8in)
Crystal system:	orthorhombic

Monazite

Monazite is a rare earth phosphate, containing up to 12 per cent ThO_2, and is feebly radioactive. It is a common accessory in granites, syenites and metamorphic gneisses, and is also found as a placer deposit, being dense and resistant to chemical weathering. Monazite is the main source of rare earths, used as a fuel-cracking catalyst in the petroleum industry, and of thorium. Thorium is important in the computer, electronic and medical industries. It is also the most efficient nuclear reactor fuel, and could potentially be used instead of uranium. Important commercial sources are Australia (with the largest reserve), Florida (USA), Madagascar, Brazil, Travancore (India), Sri Lanka and South Africa.

$(Ce,La,Nd,Th)PO_4$

Colour:	yellow to reddish brown
Lustre; opacity:	resinous to waxy, vitreous; transparent to translucent
Streak:	white
Hardness:	5–5.5
Specific gravity:	4.6–5.4
Cleavage; fracture:	good; conchoidal to uneven fracture
Habit:	crystals tabular or prismatic; usually as grains
Crystal system:	monoclinic

Amblygonite

Amblygonite is a fluorophosphate that can luminesce orange in long-wave ultraviolet light. It occurs in coarse-grained granitic rocks and in lithium- and phosphate-rich granitic pegmatites – associated, for example, with apatite, pollucite and spodumene. It is used as a source of lithium, but can also be faceted to make gems when transparent or translucent. Amblygonite being relatively soft, these tend to be of interest only to collectors. Gem-quality material occurs in pegmatites at Minas Gerais (Brazil) and at Newry (Maine, USA), and giant crystals weighing several tons have been found at Custer (Dakota, USA); it also occurs in the pegmatites of Pala (California, USA) and in Yauapa County (Arizona, USA).

(Li,Na)Al(PO$_4$)F

Colour:	milk white, yellow, bluish, greenish, pink or colourless
Lustre; opacity:	vitreous to greasy; transparent to translucent
Streak:	white
Hardness:	5.5–6
Specific gravity:	3.1
Cleavage; fracture:	two good cleavages; uneven
Habit:	crystals equant, rough, and may be twinned; also as cleavable masses, or compact
Crystal system:	triclinic

Olivenite

Named after its olive-green colour, olivenite is much in demand by mineral collectors. Although rare, it is the most common secondary copper arsenate found in the oxidized zone of hydrothermal copper deposits. It is associated with other copper minerals such as azurite, malachite, chrysocolla and chalcopyrite, and sometimes with arsenopyrite and spangolite. The type locality is Carharrak Mine in Cornwall, England; elsewhere in Britain it is found on Alston Moor (Cumbria) and Tavistock (Devon). Other localities include Cap Garonne (France), Clara Mine (near Oberwolfach, Germany), and the porphyry copper deposits of Chuquicamata in Chile. Excellent large crystals occur in Tsumeb (Namibia).

$Cu_2(AsO_4)(OH)$

Colour:	shades of olive green, dirty white, straw yellow
Lustre; opacity:	adamantine, vitreous, pearly, silky; transparent to opaque
Streak:	olive green to brown
Hardness:	3
Specific gravity:	4–4.5
Cleavage; fracture:	indistinct; forms small conchoidal fragments, very brittle
Habit:	crystals variable, may be elongated, or of short, prismatic, acicular or fibrous form; also as reniform masses
Crystal system:	monoclinic; pseudo-orthorhombic

Adamite

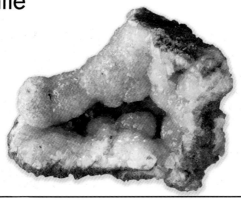

Adamite forms a series with olivenite $Cu_2(AsO_4)(OH)$. A popular mineral with collectors, adamite is generally shades of yellow, green or brown, although it may be violet to rose-coloured where cobalt has substituted for zinc. Some specimens fluoresce lemon yellow or greenish white, and some phosphoresce. Adamite is a rare secondary mineral found in the oxidized zones of ore deposits containing primary zinc- or arsenic-bearing minerals. It is associated with a variety of minerals, including malachite, reddish limonite, smithsonite or hemimorphite. It occurs at the type locality in the Atacama Desert (Chile), and also at Schwarzwald (Germany), Laurion (Greece), Durango (Mexico) and Tsumeb (Namibia).

$Zn_2(AsO_4)(OH)$

Colour:	yellow, brownish yellow to green, pink or violet
Lustre; opacity:	vitreous; transparent to translucent
Streak:	white
Hardness:	3.5
Specific gravity:	4.3–4.5
Cleavage; fracture:	good; uneven to subconchoidal
Habit:	crystals elongate tabular, platy or equant; usually forms crusts or spheroidal masses, or roughly radial aggregates
Crystal system:	orthorhombic

Lazulite

L azulite derives its name from the Arabic *azul*, meaning 'sky' or 'heaven', and the Greek *lithos*, meaning 'stone'. It has sometimes been confused with lazurite, lapis lazuli or azurite. It is a minor semi-precious stone, and may be polished, carved or tumbled to make beads or attractive ornamental stones. The rare transparent form is pleochroic, changing from blue to colourless. Lazulite occurs in high grade, quartz-rich metamorphic rocks, veins and pegmatites, associated with minerals such as andalusite, kyanite, sillimanite, and garnet. Beautiful crystals up to 5cm (2in) long occur in the Graves Mountains (Georgia, USA). Some of the best specimens occur in veins in iron-rich shales at Rapid Creek (Yukon, Canada).

$MgAl_2(PO_4)_2(OH)_2$

Colour:	deep azure blue to light blue or bluish green; mottled
Lustre; opacity:	vitreous to dull; translucent to opaque
Streak:	white
Hardness:	5.5–6
Specific gravity:	3.1–3.4 (iron rich)
Cleavage; fracture:	indistinct to good, prismatic; uneven to splintery
Habit:	crystals show steep pyramidal forms, may be tabular; also massive or granular
Crystal system:	monoclinic

Pseudomalachite

Pseudomalachite differs from malachite because it lacks the characteristic colour banding and does not effervesce in warm hydrochloric acid, whereas malachite does. It is a very rare secondary mineral found in the oxidized zone of copper deposits, associated with azurite, malachite, atacamite, limonite or chalcedony. It is sometimes used as a precious stone. Its use in paint has been recorded in an ancient tomb at the prehistoric Mayan site known as Baking Pot, in Belize. It is found at Virneberg Mine (Rheinbreitbach, Germany), Cornwall (England), Libethen (Slovakia), Harquahala Mine (Arizona, USA), Bogolo (Portugal), Manto Cuba and San Salvador Mines (Atacama, Chile) and in Nizhni Tagil (Russia).

$Cu_5(PO_4)_2(OH)_4$

Colour:	emerald green to blackish green; paler when fibrous
Lustre; opacity:	vitreous, greasy; translucent to transparent
Streak:	blue-green, paler than colour
Hardness:	4.5–5
Specific gravity:	4.3
Cleavage; fracture:	perfect; splintery
Habit:	crystals rare, rough and rounded; forms radial clusters, coatings, stalactitic, reniform or botryoidal aggregates
Crystal system:	monoclinic

Brazilianite

Brazilianite is a rare and unusual mineral, named after its discovery at Minas Gerais (Brazil) in 1945. Brazilianite occurs in phosphate-rich pegmatites, and may be associated with clay, lazulite and apatite. A striking yellow or yellowish green, it is sometimes cut for collectors, and is probably the best of the phosphates to be used as a gemstone. Crystals are fragile and easily broken, however, as they have a conchoidal fracture and a very good cleavage parallel to their length. Good crystals occur at Minas Gerais (Brazil) and Palermo Mine (New Hampshire, USA). Other localities include Roĭná (Czech Republic), Hagendorf (Germany), Etiro (Namibia) and the Dawson Mining District (Yukon Territory, Canada).

$NaAl_3(PO4)_2(OH)_4$

Colour:	colourless, light yellow to greenish yellow
Lustre; opacity:	vitreous; transparent
Streak:	white
Hardness:	5.5
Specific gravity:	3.0
Cleavage; fracture:	good; conchoidal fracture
Habit:	crystals common and well-formed, striated; also as spherical aggregates
Crystal system:	monoclinic

Clinoclase

Clinoclase forms as a secondary mineral in the oxidation zone of arsenic-rich hydrothermal copper deposits. Associated minerals include liroconite, chalcophyllite, olivenite and cornwallite ($Cu_5(AsO_4)_2(OH)_4$). The type locality is Wheal Gorland (Cornwall, England); now closed, this mine has produced many fine and rare minerals. In common with other unusual secondary copper arsenate minerals, clinoclase occurs at the Majuba Hill Mine (Nevada, USA), also closed. Blue-black crystalline aggregates from this mine have been termed beetle ore. Other localities include Sayda (near Freiberg, Germany), Cap Garonne (France), Novoveská Huta (Slovakia), Stirling Mine (New Jersey, USA) and Tintic (Utah, USA).

$Cu_3AsO_4(OH)_3$

Colour:	greenish black or bluish green
Lustre; opacity:	vitreous, pearly on cleavage; subtransparent to translucent
Streak:	blue-green
Hardness:	2.5–3
Specific gravity:	4.2–4.4
Cleavage; fracture:	very good parallel to base; uneven
Habit:	crystals small, needle-like, pseudo-orthorhombohedral, or platy; radial, fibrous, spherical aggregates or encrustations
Crystal system:	monoclinic

Descloizite

Descloizite is the zinc end member of a complete series in which copper substitutes for zinc, mottramite being the copper end member. A secondary mineral, it forms in the oxidized zone of lead, zinc and copper ore deposits, and is often associated with lead minerals such as vanadinite, cerussite, wulfenite and pyromorphite. It also forms in sandstone, having been precipitated from circulating mineralized ground water. It is a rare but important ore of vanadium. Large masses are found at Otavi (Namibia), and particularly fine crystals at Bisbee (Arizona, USA) and Lake Valley (New Mexico, USA). It also occurs at Obir (Corinthia, Austria), Bena de Padru (Sardinia, Italy) and Tsumeb (Namibia).

$PbZn(VO_4)(OH)$

Colour:	orange-red, brown, brownish-red
Lustre; opacity:	resinous to greasy; transparent to translucent
Streak:	orange to reddish brown
Hardness:	3.5
Specific gravity:	5.5–6.2
Cleavage; fracture:	poor; uneven to subconchoidal, brittle
Habit:	crystals prismatic or platy; fibrous, botryoidal masses or crusts
Crystal system:	orthorhombic

Apatite

A patite is the name for a group of minerals including fluorapatite, chlorapatite, and hydroxylapatite, in which fluorine, chlorine and hydroxyl substitute for one another, and carbonate-apatite, $Ca_5(PO_4,CO_3,OH)_3(F,OH)$. Apatite is the main inorganic constituent of bones and teeth. It is a common accessory mineral in igneous rocks, and is found in some contact metamorphic calc-silicate rocks. It also occurs in extensive and economically important bedded deposits called phosphorites. Eighty per cent of world production is used in chemical fertilizers and in the production of phosphoric acid. Major mining areas include Florida and North Carolina (USA), Morocco, and the Kola Peninsula (Russia).

$Ca_5(PO_4)_3(F,Cl,OH)$

Colour:	usually green, may be violet, red or brown
Lustre; opacity:	vitreous to sub-resinous; transparent to opaque
Streak:	white
Hardness:	5
Specific gravity:	3.1–3.2
Cleavage; fracture:	poor; conchoidal to uneven
Habit:	crystals prismatic or needle-like, sometimes tabular; also massive, concretionary, mammillated or oolitic
Crystal system:	hexagonal

Pyromorphite

Pyromorphite forms one series with mimetite ($Pb_5(AsO_4)_3Cl$) and another with vanadinite ($Pb_5(VO_4)_3Cl$). It is named after the Greek *pyr* ('fire') and *morfe* ('form'), as melted globules will recrystallize. It can luminesce yellow in ultraviolet light. Pyromorphite often occurs in rounded barrel-shaped forms known as campylite. It forms as a secondary mineral in the oxidized zone of lead veins, and rarely as a volcanic sublimate. Associated minerals include galena, wulfenite, cerussite, vanadinite and descloizite. Pyromorphite is a minor ore of lead known as green lead ore. It occurs in Cornwall (England), Cordoba (Spain), Guang Xi (China), Bunker Hill Mine (Idaho, USA), Broken Hill (NSW, Australia) and Zambia.

$Pb_5(PO_4)_3Cl$

Colour:	usually green, can be yellow, orange, brown or colourless
Lustre; opacity:	adamantine, vitreous, resinous; transparent to translucent
Streak:	white
Hardness:	3.5–4
Specific gravity:	7.1; unusually dense for a translucent mineral
Cleavage; fracture:	poor; uneven to subconchoidal
Habit:	crystals sometimes prismatic, occasionally hollow; also reniform or globular
Crystal system:	hexagonal

Mimetite

Mimetite was named from the Greek *mimetes*, the 'imitator', alluding to its resemblance to pyromorphite, with which it forms a series. It is a rare secondary mineral formed in the oxidized zone of arsenic-bearing lead deposits. It is commonly associated with pyromorphite and other lead minerals, such as vanadinite, galena and anglesite, and also with hemimorphite and arsenopyrite. It may fluoresce reddish yellow. It is soluble in hydrochloric acid, producing a very strong smell similar to garlic. Mimetite is a minor ore of lead. Well-crystallized material is found at Johanngeorgenstadt (Germany), Pfíbram (Czech Republic) and particularly at Tsumeb (Namibia).

$Pb_5(AsO_4)_3Cl$

Colour:	yellow, orange, brown, red, white, or may be colourless
Lustre; opacity:	vitreous, resinous; subtransparent to translucent
Streak:	white
Hardness:	3.5–4
Specific gravity:	7.0–7.3
Cleavage; fracture:	weak; uneven to subconchoidal, brittle
Habit:	crystals slender, acicular, or may be tabular or barrel-shaped; globular, reniform, stalactitic or granular
Crystal system:	hexagonal

Vanadinite

Vanadinite is a rare mineral that darkens on exposure and loses transparency. It shares the same structure as apatite, and hence similar crystal shapes. It forms as a secondary mineral in the oxidation zone of lead deposits, associated with wulfenite, pyromorphite, mimetite, cerussite or anglesite. Vanadinite obtained as a by-product of other mining operations can be used as an ore of vanadium, used in metal alloys such as steel, in dyes and in ceramics. The best crystals are found at Mibladen (Morocco) and Grootfontein (South Africa). The original material from the type locality of Zimapan (Hildalgo, Mexico) was lost at sea. Vanadinite also occurs in Kabwe (Zambia), Obir (Austria) and Tsumeb (Namibia).

$Pb_5(VO_4)_3Cl$

Colour:	bright reddish-orange to yellow, brown
Lustre; opacity:	resinous to adamantine; translucent to opaque
Streak:	yellowish-white to brownish-yellow
Hardness:	2.5–3
Specific gravity:	6.9–7.0
Cleavage; fracture:	none; conchoidal or uneven, brittle
Habit:	crystals stubby, hexagonal, may be hollow; as fibrous radiating masses, crusts, granular, or nodular
Crystal system:	hexagonal

Variscite

Variscite forms a series with strengite ($FePO_4 \cdot 2H_2O$). It typically forms where phosphatic meteoric water has circulated through and reacted with aluminium-rich rocks, such as feldspar-rich igneous rocks. It has also been found in caves, the source of the phosphorus being decomposing bat guano. Associated minerals include apatite, wavellite and limonite. The name comes from Variscia, the old name for Vogtland (Germany), the type locality. Nodules up to 30cm (12in) occur at Fairfield (Utah, USA), and greenish encrustations in Montgomery County (Arkansas, USA). Variscite is occasionally smoothed, polished and sold as turquoise, or it can be cut into cabochons.

$AlPO_4 \cdot 2H_2O$

Colour:	pale green, bluish green, emerald green; may be colourless
Lustre; opacity:	waxy to earthy; transparent to translucent
Streak:	white
Hardness:	4–5
Specific gravity:	2.52 (variscite)
Cleavage; fracture:	good; conchoidal or uneven to splintery
Habit:	crystals very rare; massive, encrusting or reniform
Crystal system:	orthorhombic

Strengite

Strengite is the iron-bearing end member of the strengite–variscite series. It is a secondary mineral that forms in complex granitic pegmatites, as an alteration product of primary phosphate minerals such as triphylite ($Li(Fe,Mn)PO_4$) – e.g. Bull Moose Mine, South Dakota, USA. It also occurs in limonitic iron ores and gossans, and more rarely in caves where the phosphate from bat guano has been used in the alteration of iron-rich minerals to strengite. It may be associated with dufrenite, cacoxenite, vivianite or apatite. Localities include Hagendorf (Bavaria, Germany), Bomi Hill Caves (Liberia), Indian Mountain (Alabama, USA), Minas Gerais (Brazil) and Iron Monarch Quarry (South Australia).

$FePO_4.2H_2O$

Colour:	shades of violet, pink or red, may be colourless
Lustre; opacity:	vitreous; transparent to translucent
Streak:	white
Hardness:	3.5
Specific gravity:	2.87
Cleavage; fracture:	good; conchoidal
Habit:	crystals of variable habit; botryoidal, radial, or spherical nodules
Crystal system:	orthorhombic

Phosphophyllite

Phosphophyllite is found as a primary precipitate in tin-rich hydrothermal veins in Bolivia. Elsewhere, it occurs as a secondary mineral in complex granitic pegmatites, formed by the alteration of sphalerite and iron-manganese phosphates, and in hydrothermal vein deposits. Associated minerals include vivianite, strengite, sphalerite and apatite. It is a highly prized gemstone when a pale bluish green, and may be cut into facets or cabochons. It is brittle and fragile, however, and is rarely cut, as large crystals are too valuable to break up. It is very rare. Crystals up to 10cm (4in) occur at Potosi (Bolivia); other locations include Hagendorf (Germany), North Groton (New Hampshire, USA) and Kabwe (Zambia).

$Zn_2(Fe,Mn)(PO_4)_2.4H_2O$

Colour:	bright blue green, green to colourless
Lustre; opacity:	vitreous; transparent
Streak:	white
Hardness:	3–3.5
Specific gravity:	3.1
Cleavage; fracture:	perfect; conchoidal or uneven
Habit:	crystals equant to prismatic; drusy, forming in cavities
Crystal system:	monoclinic

Vivianite

Vivianite was named after J.H. Vivian (1785–1855), the mineralogist who discovered it in St Austell (Cornwall, England). As the pigment blue ochre, it was used in Cologne in the thirteenth and fourteenth centuries, and possibly by the ancient Greeks. It is a common secondary mineral formed by the weathering of metallic ore deposits, and as an alteration product in pegmatites. Associated minerals are siderite, sphalerite and pyrite. It can replace organic matter, forming thin coatings on the damaged surfaces of fossilizing mammoth tusks or skulls (Mexico), whale bones (Richmond, Virginia, USA) or inside bivalves (Kerch iron deposits, Crimea, Russia). It also forms in lake sediments, bog iron ores and peat bogs.

$Fe_3(PO_4)_2.8H_2O$

Colour:	colourless when fresh; darkens on exposure to blue
Lustre; opacity:	vitreous, greasy or pearly; transparent to translucent
Streak:	colourless to bluish-white
Hardness:	1.5–2
Specific gravity:	2.7
Cleavage; fracture:	perfect, thin cleavage plates are flexible; fibrous
Habit:	crystals prismatic to flattened, may appear bent, acicular (needle-like) or fibrous; concretionary, earthy and encrusting masses
Crystal system:	monoclinic

Erythrite

Erythrite forms a series with annabergite, in which nickel and cobalt substitute for one another. A rare secondary mineral, it typically forms by weathering in the oxidized zone of ore deposits, as at Cobalt (Ontario, Canada). Associated minerals include silver, cobaltite and skutterudite. Its bright colour is a valuable indicator of the presence of cobalt-bearing ore deposits, hence the alternative name cobalt bloom. When heated, the powdered mineral turns lavender blue and smells of garlic. It was developed in 1803 as a natural pale to medium violet pigment, light cobalt violet, but is rarely used now because of its toxicity. Striking specimens come from Daniel Mine (Saxony, Germany) and Bou Azzer (Morocco).

$Co_3(AsO_4)_2 \cdot 8H_2O$

Colour:	violet red, light pink, purple red; grey surface alteration
Lustre; opacity:	vitreous, pearly on cleavages; translucent to transparent
Streak:	pinkish red
Hardness:	1.5–2.5
Specific gravity:	3.1–3.2
Cleavage; fracture:	perfect, small, flexible plates; uneven
Habit:	well-formed crystals rare; mostly as crusts or small reniform aggregates
Crystal system:	monoclinic

Annabergite

Annabergite is the nickel end member of the series annabergite–erythrite, in which nickel and cobalt substitute for one another. It was named in 1832 after Annaberg in Saxony. A rare secondary mineral, annabergite is also known as nickel bloom because it forms crusts or films through the surface alteration of other nickel minerals, sometimes completely replacing the original mineral. Like erythrite, it has served as a good indicator of the presence of ore, although rarely being of economic importance itself. It may also be generated during mining as part of the milling process. It occurs in Annaberg (Saxony, Germany), Cobalt (Ontario, Canada), Humboldt County (Nevada, USA) and Allemont (France).

$Ni_3(AsO_4)_2.8H_2O$

Colour:	shades of green, white, grey; may be pink
Lustre; opacity:	vitreous, pearly, dull; translucent to opaque
Streak:	light green, paler than its colour
Hardness:	1.5–2.5
Specific gravity:	3.1
Cleavage; Fracture:	perfect; uneven
Habit:	prismatic, striated crystals, small and rare; more usually as earthy crusts or powdery masses
Crystal system:	monoclinic

Cacoxenite

Cacoxenite is a rare secondary mineral formed through the alteration of other phosphates, and associated with hematite, wavellite, strengite and iron oxides. It is named from the Greek *kakos* ('wrong') and *xenos* ('guest') because the phosphorus content of cacoxenite lessened the quality of smelted iron. It is very attractive, forming golden fibrous and radiating crystals within spherical aggregates. It is also found as inclusions in quartz, particularly amethyst, which detracts from the purple of the amethyst and dulls its appearance. Cacoxenite is found at Diamond Hill Quartz Mine (South Carolina), at Ober-Rosbach (Hesse, Germany), in Indian Mountain (Alabama) and at Avant's Claim (Arkansas, USA).

$Fe_4(PO_4)_3(OH_3).12H_2O$

Colour:	yellow, ochre yellow, golden or brown
Lustre; opacity:	vitreous, silky, greasy; translucent to transparent
Streak:	straw yellow
Hardness:	3–3.5
Specific gravity:	2.3
Cleavage; fracture:	poor; fibrous, brittle
Habit:	acicular (needle-like) crystals; as spherical aggregates, internally radiating and giving a stellar appearance
Crystal system:	hexagonal

Wavellite

Wavellite is named after the English physician W. Wavell, who discovered the mineral in 1805. It forms small globular masses that exhibit a characteristic internally radiating structure of acicular crystals. A common secondary mineral, it forms as a low-temperature hydrothermal alteration product in fissures of aluminium-rich rocks, and also in pegmatites with phosphates. At one locality in Arkansas (USA), bluish-green spherulites form the cement of a phosphoritic breccia. Associated minerals are limonite, hematite, pyrolusite. Wavellite is also found in Dünsberg and Waldgermes (Germany), Cerhovice (Czech Republic), in tin veins in St Austell (Cornwall, England) and Llallagua (Bolivia).

$Al_3(PO_4)_2(OH)_3.5H_2O$

Colour:	white, yellow, greenish, brown, bluish
Lustre; opacity:	vitreous, silky, pearly; translucent
Streak:	white
Hardness:	3.5–4
Specific gravity:	2.3–2.4
Cleavage; fracture:	three good cleavages; uneven to subconchoidal
Habit:	very rare as good prismatic crystals; usually flat to spherical, radiating green to yellow-green fibrous clusters
Crystal system:	orthorhombic

Turquoise

Turquoise is a rare and valuable ornamental stone. Its name comes from the French *pierre turquoise*, meaning 'Turkish stone', having been brought from Iran to Europe via Turkey. It has been prized for thousands of years, as evidenced from its use in an Egyptian tomb of *c.* 3000 BC and in the burial mask of Tutankhamun. A secondary mineral, it forms in the alteration zone of hydrothermal porphyry copper deposits, and in veins and pockets through the alteration of volcanic rocks and phosphate-rich sedimentary rocks. It normally occurs in arid regions. The most important source of turquoise is Iran, where it has been mined for 2000 years, many being worked around Nishapur.

$CuAl_6(PO_4)_4(OH)_8.4H_2O$

Colour:	sky blue to pale blue, greenish blue, green
Lustre; opacity:	vitreous, waxy, greasy or dull; transparent to opaque
Streak:	white or pale greenish
Hardness:	5-6
Specific gravity:	2.6–2.8
Cleavage; fracture:	good; conchoidal to even fracture
Habit:	crystals rare, small, prismatic; usually massive in veins, concretions or encrustations
Crystal system:	triclinic

Chalcophyllite

Chalcophyllite is appealing to collectors because of its attractive colour and high lustre, and because it sometimes forms attractive six-sided crystals arranged into a rosette. A rare secondary mineral, it is found in the zone of oxidation of arsenic-bearing hydrothermal copper deposits. Associated minerals include azurite, chrysocolla, malachite, cuprite, spangolite, limonite and clinoclase. It occurs at Wheal Phoenix (Cornwall, England), Clara Mine (Wolfach, Germany), Bisbee (Arizona, USA) and Nizhni Tagil (Russia). It has also been found in vugs within slags derived from ancient to relatively recent copper refining at Val Varenna (Genova, Italy), which were uncovered in 1993 after severe flooding.

$(Cu,Al)_3(AsO_4,SO_4)(OH)_4.6H_2O$

Colour:	shades of green to blue
Lustre; opacity:	adamantine to vitreous, pearly; transparent to translucent
Streak:	pale green to bluish green
Hardness:	2
Specific gravity:	2.68 (light for copper minerals)
Cleavage; fracture:	perfect, basal; uneven
Habit:	crystals platy, may be striated, in rosettes; foliated, tabular or massive
Crystal system:	trigonal

Liroconite

Liroconite is named from the Greek *liros* ('pale') and *konia* ('powder'). An extremely rare secondary mineral, it forms in the oxidized zone of copper deposits, associated with other copper minerals such as clinoclase, chalcophyllite, azurite, malachite and limonite. Liroconite was discovered in the 1780s or 1790s in the old copper mines of Cornwall (England), such as Wheal Gorland. The largest known crystal is 35mm (1½in) long, and is in the Rashleigh Collection held at the Royal Cornwall Museum at Truro. Lesser quality specimens occur in Sayda (near Freiburg, Germany), Herrengrund (Slovakia), Cerro Gordo Mine (California, USA), the Khovu-Aksy deposit (Tuva, Russia), N'Kana (Zambia) and Zaire.

$Cu_2Al(As,P)O_4(OH)_4.4H_2O$

Colour:	sky blue, turquoise blue, verdigris green
Lustre; opacity:	vitreous to resinous; transparent to translucent
Streak:	pale blue, paler than colour
Hardness:	2–2.5
Specific gravity:	2.9–3.0
Cleavage; fracture:	poor; uneven to subconchoidal, slightly curving surfaces
Habit:	well-formed crystals wedge to lens-shaped, frequently appearing bent with rounded edges; also granular
Crystal system:	monoclinic

Lavendulan

Lavendulan, named after its lavender colour, is a rare secondary mineral formed in the oxidized zones of copper deposits. Associated minerals include erythrite, cuprite, malachite, covellite, brochantite and olivenite. World-class specimens of lavendulan were collected from the important former lead-silver-zinc mining district of Mazarron-Aguilas (Murcia, Spain) in 1992. Although copper mining was never significant, the area is renowned for copper arsenates and associated secondary minerals. Specimens include electric blue bladed crystals up to 4mm (⅙in), and radial groups up to 10mm (⅜in). Other localities include Jáchymov (Czech Republic), Bou Azzer (Morrocco) and Wheal Owles (Cornwall, England).

$CaNaCu_5(AsO_4)_4Cl.5H_2O$

Colour:	blue, electric blue, greenish blue or lavender
Lustre; opacity:	vitreous, waxy or satiny in aggregates; translucent
Streak:	light blue
Hardness:	2.5
Specific gravity:	3.54–3.59
Cleavage; fracture:	good; uneven
Habit:	crystals rare, generally less than 3mm, twinning common; radiating fibres or rosettes, botryoidal crusts
Crystal system:	orthorhombic

Autunite

Autunite is a hydrated calcium uranium phosphate, which loses water over time to form meta-autunite. Autunite is strongly radioactive and fluoresces a striking yellowish green. It forms through the alteration of primary uranium minerals such as uraninite, in pegmatites, hydrothermal veins and in the weathering zone of granite. Associated minerals include torbernite ($Cu(UO_2)_2(PO_4)_2.8\text{-}12H_2O$), which has similar chemistry and properties, although not forming a series with autunite. It is an important uranium ore, widely used in World War II. Named after Autun in France, it also occurs at Hagendorf (Germany), Shinkolobwe (Zaire), Minas Gerais (Brazil), Mt Spokane (Washington, USA) and Colorado (USA).

$Ca(UO_2)_2(PO_4)_2.10\text{-}12H_2O$

Colour:	shades of yellow to greenish yellow
Lustre; opacity:	vitreous to pearly; transparent to translucent
Streak:	pale yellow
Hardness:	2–2.5
Specific gravity:	3.1–3.2
Cleavage; fracture:	perfect, thin cleavage sheets are flexible; uneven
Habit:	crystals tabular, may be twinned; crusts, scaly aggregates or earthy masses
Crystal system:	tetragonal

Carnotite

Carnotite is a hydrated potassium uranium vanadate, the water content and hence specific gravity of which varies with temperature and humidity. It is strongly radioactive, but does not fluoresce. It forms through the alteration of minerals such as uraninite, occurring as an impregnation in sands and sandstones, associated with old channels or near playas. It sometimes occurs in small masses associated with vegetable matter, such as petrified trees, or is disseminated throughout, colouring the rock bright yellow. Used as an ore of vanadium and uranium, carnotite occurs in Colorado's desert regions, Utah's San Rafael Swell, Monument Valley (Arizona), New Mexico (USA), the Fergana desert (Russia) and Radium Hill (South Australia).

$K_2(UO_2)_2(VO_4)_2.3H_2O$

Colour:	canary yellow to greenish yellow
Lustre; opacity:	pearly to dull; semi-opaque to transparent
Streak:	light yellow
Hardness:	very soft, possibly 2
Specific gravity:	4.7–4.9
Cleavage; fracture:	perfect, basal
Habit:	crystals flat, pseudo-hexagonal, rare; powdered, earthy aggregates
Crystal system:	monoclinic

Willemite

Willemite is a zinc ore that became important after the discovery of huge amounts in marbles at Franklin and Sterling Hill (New Jersey, USA); it was found later in Belgium and named in honour of King Willem I of the Netherlands. Specimens from Franklin show a bright green fluorescence, contrasting beautifully with the red fluorescence of associated calcite. Willemite is a secondary mineral found in zinc-bearing limestones, associated with calcite, franklinite and zincite; specimens displaying all four minerals are popular. Other localities include Tiger (Arizona, USA), Tsumeb (Namibia), Altenberg (Belgium), Mont St Hillaire (Quebec, Canada) and Beltana (South Australia).

$ZnSiO_4$

Colour:	greenish-yellow, also dark brown to off-white, bluish
Lustre; opacity:	vitreous to resinous; transparent to opaque
Streak:	white
Hardness:	5.5
Specific gravity:	4.0
Cleavage; fracture:	perfect; conchoidal
Habit:	rare prismatic crystals; compact, granular or massive aggregates
Crystal system:	trigonal

Forsterite

Forsterite is a member of the forsterite–fayalite series (Mg_2SiO_4–Fe_2SiO_4), which comprises an important group of rock-forming minerals called olivines. Olivine is common in basic and in ultrabasic igneous rocks. Forsterite and quartz do not occur together, as they react to form enstatite ($MgSiO_3$). Forsterite is found in metamorphosed siliceous dolomites; fayalite, stable in the presence of quartz, is found in quartz-bearing rocks. Nickel-rich forsterites occur in stony meteorites. Forsterite is found on Bheinn-an-Dubhaich (Skye, Scotland), Vesuvius (Italy), Sapat (Pakistan) and Kovdor massif (Kola Peninsula, Russia). Fayalite occurs in Yellowstone Park (Wyoming, USA) and the Mourne Mountains (Ireland).

Mg_2SiO_4

Colour:	white or yellow (forsterite), brown or black (fayalite)
Lustre; opacity:	vitreous; transparent to translucent
Streak:	white
Hardness:	6.5 (forsterite); 7 (fayalite)
Specific gravity:	3.2 (forsterite); 4.4 (fayalite)
Cleavage; fracture:	poor; conchoidal, brittle
Habit:	well-formed crystals rare; granular masses, isolated grains in rock
Crystal system:	orthorhombic

Olivine

Olivine, like mica or serpentine, is not an officially recognized mineral name, but is widely used by petrologists and mineralogists. The gem variety of olivine is called peridot, having a distinctive greasy lustre and an olive- or bottle-green colour. The best peridot stones have about 15 per cent fayalite content to give sufficient green colour, plus possibly a little nickel and/or chromium to provide the best hues. Peridot was used in the Middle Ages to decorate church robes and plates. Gem-quality peridot is found at St John's Island (Zabargad) in the Red Sea, Eiffel (Germany), Mogok (Myanmar), San Carlos (Arizona, USA), Sapat (Pakistan) and Vesuvius (Italy). Olivine can make up half the mass of a stony meteorite.

$(Mg,Fe)_2SiO_4$

Colour:	green, especially olive-coloured
Lustre; opacity:	vitreous; transparent to translucent
Streak:	white
Hardness:	6.5 –7
Specific gravity:	3.2–4.4
Cleavage; fracture:	poor; conchoidal, brittle
Habit:	well-formed crystals rare; granular masses, isolated grains in igneous rocks
Crystal system:	orthorhombic

Monticellite

Monticellite is a calcium-rich member of the olivine group, and an end member of the monticellite–kirschsteinite series ($CaMgSiO_4$–$CaFeSiO_4$). It also forms a series with glaucochroite ($CaMnSiO_4$). Consequently, iron and manganese are ubiquitous impurities. It was named after the Italian mineralogist Teodoro Monticelli (1759–1845). Monticellite is formed in contact metamorphic zones between limestones and gabbros, or between granites and dolomites. Associated minerals include gehlenite, spinel, calcite, vesuvianite and apatite. It is found at Vesuvius (Italy), Isle of Muck (Scotland), Magnet Cove (Arkansas, USA), Kovdor (Kola Peninsula, Russia) and Isle Cadieux (Quebec, Canada).

$CaMgSiO_4$

Colour:	colourless, greenish grey, grey
Lustre; opacity:	vitreous; transparent
Streak:	white
Hardness:	5.5
Specific gravity:	3.2
Cleavage; fracture:	poor; subconchoidal to uneven
Habit:	crystals rare as well-formed prisms; massive, granular
Crystal system:	orthorhombic

Garnet: Pyrope

Pyrope, alongside almandine and spessartine, belongs to the pyralspite group of garnets. It is one of the less common garnets, widely used as a gem and noted for its clear red crystals. It forms a series with almandine; a member of which is the red-lavender coloured gemstone rhodolite. Unusual for a garnet, pyrope occurs mostly in igneous rocks, such as eclogites and kimberlites. These are formed at high pressures and can contain diamonds. It also occurs in related serpentinites and placer deposits. It is found at Merunice (Czech Republic), Dora-Meira (Piedmont, Italy), Kimberly (South Africa), the Umba River (Tanzania), Cowee Creek (North Carolina, USA), Buell Park (Arizona, USA) and Bingara (NSW, Australia).

$Mg_3Al_2(SiO_4)_3$

Colour:	red-orange to deep red
Lustre; opacity:	vitreous; transparent to translucent
Streak:	white
Hardness:	7–7.5
Specific gravity:	3.58
Cleavage; fracture:	absent; conchoidal
Habit:	crystals typically rhombic dodecahedra, sometimes trapezohedra; massive, granular
Crystal system:	cubic

Garnet: Almandine

Almandine, a pyralspite garnet, is darker red than pyrope and sometimes may appear black. Rare transparent specimens are facetted as gemstones, having a high lustre. Specimens having inclusions such as rutile needles are often cut *en cabochon*. Like many garnets in metamorphic rocks, their equant, well-formed crystals often stand out as porphyroblasts against a schistose matrix. It is usually found in medium-grade metamorphic rocks; less often in granites and pegmatites; and also as placer deposits. Fine orange-red crystals are found in sands in Minas Novas and Minas Gerais (Brazil) and in Sri Lanka; other locations include Sticken River (Alaska), Adirondacks (New York, USA) and the Zillertal (Austria).

$Fe_3Al_2(SiO_4)_3$

Colour:	red-violet to red-brown
Lustre; opacity:	vitreous to resinous; transparent to opaque
Streak:	white
Hardness:	7.5
Specific gravity:	4.4
Cleavage; fracture:	absent; conchoidal
Habit:	crystals typically rhombic dodecahedra, sometimes trapezohedra, embedded as separate crystals in metamorphic rocks; massive
Crystal system:	cubic

Garnet: Spessartine

Spessartine, a pyralspite garnet, is rarely found at gem quality. Forming a series with almandine, it usually contains varying amounts of manganese and iron. The crystal shapes of spessartine – i.e. the 12-sided dodecahdron with rhomb-shaped faces or the 24-sided trapezohedron with trapezium-shaped faces – are ubiquitous in the garnet group. These round-looking crystals are characteristic of garnets. Spessartine occurs in rhyolites, granites and pegmatites; in metasomatized manganous rocks; and as placer deposits. It occurs in Seriphos (Greece), the Spessart Mountains (Germany), Minas Gerais (Brazil), Lieper's Quarry (Pennsylvania, USA), Amelia (Virginia, USA) and Tsilaizina and Anjanabonoina (Madagascar).

$Mn_3Al_2(SiO_4)_3$

Colour:	yellow, orange, red, brown, black
Lustre; opacity:	vitreous; transparent to translucent
Streak:	white
Hardness:	7–7.5
Specific gravity:	4.2
Cleavage; fracture:	absent; conchoidal
Habit:	crystals as the typical rhombic dodecahedra, but more commonly 24-sided trapezohedra; massive, granular
Crystal system:	cubic

Garnet: Grossular

Grossular, alongside uvarovite and andradite, belongs to the ugrandite group of garnets, giving gem-quality specimens in many colours. Massive green grossular, such as Transvaal jade, is popular, often with black inclusions of magnetite. Some has a distinctive gooseberry colour, hence the name derived from the scientific name *R. grossularia*. Cinnamon orange-red transparent garnets called hessonite were used as gems by the ancient Greeks and Romans. Tsavorite is a transparent green variety. It occurs in metamophosed calcareous rocks and as placer deposits. Fine crystals occur at Chernyshevsk (Russia), Ala Valley (Italy), Asbestos (Canada), Ramona (California, USA), Maharitra (Madagascar) and Telemarken (Norway).

$Ca_3Al_2(SiO_4)_3$

Colour:	colourless when pure, green, brown, orange, pink, red, black
Lustre; opacity:	vitreous; transparent
Streak:	white
Hardness:	6.5–7.5
Specific gravity:	3.59
Cleavage; fracture:	none; subconchoidal
Habit:	crystals as rhombohedra or trapezohedra; granular, compact, massive
Crystal system:	cubic

Garnet: Uvarovite

Uvarovite, a ugrandite garnet, is emerald-green and much sought-after as a gemstone because of its outstanding brilliance and colour. Uvarovite is mostly found as small crystals, so much jewellery uses small druses. The best clear specimens are found in the Urals, lining cavities and fissures. Uvarovite is named after Count Sergei Semenovitch Uvarov (1765–1855), a Russian statesman and amateur mineral collector. It forms a series with grossular and often contains aluminium. It occurs in chromium-rich serpentinites. Fine crystals are found in Yerkaterinberg and the Saranovskii Mine (Urals, Russia), Outokumpu (Finland), Pico do Posets (Pyrenees, Spain), Quebec (Canada) and the Kop Mountains (Turkey).

$Ca_3Cr_2(SiO_4)_3$

Colour:	bright green to dark green
Lustre; opacity:	vitreous to adamantine; transparent to translucent
Streak:	white
Hardness:	6.5–7
Specific gravity:	3.4–3.8
Cleavage; fracture:	absent; conchoidal
Habit:	crystals as rhombic dodecahedra or trapezohedra; massive, granular
Crystal system:	cubic

Garnet: Andradite

Andradite, a ugrandite garnet, has many varieties, the most prized being the chromium-rich, emerald-green demantoid. This has a higher dispersion than diamond and is characterized by inclusions of asbestos fibres known as 'horsetails'. A yellow variety is called topazolite and the black variety melanite. It is named after the Brazilian mineralogist J.B. d'Andrada e Silva (1763–1838). Andradite is found in contact-metamorphosed limestones; schists and serpentinites; and in silica-poor igneous rocks. It occurs in Elba and Livorno (Italy), Arendal (Norway) and Franklin (New Jersey, USA); it is found as dementoid in the Urals among gold-bearing sands, and as melanite at Vesuvius and Lazio (Italy).

$Ca_3Fe_2(SiO_4)_3$

Colour:	yellow, green, brown, black
Lustre; opacity:	vitreous; transparent to opaque
Streak:	white
Hardness:	6.5–7
Specific gravity:	3.7–4.1
Cleavage; fracture:	absent; conchoidal
Habit:	crystals as rhombic dodecahedra and trapezahedra; massive, crusts showing many rhombic faces
Crystal system:	cubic

Zircon

Zircon has a similar lustre and fire to that of diamond, with colourless zircons having been used as imitation diamonds. Zircon comes in a variety of colours; the yellow variety alluded to in the name comes from the Arabic *zargun* for 'gold colour'. Heat treatment often changes the colour of zircons. Zircon that is not of gem quality is a major ore for zirconium, as well as hafnium and thorium. It occurs as an accessory mineral in acid igneous rocks, their products of metamorphism, and as placer deposits. Excellent crystals are found in river deposits at Matura (Sri Lanka) and the Ilmen Mountains (Russia). Other localities are Renfrew (Canada), Mt Ampanobe (Madagascar) and Teete (Mozambique).

$ZrSiO_4$

Colour:	colourless, yellow, red, brown, grey, green
Lustre; opacity:	vitreous to subadamantine; transparent to opaque
Streak:	white
Hardness:	6.5–7.5
Specific gravity:	3.9–4.8
Cleavage; fracture:	indistinct; conchoidal
Habit:	crystals dipyramidal or prismatic; irregular granules
Crystal system:	tetragonal

Sillimanite

Sillimanite is a polymorph of aluminium silicate, alongside kyanite and andalusite. The appearance of one of these minerals in metamorphic rocks is a key indicator of the temperatures and pressures undergone by the rock. The presence of sillimanite indicates a high temperature of formation (at least 650°C/1202°F). The slim, prismatic crystals of sillimanite distinguish it, and may be raised on weathered rocks. It is found in metamorphosed pelitic rocks and in pegmatites associated with corundum, tourmaline and topaz. Blue and violet transparent sillimanite is found in Mogok (Myanmar); grey-green varieties are found in Sri Lanka; and other locations are Minas Gerais (Brazil), Maldan (Czech Republic) and Freiberg (Germany).

Al_2SiO_5

Colour:	grey, brown, pale green
Lustre; opacity:	vitreous to subadamantine; translucent to transparent
Streak:	white
Hardness:	6.5–7.5
Specific gravity:	3.2
Cleavage; fracture:	perfect; uneven
Habit:	crystals long, slender prisms, occasionally acicular, poorly terminated; often silky, fibrous aggregates (fibrolite)
Crystal system:	orthorhombic

Andalusite

Andalusite is a polymorph of aluminium silicate, alongside kyanite and sillimanite. Its presence in a rock indicates that the rock has undergone low pressures during metamorphism (i.e. has been formed within a few kilometres of the Earth's surface). Greenish-red transparent crystals are used as gems. Twinned crystals called chiastolite form cross shapes (+ shapes) with x-shaped dark lines at the centre. Chiastolite has been used in amulets as a religious symbol. Andalusite occurs in metamophosed pelites, especially contact-metamorphosed slates, and in pegmatites. Large crystals are found at Lisenz (Austria); chiastolite is found in schists at Santiago de Compostela (Spain) and Keiva (Kola Peninsula, Russia).

Al_2SiO_5

Colour:	white, pink, pearl-grey, green, brown
Lustre; opacity:	vitreous, greasy; transparent to opaque
Streak:	white
Hardness:	7.5
Specific gravity:	3.1–3.2
Cleavage; fracture:	good; uneven
Habit:	crystals as square, stubby prisms, often twinned; granular, rod-like aggregates
Crystal system:	orthorhombic

213

Kyanite

Kyanite is a polymorph alongside andalusite and sillimanite. The presence of kyanite in rocks indicates that they have undergone moderate temperatures and medium to high pressures during metamorphism. Kyanite is named after the Greek *kyanos* for blue; it is also known as disthene, meaning 'double strength', after the hardness, which is greater across the crystal than lengthwise. Kyanite occurs in pelitic schists and gneisses, associated with garnet, staurolite and micas; and in eclogites. Large blue crystals come from Minas Gerais (Brazil) and Pizzo Forno (Switzerland); green crystals up to 30cm (12in) from Machakos (Kenya); and grey radiating crystals from Bolzano (Italy), the Tyrol (Austria) and Morbihan (France).

Al_2SiO_5

Colour:	often blue, lighter towards the margins; colourless, grey, green
Lustre; opacity:	vitreous, pearly; transparent to translucent
Streak:	white
Hardness:	4–4.5 along cleavage planes, 6–7 across cleavage planes
Specific gravity:	3.6–3.7
Cleavage; fracture:	perfect
Habit:	flat, bladed crystals in schists and gneisses, rosettes in quartz; massive aggregates
Crystal system:	triclinic

Topaz

Topaz has been a highly prized gemstone for millenia. It is found in a variety of colours and is often heat-treated and/or irradiated to give different hues; blue topaz resembling aquamarine is produced thus from colourless stones. Yellow topaz from Brazil turns pink on heating to 300°–450°C (572°–842°F). Crystals in pegmatites can be huge (up to 300kg/661lb). The Brazilian Princess, a pale blue topaz from Teofilo Otoni, is the largest cut topaz, weighing 4266g (9¼lb). Topaz is found in pneumatolytic veins in granites and pegmatites, and as a placer deposit. The largest crystals are found at Minas Gerais (Brazil) and Albashka (Siberia, Russia); it is also found in Mino Province (Japan), at Pikes Peak (Colorado, USA) and on Elba (Italy).

$Al_2SiO_4(F,OH)_2$

Colour:	colourless, yellow, pink, blue, green
Lustre; opacity:	vitreous; transparent to translucent
Streak:	opacity
Hardness:	8
Specific gravity:	3.5–3.6
Cleavage; fracture:	perfect; conchoidal, uneven
Habit:	vertically striated prismatic crystals; columnar, granular
Crystal system:	orthorhombic

Staurolite

Staurolite is a common mineral in pelitic schists and gneisses, occurring often as large porphyroblasts. Its presence is useful when assessing the degree of metamorphism, indicating that the rock has undergone medium temperature conditions. As a gem mineral, the opaque cross-shaped twins characteristic of staurolite are used in jewellery, often as amulets or religious items. The crosses occur at angles of both 90° and 60°. Fine crystals up to about 12cm (5in) are found in Pizzo Forno and Alpe Piona (Switzerland), Finistère and Morbihan (France), Mt Greiner (Austria), Keivy massif (Kola Peninsula, Russia), Franconia (New Hampshire), Blue Ridge (Georgia) and Taos County (New Mexico, USA).

$(Fe,Mg)_2(Al,Fe)_9(Si_4O_{20})(O,OH)_2$

Colour:	reddish-brown, brown, black
Lustre; opacity:	vitreous, resinous, dull; translucent to opaque
Streak:	white
Hardness:	7–7.5
Specific gravity:	3.65–3.83
Cleavage; fracture:	poor; uneven to conchoidal
Habit:	crystals prismatic or tabular, pseudohexagonal truncated diamond shapes, often twinned giving cross shapes
Crystal system:	monoclinic

Titanite

Titanite is now the more common name for sphene. Titanite displays interesting colours and strong fire, and is often faceted for display. Impurities such as iron and aluminium are always present, and rare earths such as cerium and yttrium are commonly present. Titanite is a common accessory mineral in intermediate and felsic plutonic rocks, pegmatites and veins, and is also found in gneisses and schists. Large crystals occur at St Gotthard (Switzerland), Bridgewater (Pennsylvania, USA), the Urals (Russia), Kola Peninsula (Russia), Minas Gerais (Brazil), Renfrew (Canada). Gem-quality crystals can be found at Pino Solo and La Huerta (Baja California Norte, Mexico).

$CaTiSiO_5$

Colour:	yellow, green, red, reddish brown, brown, black
Lustre; opacity:	vitreous to subadamantine; transparent to opaque
Streak:	white
Hardness:	5–5.5
Specific gravity:	3.4–3.6
Cleavage; fracture:	distinct; conchoidal, brittle
Habit:	crystals usually flattened, prismatic, wedge-shaped; compact or lamellar masses, disseminated grains
Crystal system:	monoclinic

Chloritoid

Chloritoid is found as small, platy crystals or occasionally as porphyroblasts in metamorphic rocks. Two related minerals are the magnesium-rich sismondine and the manganese-rich otrellite. Chloritoid indicates that the rock formed under low to medium pressures and temperatures. It shows a strong pleochroism, dark green to light green or yellow in varying orientations. Chloritoid occurs in metamorphic pelites, especially schists and marbles, associated with muscovite, staurolite, garnet, chlorite, kyanite and quartz. It is found at Kosoi Brod (Urals, Russia), Svalbard (Norway) and Natick (Rhode Island, USA). Sismondine is found at Zermatt (Switzerland) and Pregatten (Austria); otrellite at Ottré (Belgium).

$(Fe,Mn)_2Al_4Si_2O_{10}(OH)_4$

Colour:	dark grey, greenish-grey
Lustre; opacity:	vitreous, pearly
Streak:	white, pale grey, pale green
Hardness:	6.5
Specific gravity:	3.56–3.61
Cleavage; fracture:	perfect; brittle
Habit:	pseudohexagonal tabular crystals; commonly compact, foliated aggregates, massive
Crystal system:	monoclinic or triclinic (but always pseudohexagonal)

Datolite

Datolite is an ore of boron when available in sufficient quantities. It forms complex crystals that can superficially look like cubic forms such as dodecahedra or trapazohedra, but on closer inspection are revealed as lacking in symmetry. Datolite occurs as a secondary mineral in basalts, serpentinites and in hydrothermal deposits. It is often found in vesicles, associated with zeolites, prehnite and calcite. Datolite can be found in Kratzenberg (Austria), Serra dei Zanchetti and Alpe de Siusi (Italy), St Andreasberg (Harz, Germany), Arendal (Norway), Lane (Massachusetts, USA) and Prospect Park (New Jersey, USA).

$CaBSiO_4(OH)$

Colour:	colourless, white, sometimes light green or yellow
Lustre; opacity:	vitreous, greasy; transparent to translucent
Streak:	white
Hardness:	5–5.5
Specific gravity:	2.9–3
Cleavage; fracture:	imperfect; irregular, subconchoidal
Habit:	crystals as plates or prisms, sometimes large and complex; granular to compact, crusty aggregates with radiating fibres called botryolites
Crystal system:	monoclinic

219

Gadolinite

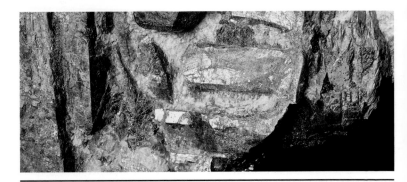

Gadolinite has varying amounts of other rare-earth elements such as cerium, lanthanum and neodymium substituting for yttrium; it also contains varying amounts of thorium and uranium, making it radioactive, and the crystals often metamict. Gadolinite is an ore of thorium and the rare-earth elements, which are finding an increasing variety of uses. It is named after the Finnish chemist Johan Gadolin (1760–1852), who discovered yttrium. Gadolinite occurs in granite and syenite pegmatites, and in veins in metamorphic rocks. Crystals (up to 500kg/1100lb) occur in Hitterö, Iveland and Hundholmen (Norway). Other localities include Finbo and Ytterby (Sweden), Llano County (Texas, USA) and Novara (Italy).

$Be_2FeY_2Si_2O_{10}$

Colour:	green, greenish-brown, black
Lustre; opacity:	vitreous; transparent to opaque
Streak:	grey–green
Hardness:	6.5–7
Specific gravity:	4.0–4.5
Cleavage; fracture:	none; conchoidal, splintery
Habit:	rare prismatic crystals of diamond-shape cross-section; microgranular, earthy masses
Crystal system:	monoclinic

Euclase

Euclase forms extremely attractive hard crystals; blue and green varieties are popular. The name is derived from the Greek *eu* for 'well' and *klasis* for 'break', alluding to its facile cleavage. The distinctive crystal shapes can be confused only with barite or celestite, minerals which do not occur in the same environment as euclase. Euclase is formed as an alteration product of beryl in pegmatites and veins, and may be found as a placer deposit. The best crystals are found at Ouro Prêto (Minas Gerais, Brazil) as colourless, blue and green gem-quality stones. Other locations include the Sanarka River (Russia), Park County (Colorado, USA) and Las Cruces (Chivor, Colombia).

$AlBeOHSiO_4$

Colour:	colourless, white, green, blue
Lustre; opacity:	vitreous to adamantine; transparent to translucent
Streak:	white
Hardness:	7.5
Specific gravity:	3.1
Cleavage; fracture:	perfect; conchoidal
Habit:	stubby prismatic crystals with non-symmetrical slanted terminations; reniform, stalactitic, mammilliary masses
Crystal system:	monoclinic

Humite

Humite is part of a rare related group of minerals (called the humite group), which also includes norbergite, chondrodite, clinohumite and titanclinohumite. The structure of humites comprises alternating layers of brucite ($Mg(OH)_2$) and olivine (Mg_2SiO_4). It was named after Sir Abraham Hume (1749–1838), English collector of art, gems and minerals. Humite occurs in hydrothermal veins and contact-metamorphosed limestones and dolomites. Associated minerals include magnetite, diopside, spinel, biotite, serpentine, olivine and calcite. It is found at Vesuvius (Italy), Los Llanos de Januar (Spain), Pargas (Finland), Varmland (Sweden), Franklin (New Jersey, USA) and the Tilly Foster mine (New York, USA).

$Mg_7(OH,F)_2(SiO_4)_3$

Colour:	often yellow or yellow-green; white, brown, orange
Lustre; opacity:	vitreous; transparent to translucent
Streak:	white
Hardness:	6
Specific gravity:	3.2–3.3
Cleavage; fracture:	poor; conchoidal
Habit:	small prismatic crystals; granular, grains embedded in matrix
Crystal system:	orthorhombic

Braunite

Braunite, named after K. Braun (1790–1872), Adviser of the Chambers, of Gotha (Germany), is sometimes referred to as braunite-I to differentiate it from the calcium-bearing variant braunite-II (Ca(Mn,Fe)$14Si_2O_{24}$). Crystals are usually small and rarely well formed, but they can have an attractive metallic grey sheen. Braunite is formed by metamorphism of manganese-rich silicates and by weathering. Associated minerals are calcite, quartz, pyrolusite, hausmannite, rhodonite and spessartine. It is found at Friedrichroda (Thuringia, Germany), Långban and Jacobsberg (Sweden), Sitapar (India), Postmasburg (South Africa), Val d'Aosta (Italy), Batesville (Arkansas, USA) and Mason County (Texas, USA).

$Mn^{2+}Mn^{3+}_6SiO_{12}$

Colour:	black, brownish-black, greyish-black
Lustre; opacity:	metallic, greasy; opaque
Streak:	brownish-black
Hardness:	6–6.5
Specific gravity:	4.7–4.8
Cleavage; fracture:	perfect; uneven; subconchoidal
Habit:	crystals pseudooctahedral and bipyramidal up to 5cm (2in), rare; granular or massive aggregates
Crystal system:	tetragonal

Dumortierite

Dumortierite is a borosilicate second only in abundance to tourmaline. It forms attractive violet to blue masses used for items such as cabochons, beads and sculptures. It was named after Eugène Dumortier (1802–1873), a French palaeontologist. Dumortierite can be mistaken for sodalite, the latter being less dense and usually including more white minerals. Dumortierite is found in aluminium-rich metamorphic rocks, pegmatites and contact metamorphosed rocks. Large deposits are found in Oreana and Rochester (Nevada, USA), Dehesa (California, USA), Arizona (USA); it is also found at the type locality Chaponost (Rhône-Alps, France) and at Minas Gerais (Brazil), Sondria (Italy), Mogra (India) and Madagascar.

$Al_7(BO_3)(SiO_4)_3O_3$

Colour:	blue or violet crystals, sometimes reddish-brown
Lustre; opacity:	silky, vitreous; translucent
Streak:	white
Hardness:	7–8.5
Specific gravity:	3.2–3.4
Cleavage; fracture:	good; uneven
Habit:	rare prismatic or acicular needles; columnar or fibrous, radiating aggregates
Crystal system:	orthorhombic

Gehlenite-Åkermanite

The gehlenite–åkermanite group is a solid solution series where magnesium and silicon replace aluminium; melilite is a term applied to intermediate minerals. Specimens can be attractive, varying from transparent white or reddish to opaque grey-green, especially as micromounts. Gehlenite and åkermanite are found in contact-metamorphosed limestones and dolomites, respectively. Melilite occurs in silica-deficient lavas, associated with nepheline or leucite. Åkermanite is found on Vesuvius (Italy) and often in blast furnace slag; gehlenite in Trento (Italy) and Oravita (Banat, Romania); melilite at Mount Monzoni and Canzocoli (Italy), Isle of Muck and Ardnamurchan (Scotland), and Belerberg (Eiffel, Germany).

$Ca_2Al_2SiO_7 - Ca_2MgSi_2O_7$

Colour:	white, grey, yellow, grey-green, brownish-red
Lustre; opacity:	vitreous, greasy; transparent to opaque
Streak:	white
Hardness:	5
Specific gravity:	2.95–3.05
Cleavage; fracture:	good; conchoidal, uneven
Habit:	crystals as small, stubby, square prisms or plates; granular and massive aggregates
Crystal system:	tetragonal

Ilvaite

Ilvaite, named from Ilva, the old Latin name for Elba, is a blackish, almost opaque mineral. It forms attractive large black crystals and fibrous radiating aggregates, often with calcite, hedenbergite, magnetite, andradite and pyrite. It melts readily in a flame, leaving a magnetic residue. Ilvaite appears translucent on freshly exposed surfaces, but almost opaque on old specimens. It forms in contact metasomatic rocks and metamorphosed iron-rich limestones, and less commonly in syenites. It is found at Livorno and Elba (Italy), Seriphos (Greece), Julianahaab (Greenland), Dal'negorsk (Primorskiy Kray, Russia), Laxey (Idaho, USA), Dragoon Mountains (Arizona, USA), Kamioka (Oita, Japan) and Thyrill (Iceland).

$CaFe_3(SiO_4)_2(OH)$

Colour:	brownish-black, black
Lustre; opacity:	submetallic, vitreous, greasy; semi-opaque
Streak:	black, greeny-black
Hardness:	5.5–6
Specific gravity:	3.8–4.1
Cleavage; fracture:	good; conchoidal
Habit:	crystals as striated prisms; radiating fibrous, massive, granular
Crystal system:	orthorhombic

Hemimorphite

Hemimorphite is a common mineral in all lead, zinc and silver deposits. It forms part of the altered top of sphalerite deposits called 'gossan', or 'iron cap'. It is named from the hemimorphism displayed by its crystals; the prisms usually have different terminations (ends), one being rather blunt and the other being pyramidal. It has long been mined alongside smithsonite as a zinc ore. It forms in the oxidized zones of zinc-bearing veins below 240°C (464°F), above which willemite is formed. It is found at Chihuahua (Mexico), Moresnet (Belgium), Cumberland and Derbyshire (England), Carinthia (Austria), Nerchinsk (Siberia, Russia), Franklin (New Jersey) and Granby (Missouri, USA) and Baita (Romania).

$Zn_4Si_2O_7(OH)_2.H_2O$

Colour:	colourless, white, tinged yellow, blue or green, grey, brown
Lustre; opacity:	vitreous, pearly, silky; transparent to translucent
Streak:	white
Hardness:	5
Specific gravity:	4.0
Cleavage; fracture:	perfect; conchoidal, splintery
Habit:	crystals hemimorphic prismatic or plates; stalactitic, encrustations, massive granular, fan-shaped aggregates
Crystal system:	orthorhombic

Clinozoisite

Clinozoisite forms a series with the iron-bearing mineral epidote, but tends to be paler. Attractive specimens are transparent rodlike crystals, clusters of randomly orientated prisms and pinkish radiating fibrous aggregates. Clinozoisite occurs in low- to medium-grade metamorphic rocks, in contact-metamorphosed calcium-rich sediments and as altered calcium-rich plagioclase. Associated minerals include amphiboles, plagioclase and quartz. It is found at Goslarwand (Tirol, Austria), Camaderry Mountain (Wicklow, Ireland), Amborompotsy (Madagascar), Chiampernotto (Turin, Italy), Belvidere Mountain (Vermont, USA), Spade Spring Canyon (California, USA), Nightingale (Nevada, USA) and Baja California (Mexico).

$Ca_2Al_3Si_3O_{12}(OH)$

Colour:	grey, yellow, greenish, light rose
Lustre; opacity:	vitreous; transparent to opaque
Streak:	white
Hardness:	6.5
Specific gravity:	3.2
Cleavage; fracture:	perfect; uneven
Habit:	prismatic crystals; granular, massive, fibrous aggregates
Crystal system:	monoclinic

Epidote

Epidote is noted for its characteristic green colour, sometimes described as 'pistachio'. It is a common secondary mineral in a number of rocks. Epidote is pleochroic, displaying greens, yellows and browns. Rock composed mostly of epidote may be polished or tumbled as unakite. Epidote occurs in low- to medium-grade metamorphic rocks of mafic composition. Associated minerals depend on the degree of metamorphism and the bulk composition of the rock. It is found at Bourg d'Oisans (France), Arendal (Norway), Traversella (Piedmont, Italy), Knappenwand (Austria), Sulzer (Prince of Wales Island, Alaska, USA), Seven Devils (Idaho, USA) and San Quentin (Baja California, Mexico).

$Ca_2(Al,Fe)_3Si_3O_{12}(OH)$

Colour:	dark green to yellow-green
Lustre; opacity:	vitreous; translucent
Streak:	grey
Hardness:	6.7
Specific gravity:	3.3–3.5
Cleavage; fracture:	perfect; uneven
Habit:	crystals as columnar prisms, finely striated lengthwise; granular, massive, fibrous aggregates
Crystal system:	monoclinic

Piemontite

Piemontite or piedmontite is a member of the epidote group. Named after the Piedmont area in northwest Italy, it is also referred to as manganiferous epidote. The manganese content gives piemontite a red coloration, which can be difficult to distinguish from red varieties of clinozoisite. Piemontite is widespread and found in low- to medium-grade metamorphic rocks, metasomatized manganese deposits and hydrothermal veins. It is associated with epidote, glaucophane, quartz, orthoclase and calcite. It is found at Saint Marcel (Piedmont, Italy), Ceres (Turin, Italy), Groix (France), Shikoku (Japan), Old Bookoomata (South Australia), Garnet Lake (California, USA), Tucson Mountains (Arizona, USA) and Tachgagalt (Morocco).

$Ca_2(Al,Mn,Fe)_3Si_3O_{12}(OH)$

Colour:	red, purple, reddish brown, reddish black
Lustre; opacity:	vitreous; translucent to nearly opaque
Streak:	red
Hardness:	6
Specific gravity:	3.4–3.5
Cleavage; fracture:	perfect; splintery
Habit:	blocky, equant, prismatic bladed or acicular crystals; massive, granular
Crystal system:	monoclinic

Allanite

Allanite is a member of the epidote group, calcium being replaced by rare earth elements and thorium. When thorium-rich, allanite can be sufficiently radioactive to be metamict. Allanite–(Ce), –(La) and –(Y) are minerals particularly rich in cerium, lanthanum or yttrium. It is named after the Scottish mineralogist Thomas Allan (1777–1833), its discoverer. It occurs as an accessory in granites and in pegmatites; and rarely in schists, gneisses and contact-metamorphic limestones. Associated minerals are epidote, muscovite and fluorite. Allanite is found at Qáqassuatsiaq (Aluk Island, Greenland), Ytterby and Finbo (Sweden), Miask (Urals, Russia), Franklin (New Jersey, USA) and Barringer Hill (Texas, USA).

$(Ce,Ca,Y)_2(Al,Fe)_3Si_3O_{12}(OH)$

Colour:	brown to black
Lustre; opacity:	vitreous, resinous, submetallic; translucent to opaque
Streak:	grey
Hardness:	5.5–6
Specific gravity:	3.5–4.2
Cleavage; fracture:	imperfect; conchoidal, uneven
Habit:	tabular, prismatic or acicular crystals; massive, granular
Crystal system:	monoclinic

231

Zoisite

Zoisite is an epidote noted for its gem and ornamental use. The violet-blue variety tanzanite is very popular, having more fire than tourmaline or peridot. It shows a distinct purple to blue to slate-grey pleochroism and can appear more violet in incandescent light. Massive green zoisite containing rubies is a popular rock among collectors and can be polished or carved. A massive manganous pink variety called thulite is also polished and carved. Zoisite occurs in high-grade metamorphic rocks, in hydrothermal veins and as altered calcic plagioclase. Excellent tanzanite is found at Merelani Hills (Letatina Mountains, Tanzania); thulite is found at Telemark (Norway) and in Tennessee and South Carolina (USA).

$Ca_2Al_3Si_3O_{12}(OH)$

Colour:	white, blue, pale green, pink, violet-blue
Lustre; opacity:	vitreous, pearly; transparent to translucent
Streak:	white
Hardness:	6–6.5
Specific gravity:	3.15–3.36
Cleavage; fracture:	perfect; uneven
Habit:	finely striated, elongated, prismatic crystals, usually poorly terminated; grains, granular masses, aggregates
Crystal system:	orthorhombic

Vesuvianite

Also known as idocrase, vesuvianite is a popular mineral that forms fine transparent crystals and is often confused with other gemstones. It is commonly cut for collections, but not for wearing. Varieties include green californite or california jade, blue cyprine, yellow-green xanthite and pale-green or whitish wiluite. Vesuvianite is formed by metamorphism of limestones and metasomatism of serpentinized ultrabasic rocks. Associated minerals are grossular, andradite, wollastonite and diopside. Vesuvianite is found at Vesuvius (Italy), California (USA), Morelos and Chiapas (Mexico), Zermatt (Switzerland), Arendal (Norway), the Akhmatovsk mine (Urals, Russia) and Chernyshevsk (Yakutia, Russia).

$Ca_{10}Mg_2Al_4(SiO_4)_5(Si_2O_7)_2(OH)_4$

Colour:	yellow, green, brown, colourless, white, blue, violet, red , black
Lustre; opacity:	vitreous to resinous; transparent to translucent
Streak:	white
Hardness:	6.5
Specific gravity:	3.27–3.45
Cleavage; fracture:	poor; subconchoidal to irregular
Habit:	stubby, prismatic crystals, rare pyramidal terminations; compact granular masses
Crystal system:	tetragonal

Benitoite

Benitoite is a rare mineral, until recently known only from Diablo Range (San Benito, California). Discovered in 1906, it was mistaken for sapphire. Benitoite shows a high dispersion, similar to diamond, and is strongly pleochroic (i.e. it displays different colours in different orientations), changing from blue to colourless. It gives a pale blue fluorescence under ultraviolet light. Clusters of blue benitoite and black-red neptunite on a matrix of white natrolite are appealing and rare specimens. Benitoite is formed in hydrothermal veins cutting serpentinites associated with natrolite, albite, neptunite and joaquinite. Other localities include Magnet Cove (Arkansas, USA), Ohmi (Niigata, Japan) and Broken Hill (NSW, Australia).

BaTiSi$_3$O$_9$

Colour:	sapphire-blue, white, colourless
Lustre; opacity:	vitreous; transparent to translucent
Streak:	white
Hardness:	6–6.5
Specific gravity:	3.65
Cleavage; fracture:	poor; conchoidal
Habit:	crystals stubby, prismatic, dipyramidal
Crystal system:	hexagonal

Axinite

Axinite is a complex mineral in the ferroaxinite-manganaxinite series. Iron-rich axinite is brown to black whereas manganese-rich specimens are yellow-orange. Magnesioaxinite is a blue-grey magnesium-rich analogue and tinzenite is a low-calcium variety. Axinite is named after the sharp axe- or spearhead-shaped crystals. Axinite occurs in cavities in granites and adjacent zones, associated with diopside, andradite, quartz, calcite, scheelite and prehnite. It is found at Bourg d'Oisans (France), St Just (Cornwall, England), Obira (Japan) and Luning and Pala (California, USA); manganaxinite is found at Franklin (New Jersey, USA), Tinzon (Switzerland) and Liguria (Italy); and ferroaxinite at Baveno (Italy).

$Ca_2(Fe,Mn)Al_2BO_3(OH)Si_4O_{12}$

Colour:	yellow to brown, violet, grey
Lustre; opacity:	vitreous; transparent to translucent
Streak:	white
Hardness:	6.5–7
Specific gravity:	3.25
Cleavage; fracture:	perfect; conchoidal
Habit:	sharp-edged crystals, varying shape; granular and platy masses
Crystal system:	triclinic

Beryl

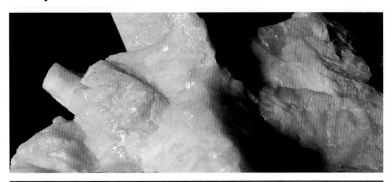

Beryl is famous for forming a wide variety of gemstones. It has been known since ancient times, the Greeks naming it after the colour of the sea, and druids believing it to aid psychic powers. Beryl is also of great importance as the main ore of beryllium, one of the lightest metals known. Beryl is formed in granites and pegmatites, where crystals can grow to enormous sizes (6m, 18 tonnes). It is also found in metamorphic rocks, hydrothermal veins and placer deposits. The availability varies enormously; non-precious varieties are easily obtained whereas emerald is so expensive that even small crystals are much sought-after. It is mined in Brazil, USA, Madagascar, Germany, Czech Republic, Russia and India.

$Be_3Al_2Si_6O_{18}$

Colour:	colourless, white, blue, green to yellows, rose, peach, red
Lustre; opacity:	vitreous, resinous; transparent to translucent
Streak:	white
Hardness:	7.5–8
Specific gravity:	2.63–2.97
Cleavage; fracture:	imperfect; conchoidal
Habit:	hexagonal prismatic crystals, often without clear terminations; rarely in druses or compact masses
Crystal system:	hexagonal

Beryl: Gem varieties

Emerald is the most famous and long-prized variety of beryl, having a beautiful green colour, derived from the presence of chromium and vanadium. It is rarely faultless, making good specimens very expensive, and stones are often oiled to disguise faults. The best emeralds come from Muso and Chivor (Colombia). Aquamarine is usually sky-blue to dark-blue, but in the nineteenth century a sea-green variety was favoured (the name meaning 'sea water'). A dark blue variety comes from Madagascar, often enhanced by heat treatment. The best-quality yellow to golden yellow heliodor comes from the Urals (Russia). Often associated with the Sun, heliodor is found only rarely as gem-quality material.

$Be_3Al_2Si_6O_{18}$

Colour:	colourless, white, blue, green to yellows, rose, peach, red
Lustre; opacity:	vitreous, resinous; transparent to translucent
Streak:	white
Hardness:	7.5–8
Specific gravity :	2.63–2.97
Cleavage; fracture:	imperfect; conchoidal
Habit:	hexagonal prismatic crystals, often without clear terminations; rarely in druses or compact masses
Crystal system:	hexagonal

Cordierite

Cordierite is a common mineral in metamorphic rocks, frequently occurring as porphyroblasts. The most interesting variety is the transparent violet-blue stone often known as iolite or water-sapphire. A black iron-rich variety called sekaninaite occurs only in Dolni Bory (Moravia, Czech Republic). Cordierite is named after the French geologist Pierre Cordier (1777–1861), who first described it, although it had been used as a gem long before this. Associated minerals include sillimanite, potassium feldspar, muscovite, biotite and andalusite. It is found at Bodenmais (Germany), Orijärvi (Finland), Kragerö (Norway), Bity (Madagascar), Tamil Nadu (India) and Thompson (Manitoba, Canada).

$Mg_2Al_4Si_5O_{18}$

Colour:	grey, rarely blue
Lustre; opacity:	vitreous; transparent to translucent
Streak:	white
Hardness:	7
Specific gravity:	2.6–2.66
Cleavage; fracture:	fair; subconchoidal
Habit:	stubby, pseudohexagonal, twinned crystals; granular, compact, massive
Crystal system:	orthorhombic

Dioptase

Dioptase is a beautiful deep green mineral, mistaken for emerald when first discovered in the eighteenth century. The strong green colour can hide the fire and transparency of some crystals. The crystal faces are usually very reflective and sparkling in clusters. The name comes from the Greek for 'see through', after clearly visible cleavages in crystals. Dioptase is found in the oxidized zones of copper deposits, associated with chrysocolla, malachite, wulfenite, cerrusite and quartz. It was first found at Altin-Tyube (Kirghiz Steppe, Kazakhstan); other locations include Tsumeb (Namibia), Copiapo and Atacama (Chile), Mindouli (Congo), Mammoth Mine (Tiger, Arizona, USA) and Baita (Romania).

$CuSiO_2(OH)_2$

Colour:	emerald-green, blue-green
Lustre; opacity:	vitreous; transparent to translucent
Streak:	green
Hardness:	5
Specific gravity:	3.3
Cleavage; fracture:	perfect; conchoidal to uneven
Habit:	crystals short, six-sided prisms with rhombohedral ends; granular, massive
Crystal system:	hexagonal

Tourmaline

Tourmalines are a group of aluminoborosilicates, notably elbaite (lithium-rich), dravite (magnesium-rich), schorl (iron- and manganese-rich) and uvite (iron- and magnesium-rich). Tourmalines crystallize as characteristic long prisms of triangular cross-section. All are strongly piezoelectric and pyroelectric, finding use in high-pressure gauges. Tourmalines are found in granites, pegmatites and quartz veins, and as an accessory in schists and gneisses. Elbaite occurs in Elba (Italy), the Urals (Russia), Sri Lanka, at Pala and Ramona (California, USA), and Newry (Maine, USA); uvite is found at Franklin (New Jersey, USA) and Gouverneur (New York, USA); and dravite at Yinniethara (Australia) and New York (USA).

$NaAl_9(BO_3)_3Si_6O_{18}(OH)_4$

Colour:	colourless, yellow, blue, olive-green, brown, black
Lustre; opacity:	vitreous; transparent to translucent
Streak:	white
Hardness:	7–7.5
Specific gravity:	3.01–3.26
Cleavage; fracture:	poor; uneven to conchoida
Habit:	prismatic crystals, often striated and elongated
Crystal system:	trigonal

Tourmaline: Gem Varieties

Of the tourmaline gem varieties, dravite is usually brown, with strong dichroism. Rubellite is a pink or red variety. Fibrous rubellite is cut *en cabochon*, for a cat's-eye effect. Watermelon tourmaline is coloured pink and green, resembling contrasting flesh and rind of the watermelon. Tourmaline may vary throughout the crystal, with up to 10 or more colours or shades. Achroite is a rare colourless variety of elbaite; it is easier to cut than other tourmalines, the strong dichrosim of which dictates the direction of cuts. Indicolite is a rare deep blue variety; paraiba tourmaline is deep blue to bluish green. The black opaque iron-rich variety schorl can be found as crystals up to several metres long. Verdelith is a yellow-green.

$NaAl_9(BO_3)_3Si_6O_{18}(OH)_4$

Colour:	colourless, yellow, blue, olive-green, brown, black
Lustre; opacity:	vitreous; transparent to translucent
Streak:	white
Hardness:	7–7.5
Specific gravity:	3.01–3.26
Cleavage; fracture:	poor; uneven to conchoida
Habit:	prismatic crystals, often striated and elongated
Crystal system:	trigonal

Milarite

Milarite is a rare mineral and the end member of the milarite–osumilite group of minerals. Osumilite contains magnesium and iron, but no beryllium. Crystals of milarite are rarely transparent and usually small, but are excellent as micromounts. Crystals are often well-shaped hexagonal prisms in muted shades of green to yellow. It has been called giufite, but milarite is now the preferred name. Milarite is found in hydrothermal veins, classically in Val Giuf and Val Striem (Grissons, Switzerland), as crystals up to 3cm (1¼in); it is also found on the Kola Peninsula (Russia) and in Valencia Guanajuato (Mexico) and Jaguaraçú (Minas Gerais, Brazil). Osumilite is found on the volcano Sakurajima near Osumi (Japan).

$K_2Ca_4Be_4Al_2Si_{24}O_{60}.H_2O$

Colour:	colourless, green to yellow
Lustre; opacity:	vitreous; transparent to translucent
Streak:	white
Hardness:	5.5–6
Specific gravity:	2.4–2.6
Cleavage; fracture:	absent; conchoidal, uneven
Habit:	hexagonal prismatic crystals; radial-fibrous aggregates and intergrowths
Crystal system:	hexagonal

Eudialite

Eudialite is a rare mineral popular for the attractive colours of some specimens. Crystals are commonly red and usually found embedded in a matrix of host rock. The site on the Kola peninsula is well known for pegmatites rich in sodium and some rarer elements giving rise to unusual minerals such as eudialite. The name comes from the Greek, alluding to its ready solubility in acids. Eudialite occurs in nepheline syenites, granites and associated pegmatites, associated with quartz, albite, nepheline, aegerine and natrolite. Notable localties are the Kangerdluarssuk Plateau (Greenland), Magnet Cove (Arkansas, USA), Pajarito Mountain (New Mexico, USA) and the Lovozero and Khibiny massifs (Kola Peninsula, Russia).

$(Na,Ca,Fe)_6Zr(OH,Cl)(Si_3O_9)_2$

Colour:	pink, red, yellow-brown, violet
Lustre; opacity:	vitreous; translucent
Streak:	white
Hardness:	5–5.5
Specific gravity:	2.8–3.0
Cleavage; fracture:	imperfect; conchoidal, uneven, splintery
Habit:	crystals as plates, rarely well formed; granular, massive aggregates
Crystal system:	hexagonal

Hedenbergite

The pyroxenes are an important group of rock-forming minerals. Their structure is characterized by straight chains of linked silicon oxide (SiO_4) tetrahedra. They can be distinguished from the amphiboles by the angle between their two good cleavages (about 90° rather than 120°). Hedenbergite and diopside ($CaMgSi_2O_6$) are monoclinic clinopyroxenes, forming a calcium-bearing series in which magnesium and iron substitute for one another. Hedenburgite is common in iron-magnesium skarns at the contact of granitic rocks with limestones, and in contact-metamorphosed iron-rich sediments. Good hedenburgite crystals have been found in the Skardu area (Pakistan), Broken Hill (NSW, Australia) and the Harstig Mine (Varmland, Sweden).

$CaFeSi_2O_6$

Colour:	brownish-green, black; almost opaque
Lustre; opacity:	vitreous or dull; translucent to opaque
Streak:	white, grey
Hardness:	5.5–6.5
Specific gravity:	3.6
Cleavage; fracture:	good in two directions at about 90°; uneven to subconchoidal
Habit:	crystals short, prismatic; radiating aggregates, granular, massive
Crystal system:	monoclinic

Diopside

Diopside is a calcium-bearing clinopyroxene that forms a solid solution series with hedenbergite ($CaFeSi_2O_6$). Pure diopside is common in contact-metamorphic siliceous magnesian limestones, associated with calcite, dolomite and sometimes forsterite, as on Skye (Scotland). There are several varieties of diopside: chrome diopside is a chromium-rich gem variety found in Burma, Siberia (Russia), Pakistan and South Africa. Some specimens have inclusions (probably of rutile) that may form a 'cat's eye' effect when polished. Violane is a rare blue manganese-bearing variety from Italy. One dark green to black variety from southern India, known as star diopside, shows a four-rayed star when cut *en cabochon*.

$CaMgSi_2O_6$

Colour:	pale to dark green or black, may be colourless
Lustre; opacity:	vitreous or dull; transparent to opaque
Streak:	white, grey, grey-green
Hardness:	5.5–6.5
Specific gravity:	3.29
Cleavage; fracture:	good in two directions at about 90°; uneven to subconchoidal
Habit:	prismatic crystals; granular, columnar, massive
Crystal system:	monoclinic

Jadeite

Jadeite is the characteristic pyroxene of high-pressure metamorphic rocks, in which it may be associated with glaucophane, aragonite and quartz, and also of eclogites, along with garnet. It occurs in the high-pressure metamorphic belt in California (USA) and at Shibukawa (Japan). Jadeite is used as an ornamental stone for carving, and as a precious stone in jewellery. Of the two minerals known as jade, (the other being nephrite, an amphibole), it is less common. The emerald green chromium-bearing variety is most prized, and known as imperial jade. For more than two centuries, China has been supplied with such jade from Burma, whereas the Central American Indians used jadeite from Guatemala.

$Na(Al,Fe)Si_2O_6$

Colour:	shades of light and dark green; rarely white or violet
Lustre; opacity:	subvitreous, pearly on cleavages; translucent
Streak:	white
Hardness:	6–7
Specific gravity:	3.2–3.4
Cleavage; fracture:	good in two directions at about 90°; splintery
Habit:	crystals rare, prismatic; commonly massive, granular, compact felty masses
Crystal system:	monoclinic

Spodumene

The pyroxene spodumene was named from the Greek for 'ash', in reference to its colour. Spodumene comes in other colours: gem varieties include the lilac pink kunzite, coloured by manganese, and bright emerald green hiddenite, coloured by chromium. Well-cut stones display strong pleochroism from colourless to two shades of body colour. Crystals can weigh up to 65 tonnes (72 tons). Spodumene is mined as a raw material for lithium compounds and ceramics. It is commonly found in lithium-rich granitic pegmatites and also occurs in gneisses. Associated minerals include quartz, albite, petalite, lepidolite and beryl. Notable localities include the Black Hills (South Dakota, USA) and Sterling and Chesterfield (Massachussetts, USA).

$LiAlSi_2O_6$

Colour:	commonly yellowish-grey, but shows a range of colours
Lustre; opacity:	vitreous, dull; transparent to translucent
Streak:	white
Hardness:	6.5–7
Specific gravity:	3.1–3.2
Cleavage; fracture:	good in two directions at about 90°; uneven to subconchoidal
Habit:	crystals prismatic, may be striated; commonly massive
Crystal system:	monoclinic

247

Aegirine

First described from Norway, aegirine is named after Aegir, a Scandinavian god of the sea. It is also known as acmite, from the Greek for 'point', in reference to its unusually sharply pointed crystals. This helps to distinguish it from augite, which is very similar. Aegerine occurs in alkaline rocks such as sodium-rich nepheline syenites, carbonatites and pegmatites, and rarely in regionally metamorphosed schists, gneisses and iron formations. Associated minerals include analcime, nepheline, clinochlore, eudialyte and rhodochrosite. Good crystals occur at Malosa (Zomba District, Malawi), Mont Saint-Hilaire (Quebec, Canada), Khibiny Massif (Kola Peninsula, Russia) and Magnet Cove, Arkansas (USA).

$NaFeSi_2O_6$

Colour:	dark green, greenish-black, reddish-brown
Lustre; opacity:	vitreous to resinous; translucent to opaque
Streak:	yellowish-grey
Hardness:	6
Specific gravity:	3.5
Cleavage; fracture:	good in two directions at about 90°; uneven, brittle
Habit:	acicular crystals terminated by a steep pyramid, striated; disseminated grains, fibrous, in radiating aggregates
Crystal system:	monoclinic

Augite

Augite is a calcium-bearing clinopyroxene, similar to the diopside-hedenbergite series, but with aluminium, titanium and sodium incorporated into the structure. The commonest pyroxene, it can be distinguished from hornblende by the angle between the cleavages. It is an important rock-forming mineral in basic igneous rocks (basalt and gabbro) and some syenite, and in ultrabasic rocks (pyroxenites and peridotites). It also forms in high-temperature metamorphic rocks such as pyroxene gneisses and pyroxene granulites. It is commonly associated with plagioclase feldspars, amphiboles, olivine and orthopyroxenes. Rare well-formed crystals occur at Vesuvius and Stromboli (Italy), and at Eifel (Germany).

$(Ca,Na)(Mg,Fe,Al,Ti)(Si,Al)_2O_6$

Colour:	dark green, brown, black, rarely cream
Lustre; opacity:	vitreous, resinous to dull; transparent to opaque
Streak:	greyish-green
Hardness:	5.5–6
Specific gravity:	3.2–3.6
Cleavage; fracture:	good in two directions at about 90°; uneven to subconchoidal
Habit:	short prismatic crystals of four- or eight-sided cross-section, often twinned; compact, granular
Crystal system:	monoclinic

Enstatite

Enstatite and ferrosilite ($Fe_2Si_2O_6$) form a series in which iron and magnesium substitute for each other. The intermediate composition, $MgFeSi_2O_6$, is called hypersthene. Minerals of this series are calcium-poor and are mostly orthorhombic. These orthopyroxenes are important rock-forming minerals, and occur worldwide in basic and ultrabasic igneous rocks, and high-temperature metamorphic rocks such as pyroxene gneisses. Enstatite is also found in meteorites. Crystals are quite rare, but rolled pebbles are sometimes faceted as gems. The variety chrome-enstatite is emerald green; a dark brown six-rayed star enstatite occurs in Mysore (India); and unique gem-quality colourless crystals are found at Embilipitiya (Sri Lanka).

$Mg_2Si_2O_6$

Colour:	pale to dark brownish-green, white, greyish, yellowish
Lustre; opacity:	vitreous; transparent to opaque
Streak:	white to greyish
Hardness:	5–6
Specific gravity:	3.2–3.4
Cleavage; fracture:	good in two directions at about 90°; uneven
Habit:	crystals stumpy, may be twinned; as grains, lamellar or massive
Crystal system:	orthorhombic

Bronzite

The orthopyroxene bronzite is characterized by a bronze sheen on its cleavage surface; however, it is now regarded as an iron-bearing enstatite, and is no longer recognized as a separate species. It is affected by partial alteration of a type called 'schillerization', which gives rise to the bronzelike submetallic lustre. Bronzite is sometimes cut and polished for ornaments, and may have a fibrous structure. It occurs in the basic igneous rock norite in the Fichtelgebirge (Germany) and in the serpentine of Kraubat (Styria, Austria). Relict crystals of more highly altered bronzite or enstatite occurring in serpentinite are known as bastite, after the locality Baste (Harz, Germany); it also occurs in the Lizard (Cornwall, England).

$(Mg,Fe)_2Si_2O_6$

Colour:	grey-green or bronze-brown
Lustre; opacity:	vitreous, pearly, bronze sheen on cleavage; translucent to opaque
Streak:	white to greyish
Hardness:	5–6
Specific gravity:	3.2–3.3
Cleavage; fracture:	good in two directions at about 90°; scaly; uneven
Habit:	coarse prismatic crystals; massive, granular
Crystal system:	orthorhombic

Cummingtonite

Cummingtonite belongs to the amphibole group, whose structure is characterized by double chains of silicon oxide (SiO_4) tetrahedra. It is in the middle compositional range of a magnesium–iron series, the end members of which are magnesiocummingtonite and grunerite. Cummingtonite cannot be distinguished from anthophyllite, its polymorph, with the naked eye. It forms in medium-grade regionally metamorphosed rocks. The only commercial source of brown asbestiform amosite was the Precambrian ironstone formations of Transvaal (South Africa). Grunerite is found in the iron deposits of Lake Superior (USA); cummingtonite is found at Cummington (Massachusetts, USA) and Val d'Ossola (Italy).

$(Mg, Fe)_7Si_8O_{22}(OH)_2$

Colour:	dark green, pale to dark brown depending on iron content
Lustre; opacity:	vitreous to silky; translucent to transparent
Streak:	white
Hardness:	5–6
Specific gravity:	3.1–3.6 (increasing with iron content)
Cleavage; fracture:	two good cleavages at about 120°; splintery
Habit:	individual crystals rare; aggregates of rodlike or fibrous crystals, often radiating
Crystal system:	monoclinic

Glaucophane

Glaucophane is named after the Greek *glaucos* ('blue') and *fanos* ('appearing'). It is an alkali-bearing amphibole which, like riebeckite, has a characteristic blue-grey colour, and typical 120° prismatic amphibole cleavage. It characteristically forms in subduction zones under high-pressure, low-temperature conditions, and may eventually be exposed at the surface. Glaucophane-rich schists are known as blueschists. Glaucophane may be associated with the high-pressure minerals lawsonite, jadeite and aragonite, and also with garnet, epidote or pumpellyite. Glaucophane occurs in the Franciscan belt of California (USA), in Shikoku (Japan), Anglesey (Wales), Zermatt (Switzerland) and Euboea Island (Greece).

$Na_2(Mg,Fe)_3Al_2Si_8O_{22}(OH)_2$

Colour:	grey-blue to lavender blue
Lustre; opacity:	vitreous, silky in fibrous varieties; translucent
Streak:	blue–grey
Hardness:	6
Specific gravity:	3.08–3.22
Cleavage; fracture:	good in two directions at about 120°; uneven to subconchoidal
Habit:	good crystals rare, slender prismatic to acicular; sometimes massive, fibrous, columnar or granular
Crystal system:	monoclinic

Riebeckite

Named after the German traveller E. Riebeck, riebeckite is a blue-grey alkali-amphibole formed in alkali granites, syenites, more rarely in granite pegmatites and in some volcanic rocks. The asbestiform variety crocidolite is found in iron formations and was formerly mined in South Africa and Australia. The associated health risk means that specimens should be professionally sealed in boxes or bags. Associated minerals include nepheline, albite, aegerine, tremolite, magnetite, hematite or siderite. Crocidolite inclusions in quartz, and their pseudomorphing by quartz, have given rise to the gems tiger's eye and hawk's eye. The type locality for riebeckite is Socotra Island (Indian Ocean, Yemen).

$Na_2(Mg,Fe)_5Si_8O_{22}(OH)_2$

Colour:	dark blue to black
Lustre; opacity:	vitreous or silky; translucent
Streak:	none determined
Hardness:	5
Specific gravity:	3.32–3.38
Cleavage; fracture:	two good cleavages at about 120°; conchoidal to uneven
Habit:	long, prismatic and striated crystals; massive, fibrous or asbestiform
Crystal system:	monoclinic

Tremolite

Tremolite is a calcic amphibole that forms a series with the iron-bearing ferroactinolite. Pure tremolite is creamy white, but grades to green with increasing iron content. It may fluoresce yellow or pink. It forms through the contact metamorphism of calcium- and magnesium-bearing siliceous sediments, impure dolomitic limestones, or ultramafic rocks. It can be associated with calcite, dolomite, garnet, wollastonite, talc, diopside or forsterite. A fibrous variety of tremolite has been mined for use as asbestos. It is named after Tremola (Italy). Other notable localities include St Marcel (Piedmont, Italy), Franklin (New Jersey, USA), Wilberforce (Ontario, Canda) and Brumada Mine (Bahia, Brazil).

$Ca_2Mg_5Si_8O_{22}(OH)_2$

Colour:	colourless, white, grey, can be green, pink or brown
Lustre; opacity:	vitreous or silky to dull; transparent to translucent
Streak:	white
Hardness:	5–6
Specific gravity:	2.9–3.2
Cleavage; fracture:	two good cleavages at about 120°; uneven to subconchoidal
Habit:	crystals long, bladed, often twinned; columnar, fibrous, plumose aggregates, radiating or granular
Crystal system:	monoclinic

Actinolite

Actinolite is a fairly common calcic amphibole within the series tremolite–ferroactinolite. It occurs in relatively low-grade metamorphosed mafic, ultramafic or magnesian carbonate rocks, and also in glaucophane-bearing blue schists. It is a common alteration product of primary pyroxenes in gabbro or dolerite, and used to be called uralite. Byssolite is an asbestiform type. The variety nephrite is the more common of the two types of green gemstone known as jade; it is tough, and takes a good polish. Beautiful nephrites from New Zealand are known as Maori stone because of their widespread use in traditional Maori art. Actinolite is found in the Zillertal (Austria), Val Malenco (Italy) and the Urals (Russia).

$Ca_2(Mg,Fe)_5Si_8O_{22}(OH)_2$

Colour:	bright to greyish green
Lustre; opacity:	vitreous, silky when fibrous; transparent to translucent
Streak:	white
Hardness:	5–6
Specific gravity:	3.0–3.24
Cleavage; fracture:	two good cleavages at about 120°; uneven or splintery
Habit:	long bladed or prismatic crystals, may be bent; columnar, radiating fibrous to asbestiform, granular, massive
Crystal system:	monoclinic

Hornblende

Hornblende is a calcic amphibole and a very important rock-forming mineral. It is widespread in igneous rocks (granodiorites, diorites, syenites and some gabbros) and in medium-temperature metamorphosed basalts (amphibolites) accompanied by plagioclase feldspar and sometimes garnet. It is occasionally found in metamorphosed impure dolomitic limestones and ironstones. It forms complex solid-solution series with several other amphiboles. The variety edenite is pale green, iron-poor and found at Edenville (New York, USA); pargasite is dark green, iron-rich and found in Pargas (Finland); hastingsite is richer in sodium and aluminium, and occurs in gabbros in Ontario (Canada).

$(Ca,Na)_{2-3}(Mg,Fe,Al)_5(Al,Si)_8O_{22}(OH)_2$

Colour:	black to dark green
Lustre; opacity:	vitreous to dull; opaque
Streak:	brown to grey
Hardness:	5–6
Specific gravity:	3.28–3.41
Cleavage; fracture:	two good cleavages at about 120°; uneven
Habit:	crystals long and thin, or short and stumpy, generally six-sided; granular, may form radiating aggregates
Crystal system:	monoclinic

Anthophyllite

Anthophyllite is the orthorhombic polymorph of monoclinic cummingtonite, from which it can be distinguished only through optical, density or x-ray study. It occurs in medium-grade metamorphic rocks, derived from mafic or ultramafic igneous rocks, or dolomitic sedimentary rocks. It can form through the hydration of olivine. It is associated with talc, cordierite, chlorite, mica, hornblende or olivine. Anthophyllite has been commercially used in asbestos, with the associated health hazards. It is named from the Latin *anthophyllum*, meaning 'clove', in reference to its distinctive colour. Good crystals occur at Köngsberg (Norway); it is also found at Fahlan (Sweden), Orijarvi (Finland) and the Isle of Elba (Italy).

$(Mg,Fe)_7Si_8O_{22}(OH)_2$

Colour:	grey-brown, cinnamon or clove brown, green
Lustre; opacity:	vitreous, silky when fibrous; translucent to translucent
Streak:	white
Hardness:	5.5–6
Specific gravity:	2.85–3.57
Cleavage; fracture:	two good cleavages at about 120°; uneven, brittle
Habit:	individual crystals rare; usually aggregates of prismatic crystals, fibrous, asbestiform
Crystal system:	orthorhombic

Wollastonite

Wollastonite, formerly known as table spar, is found in thermally metamorphosed siliceous limestones at igneous contacts, or in regionally metamorphosed rocks, associated with calcium-rich garnets, diopside, epidote and tremolite. It also occurs in xenoliths in igneous rocks. There are many different polymorphs, generally called wollastonite. Although common, commercially viable deposits are unusual, occurring in New York (USA), Finland, Mexico, China, India and Africa. Wollastonite has many applications, e.g. in refractory ceramics, plastics, and sealants. It was named after the British mineralogist and chemist William H. Wollaston (1766–1828), who also discovered palladium and rhodium.

$CaSiO_3$

Colour:	white or greyish
Lustre; opacity:	vitreous, may be silky; subtransparent to translucent
Streak:	white
Hardness:	4.5–5
Specific gravity:	2.87–3.09
Cleavage; fracture:	perfect; splintery, uneven
Habit:	crystals tabular to short prismatic, twinning common; fibrous masses, radial, granular, compact
Crystal system:	triclinic (or monoclinic)

Pectolite

Pectolite must be handled carefully, as the fine white needles are sharp and easily puncture the skin and can become embedded. Some specimens are triboluminescent. A variety called larimar is a fine pale blue and green, and comes from the Dominican Republic. Pectolite occurs as a primary mineral in nepheline syenites, but is also formed by hydrothermal processes, filling cavities in basalts, associated with zeolites, datolite, prehnite and calcite. It is also found filling fractures in serpentinites and peridotites, and in contact-metamorphosed limestones. Notable occurrences include Lake County (California, USA), Franklin (New Jersey, USA), Bahamas, Mt Baldo (Trento Province, Italy) and England.

$NaCa_2Si_3O_8(OH)$

Colour:	colourless, white or grey, larimar is pale blue
Lustre; opacity:	vitreous to silky; translucent to opaque
Streak:	white
Hardness:	4.5–5
Specific gravity:	2.84–2.90
Cleavage; fracture:	perfect in two directions close to 90°; uneven, splintery
Habit:	crystals tabular or acicular; fibrous tufts or fibrous-radiating spherical aggregates or compact masses
Crystal system:	triclinic

Rhodonite

Rhodonite has a similar structure to wollastonite and pectolite, built of chains of silicon oxide (SiO_4) tetrahedra with bases that do not lie in a plane. It has a distinctive pink colour and can be distinguished from rhodochrosite (which is also pink) because it is harder and does not dissolve in dilute hydrochloric acid. It often occurs in association with manganese ore deposits in hydrothermal veins, or in contact and regional metamorphosed manganese-bearing sedimentary rocks. Rhodonite from Sverdlovsk in the Urals (Russia) has been used as an ornamental stone. It also occurs at Långban (Sweden) in iron ore, at Broken Hill (NSW, Australia) and at Franklin and Sterling Hill (New Jersey, USA).

(Mn,Fe,Mg)SiO$_3$

Colour:	deep pink, with brown or black surface oxidation
Lustre; opacity:	vitreous, pearly on cleavages; transparent to translucent
Streak:	white
Hardness:	5.5–6.5
Specific gravity:	3.57–3.76
Cleavage; fracture:	perfect in two directions close to 90°, with a third good cleavage; conchoidal to uneven
Habit:	crystals rare, tabular or prismatic; massive or granular
Crystal system:	triclinic

Babingtonite

Babingtonite is a rare mineral named after the Irish physicist and mineralogist W. Babington (1757–1833). Relatively recently discovered, it contains both ferrous iron (Fe^{2+}) and ferric iron (Fe^{3+}) (which takes the place of aluminium typical of many silicates), and is weakly magnetic. It tends to grow in cavities in mafic volcanic rocks and gneisses, which enables crystals to grow freely; it is also found in veins cross-cutting granite and diorite, and in skarns. It is commonly associated with prehnite, epidote, pyrite and quartz, and with zeolites. Notable occurrences include Poona (India), Devon (England), Herbornseelbach (Germany), Baveno (Italy), Yakubi Mine (Japan) and Massachusetts (USA), where it is the official mineral emblem.

$Ca_2(Fe,Mn)FeSi_5O_{14}(OH)$

Colour:	dark greenish-black to brownish-black
Lustre; opacity:	brilliantly vitreous; opaque to translucent
Streak:	green-grey
Hardness:	5–6
Specific gravity:	3.0–3.2
Cleavage; fracture:	good in two directions at about 90°; uneven to subconchoidal
Habit:	crystals short, columnar, often striated; as platey, radial or fan-shaped aggregates
Crystal system:	triclinic

Neptunite

Neptunite is a rare mineral that forms as an accessory in intermediate plutonic igneous rocks such as nepheline syenite, and its pegmatites. It is also found in serpentinites associated with benitoite and natrolite. Its name is derived from Roman mythology, Neptune being the god of the sea, and alludes to its close association at its type locality with aegerine, named after Aegir, the Scandinavian sea god. Excellent crystals occur at San Benito (California, USA), where it is found in natrolite veins in a serpentinite body, along with blue benitoite; other localities include Barnavave (Ireland), Kola Peninsula (Russia), Igaliko (Greenland) and Mont St Hilaire (Quebec, Canada).

$Na_2KLi(Fe,Mn)_2Ti_2Si_8O_{24}$

Colour:	black with deep reddish-brown internal reflections
Lustre; opacity:	vitreous to submetallic; opaque to translucent
Streak:	dark reddish-brown
Hardness:	5–6
Specific gravity:	3.19–3.23
Cleavage; fracture:	perfect; conchoidal
Habit:	elongate prismatic crystals of square cross-section and pointed terminations
Crystal system:	monoclinic

Bavenite

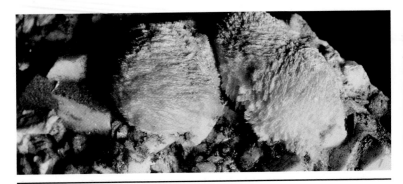

Bavenite is an extremely rare mineral, named in 1901 after Baveno (Italy) where it was discovered. It is formed by the alteration of beryl and other beryllium-bearing minerals, and occurs as druses in cavities in granite and associated pegmatites. It is also found in hydrothermal veins and skarns. Associated minerals include orthoclase, fluorite, albite and beryl. It has a distinctive habit, with very thin tabular crystals forming diverging groups like the pages of a book. It can also form fibrous crystals, as in the granitic pegmatite bodies of Bustarviejo (Madrid, Spain). Other localities include Shap Pink Quarry (Cumbria, England), Mont St Hilaire (Quebec, Canada), Strzegom (Poland) and Londonderry (Australia).

$Ca_4Al_2Be_2(OH)_2Si_9O_{26}$

Colour:	colourless, white, green, pink, brown
Lustre; opacity:	silky, vitreous, pearly; transparent to translucent
Streak:	white
Hardness:	5.5–6
Specific gravity:	2.7
Cleavage; fracture:	perfect; uneven
Habit:	very thin tabular crystals, may be twinned; lamellar, rose-shaped aggregates, radiating tufts of needle-like crystals
Crystal system:	orthorhombic

Prehnite

Prehnite is found in hollows in basaltic lavas, often associated with zeolites; it also occurs in very low grade metamorphic rocks, and through the decomposition of plagioclase feldspar. Associated minerals include datolite, epidote or calcite. It was the first mineral to be named after a person, the Dutchman Colonel Hendrick von Prehn (1733–1785), who discovered it at the Cape of Good Hope. It occasionally forms epimorphs (crystal growth over another mineral surface) on laumontite, which may subsequently dissolve, leaving the prehnite as a crust. Yellowish brown prehnite is occasionally cut *en cabochon*. Pale green masses are found in Scotland, and dark green or greenish-brown prehnite in Australia.

$Ca_2Al_2Si_3O_{10}(OH)_2$

Colour:	pale watery or oily green, sometimes white or yellowish
Lustre; opacity:	vitreous, pearly; transparent to translucent
Streak:	white
Hardness:	6–6.5
Specific gravity:	2.8–2.9
Cleavage; fracture:	distinct; uneven, rather brittle
Habit:	crystals rare, prismatic, tabular or pyramidal; usually botryoidal or globular masses with a radiating structure
Crystal system:	orthorhombic

Apophyllite

Apophyllite is the name given to a group of minerals including fluoroapophyllite and hydroxyapophyllite. Its name, from the Greek, roughly means 'to flake off' because it peels apart in leafs on heating as water is driven off. Apophyllite is quite abundant and popular, forming some attactive clear, colourless crystals and some of pale pastel shades; the most famous is an emerald-green variety from India. Apophyllite occurs in cavities in basalts, associated with stilbite, scolecite, calcite, prehnite and analcime. Excellent crystals occur at Bolzano (Italy), Poona (India), Mont St Hilaire (Canada), St Andreasburg (Harz, Germany), Paterson (New Jersey, USA) and Rio Grande do Sul (Brazil).

$KCa_4Si_8O_{20}(F,OH).8H_2O$

Colour:	colourless, white, pale pink, green or yellow
Lustre; opacity:	vitreous, pearly; transparent to translucent
Streak:	white
Hardness:	4.5–5
Specific gravity:	2.33–2.37
Cleavage; fracture:	perfect; irregular
Habit:	tabular pseudocubic or pseudooctahedral crystals, sometimes dipyramidal or platy; aggregates
Crystal system:	tetragonal

Pyrophyllite

Pyrophyllite is named after the Greek for 'fire' and 'leaf', after its behaviour on heating, when water is driven off and the mineral peels into flakes. Breaking into thin flakes like this is typical of phyllosilicates and reflects a sheetlike structure. Pyrophyllite has a greasy feel and is used as a dry lubricant like talc, from which it is almost indistinguishable. The variety agalmatolite has been used for carved ornaments in China. Pyrophyllite occurs in hydrothermal veins and schists, associated with kyanite, andalusite, topaz, mica and quartz. It is found in Orange County (North Carolina), Graves Mountain (Georgia) and La Paz County (Arkanas, USA), Ibitiara (Bahia, Brazil), Zermatt (Switzerland) and Krassik (Urals, Russia).

$Al_2Si_4O_{10}(OH)_2$

Colour:	yellowish-white, grey, green
Lustre; opacity:	pearly, dull; translucent to opaque
Streak:	white
Hardness:	1–1.5
Specific gravity:	2.8
Cleavage; fracture:	perfect; uneven or splintery
Habit:	lamellar or radiating foliated aggregates, granular, massive
Crystal system:	monoclinic

Talc

Talc, a member of the mica group, is noted for being soft, by definition 1 on Moh's scale. It is used in many industries, including paper, paints, personal care and roofing. Massive steatite, or soapstone (after its greasy feel), is easily carved or turned on a lathe. Talc is a good electrical and heat insulator, and repels water. The name comes from the Persian or Arabic *talq*. Talc occurs in schists and hydrothermally altered mafic rocks, associated with serpentine, actinolite, chlorite, vermiculite, dolomite and calcite. It is found at Mt Greiner (Austria), Zermatt (Switzerland), Pfitschal (Italy), Trimouns (Ariège, France), Onotosk (Siberia, Russia), Yellowstone mine (Montana, USA) and Delta (Pennsylvania, USA).

$Mg_3Si_4O_{10}(OH)_2$

Colour:	white, brown, green
Lustre; opacity:	greasy when massive, pearly on cleavage; translucent
Streak:	white
Hardness:	1
Specific gravity:	2.58–2.83
Cleavage; fracture:	perfect; uneven, lamellar
Habit:	very rare crystals as pseudohexagonal plates; usually scaly, foliated aggregates
Crystal system:	monoclinic

Muscovite

Muscovite is the most common mica, well known for its cleavage into thin, flexible lamellae – a manifestation of its underlying two-dimensional sheetlike structure. The name comes from muscovy glass, after its former use in windows. Muscovite is used for electrical and heat insulation, as a dry lubricant and in paper, rubber, paints and plastics. The variety fuchsite is emerald green and popular among collectors. Muscovite is a very common rock-forming mineral in granites, pegmatites, phyllites, schists and gneisses. Crystals up to 30–50 m² (323–538 sq ft) are found in pegmatites in Methuen and Calvin (Ontario, Canada), Nellore (Andhra Pradesh, India), Custer (South Dakota, USA) and Minas Gerais (Brazil).

$KAl_2(AlSi_3)O_{10}(OH)_2$

Colour:	white, grey, yellow, brown, greenish; bright green
Lustre; opacity:	vitreous, pearly, silky; transparent to translucent
Streak:	white
Hardness:	2.5
Specific gravity:	2.76–2.88
Cleavage; fracture:	perfect; uneven, lamellar
Habit:	pseudo-hexagonal tabular crystals, deep striations on prism faces; foliated, scaly, lamellar masses
Crystal system:	monoclinic

Phlogopite

Phlogopite forms a series with the more common mica biotite, which is iron-rich and generally of darker colour. It is used in electronics for its excellent heat- and electrical-insulating properties. Phlogopite forms tabular crystals up to 2m (6½ft) across, some strikingly transparent. It decomposes in concentrated sulphuric acid, distinguishing it from muscovite. The name comes from the Greek *flogopos*, meaning 'to resemble fire'. It occurs in metamorphosed dolomites and in ultramafic rocks, associated with dolomite, calcite, diopside and epidote. It is found at Campolungo (Switzerland), Frontenac (Ontario, Canada), Ødegården (Norway), Franklin (New Jersey, USA), Saharakara (Madagascar) and Anxiety Point (New Zealand).

$KMg_3(Al,Fe)Si_3O_{10}(F,OH)_2$

Colour:	brown, reddish-brown, yellow, green
Lustre; opacity:	pearly; transparent to translucent
Streak:	white
Hardness:	2–3
Specific gravity:	2.86
Cleavage; fracture:	perfect; uneven, lamellar
Habit:	crystals as six-sided plates; foliated, granular aggregates
Crystal system:	monoclinic

Biotite

Biotite, named after the French mineralogist J.-B. Biot (1774–1862), is also called black mica, as opposed to white mica (muscovite). It can form enormous crystals in pegmatites, but is most common as small dark plates in granites, contrasting with larger, paler quartz and feldspar. Biotite in exposed surfaces of rocks can sparkle in sunlight, occasionally with a golden sheen. It is an important mineral in schists, gneisses, granites, nepheline syenites and contact-metamorphosed rocks. Associated minerals include muscovite, pyroxenes, amphiboles, andalusite and cordierite. Good crystals occur at Vesuvius (Italy), Miass (Ilmen Mountains, Russia), Arendal (Norway), Franklin (New Jersey, USA) and Bancroft (Ontario, Canada).

$K(Mg,Fe)_3(AlSi_3)O_{10}(F,OH)_2$

Colour:	brown, black, dark green, greyish-yellow
Lustre; opacity:	vitreous, pearly on cleavage; transparent to translucent
Streak:	grey-white
Hardness:	2.5–3
Specific gravity:	2.8–3.2
Cleavage; fracture:	perfect; irregular, lamellar
Habit:	rare tabular or short prismatic pseudohexagonal crystals; foliated masses or aggregates, disseminated grains
Crystal system:	monoclinic

Lepidolite

Lepidolite, or lithium mica, named after the Greek for 'scale', is an ore of lithium and a source of rubidium and caesium. The pink colour is fairly characteristic, but may be confused with pink muscovite. Like other micas, lepidolite can be found as 'books' of crystal sheets. Large forms are carved, shaped or used as polished stones. Attractive specimens from Brazil comprise lepidolite with red tourmaline and colourless quartz. Lepidolite occurs in lithium-rich pegmatites and altered granites, associated with quartz, feldspars, spodumene and tourmaline. It is found at Pala (California, USA), Mount Mica (Maine, USA), Rozna (Czech Republic), Alabashka (Urals, Russia) and Virgem de Lape (Minas Gerais, Brazil).

$K(Li,Al)_3(Si,Al)_4O_{10}(F,OH)_2$

Colour:	pink to lilac
Lustre; opacity:	pearly; transparent to translucent
Streak:	white
Hardness:	2.5–4
Specific gravity:	2.8–2.9
Cleavage; fracture:	perfect; uneven, lamellar giving flexible sheets
Habit:	six-sided tabular crystals; fine, platy aggregates, massive
Crystal system:	monoclinic

Zinnwaldite

Zinnwaldite is a widespread but rare mica, difficult to distinguish from other micas but for the environment in which it occurs. Its colour is usually darker than muscovite, but lighter than biotite. Zinnwaldite forms six-sided crystals up to 20cm (8in). As with other micas, zinnwaldite occurs as 'books' of pseudohexagonal crystals. Zinnwaldite occurs in tin-bearing pneumatolytic deposits in greisens and rarely in pegmatites and quartz veins, associated with topaz, cassitrerite, lepidolite, wolframite and spodumene. It is found at Cínovec (Zinnwald, Czech Republic), Altenberg and Waldstein (Germany), St Just (Cornwall, England), Amelia (Virginia, USA), Pala (California, USA) and Antaboaka (Madagascar).

$KLiFeAl(Al_2Si_2)O_{10}(F,OH)_2$

Colour:	grey-brown, yellow-brown, pale violet, dark green
Lustre; opacity:	vitreous, pearly; transparent to translucent
Streak:	white
Hardness:	2.5–4
Specific gravity:	2.9–3.2
Cleavage; fracture:	perfect; uneven, lamellar giving flexible, elastic scales
Habit:	short prismatic or tabular crystals; rosettes, lamellar or scaly aggregates, disseminated
Crystal system:	monoclinic

Margarite

Margarite is a brittle mica named from the Greek for 'pearl', after its lustre. Brittle micas differ from common micas in that the thin platy crystals are inflexible and readily broken. Beryllian margarite is a related brittle mica, rich in beryllium. Brittle micas have doubly charged ions positioned between the aluminosilicate sheets, whereas common micas have singly charged ions. It occurs in low- to medium-grade metamorphic rocks; commonly in emery deposits, associated with corundum, diaspore, tourmaline, chlorite, andalusite, calcite and quartz. It is found at Mt Greiner (Austria), Naxos (Greece), Smyrna (Turkey), Chester (Massachusetts, USA), Glen Esk (Scotland) and Sverdlovsk (Russia).

$CaAl_2(Al_2Si_2)O_{10}(OH)_2$

Colour:	pale pink, white, grey, green
Lustre; opacity:	pearly
Streak:	white
Hardness:	3.5–4.5
Specific gravity:	2.99–3.1
Cleavage; fracture:	perfect; uneven, giving brittle lamellae
Habit:	rare, tabular, pseudohexagonal crystals; usually foliated aggregates
Crystal system:	monoclinic

Vermiculite

Vermiculite, when heated to 300°C (572°F), quickly loses water and expands 18 to 25 times its original volume, giving long, twisted forms of a golden yellow colour; hence the name, from the Greek for breeding worms. The expanded material is used extensively as packing and insulating material, in soil conditioners, hydroponics, fireproof fillings, and many others. It is formed by hydrothermal alteration of biotite or phlogopite, especially at the contact between felsic intrusions in ultramafic rocks. Associated minerals include corundum, serpentine and talc. Large masses are found at Bulong (Western Australia), Palabora (South Africa), Libby (Montana, USA), Macon (North Carolina, USA) and Ajmer (Rajasthan, India).

$(Mg,Fe,Al)_3(Al,Si)_4O_{10}(OH)_2.4H_2O$

Colour:	white, yellow, green, brown
Lustre; opacity:	vitreous, pearly; transparent to opaque
Streak:	white
Hardness:	1.5
Specific gravity:	2.2–2.6
Cleavage; fracture:	perfect; uneven
Habit:	large crystalline plates, possibly pseudohexagonal; scaly aggregates
Crystal system:	monoclinic

Pennine

Pennine, or penninite, is a pseudotrigonal variety of clinochore. It is a silica-rich member of the chlorite group and occurs as complex admixtures with chromium-rich kammererite. Pennine forms hexagonal platy crystals and it flakes on heating. It is named after the Pennine Alps in Switzerland, where it was first found. Massive pennine is easily carved and is used for ornaments. Pennine occurs in low-grade metamorphic and hydrothermally altered rocks, associated with actinolite, epidote and garnet. It is an essential component of some massive chlorite schists. Pennine is found at Rimpfischwäge and Zermatt (Switzerland), Zillertal (Austria), Val d'Ala (Turin, Italy) and Harstigen (Värmland, Sweden).

$(Mg,Fe)_5Al(Si,Al)_4O_{10}(OH)_8$

Colour:	white, yellow, green
Lustre; opacity:	vitreous, pearly; translucent
Streak:	white
Hardness:	2–2.5
Specific gravity:	2.5–2.6
Cleavage; fracture:	perfect; uneven
Habit:	crystals as plates or rhombohedra; scaly, massive, granular aggregates
Crystal system:	monoclinic

Kämmererite

Kämmererite is a rare chromium-rich variety of clinochlore, displaying a variety of shades of red. The red colours contrast with the green colours produced by iron substitution in clinochlore. It is usually a mixture with pennine, which is regarded by some as a green variety of kämmererite. Particularly good clusters of deep magenta crystals come from Kop Daglari (Erzerum Province, Turkey), many of the crystals being clear, well-formed rhombohedra. Kämmererite occurs in metamorphic rocks, especially serpentinized rocks containing chromium-rich olivines. It is associated with serpentine, chromite and uravorite. It is found at Miass (Ilmen Mountains, Russia), Texas (USA) and the Shetland Islands (UK).

$(Mg,Cr)_3(OH)_2AlSi_3O_{10}Mg_3(OH)_6$

Colour:	red to purple
Lustre; opacity:	vitreous; transparent to translucent
Streak:	reddish
Hardness:	2–2.5
Specific gravity:	2.64
Cleavage; fracture:	perfect; uneven
Habit:	pseudohexagonal crystals; scaly aggregates
Crystal system:	monoclinic

Sepiolite

Sepiolite, also called meerschaum or sea foam, is a light and porous clay mineral named after the Greek for 'cuttle-fish'. It has been traditionally used in tobacco pipes, and is now used to absorb oil on workshop floors or as a filler, binder or free-flow agent in formulations. Thousands of tons per year are imported to the United Kingdom from Spain. Possible asbestos contamination requires monitoring of its use. It is a sedimentary clay mineral formed by surface alteration of magnesite or serpentine. It is associated with opal and dolomite. Sepiolite is found at Eskiflhahir (Turkey), Vallecas and Cabañas (Spain), Middletown (Pennsylvania, USA), Amarillo (Texas, USA), Hobbs (New Mexico, USA) and Cerro Mercado (Mexico).

$Mg_4(OH)_2Si_6O_{15}.2\text{-}4H_2O$

Colour:	white, greyish-, yellowish- or reddish-white
Lustre; opacity:	dull; opaque
Streak:	white
Hardness:	2–2.5
Specific gravity:	1–2
Cleavage; fracture:	unknown; conchoidal
Habit:	compact, nodular, earthy, massive
Crystal system:	orthorhombic

Clinochlore

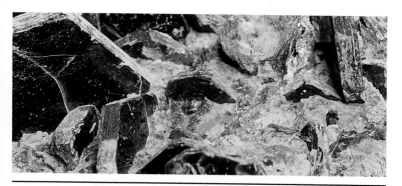

Clinochlore is a common mineral of the chlorite group, the members being difficult to distinguish and usually referred to simply as chlorite by petrologists. Clinochlore is magnesium-rich and forms a series with the iron-rich chamosite. Massive green clinochlore, often banded with a silvery iridescence, is used for carving and polishing as seraphinite. It is an essential mineral in chlorite and talc schists and is found in serpentinites and marbles, associated with serpentine, biotite, actinolite, plagioclase, calcite and dolomite. It is found at Val d'Ala (Turin, Italy), Val Malenco (Sondrio, Italy), Chester (Massachusetts, USA,) Chester County (Pennsylvania, USA), the Zillertal (Austria) and Akhmatovsk (Urals, Russia).

$(Mg,Fe)_5Al(Si,Al)_4O_{10}(OH)_8$

Colour:	grass-green, olive-green, white, pink, yellowish
Lustre; opacity:	pearly, greasy, dull; transparent to translucent
Streak:	greenish-white to white
Hardness:	2–2.5
Specific gravity:	2.6–2.8
Cleavage; fracture:	perfect; uneven, giving inelastic lamellae
Habit:	pseudohexagonal crystals with tapering prismatic faces; foliated, granular, earthy, massive
Crystal system:	monoclinic

Chrysotile

Chrysotile is a fibrous variety of serpentine and the main type of asbestos. Formerly widely used for thermal and electrical insulation, it is now being removed because of the carcinogenic properties of some fibrous forms. The fibres are flexible enough to form woven products, such as fire blankets. It is produced by low-grade metamorphism of ultrabasic rocks in water-rich environments (especially the ocean floor). It is associated with antigorite and lizardite in serpentinites, often filling veins, with the fibres aligned in the direction of the vein. Large masses are found in Asbestos (Quebec, Canada), Eden Mills (Vermont, USA), Brewster (New York, USA), the Urals (Russia), the Alps and Cyprus.

$Mg_3Si_2O_5(OH)_4$

Colour:	greyish-white, green, yellow, brown
Lustre; opacity:	silky; translucent to opaque
Streak:	white
Hardness:	2.5–4
Specific gravity:	2.55
Cleavage; fracture:	none; splintery
Habit:	microcrystalline; aggregates of flexible fibres
Crystal system:	monoclinic (clinochrysotile) and orthorhombic (orthochrysotile)

Chrysocolla

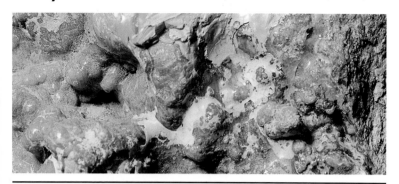

Chrysocolla is an attractive blue-green copper silicate used as an ornamental stone. It is softer than tourquoise or chalcedony, with which it can be confused. Specimens of chrysocolla with quartz not only go well together but also tend to be harder ('agatized') and are more suitable for use in jewellery. It is named from the Greek for 'gold' and 'glue', after its use in soldering gold in ancient times. It occurs in the oxidized zones of copper deposits, associated with malachite, tenorite, quartz and cuprite. Chrysocolla is found at Chuquicamata (Chile), Nizhni Tagil (Urals, Russia), Lubumbashi (Katanga, Congo), Timna (King Solomon's) mine (Israel), Santa Rita (New Mexico, USA) and Clifton (Arizona, USA).

$CuSiO_3.nH_2O$

Colour:	green, blue-green
Lustre; opacity:	vitreous, greasy, earthy; translucent to opaque
Streak:	light green
Hardness:	2–4
Specific gravity:	2.0–2.4
Cleavage; fracture:	none; conchoidal
Habit:	rare acicular crystals in radiating clusters; concretions, porcellanous masses, earthy
Crystal system:	monoclinic

Serpentine

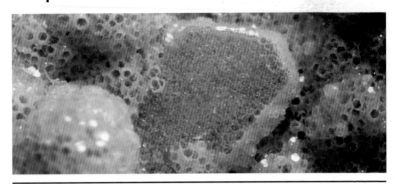

Serpentine is a group of closely related minerals: antigorite, lizardite and chrysotile, the latter most likely to be fibrous and form asbestos. Fibrous serpentine is now regarded as a health hazard. Antigorite and lizardite are the main components of the rock serpentinite, which is formed under oceans by low-grade metamorphism of ultrabasic rocks. Such rocks may be cut by veins of fibrous chrysotile. Attractive, multi-hued and easily worked, serpentine is popular for ornaments; some varieties (bowenite and williamsite) are used for jewellery. The name refers to the snakelike appearance of massive forms. It is found widely, including at Stillwater (Montana, USA), Val Malenco (Italy), Lizard Peninsula (England) and Troodos Mountains (Cyprus).

$(Mg,Fe)_3Si_2O_5(OH)_4$

Colour:	mostly green, also yellow, brown, black, red
Lustre; opacity:	greasy, dull; translucent to opaque
Streak:	white, grey
Hardness:	2.5–4
Specific gravity:	2.5–2.6
Cleavage; fracture:	none; conchoidal, splintery
Habit:	microcrystalline; lamellar (antigorite), fibrous (chrysotile), scaly (lizardite)
Crystal system:	monoclinic/orthorhombic/hexagonal

Nepheline

Nepheline is a major rock-forming mineral found in igneous rocks that are low in silica and contain no quartz, such as nepheline syenites. It is a feldspathoid, a group of minerals like alkali feldspars, but poorer in silica. Although abundant, nepheline rarely forms attractive specimens. Nepheline is especially liable to alteration; it becomes cloudy when treated with acids, hence the name from the Greek for 'cloud'. Nepheline is often altered to zeolites, often natrolite. It occurs in alkaline rocks and pegmatites associated with feldspar, leucite, augite and olivine. Some of the many locations include Vesuvius (Italy), Kola Peninsula and Urals (Russia), Bancroft (Canada) and Litchfield (Maine, USA).

$(Na,K)AlSiO_4$

Colour:	off-white, grey, brown, other tints
Lustre; opacity:	greasy; translucent
Streak:	white
Hardness:	5.5–6
Specific gravity:	2.56–2.66
Cleavage; fracture:	poor; conchoidal
Habit:	prismatic crystals; compact, granular aggregates
Crystal system:	hexagonal

Petalite

Petalite is one of the less common feldspathoids and an ore of lithium. It is too fragile and too rarely of high quality to be commonly used in jewellery, but is sometimes cut for collectors. When heated, petalite slowly gives a blue phosphorescence; it is easily melted and colours a flame crimson (due to lithium). Petalite is named from the Greek for 'leaf', after its perfect cleavage. Petalite occurs in lithium-rich granite pegmatites associated with spodumene, lepidolite, tourmaline, topaz, albite, microcline and quartz. Good localities for petalite include Varaträsk (Sweden), Bikita (Zimbabwe), Norwich (Massachusetts, USA), Oxford County (Maine, USA) and Minas Gerais (Brazil).

$LiAlSi_4O_{10}$

Colour:	colourless or white, grey to reddish-grey
Lustre; opacity:	vitreous to pearly; transparent to translucent
Streak:	white
Hardness:	6–6.5
Specific gravity:	2.4
Cleavage; fracture:	perfect; conchoidal
Habit:	tabular or columnar crystals; usually massive
Crystal system:	monoclinic

Leucite

Leucite is a feldspathoid found abundantly in potassium-rich, silica-poor volcanic rocks. The usual trapezohedral crystal form of leucite is pseudomorphic after the cubic ,-leucite, which forms above 605°C (1121°F). Leucite usually occurs as single crystal grains embedded in host rock. The garnets and analcime are the only other common minerals to crystallize in the form of a trapezohedron. The name is derived from the Greek for 'white', and leucite has been known as white garnet. Good crystals occur in many localities, including Alban Hills and Vesuvius (Italy), Leucite Hills (Wyoming, USA), Bear Paw Mountains (Montana, USA), Mount Nyiragongo (Congo) and Kaiserstuhl (Germany).

KAlSi$_2$O$_6$

Colour:	white
Lustre; opacity:	vitreous; transparent to translucent
Streak:	white
Hardness:	5–5.5
Specific gravity:	2.24–2.29
Cleavage; fracture:	absent; conchoidal, brittle
Habit:	trapezohedral crystals; round grains
Crystal system:	tetragonal

285

Sanidine

Sanidine is a potassium-feldspar formed at high temperatures, above 900°C (1652°F) (orthoclase and microcline are polymorphs formed at medium/high and low temperatures, respectively). It forms a partial series with albite, $NaAlSi_3O_8$, which can form up to 20 per cent of its composition. Sanidine is found in felsic igneous rocks, potassium-rich contact metamorphic rocks and hydrothermal veins. Associated minerals include quartz, sodic plagioclase, biotite, muscovite, hornblende and magnetite. Locations include Vesuvius and Monte Cimine (Italy), Drachenfels and Hohenfels (Germany), Daichi (Japan), Tooele (Utah), Cottonwood Canyon (Arizona) and Rabb Canyon (New Mexico, USA), and Kanchin-do (Korea).

$KAlSi_3O_8$

Colour:	colourless, white, greyish, yellowish
Lustre; opacity:	vitreous, pearly; transparent to translucent
Streak:	white
Hardness:	6
Specific gravity:	2.5
Cleavage; fracture:	perfect; uneven,conchoidal
Habit:	tabular, prismatic crystals, often twinned
Crystal system:	monoclinic

Orthoclase

Orthoclase is abundant as an essential mineral in granitic rocks, forming between about 500° and 900°C (932° and 1652°F). It also occurs in pegmatites and syenites; cavities in basalts; high-grade metamorphic rocks; and hydrothermal veins. Adularia is a transparent colourless gem variety with a bluish sheen. Yellow orthoclase is often faceted when transparent or cut *en cabochon* when displaying a cat's-eye effect. Moonstone is opalescent with a blue or white sheen resembling moonlight. Locations include Madagascar, Sri Lanka and Burma for gem varieties; fine crystals are found at Baveno and Elba (Italy), Carlsbad (Czech Republic), Kirkpatrick (Scotland) and Sverdlovsk (Urals, Russia).

$KAlSi_3O_8$

Colour:	colourless or white, pale yellow, pink, blue or grey
Lustre; opacity:	vitreous, pearly; transparent to translucent
Streak:	white
Hardness:	6
Specific gravity:	2.55–2.63
Cleavage; fracture:	perfect; conchoidal
Habit:	prismatic, columnar or tabular crystals; compact, granular masses
Crystal system:	monoclinic

Microcline

Microcline is a potassium feldspar that forms at low temperatures (usually less than 400ºC/752°F), abundant in granites and related rocks. Amazonite is a blue-green semi-opaque stone resembling jade or turquoise. Perthite is microcline or orthoclase containing undulating layers or intergrowths of plagioclase feldspar, which have separated out on cooling. Microcline is found in felsic plutonic igneous rocks, metamorphic rocks and hydrothermal veins; it is associated with quartz, sodic plagioclase, biotite, muscovite and hornblende. Localities include Arendal (Norway), Ilmen Mountains (Russia), Cala Francese (Sardinia, Italy); amazonite is found at Pikes Peak (Colorado, USA), Bancroft (Canada) and Minas Gerais (Brazil).

$KAlSi_3O_8$

Colour:	white, pink, red, yellowish, blue-green
Lustre; opacity:	vitreous; transparent to translucent
Streak:	white
Hardness:	6–6.5
Specific gravity:	2.55–2.63
Cleavage; fracture:	perfect; conchoidal, uneven
Habit:	prismatic crystals, often twinned; compact aggregates
Crystal system:	triclinic

Hyalophane

Hyalophane is an unusual barium-rich feldspar forming a series with orthoclase and celsian ($BaAl_2Si_2O_8$). The clarity of the crystals, which may be found up to 20cm (8in) long, inspires the name from the Greek for 'glassy'. Hyalophane often shows a weak violet fluorescence under ultraviolet light. Found in metamorphosed manganese-rich rocks and manganiferous deposits, it is associated with epidote, rhodonite, plagioclase and analcime. Some excellent crystals have come from the Zagradski Potok region of Bosnia. Other locations include Lengenbach (Switzerland), Långban and Värmland (Sweden), Slyudyanka (Baikal, Russia), Broken Hill (NSW, Australia) and Kaso (Japan).

(K,Ba)Al(Al,Si)Si$_2$O$_6$

Colour:	colourless, white, yellow
Lustre; opacity:	vitreous; transparent to translucent
Streak:	white
Hardness:	6–6.5
Specific gravity:	2.6–2.8
Cleavage; fracture:	perfect; uneven, conchoidal
Habit:	crystals as prisms or rhombohedra; compact, massive
Crystal system:	monoclinic

Albite

Albite is a plagioclase feldspar in the series albite–anorthite ($NaAlSi_3O_8$–$CaAl_2Si_2O_8$) containing less than 10 per cent anorthite. It forms a high-temperature series with sanidine, but separates at lower temperatures to give intergrowths (perthites). A thin, platy and often transparent variety often associated with tourmaline is called cleavelandite. Albite is abundant in many acid igneous rocks; it is also found in mica schists and gneisses, and in rocks altered ('albitized') by sodium-rich fluids (sea-floor metamorphism). It is associated with orthoclase, quartz, muscovite and biotite. Good crystals are found at St Gotthard (Switzerland), Baveno (Italy), Cazadero (California, USA) and Rio Grande do Sul (Brazil).

$NaAlSi_3O_8$

Colour:	colourless or white
Lustre; opacity:	vitreous; transparent to translucent
Streak:	white
Hardness:	6–6.5
Specific gravity:	2.62
Cleavage; fracture:	perfect; conchoidal
Habit:	crystals as prisms or plates, often twinned; granular and massive aggregates
Crystal system:	triclinic

Anorthite

Plagioclase feldspars range from sodium-rich albite, through oligoclase, andesine, labradorite and bytownite, to calcium-rich anorthite (contains up to 10 per cent albite). Sunstone is a variety of oligoclase containing reflective platy inclusions, red, orange or green in colour, which give a metallic glitter. Labradorite (50–70 per cent anorthite) is used as a gem or ornamental stone, displaying a play of colours produced by internal interference of light. Plagioclase occurs in igneous and metamorphic rocks, becoming more calcium-rich as the rocks become more mafic. Labradorite occurs in Larvik-Tvedalan (Norway) and Labrador (Canada). Anorthite is found at Vesuvius (Italy), Hokkaido (Japan) and Tunaberg (Sweden).

$CaAl_2Si_2O_8$

Colour:	white to grey, yellowish, greenish, reddish
Lustre; opacity:	vitreous, pearly; transparent to translucent
Streak:	white
Hardness:	6
Specific gravity:	2.62–2.76
Cleavage; fracture:	good; uneven
Habit:	crystals as prisms or plates, often twinned; granular or massive aggregates
Crystal system:	triclinic

Hauyne

Hauyne, or hauynite, is a feldspathoid and a member of the sodalite group, where calcium partially substitutes for sodium, and sulphate groups replace chlorine atoms. It is popularly collected as bright blue grains in a matrix and occasionally faceted, but is difficult to cut because of its perfect cleavage. Hauyne is found in phonolites and related silica-poor rocks, associated with nepheline, leucite, augite, sanidine, biotite and apatite. It is named after the French crystallographer Abbé René Haüy (1743–1822), who discovered it on Vesuvius (Italy); other localities are Laacher See and Niedermendig (Germany), Winnett (Montana, USA), Edwards Mine (New York, USA), Tasmania (Australia) and Nanjing (Jiangsu, China).

$(Na,Ca)_{4-8}(Al_6Si_6)O_{24}(SO_4,Cl)_{1-2}$

Colour:	bright blue to greenish-blue, white, grey, brown
Lustre; opacity:	vitreous, greasy; transparent to translucent
Streak:	white
Hardness:	5.5
Specific gravity:	2.44–2.5
Cleavage; fracture:	perfect; conchoidal
Habit:	rare octahedral or dodecahedral crystals; usually rounded grains
Crystal system:	cubic

Sodalite

Sodalite is a feldspathoid mostly notable for its occurrence in a bright blue, massive form, used as a semi-precious stone for carvings and jewellery. It is a component of lapis lazuli, but, unlike it, sodalite rarely contains contrasting golden yellow pyrite and has a lower density. Crystals of sodalite are rare and usually small. Sodalite is found in nepheline syenites, phonolites and related rocks, and in metasomatic calcareous rocks. Associated minerals include nepheline, microcline and sanidine. Excellent massive sodalite is found at Bancroft (Canada), Litchfield (Maine, USA), Magnet Cove (Arkansas, USA), Ilimaussaq (Greenland) and Cerro Sapo (Bolivia); clear crystals are found in the calcium-rich lavas of Vesuvius (Italy).

$Na_8Cl_2(AlSiO_4)_6$

Colour:	bright blue, white, grey, green
Lustre; opacity:	vitreous, greasy; transparent to opaque
Streak:	white
Hardness:	5–6
Specific gravity:	2.3
Cleavage; fracture:	perfect; conchoidal, uneven
Habit:	very rare dodecahedral crystals; compact masses
Crystal system:	cubic

Lazurite

Lazurite is most commonly found as a component of lapis lazuli, the semi-precious massive stone used for jewellery and carved ornaments. Lapis lazuli is mostly lazurite, but contains blue sodalite and hauyne, white calcite and brassy pyrite. The deposits in Afghanistan were worked 6000 years ago for high-quality lapis lazuli used, for example, for the mask of Tutunkhamun. Lazurite is found in contact metamorphosed limestones. Exceptional crystals and fine lapis lazuli are found at Sar-e-Sang (Kokscha Valley, Afghanistan); other locations include Vesuvius and the Alban Hills (Italy), Sayan Mountains (Russia), Baffin Island (Canada) and California and Colorado (USA).

(Na,Ca)$_8$(Al,Si)$_{12}$O$_{24}$(S,SO$_4$)

Colour:	deep blue, blue-green
Lustre; opacity:	vitreous; translucent to opaque
Streak:	light blue
Hardness:	5–5.5
Specific gravity:	2.38–2.42
Cleavage; fracture:	imperfect; uneven
Habit:	very rare octahedral crystals; compact masses
Crystal system:	cubic

Scapolite

Scapolite is a member of the complex scapolite series, between meionite $(Ca_4(Si,Al)_{12}O_{24}(CO_3,SO_4))$ and marialite $(Na_4(Si,Al)_{12}O_{24}Cl)$. It is formed by alteration of plagioclase feldspars, its composition generally reflecting the sodium/calcium ratio of the parent mineral. Clear crystals may be faceted; those containing inclusions are cut *en cabochon*. Scapolite is named from the Greek for 'rod', alluding to the crystal shape. Scapolite is found in metamorphosed limestones and hydrothermally altered basic rocks. Crystals up to 50cm (20in) are found at Rossie and Pierrepoint (New York, USA); other localities are Renfrew (Ontario) and Grenville (Quebec, Canada), Lake Tremorgio (Switzerland) and Minas Gerais (Brazil).

$(Na,Ca)_8(Cl_2,SO_4,CO_3)(AlSi_3O_8)_6$

Colour:	white, bluish, grey, pink
Lustre; opacity:	vitreous; transparent to translucent
Streak:	white
Hardness:	5–6.5
Specific gravity:	2.54–2.77
Cleavage; fracture:	poor; conchoidal
Habit:	columnar prismatic crystals; fibrous or massive aggregates
Crystal system:	tetragonal

Natrolite

Natrolite is a zeolite forming attractive sprays of radiating needles. It is named from the Greek *natron* for 'soda', alluding to its sodium content. Being a typical molecular sieve, it holds water in voids in its structure. On heating to 300°C (572°F), it loses water without changing crystal structure, which it will reabsorb from the atmosphere when cooled. Natrolite may also exhibit orange fluorescence under ultraviolet light. It is found in hydrothermal veins in basalts, associated with nepheline, sodalite, quartz and other zeolites. Good crystals come from Nova Scotia and British Columbia, and specimens up to 1m (3¼ft) long from Asbestos (Canada); other localities are Puy-de-Dôme (France) and White Head (Antrim, Northern Ireland).

$Na_2(Al_2Si_3)O_{10}.2H_2O$

Colour:	colourless, white, pink, yellowish
Lustre; opacity:	vitreous; transparent to translucent
Streak:	white
Hardness:	5.5–6
Specific gravity:	2.2–2.4
Cleavage; fracture:	good; conchoidal
Habit:	acicular needles; globular aggregates of radiating needles, rarely compact
Crystal system:	orthorhombic

Mesolite

Mesolite forms characteristic sprays of transparent acicular crystals. A typical zeolite, it forms only as a secondary mineral from the breakdown of feldspars and fills voids in altered igneous rocks. It is a molecular sieve and easily loses and absorbs water. It is named from the Greek for 'middle', having a composition between natrolite and scolecite. Mesolite is found in cavities in volcanic rocks, especially basalt, and hydrothermal veins. It is associated with natrolite and scolecite (from which it is difficult to distinguish), and other zeolites and calcite. Locations include Nova Scotia (Canada), Grant County (Oregon, USA), Faeroe Islands (Norway), Puy-de-Dôme (France) and Poona (Maharashtra, India).

$Na_2Ca_2(Al_2Si_3O_{10})_3.8H_2O$

Colour:	white, grey, yellow
Lustre; opacity:	vitreous, silky; transparent to translucent
Streak:	white
Hardness:	5–5.5
Specific gravity:	2.2–2.4
Cleavage; fracture:	perfect; conchoidal, uneven
Habit:	acicular needles; massive, spherulitic aggregates, earthy
Crystal system:	monoclinic

297

Thomsonite

Thomsonite is a rare zeolite, difficult to distinguish from natrolite. The name, after the Scottish chemist Thomas Thomson, is applied to the series thomsonite-Ca to thomsonite-Sr. The strontium-rich minerals are visually indistinguishable and rarer. Thomsonite is found filling voids in mafic igneous rocks, especially basalts, and in hydrothermal veins in other igneous rocks. Associated minerals are calcite, prehnite, quartz and other zeolites. Localities include Mount Monzoni (Italy), Faeroe Islands (Norway), Old Kilpatrick and Bishopton (Scotland), Disko Island (Greenland) and Springfield (Oregon, USA). Attractively banded specimens of massive nodules are found on the southern shore of Lake Superior.

$NaCa_2(Al_5Si_5)O_{20}.6H_2O$

Colour:	white, brown
Lustre; opacity:	vitreous, pearly; transparent to translucent
Streak:	white
Hardness:	5–5.5
Specific gravity:	2.2–2.4
Cleavage; fracture:	perfect; uneven
Habit:	individual crystals rare; radiating globular clusters
Crystal system:	monoclinic

Scolecite

Scolecite is a rare zeolite occurring as sprays of white needles. It is closely related to natrolite and mesolite, and difficult to distinguish from these. The name, from the Greek *skolec* for 'worm', is an allusion to its behaviour in a blowpipe flame, when it curls into wormlike forms before melting. Scolecite is found lining cavities in lavas, especially basalts; and also in veins in contact-metamorphosed pelites and limestones. Associated minerals are calcite, prehnite and other zeolites. Excellent crystals are found at Teigarhorn (Iceland), Ben More, Isle of Mull and Talisker Bay, Isle of Skye (Scotland) and the Faeroe Islands (Norway); other locations include the Deccan Traps (Poona, India) and Vesuvius (Italy).

$Ca(Al_2Si_3)O_{10}.3H_2O$

Colour:	colourless or white
Lustre; opacity:	vitreous, silky; transparent
Streak:	white
Hardness:	5–5.5
Specific gravity:	2.2–2.4
Cleavage; fracture:	perfect; fragile
Habit:	prismatic crystals; fibrous, radiating masses
Crystal system:	orthorhombic

Mordenite

Mordenite is a rare but widespread zeolite, found often as clusters of white or pinkish needles. It is used in catalysis, petrochemicals and fine chemicals, and is synthetically produced for such applications. Mordenite, like other zeolites, has holes in its molecular framework of exact sizes, making it specific in interactions for absorption and catalysis. It is a secondary mineral, found in veins and cavities in igneous rocks, associated with calcite, kaolinite, glauconite and other zeolites; it is also deposited among some sediments. Locations include Morden (Nova Scotia, Canada), Berufjord (Iceland), Elba (Italy), Isle of Mull (Scotland), Custer County (Colorado, USA) and Hoodoo Mountains (Wyoming, USA).

$(Ca,K_2,Na_2)(AlSi_5O_{12})_2.6H_2O$

Colour:	colourless, white, yellowish, pinkish
Lustre; opacity:	vitreous, pearly; transparent to translucent
Streak:	white
Hardness:	4.5
Specific gravity:	2.1
Cleavage; fracture:	perfect; uneven
Habit:	crystals as striated needles; fibrous, reniform aggregates
Crystal system:	orthorhombic

Laumontite

Laumontite crystals can be impressive when large, almost acicular prisms. It is a zeolite named after the Frenchman François de Laumont (1747–1834), who found the mineral. Specimens exposed to dry air become opaque and powdery due to partial dehydration; this can be avoided by using a sealant or airtight containment. Laumontite is found in veins and cavities in most rock types and as a cement in sandstones. Associated minerals include datolite, calcite, chlorite and other zeolites. Crystals up to 30cm (12in) are found at Bishop (California, USA); good crystals are also found in the eastern Pyrenees (France), St Gotthard (Switzerland), the Tirol (Austria), Baveno (Italy) and Poona and Khandivali (Maharashtra, India).

$Ca(Al_2Si_4)O_{12}.4H_2O$

Colour:	white to grey, yellowish, brownish, pink
Lustre; opacity:	vitreous, pearly; transparent to translucent
Streak:	white
Hardness:	3–4
Specific gravity:	2.2–2.3
Cleavage; fracture:	perfect; uneven
Habit:	square, prismatic crystals often elongated; columnar, fibrous and radiating aggregates
Crystal system:	monoclinic

Heulandite

Heulandite is a common zeolite forming attractive aggregates of tabular crystals in a range of different hues. Distinguished from stilbite in 1818, it was first named euzeolite, meaning 'beautiful zeolite', but eventually named after the English mineral collector John Heuland (1778–1856). Heulandite is now a term referring to a series of calcium-, sodium-, potassium- or strontium-rich rocks. It is found in cavities in volcanic rocks, in veins in schists and gneisses, and disseminated in sedimentary rocks; it is usually associated with stilbite and chabazite. Good crystals are found at Paterson (New Jersey, USA), Poona (Maharashtra, India), Berufjord (Iceland), Stirlingshire (Scotland), Nova Scotia (Canada) and Hawaii (USA).

$(Na,Ca)_{4-6}Al_6(Al,Si)_4Si_{26}O_{72}.24H_2O$

Colour:	colourless, yellow, green, reddish-orange
Lustre; opacity:	vitreous, pearly; transparent, usually translucent
Streak:	white
Hardness:	3.5–4
Specific gravity:	2.2
Cleavage; fracture:	perfect; uneven
Habit:	tabular crystals; aggregates
Crystal system:	monoclinic

Stilbite

Stilbite is a common mineral and one of the most popularly collected zeolites. It is named from the Greek for 'shine', after the pearly effect on the cleavage plane. The structure of stilbite contains channels that hold certain sizes of molecules and ions. This is put to use in the separation of hydrocarbons in petroleum refining and in ion-exchange processes. Stilbite is soluble in acids and melts easily, giving a white glass. It is found in hydrothermal veins in basalts, associated with calcite and other zeolites, especially heulandite. Fine crystals are found at Paterson (New Jersey, USA), Kilpatrick and Isle of Skye (Scotland), Nova Scotia (Canada), Rio Grande do Sul (Brazil), Teigarhorn (Iceland) and Poona (India).

$NaCa_2(Al_5Si_{13})O_{36}.14H_2O$

Colour:	white, grey, reddish-brown
Lustre; opacity:	vitreous, pearly; transparent to translucent
Streak:	white
Hardness:	3.5–4
Specific gravity:	2.1
Cleavage; fracture:	perfect; conchoidal, fragile
Habit:	prismatic crystals, usually in sheaflike aggregates, cruciform twins; radiating fibrous masses
Crystal system:	monoclinic

Harmotome

Harmotome is a rare zeolite, but popular for its twinned crystals. Single twinning gives crosses of interpenetrating prisms, with the ends resembling blunt Phillips head screwdrivers; more uncommonly, double twinning can result in three prisms interpenetrating at 90° to each other. Harmotome is named from the Greek for 'joint' and 'cut', after the easily separated twinned crystals. Harmotome occurs in hydrothermal veins in basalts and other volcanic rocks, in gneisses and in some ore veins; it is associated with calcite, barite and quartz. Good crystals are found in St Andreasburg (Germany), Argyll (Scotland), Pribram (Czech Republic), Thunder Bay (Ontario, Canada) and Manhattan (New York City, USA).

$BaAl_2Si_6O_{16}.6H_2O$

Colour:	white, grey, yellow, red, brown
Lustre; opacity:	vitreous, pearly; transparent to translucent
Streak:	white
Hardness:	4.5
Specific gravity:	2.44–2.5
Cleavage; fracture:	distinct; uneven, subconchoidal
Habit:	prisms, plates, often twins; aggregates
Crystal system:	monoclinic

Gismondine

Gismondine is a rare zeolite, often forming clear colourless pseudotetragonal crystals. It was named after Professor Carlo Gismondi (1762–1824), the Italian mineralogist who first examined it. A similar mineral gismondine-Ba has been found on lead-rich slags. Gismondine is easily melted in a flame and soluble in hydrochloric acid. It occurs in hydrothermal veins in nepheline and olivine basalts, associated with calcite, quartz, chlorite and other zeolites. It is found at Capo di Bove (Lazio, Italy), Bühne (Westphalia), Fulde (Hesse) and Arensberg (Eiffel, Germany), Round Top (Oahu) and Alexander Dam (Kauai, Hawaii), Antrim (Northern Ireland) and Reydarfjord and Fáskrúdsfjord (Iceland).

$CaAl_2Si_2O_8 \cdot 4H_2O$

Colour:	white, grey, bluish, reddish
Lustre; opacity:	vitreous; transparent to translucent
Streak:	white
Hardness:	4.5
Specific gravity:	2.26
Cleavage; fracture:	imperfect; conchoidal
Habit:	bipyramidal crystals; stellate or radiating spherulitic aggregates
Crystal system:	monoclinic

Chabazite

Chabazite is a lesser known zeolite forming rhombohedral crystals that appear almost like cubes, with angles close to 90°. Chabazite-Na, -K and -Sr are varieties rich in sodium, potassium or strontium. It is used as an acid-resistant absorbant in natural gas production and to remove heavy metals from waste streams. Chabazite occurs in hydrothermal veins in basalts and andesites; and rarely in limestones, schists and ore veins. Associated minerals include nepheline, olivine, pyroxenes, tridymite, calcite and dolomite. Chabazite is found at Idar-Oberstein (Germany), Repcice (Czech Republic), Kilmalcolm (Scotland), Breidhdalsheidhi (Iceland), Bowie (Arizona, USA) and Paterson (New Jersey, USA).

$Ca(Al_2Si_4)O_{12}.6H_2O$

Colour:	colourless, white, yellow, pink, red
Lustre; opacity:	vitreous; transparent to translucent
Streak:	white
Hardness:	4.5
Specific gravity:	2.08
Cleavage; fracture:	distinct; uneven
Habit:	crystals as rhombohedra, twins; druses, granular, massive
Crystal system:	trigonal

Gmelinite

Gmelinite is a rare zeolite that forms characteristic crystals in the form of shallow six-sided double pyramids, which have been described as 'flying saucers'. It can also occur as thin plates and attractive rosettes of such crystals. Varieties are rich in calcium, sodium or potassium. It is named after Christian Gmelin (1792–1860), a German chemist. Gmelinite occurs in basalts, related rocks and pegmatites, associated with calcite, aragonite, quartz and other zeolites, especially chabazite and analcime. Locations include Montecchio Maggiore (Vicenza, Italy), Glenarm (Antrim, Northern Ireland), Pyrgos (Cyprus), Paterson (New Jersey, USA), Sarbay-Sokolov (Kazakhstan) and Bekiady (Madagascar).

(Na,Ca)Al$_2$Si$_4$O$_{12}$.6H$_2$O

Colour:	white, yellow, pink, reddish
Lustre; opacity:	vitreous; transparent to opaque
Streak:	white
Hardness:	4.5
Specific gravity:	2.03
Cleavage; fracture:	imperfect; uneven
Habit:	crystals as plates, bipyramids, rhombohedra, twins; druses, rarely radiating aggregates or granular
Crystal system:	hexagonal

Analcime

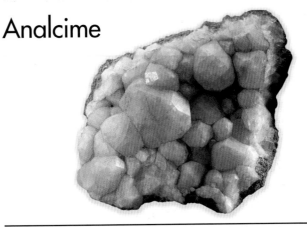

Analcime, or analcite, is a popular zeolite forming distinctive trapezohedral crystals, a shape commonly found only in garnets (harder and strongly coloured) and leucite (lower specific gravity). It has been classified as a feldspathoid, but has an open structure, typical of zeolites. It is named from the Greek for 'weak', after the weak static charge produced on heating or rubbing. Analcime occurs in basalts and phonolites, associated with prehnite, glauconite, quartz and other zeolites. It is found at Kotchechovmo (Krasnoyarski, Russia), St Keverne (Cornwall, England), Breidhdalsheidhi (Iceland), Lake Superior (Michigan, USA), Bergen Hill (New Jersey, USA) and Mt Saint Hilaire (Quebec, Canada).

$NaAlSi_2O_6.2H_2O$

Colour:	white, colourless, grey, pink, greenish
Lustre; opacity:	vitreous, dull; transparent to translucent
Streak:	white
Hardness:	5.5
Specific gravity:	2.2–2.3
Cleavage; fracture:	perfect; uneven, conchoidal
Habit:	trapezohedral crystals; granular, compact, massive, showing concentric structure
Crystal system:	cubic

Pollucite

Pollucite is a zeolite and the major ore of caesium. Typically pollucite contains about 28 per cent caesium, and it may also be a source of rubidium, which is present at about 1–2 per cent. Caesium compounds are used in catalyst promoters, special glasses and radiation monitoring equipment. The name comes from its association with petalite (castorite) from classical mythology, Pollux being the brother of Castor. Pollucite occurs in kilotonne quantities in lithium-rich granite pegmatites, associated with spodumene, lepidolite, potassium feldspar, microcline and quartz. Pollucite is found at Varutrask (Sweden), Elba (Italy), Bikita (Zimbabwe), Bernic Lake (Manitoba, Canada) and Shengus, Skardu and Gilgit (Pakistan).

$(Cs,Na)AlSi_2O_6.H_2O$

Colour:	colourless, white, grey, pale pink, pale blue
Lustre; opacity:	vitreous; transparent to translucent
Streak:	white
Hardness:	6.5
Specific gravity:	2.44–2.5
Cleavage; fracture:	absent; brittle, uneven
Habit:	cubic crystals; grains, granular and massive aggregates
Crystal system:	cubic

Amber

Amber is a fossil tree resin that, although not mineralized, is widely used in jewellery. It often has small insects preserved inside, but it can also contain plant remains, moss, pine needles, spiders or even frogs and lizards. Amber will become electrically charged when rubbed. Unlike most minerals, amber will melt and burn on heating, being chiefly of polymerized hydrocarbons called 'terpenes'. Most amber is *c.* 30–90 million years old; semi-fossilized resin is called copal. The best known amber deposits, known as succinite, are in the Baltic region, along the coasts of Poland and Russia, especially the Sambia Peninsula near Kaliningrad. Burmese amber is called burmite and amber from Sicily is known as simetite.

Formula

Colour:	golden yellow/orange, brown, reddish, green, violet, black
Lustre; opacity:	resinous; transparent to translucent
Streak:	white
Hardness:	1.5–3
Specific gravity :	1.05–1.1
Cleavage; fracture:	absent; conchoidal
Habit:	irregular nodules in marine sands and clays; often a placer deposit washed up on shore
Crystal system:	amorphous

Petrified wood

Petrified wood is really a fossil, but is included here, as it is very popular as ornaments and polished articles. Like opal, it is usually composed of hydrated silica, which has slowly replaced the lignin and cellulose of the wood as they have decomposed. The original structure of the wood is often preserved in great detail, down to microscopic levels. Various colours can be produced by small amounts of impurities such as iron, manganese and copper, which can help to pick out the original structure. The most famous petrified forests are those of Santa Cruz (Patagonia, Argentina), Navá Paka (Czech Republic), Lesbos (Greece) and the Petrified Forest National Park (Arizona, USA).

$SiO_2.nH_2O$

Colour:	various
Lustre; opacity:	translucent to opaque
Streak:	white
Hardness:	5–7
Specific gravity :	1.9–2.5
Cleavage; fracture:	absent; conchoidal
Habit:	best examples are intact tracts of forest, with trees up to 3m (10ft) diameter and 30m (98ft) long and complete root systems
Crystal system:	amorphous

Jet

Jet is a hard gem variety of lignite (a type of coal), which is often carved or faceted, and takes a good polish. Occasionally it has pyrite inclusions displaying a brassy, metallic lustre. Used in jewellery since ancient times, jet has been traditionally made into rosaries for monks. As part of her mourning attire, Queen Victoria wore jet from Whitby (England), famous at the time for the mining and crafting of jet. Unlike higher grade coals, jet retains a microscopic, woody structure. Jet is found at Pictou (Nova Scotia, Canada), Cabo Mondego (Coimbra, Portugal), Wet Mountain Valley (Colorado, USA), Acoma (New Mexico, USA), Anne Arundel County (Maryland, USA) and Henry Mountains and Coaly Basin (Utah, USA).

Formula

Colour:	dark brown, black
Lustre; opacity:	velvety, waxy; opaque
Streak:	brown
Hardness:	2.5
Specific gravity :	1.33
Cleavage; fracture:	none; conchoidal
Habit:	masses in bituminous shales
Crystal system:	amorphous

Pearl

Once thought to be the tears of the gods, pearls are used widely in jewellery and decorations. They form in modern shellfish, such as mussels, as protection against a natural or emplaced irritant (e.g. a piece of sand) within the soft parts. Layers of nacre are secreted around the foreign body at a rate of up to 0.6mm per year. The nacre consists of aragonite crystals held together by conchiolin, a hornlike substance, and water. Light reflecting from overlapping layers produces the characteristic iridescent lustre. The most important source of creamy white pearls is the Persian Gulf, attracting the highest prices; they have also been collected from the Gulf of Manaar (Sri Lanka) and the Red Sea for thousands of years.

Formula

Colour:	pale yellow, white, salmon pink to grey, brown or black
Lustre; opacity:	pearly, iridescent; translucent to opaque
Streak:	generally white
Hardness:	3
Specific gravity :	2.71
Cleavage; fracture:	pearls are soft and easily damaged
Habit:	roughly spherical, but a variety of shapes depending on original irritant and whether the pearl has moved
Crystal system:	orthorhombic (aragonite)

Glossary

accessory mineral	minor component of a rock; not necessary to define the rock type
acid	(of igneous rock) containing at least 10% quartz and chemically, more than 65% SiO_2
amorphous	non-crystalline, having no regular microscopic structure
aqua regia	a mixture of 3:1 nitric and hydrochloric acids
basic	(of igneous rock) containing 45–55% SiO_2
cation	a positively charged ion such as $Na+$ or Fe_2+
columnar	elongated prismatic form of a crystal
druse	fine crystalline coating on a matrix or filling a cavity
en cabochon	a rounded, convex cut of a stone
essential mineral	one that is necessary for the definition of a rock
euhedral	(of a mineral) having well-formed crystal faces
evaporite	sedimentary rock formed by evaporation of an igneous rock
felsic	(of rocks) composed mostly of light-coloured minerals, especially feldspar and quartz or feldspathoid
fluorescence	the emission of visible light on absorption of invisible ultraviolet (uv) light, giving an apparent 'glow-in-the-dark' effect
foliated	made up of thin aligned plates that flake easily
fumarole	a vent for volcanic gasses
gangue	the waste component of a mineral deposit
grade, metamorphic	degree of metamorphism, mostly related to temperatures and pressures experienced by a rock
hydrothermal vein	a vein produced by hot, mineral-rich waters of igneous origin
igneous rock	a rock produced from magma or lava
iridescence	a play of colours on a surface produced by light interference
lamellar	made of crystals mostly developed in parallel to each other
mafic	(of rocks) composed mostly of dark, ferromagnesian minerals, especially olivine, pyroxene and amphibole

magma	hot, molten rock material, called lava on reaching the surface
matrix	fine-grained background mass of a rock
metamict	pertaining to a mineral whose structure has been broken down by radiation damage, typically powdery
metamorphic rock	rock produced by increased temperature and/or pressure when minerals are altered without substantial melting
metasomatism	the alteration of a mineral or rock by hot fluids, either removing or adding chemical elements
oxidation	chemical process involving addition of oxygen
pelitic	(of a rock or sediment) of high aluminium and low calcium content, especially of mudstones and clays
phosphorescence	the emission of light after illumination (by visible or ultaviolet light) has ceased
placer deposit	concentration of residual resistant, heavy minerals found in river gravels after removal of other components of a weathered and eroded rock
pneumatolysis	the reaction of magma-derived hot gasses with the surrounding rock
porphyroblast	large, well-developed crystal found in a metamorphic rocks
schist	metamorphic rock containing platy minerals, usually mica, in a roughly parallel arrangement
secondary	(of minerals) one produced by alteration of pre-existing minerals
sedimentary rock	rock formed by sedimentary processes such as weathering, erosion, transport, deposition, compaction and cementation
solid solution	(of minerals) a product of two or more minerals mixing at a molecular level while retaining the same structure to give intermediate compositions
sublimate	a material condensing as a solid from a gas
thermoluminescence	luminescence produced by heating
triboluminescence	luminescence produced by rubbing
ultrabasic	(of rocks) containing <45% SiO_2
ultramafic	(of rocks) containing >90% mafic minerals
vein	an irregular but essentially tabular intrusion, of a width between a millimetre and tens of centimetres

Index

acanthite 29, 59
acmite 248
actinolite 256
adamite 177
aegirine 248
agate 104, 105
aikinite 48
åkermanite 225
alabaster 151
albite 286, 290, 291
alexandrite 93
allanite 231
almandine 205, 206, 207
alum 158
alumina 94
aluminium
  borosilicates 224, 240–1
  oxides 116
  phosphates 187, 194
  silicates 212–15
alunite 158
amber 310
amblygonite 175
amethyst 101, 193
ammonium chloride 76
amphiboles 252–7
analcime 307, 308
anatase 110
andalusite 213
andradite 210
anglesite 154
anhydrite 152
ankerite 133
annabergite 41, 191, 192
anorthite 290, 291
anthophyllite 252, 258
antimonite 46
antimony 21, 22
  sulphides 32, 46,
    51, 67, 68

antimony bloom 97
apatite 9, 183, 186
apophyllite 266
aragonite 135, 313
argentite 29, 58
arsenates 176–7, 181,
    185, 191–2, 196–8
arsenic 20, 21, 22
  sulphides 36, 53, 64–5, 67
arsenolite 96
arsenopyrite 53
asbestos 252, 254–6,
    258, 278, 280, 282
atacamite 79
augite 248, 249
aurichalcite 142
automolite 87
autunite 199
axinite 235
azurite 27, 139, 159, 178

babingtonite 262
barite 134, 153, 155
barium 125, 134, 155, 289
bauxite 116
bavenite 264
benitoite 234, 263
beryl 221, 236–7, 264
beryllium 93, 236, 264, 274
betafite 124
billietite 122
biotite 270, 271, 275
bismuth 22, 48, 69
bismuthinite 48
boleite 81
borates 13, 146–9
borax 146, 149
bornite 28
boron 219
borosilicates 224, 240–1

bort 24, 25
botryogen 163
boulangerite 63, 68
bournonite 61
brass 30
braunite 223
brazilianite 180
brochantite 157
bromargyrite 72
bronze 14, 108
bronzite 251
brookite 111
brucite 114, 222
buckminsterfullerene 23

cacoxenite 193
cadmium 30, 35
caesium 272, 309
calcite 9, 55,
    123, 127, 135, 201, 303
calcium
  carbonates 127, 132–3,
    135, 145
  silicates 204, 208–10,
    228–30, 244–5, 249,
    255–7, 259, 264–5, 291
  sulphates 151, 152,
    160, 166
  uranium phosphate 199
carbonado 24
carbonates 13, 127–45
carbonatite 127, 136
carnallite 78
carnelion 104, 105
carnotite 200
cassiterite 108
celestine 136, 153
celestite 136, 153
cerium 85, 220, 231
cerussite 92, 137

chabazite 302, 306
chalcedony 104–5, 281
chalcocite 27
chalcophyllite 196
chalcopyrite 27, 31
chalcosine 27
chiastolite 213
chlorargyrite 72
chlorides 70, 72, 80
chlorite 279
chloritoid 218
chromates 169
chromite 90
chromium 90, 209, 277
chrysoberyl 93
chrysocolla 86, 281
chrysoprase 104, 105
chrysotile 280, 282
cinnabar 38, 64, 65
classification of minerals 12–13
clinochlore 276, 277, 279
clinoclase 181
clinozoisite 228, 230
cobalt 47, 54, 55, 191
cobaltite 47
colemanite 148
colour 6
columbite 112
copper 14, 20
  arsenates 176, 181, 196–8
  carbonates 138–9, 141–2
  halogenides 79
  oxides 84, 86
  phosphates 179, 195, 197
  silicates 239, 281
  sulphates 156–7, 159, 162, 164, 166
  sulphides 27–8, 31–3, 36, 43
cordierite 238
corundum 9, 94–5

cosalite 69
cotunnite 80
covellite 43
cristobalite 102
crocoite 169
cryolite 77
crystal system 11
cummingtonite 252, 258
cuprite 84
cyanotrichite 162
cyclosilicates 13, 234–43

datolite 219
descloizite 182
devilline 166
diaboleite 80, 82
diamond 9, 23, 24–5, 170, 210, 211, 234
diaspore 116
diopside 244, 245
dioptase 239
disthene 214
dolomite 67, 128, 132, 133
dumortierite 224

elbaite 240, 241
emerald 236, 237, 239
enargite 36
endellionite 61
enstatite 202, 250, 251
epidote 228, 229, 230, 231, 232
epsomite 167
erythrite 47, 54, 55, 191, 192
euclase 221
eudialyte 243

fayalite 202, 203
feldspar 158, 187, 283–6, 288–93, 295, 297
ferrosilite 250
flint 104

fluorides 71, 74–5
fluorite 9, 74–5
fluorspar 74–5
fool's gold 44
forsterite 202
franklinite 91
furgusonite 113

gadolinite 220
gahnite 87
galena 29, 30, 37, 58, 92
garnet 205–10, 308
gaylussite 145
gehlenite-åkermanite 225
gersdorfite 51, 66
giekielite 98
gismondine 305
glauberite 160
glaucodot 54
glaucophane 253
gmelinite 307
goethite 115, 117
gold 16, 31, 44, 49
graphite 23, 56
greenockite 35
grossular 208, 209
gypsum 9, 151, 152

halite 70
halogenides 70–82
halotrichite 161, 162, 163
hardness 8–9, 24
harmotone 304
hauerite 50
hauyne 292
hedenbergite 244, 245
hematite 99
hemimorphite 227
heulandite 302, 303
hornblende 249, 257
humite 222
hutchinsonite 62

INDEX

hyalophane 13, 289
hydroxides 114–19, 122, 124–5
hydrozincite 140
hypersthene 250

idocrase 233
ilmenite 98
ilvaite 226
indigo copper 43
inosilicates 13, 244–65
iodargyrite 72
iron
  carbonates 129, 133
  hydroxides 115, 117
  oxides 84, 89, 99
  phosphates 188, 190, 193
  silicates 235, 251–8, 262
  sulphates 161, 163, 168
  sulphides 44–5, 52
  tungstates 171
  see also pyrite
iron-nickel 18, 39, 42

jade 246, 256
jadeite 246
jamesonite 63, 68
jet 312
jordanite 67

kainite 165
kämmererite 276, 277
kernite 149
kröhnkite 164
kyanite 214

lanthanum 220, 231
lapis lazuli 178, 294
laumontite 265, 301
laurionite 82
lavendulan 198
lazulite 178

lazurite 6, 178, 294
lead
  arsenates 185
  carbonates 137, 143
  chloride 80
  chromate 169
  molybdate 172
  oxides 92, 123
  phosphate 184
  sulphate 154, 159
  sulphides 37, 63, 67, 69
  vanadate 182, 186
lepidolite 272
leucite 285, 308
limestone 127, 128, 132
limonite 115, 117
linarite 159
liroconite 197
lithiophyllite 173
lithium 175, 247, 272, 284
lodestone 89
loparite 85
lustre 7

magnesite 128, 278
magnesium 78
  carbonate 128, 132
  oxides 114, 128, 132
  silicates 202–5, 222, 250–8, 268, 279
  sulphate 167
magnetite 89, 90
malachite 27, 138, 179
manganese 114, 118, 125, 171
  carbonate 131
  disulphide 50
  iron phosphates 173
  oxide 107
  silicates 207, 223, 230, 235, 261
manganite 107, 118

marble 114, 127, 132, 201
marcasite 45, 52
margarite 274
melaconite 86
menaccanite 98
mercury 17, 38
mesolite 297, 299
metalloids 21
meteorites 18, 39, 42, 203
miargyrite 40
mica 268–74
microcline 286, 288
microlite 119
milarite 242
millerite 42, 68
mimetite 184, 185
minium 92
mirabilite 150
Mohs' scale of hardness 8–9
molybdenite 56
molybdenum 172
monazite 174
monticellite 204
mordenite 300
mottramite 182
muscovite 12, 269, 270, 272

native elements 13, 14–26
natrocalcite 145
natrolite 234, 263, 283, 296, 297, 298, 299
neosilicates 13, 201–24
nepheline 283
nephrite 246, 256
neptunite 234, 263
nickel 18
  sulphides 39, 41–2, 51, 55, 66
nickel bloom 192
nickeline 41
niobium 85, 112, 124

nitrates 13, 126
nitratine 126
nitre 126
'noble' metals 15, 16

ochre 115, 117, 190
olivenite 176, 177
olivine 202, 203, 204, 222, 258
onyx 105, 135
opacity 8
opal 102, 104, 106, 311
orpiment 64, 65
orthoclase 9, 286, 287, 288, 289
oxides 13, 83–113, 116, 120–1, 123, 128, 132

palladium 19
peacock ore 28
pearl 313
pectolite 260
pennine 276, 277
penninite 276
peridot 203
perovskite 85
petalite 284, 309
petrified wood 311
petzite 49
phlogopite 270, 275
phosgenite 143
phosphates 13, 173–5, 178–80, 183–4, 187–90, 193–5, 197, 199
phosphophyllite 189
phosphorites 183
phyllosilicates 13, 266–82
pickeringite 161
piemontite 230
pitchblende 121
platinum 19
plattnerite 123

pollucite 309
polybasite 60
potash 78
potassium 78
nitrate 126
silicates 285–8
uranium vanadate 200
prehnite 265
proustite 57, 58
pseudoboleite 82
pseudomalachite 179
psilomelane 107, 118, 125
purpurite 173
pyragyrite 40, 57, 58
pyralspite garnets 205–7
pyrite 31, 39, 44–5, 47, 51–3
see also iron
pyritohedra 45, 52
pyrochlore 119
pyrolusite 107, 118
pyromorphite 184, 185
pyrope 205, 206
pyrophanite 98
pyrophyllite 267
pyroxenes 244–51
pyrrhotite 31, 39

quartz 7, 9, 100–1, 103, 104, 109, 202, 281

radium 121
rammelsbergite 66
rare earths 85, 113, 119, 124, 174, 220, 231
realgar 64, 65
red lead 92, 169
red oxide of iron 99
rhodochrosite 131, 261
rhodonite 261
riebeckite 253, 254
rock crystal 101

rock salt 70
romanechite 125
römerite 168
rosasite 141
rubidium 272, 309
ruby 88, 94, 95, 232
ruby copper 84
ruby silver 57, 58, 60
ruby zinc 83
rutile 109, 110

sal-ammoniac 11, 76
sanidine 286, 290
sapphire 88, 95, 234
satin spar 151
scapolite 295
scheelite 170
scolecite 297, 299
senarmontite 97
sepiolite 278
serpentine 128, 263, 278, 280, 282
siderite 129
silicates 13
cyclosilicates 13, 234–43
inosilicates 13, 244–65
neosilicates 13, 201–24
phyllosilicates 13, 266–82
sorosilicates 13, 225–33
tectosilicates 13, 283–309
silicon oxides 100–6
sillimanite 212
silver 15, 31
halogenides 72–3
sulphides 29, 37, 40, 49, 57–60
skutterudite 55
smaltite 55
smithsonite 130, 140, 227
'snowflake' twins 137

sodalite	224, 292, 293
sodium	
carbonate	144, 145
chloride	70
fluoride	71
nitrate	126
silicates	248, 253–4, 290, 296
sulphates	150, 160, 164
sorosilicates	13, 225–33
spangolite	156
specific gravity	9
spessartine	207
sphalerite	30, 34, 35, 140
sphene	217
spinel	88, 90, 91
spodumene	247
stannite	33
staurolite	216
stephanite	59
stibnite	21, 46, 48, 64
stilbite	302, 303
	streak 7
strengite	187, 188
strontianite	134, 136
strontium	136, 153, 298
sublimate	76
sulphates	13, 150–68
sulphides	13, 27–69
native elements from	21, 22
sulphur	26, 44, 64
sulphuric acid	26, 44, 52, 152
sylvanite	49
talc	9, 267, 268
tantalite	112
tantalum	124
tectosilicates	13, 283–309
tellurium	49
tennanite	32

tenorite	86
tetrahedrite	32
thallium	62
thenardite	150
thomsonite	298
thorianite	120
thorium	120, 124, 174, 211, 220, 231
tin	33, 108
tincalconite	146
titanite	217
titanium	98, 109, 110, 111
topaz	9, 215
tourmaline	240–1, 290
transparency	8
tremolite	255
tridymite	103
trona	144
tungstates	170–1
turquoise	195, 281
ugrandite garnets	208–10
ulexite	147, 148
ullmannite	51
uraninite	121, 122, 199, 200
uranium	119, 121, 122, 124, 199, 200, 220
uvarovite	209
valentinite	97
vanadates	182, 186, 200
vanadinite	184, 186
variscite	187, 188
vermiculite	275
vesuvianite	233
villiaumite	71
vivianite	190
wad	107, 125
wavellite	194
white arsenic	96
willemite	201, 227

witherite	134
wolframite	171
wollastonite	259
wulfenite	123, 172
wurtzite	34
yellow lead	172
yttrium	74, 113, 220, 231
zeolites	265, 283, 296–309
zinc	
arsenate	177
carbonates	130, 140–2
copper deposits	141–2
oxides	83, 87
silicates	201, 227
sulphides	30, 34, 62–3
vanadate	182
zincite	83, 140
zinnwaldite	273
zircon	211
zoisite	232

HAPPY AS TWO THIEVES

"Kristin—what is it?" he asked fearfully.

She made no answer, but stood as if she were listening for something. Her eyes were far away and strange.

Now she felt it again. Deep down within her she felt as though a fish moved its tail. And again it was as if the whole earth swayed around her, and she gr— dizzy and weak, but less now than at first.

"W— ld that grew quick within you?" he — w voice, touching her shoulder.

— d with shining eyes up in his wife's — ned and happy, Kristin bent her head — d to hide from him her smile and her eyes.

— ow," he said gaily, "now will we ride home Husaby, my Kristin, and be as happy as two thieves!"

Bantam Books by Sigrid Undset
Ask your bookseller for the books you have missed

KRISTIN LAVRANSDATTER I:
   THE BRIDAL WREATH
KRISTIN LAVRANSDATTER II:
   THE MISTRESS OF HUSABY
KRISTIN LAVRANSDATTER III:
   THE CROSS

KRISTIN LAVRANSDATTER

by
Sigrid Undset

II

The Mistress of Husaby

Translated by Charles Archer

BANTAM BOOKS
TORONTO · NEW YORK · LONDON · SYDNEY · AUCKLAND

KRISTIN LAVRANSDATTER
II: THE MISTRESS OF HUSABY

*A Bantam Book / published by arrangement with
Alfred A. Knopf, Inc.*

PRINTING HISTORY

*Original title: Husfrue; copyright 1921 by
H. Aschehoug & Company, Oslo
Translated by Charles Archer*

*Knopf edition published March 1925
18 printings through October 1976*

Bantam edition / October 1978
2nd printing .. December 1978   4th printing ..... March 1982
3rd printing ...... May 1979   5th printing ...... April 1983
6th printing .. November 1984

Printed in Canada

H   15 14 13 12 11 10 9 8 7 6

THE MISTRESS OF HUSABY

PART ONE

THE FRUIT OF SIN

THE EVENING before Simon's Mass,* Baard Peters-sön's galleass lay in to the landing-place at Birgsi. Abbot Olav of Nidarholm had himself ridden down to the strand to greet his kinsman, Erlend Nikulaus-sön, and bid welcome to the young wife he was bringing with him home. The pair were to be the Abbot's guests, and to sleep at Vigg that night.

It was a deathly pale, woebegone young wife that Erlend led shoreward from the pier. The Abbot spoke jestingly of the pains of the sea-voyage; Erlend laughed and said he well believed his wife longed for nothing so much as to lie once more in a bed well fixed into a house-wall. And Kristin strove to smile; but within herself she thought that never, so long as she lived, would she willingly set foot on shipboard again. She turned sick if Erlend so much as came near her, he smelt so of the ship and of the sea — his hair was all matted and sticky with sea-water. He had been crazy with joy all the time on board — and Sir Baard had laughed: at his home in Möre, where Erlend had grown up, the boys had been out in the boats sailing or rowing late and early. 'Twas true they had been a little sorry for her, Erlend and Sir Baard, but not so sorry, Kristin thought, as her wretchedness deserved. They kept on saying the sea-sickness would pass when she grew used to being aboard ship. But from first to last her misery had not abated.

Even the next morning she felt as if she were still sailing, as she rode up through the settled lands. Uphill and down-hill, their road led over great, steep clay ridges, and when she tried to fix her eyes on some spot on the hills far ahead, it was as though the whole country-side were dipping, then rising, in waves cast up against the shining, blue-white win-ter-morning sky.

---

* 28th October.

A whole troop of Erlend's friends and neighbours had come to Vigg in the early morning to attend the bridal pair, so that they rode in a great company. The ground sounded hollow under the horses' hoofs, for the earth was as hard as iron with the black frost. The air was full of steam from the men and horses; the bodies of the beasts and the men's hair and furs were white with rime. Erlend seemed as white-haired as the Abbot; his face glowed from his morning draught and the biting wind. He wore his bridegroom's dress to-day; youth and gladness seemed to shine out of him, and joy and wantonness welled out in the tones of his mellow, supple voice as he rode, shouting and laughing, amidst his guests.

Kristin's heart began to tremble strangely — with sorrow, with tenderness and with fear. She was still sick from the voyage; she had the burning pain at her breast that came now whenever she had eaten or drunk never so little; she was bitterly cold; and deep down in her mind was a dull, dumb spot of anger with Erland, that he could be so gay and care-free. . . . And yet, now that she saw his child-like pride and sparkling happiness in bringing her home as his wife, a bitter regret welled up in her; her breast ached with pity for him. She wished now that she had not hearkened to the counsel of her own self-will, but had let Erlend know when he was at her home last summer — let him know what made it most unfit that their wedding should be held with too great pomp. She saw now that she had wished he should be made to feel — he too — that they could not escape unhumbled from what they had done.

— And she had been afraid of her father, too. And she had thought in her mind: when once their bride-ale had been drunk, they were to journey so far off; 'twas like she would not see her home-country again for a long, long time — not till all talk about her had had good time to die away. . . .

Now she saw that things here would be much worse than she had deemed. True, Erlend had spoken of the great house-warming he would hold at Husaby, but she had not thought it would be like a second wedding-feast. And the guests here were the folk that Erlend and she were to live among — it was their respect and friendship they had need to win. It was these folk that had had

Erlend's folly and evil fortunes before their eyes all these years. Now he himself believed that he had redeemed himself in their judgment, that now he could take the place among his fellows that was due to his birth and fortune. And now 'twas like he would be a laughing-stock through this whole country-side when it came out that he had done amiss with his own betrothed bride.

The Abbot leant over towards her from his horse:

"You look so sad, Kristin Lavransdatter; are you not quit of your sea-sickness yet? Or is it, perhaps, that you are home-sick for your mother?"

"Even so, sir," said Kristin softly; " 'tis of my mother I am thinking."

They had come up into Skaun, and were riding high on a hill-side. Below them in the valley bottom the woods stood white and shaggy with rime; everywhere the sunlight glittered, and a small lake in the midst flashed blue. Then all at once the troop passed out from a little pinewood, and Erlend pointed ahead:

"There lies Husaby,* Kristen. God grant you many happy days there, my own wife!" he said, with a thrill in his voice.

Before them stretched broad plough-lands, white with rime. The manor stood, as it were, on a broad shelf midway on the hill-side — nearest them lay a small church of light-coloured stone, and just south of it were the clustered houses; they were many and great; the smoke whirled up from their smoke-vents. Bells began ringing from the church, and many folk came streaming from the courtyard to meet them, with shouts of greeting. The young men in the bridal trains clashed their weapons one on another — and with a great clattering and the thunder of hoofs and joyous uproar the troop swept forward towards the new-married man's abode.

They stopped before the church. Erlend lifted his bride from her horse, and led her forward to the church-door, where a little crowd of priests and clerks stood waiting to

---

* See Note 1.

welcome them. Within, it was bitterly cold, and the daylight, sifting in through the small round-arched windows of the nave, dulled the shine of the tapers burning in the choir.

Kristin felt lost and afraid when Erlend loosed her hand and went over to the men's side, while she herself took her place among the throng of strange women, all in festal dress. The service was most goodly. But Kristin was very cold, and it seemed as though her prayers were blown back upon her when she tried to free her heart and lift it upwards. She thought maybe it was no good omen that this should be St. Simon's day — the guardian saint of the man by whom she had done so ill.

From the church all the people went in procession down to the manor, the priests first, then Kristin and Erlend hand in hand, then their guests pair by pair. Kristin was not enough herself to see much of the manor buildings. The courtyard was long and narrow; the houses lay in two rows, south and north of it. They were big and built close together; but they looked old and ill-tended.

The procession halted at the door of the hall-house, and the priests blessed it with holy water. Then Erlend led her through a dark outer room. On her right a door was thrown open, letting out a flood of light. She bent, passing through the doorway, and stood with Erlend in his hall.

It was the greatest room* she had ever seen in any man's dwelling-house. There was a hearth-place in the midst of the floor, and it was so long that there were two fires on it, one at each end; and the room was so broad that the crossbeams were borne up on carven pillars — it seemed to her more like the body of a church or a king's hall than a room in a manor-house. Up by the eastern gable-end, where the high-seat stood in the middle of the wall-bench, closed box-beds were built in between the timber pillars.

And what a mass of lights were burning in the hall — on the tables, that groaned with costly cups and vessels, and on sconces fastened to the walls! After the fashion of the old age, weapons and shields hung amidst the stretched-out tapestries. Behind the high-seat the wall was covered with a velvet hanging, and against it a man was even now fas-

---

* See Note 2.

tening up Erlend's gold-mounted sword and his white shield with the red lion salient.

Serving-men and women had taken the guests' outer garments from them. Erlend took his wife by the hand and led her forward to the hearth, the guests standing in a half-ring just behind them. A fat lady with a gentle face came forward and shook out Kristin's head-linen, where it had been crumpled by the hood of her cloak. As she stepped back into her place, she nodded to the young couple and smiled; Erlend nodded and smiled back to her, and looked down at his wife — his face, as he looked, was beautiful. And again Kristin felt her heart sink — with pity for him. She knew what he was thinking now, as he saw her standing there in his hall with the long snow-white linen coif over her scarlet bridal dress. And this morning she had had to wind a long woven belt tight around her waist under her clothes before she could get the dress to fit upon her, and she had rubbed upon her cheeks some of a red salve that Lady Aashild had given her; and while she thus bedecked herself, she had thought in sorrow and bitterness that Erlend must look but little upon her, now that he had her safely his own — since he still saw and knew nothing. Bitterly she repented now that she had not told him.

While the married pair stood thus, hand in hand, the priests were walking the round of the hall, blessing house and hearth and bed and board.

Next a serving-woman bore forth the keys of the house to Erlend. He hooked the heavy bunch on Kristin's belt — and looked, as he did so, as though he had been fain to kiss her where she stood. A man brought a great horn ringed about with golden rings — Erlend set it to his lips and drank to her:

"Hail, and welcome to thy house, Lady of Husaby!"

And the guests shouted and laughed while she drank with her husband and poured out the rest of the wine on the hearth-fire.

Then the minstrels struck up their music, as Erlend Nikulaussön led his wedded wife to the high-seat, and the wedding guests took their seats at the board.

On the third day the guests began to break up, and by the hour of nones on the fifth day the last of them were gone, and Kristin was alone with her husband at Husaby.

The first thing she did was to bid the serving-folk take all the bed-gear out of the bed, wash it and the walls round about it with lye, and carry out and burn up the straw. Then she had the bedstead filled with fresh straw, and above it made up the bed with bed-clothes from the store she had brought with her. It was late in the night before this work was at an end. But Kristin gave order that the same should be done with all the beds on the place, and that the skin rugs should be well baked in the bath-house — the maids must set to the work in the morning the first thing, and get as much done towards it as they could before the Sunday holiday. Erlend shook his head and laughed — she was a housewife indeed! But he was not a little ashamed.

For Kristin had not had much sleep the first night, even though the priests had blessed her bed. 'Twas spread above with silken pillows, with sheets of linen and the bravest rugs and furs; but beneath was dirty, mouldy straw, and there were lice in the bed-clothes and in the splendid black bearskin that was spread over all.

Many things had she seen already in these few days. Behind the costly tapestry hangings, the unwashed walls were black with dirt and soot. At the feast there had been masses of food, but much of it spoilt with ill dressing and ill service. And to make up the fires they had had naught but green and wet logs, that would scarce catch fire, and that filled the hall with smoke.

Everywhere she had seen ill husbandry, when on the second day she went round with Erlend and looked over the manor and farm. By the time the feasting was over, little would be left in barn and storehouse; the corn-bins were all but swept clean. And she could not understand how Erlend could think to keep all the horses and so many cattle through the winter on the little hay and straw that was in the barns — of leaf-fodder there was not enough even for the sheep and goats.

But there was a loft half full of flax that had been left lying unused — there must have been the greatest part of many years' harvest. And then a storehouse full of old, old unwashed and stinking wool, some in sacks and some lying loose in heaps. When Kristin took up a handful, a shower of little brown eggs fell from it — moth and maggots had got into it.

The cattle were wretched, lean, galled and scabby; and never had she seen so many aged beasts together, in one place. Only the horses were comely and well-tended. But, even so, there was no one of them that was the equal of Guldsveinen or of Ringdrotten, the stallion her father had now. Slöngvanbauge, the horse he had given her to take along with her from home, was the fairest beast in the Husaby stables. When she came to him, she had to go and throw her arms round his neck and press her face against his cheek.

She thought on her father's face, when the time came for her to ride away with Erlend and he lifted her to the saddle. He had put on an air of gladness, for many folk were standing round them; but she had seen his eyes. He stroked her arm downwards, and held her hand in his for farewell. At the moment, it might be, she had thought most how glad she was that she was to get away at last. But now it seemed to her that as long as she lived her soul would be wrung with pain when she remembered her father's eyes at that hour.

And so Kristin Lavransdatter began to guide and order all things in her house. She was up at cock-crow every morning, though Erlend raised his voice against it, and made as though he would keep her in bed by force — surely no one expected a newly-married wife to rush about from house to house long before 'twas daylight.

When she saw in what an ill way all things were here, and how much there was for to set her hand to, a thought shot through her clear and hard; if she had burdened her soul with sin that she might come hither, let it even be so — but 'twas no less sin to deal with God's gifts as they had been dealt with here. Shame upon the folk that had had the guidance of things here, and on all them that had let Erlend's goods go so to waste! There had been no fit steward at Husaby for the last two years; Erlend himself had been much away from home in that time, and besides, he understood but little of the management of the estate. 'Twas no more than was to be looked for, then, that his bailiffs in the outlying parishes should cheat him, as she was sure they did, and that the serving-folk at Husaby should work only as much as they pleased, and when and how it chanced to suit them. 'Twould be no light task for her to put things right again.

One day she talked of these things with Ulf Haldorssön, Erlend's own henchman. They ought to have had the threshing done by now, at least of the corn from the home farm — and there was none too much of it either — before the time came to slaughter for winter meat. Ulf said:

"You know, Kristin, that I am not a farm-hand. It has been our place to be Erlend's arms-bearers — Haftor's and mine — and I have no skill in husbandry any longer."

"I know it," said the mistress of the house. "But so it is, Ulf, that 'twill be no easy task for me to guide things here this winter, a newcomer as I am here north of Dovre, and with no knowledge of our folks. 'Twould be a friend's turn of you if you would help and counsel me."

"I can well believe it, Kristin — that you will have no easy task this winter," said the man, looking at her with a little smile — the strange smile that was always on his face when he spoke with her or with Erlend. It was bold and mocking, and yet there were both kindness and a sort of respect for her in his bearing. Nor, it seemed to her, had she a right to take offence that Ulf should bear himself more forwardly towards her than might have been seemly otherwise. She herself and Erlend had made this serving-man a party to their wanton and deceitful doings; and she could see that he knew, too, how things stood with her now. She must let this pass — and indeed she saw that Erlend put up with anything Ulf might say or do, and that the man showed but little reverence for his master. True, they had been friends from childhood; Ulf came from Möre, and was son to a small farmer that lived near Baard Petersson's manor. He called Erlend by his name; and her, too, now — it was true that this way of speech was commoner here north of Dovre than in her own country.

Ulf Haldorssön was a proper man, tall and dark, with sightly eyes; but his mouth was ugly and coarse. Kristin had heard ugly tales of him from the maids on the place — when he was in at the city he drank beyond measure and spent his time in revel and roistering in the lowest houses of call — but at home at Husaby he was the best man to have at one's beck, the fittest, the hardest worker, and the shrewdest. Kristin had come to like him well.

" 'Twere no easy thing for any woman," he went on, "to come hither to this house — after all that has come and gone. And yet, Mistress Kristin, I deem that you will win through it better than most could have done. You are not the woman to sit down and moan and whimper; but you will set your thoughts on saving your children's inheritance yourself, since none else here takes thought for such things. And methinks you know that you can trust me, and that I will help you as far as in me lies. You must bear in mind that I am unused to farm work. But if you will take counsel with me and let me come to you for counsel, I trow we will tide over this winter none so ill."

Kristin thanked Ulf, and went into the house.

She was heavy at heart with unrest and fear, but she tried to forget it in work. One thing was that she understood not Erlend — even now he seemed to suspect nothing. But another and a worse trouble was that she could feel no life in the child she bore within her. At twenty weeks it should quicken, she knew — and now more than three weeks over the twenty had gone by. She lay awake at night and felt the burden within her that grew greater and heavier, but was still as dull and lifeless as ever. And there floated through her mind all she had heard of children that were born crippled, with sinews stiff as stone, of births that had come to the light without limbs — with scarce a semblance of human shape. Before her tight-shut eyes would pass pictures of little infants, dreadfully misshapen; one shape of horror melting into another still worse. Southward in the dale at home, at Lidstad, the folks had a child — nay, it must be grown up now. Her father had seen it, but would never speak of it; she had marked that he grew ill at ease if anyone but named aught of it. What did it look like? — Oh, no! Holy Saint Olav, pray for me! — She must needs trust firmly on the holy King's tender mercy; had she not placed her child under His ward? She would suffer for her sins in meekness, and with her whole heart have faith that there would be help and mercy for the child. It must be the Enemy himself that tempted her with these ugly visions, to drive her to despair. But her nights were evil. . . . If a child had no limbs, if it were palsied, like enough the mother would feel no sign of life within her. . . . Erlend, half waking, marked that his wife was restless, drew her closer into his arms, and laid his face against the hollow of her throat.

But by day she showed no sign of trouble. And every morning she dressed her body with care so as to hide from the house-folk yet a little longer that she bore another life about with her.

It was the custom at Husaby that after the evening meal the serving-folk went off to the houses where they slept; so that she and Erlend were left alone in the hall. Altogether the ways of this manor were more as they had been in the ancient days, when folks kept thralls and bonds-women for the household work. There was no fixed table in the hall, but morning and evening the meals were spread on a great board that was laid on trestles, and after the meal it was hung up again on the wall. At the other meals folks took their food over to the benches and sat and ate it there. Kristin knew such had been the custom in former times. But nowadays, when 'twas hard to find men to serve at table, and all folks had to content them with maids for indoor work, it fitted the times no longer — the women were loath to break their backs lifting the heavy tables. Kristin remembered her mother telling how at Sundbu they had put a fixed table into the hall when she was but eight winters old, and that the women thought it in every way the greatest boon — they need no longer take all their sewing out to the women's house, but could sit in the hall and clip and cut out — and it made such a goodly show to have candlesticks and a few costly vessels standing out in view. Kristin thought: next summer she would pray Erlend to have a fixed table set up along the northern wall.

So it was at her home, and there her father had his high-seat at the board's end — but then the beds there were by the entrance wall. At home her mother sat highest up on the outer bench, so that she could go to and fro and keep an eye on the service of the food. Only when there was a feast did Ragnfrid sit by her husband's side. But here the high-seat stood in the middle of the eastern gable-end, and Erlend would have her always sit in it with him. At home

her father always placed God's servants in the high-seat,
if any such were guests at the manor, and he himself and
Ragnfrid served them while they ate and drank. But
Erlend would have none of this, unless they were high of
station. He was no great lover of priests and monks — they
were costly friends, he was used to say. Kristin could not
but think of what her father and Sira Eirik always said,
when folk complained of the churchmen's greed of
money: men forgot the sinful joys they had snatched for
themselves when the time came to pay for them.

She questioned Erlend about the life here at Husaby in
ancient days. But he knew strangely little. Things were
thus and thus, he had heard; but he could not remember so
nicely. King Skule had owned the manor and built on it
— 'twas said he had meant to make Husaby his dwelling-
place, when he gave away Rein for a nunnery. Erlend was
right proud of his descent from the Duke, whom he always
called King, and from Bishop Nikulaus; the Bishop was
the father of his grandfather, Munan Bishopsson. But it
seemed to Kristin that he knew no more of these men than
she herself knew already from her father's tales. At home
it was otherwise. Neither her father nor her mother was
overproud of the power of their forbears and the high
esteem they had enjoyed. But they spoke often of them;
held up the good that they knew of them as a pattern, and
told of their faults and the evil that had come of them as
a warning. And they had little tales of mirth too — of Ivar
Gjesling the Old and his quarrel with King Sverre; of
Ivar Provst's quick and witty sallies; of Haavard Gjesling's
huge bulk; and of Ivar Gjesling the Young's wonderful
luck in the chase. Lavrans told of his grandfather's brother
that carried off the Folkunga maid from Vreta cloister; of
his grandfather's mother Ramborg Sunesdatter, who
longed always for her home in Wester Gothland and at
last went through the ice and was lost, when driving on
Lake Vener one time she was staying with her brother at
Solberga. He told of his father's prowess in arms, and of
his unspeakable sorrow over his young first wife, Kristin
Sigurdsdatter, that died in childbirth when Lavrans was
born. And he read, from a book, of his ancestress the holy
Lady Elin of Skövde, who was given grace to be one of
God's blood-witnesses. Her father had often spoken of

making a pilgrimage with Kristin to the grave of this holy widow. But it had never come to pass.

In her fear and distress, Kristin tried to pray to this saint that she herself was linked to by the tie of blood. She prayed to St. Elin for her child, kissing the reliquary that she had had of her father; in it was a shred of the holy lady's shroud. But Kristin was afraid of St. Elin, now when she had brought such shame on her race. When she prayed to St. Olav and St. Thomas for their intercession, she often felt that her complaints found a way to living ears and merciful hearts. These two martyrs for righteousness her father loved above all other saints; above even St. Laurentius himself, though this was the saint he was called after, and in honour of whose day in the late summer he always held a great drinking-feast and gave richly in alms. St. Thomas her father had himself seen in his dreams one night when he lay wounded outside Baagahus. No tongue could tell how lovely and venerable he was to look on, and Lavrans himself had been able to say naught but "Lord! Lord!" But the radiant figure in the Bishop's raiment had gently touched his wounds and promised that he should have his life and the use of his limbs, so that he should see again his wife and his daughter, according to his prayer. But at that time no man had believed that Lavrans Björgulfsön could live the night through.

Aye, said Erlend. One heard of such things. Naught of the kind had ever befallen him, and to be sure 'twas not like that it should — for he had never been a pious man, such as Lavrans was.

Then Kristin asked of all the folk who had been at their homecoming feast. Erlend had not much to say of them either. It seemed to Kristin that her husband was not much like the folks of this country-side. They were comely folk, many of them, fair and ruddy of hue, with round hard heads and bodies strong and heavily built — many of the older folks were hugely fat. Erlend looked like a strange bird among his guests. He was a head taller than most of the men, slim and lean, with slender limbs and fine joints. And he had black silky hair and was pale brown of hue — but with light-blue eyes under coal-black brows and long black eyelashes. His forehead was high and narrow, the temples hollow, the nose somewhat too great and the

mouth something too small and weak for a man — but he
was comely none the less; she had seen no man that was
half so fair as Erlend. Even his mellow, quiet voice was
unlike the others' thick full-fed utterance.

Erlend laughed and said his forbears were not of these
parts either — only his grandfather's mother, Ragnfrid
Skulesdatter. Folks said he was much like his mother's
father, Gaute Erlendssön of Skogheim. Kristin asked what
he knew about this grandfather. But it proved to be almost
nothing.

One night Erlend and Kristin were undressing in the
hall. Erlend could not get his shoe-latchet unloosed; as he
cut it, the knife slipped and gashed his hand. He bled
much and swore savagely. Kristin fetched a piece of linen
from her chest. She was in her shift. As she was binding
up his hand, Erlend passed his other arm around her waist.

Of a sudden he looked down into her face with fear and
confusion in his eyes, and his face grew red as fire. Kristin
bowed her head. Erlend took away his arm, saying nothing
— then Kristin went off in silence and crept into the bed.
Her heart beat with hard dull strokes against her ribs.
Now and again she looked over at her husband. He had
turned his back to her, and was slowly drawing off one
garment after another. At last he came to the bed and lay
down.

Kristin waited for him to speak. She waited so, that at
times 'twas as though her heart no longer beat, but only
stood still and quivered in her breast.

But Erlend said no word. Nor did he take her in his
arms. At last, falteringly, he laid a hand across her breast
and pressed his chin down on her shoulder so strongly that
the stubble of his beard pricked her skin. As he still spoke
not a word, Kristin turned to the wall.

It was as though she were sinking, sinking. Not a word
could he find to give her — now when he knew that she
had borne his child within her all this long weary time.
She clenched her teeth hard in the dark. Never would she
beg and beseech — if he chose to be silent, she would be
silent too, even, if need be, till the day she bore his child.
Bitterness surged through her heart; but she lay stock-still
against the wall. And Erlend too lay still in the dark. Hour

after hour they lay thus, and each knew that the other was not sleeping. At last she heard by his even breathing that he had fallen asleep, and then she let the tears flow as they would, in sorrow and bitterness and shame. Never, it seemed to her, could she forgive him this.

For three days Erlend and Kristin went about thus — he like a wet dog, the young wife thought. She was hot and hard with wrath — she grew wild with rage when she marked that he would look searchingly at her and then hastily look away again if she turned her eyes towards him.

On the morning of the fourth day, as she sat in the hall, Erlend came in through the doorway, dressed for riding. He said he was going westward to Medalby; maybe she would come with him and see the place; it was one of the farms that fell under her morning-gift. Kristin said yes; and Erlend helped her himself to put on her long shaggy boots and the black sleeve-cloak with the silver clasps.

In the courtyard were four horses ready saddled, but Erlend said now that Haftor and Egil might stay at home and help with the threshing. Then he helped his wife up into the saddle. Kristin felt that 'twas in Erlend's mind to speak now of what lay between them unuttered. Yet he said naught as they rode slowly southward towards the woods.

It was far on now in the early winter, but no snow had yet fallen in this country-side. The day was fresh and fair, the sun just risen, and the white rime glittered in silver and gold everywhere, on the fields and on the trees. They were riding over the Husaby lands. Kristin saw that there was little sown or stubble land, but mostly fallows left for grass, and old meadow-land, uneven, moss-grown, and choked with alder-shoots. She spoke of this.

Her husband answered jauntily:

"Know you not, Kristin, you that have such skill in guiding goods and gear, that it profits not to raise corn so near to a great market? — a man does better by bartering his wool and butter for the outland merchants' corn and flour — "

"Then should you have bartered away all that wool that lies now in your lofts and is long since spoiled," said

Kristin. "But so much I know, that the law says every man that leases land shall sow corn on three parts of it, and let the fourth part lie fallow for grass. And 'tis not fit that the landlord's manor should be worse cared for than the tenants' farms — so my father always said."

Erlend laughed a little, and answered:

"I have never searched out the law in that matter — so long as I have my dues, my tenants can till their farms as likes them best, and, for Husaby, I manage it as seems to me best and fittest."

"Would you be wiser, then," asked Kristin, "than our fathers that went before us, and St. Olav and King Magnus that made these laws?"

Erlend laughed again and said:

" 'Tis a matter I have never thought on — but the devil and all must be in it, Kristin, that you have the laws of the land so at your finger-ends — "

"I know a little of these things," said Kristin, "because my father often prayed Sigurd of Loptsgaard to say over the laws to us when he came to visit us and we sat at home of an evening. Father deemed it profitable for the servants and the young folk to learn somewhat of such things; and so Sigurd would repeat one passage or another."

"Sigurd — " said Erlend. "Aye, now I remember seeing him at our wedding. He was that long-nosed toothless old fellow that wept and drivelled and patted you on the breast — he was drunk as an owl even the next morning, when the folks came up to see me set the linen coif on your head — "

"He has known me from before I can remember," said Kristin angrily. "He used to take me on his lap and play with me when I was a little maid — "

Erlend laughed again:

"Well, 'twas a strange pastime enough — for you all to sit there and listen to that old fellow chanting out the laws, part by part. Sure Lavrans is in all ways unlike other men — others are used to say that if the peasant knew the laws of the land in full, and the stallion knew his strength, 'twould take the devil to be a knight."

Suddenly, with a cry, Kristin struck her horse on the quarter and dashed on, leaving Erlend gazing after his wife in wonderment and anger.

Of a sudden he put spurs to his horse. Christ — the fjord — there was no crossing over it now — the clay bank had slipped in the autumn.

Slöngvanbauge stretched himself to gallop the harder when he heard the other horse behind him. Erlend was in deadly fear — how she was dashing down the steep hillsides too! At last he tore past her through the undergrowth, then swung into the road where it ran level for a little way, and stood so that she must needs stop. When he came alongside her, he saw that she seemed a little frightened herself now.

Erlend leant forward towards his wife and struck her a ringing blow under the ear—so that Slöngvanbauge leapt aside and reared in fright.

"Aye, and you deserved it," said Erlend in a shaken voice, when the horses had quieted down and they were riding side by side again. "To carry on so — clean crazy with rage. You frighted me — "

Kristin held her head so that he could not see her face. Erlend was wishing that he had not struck her. But he said again:

"You made me afraid, Kristin — to behave so! And of all times, now — " he added in a low voice.

Kristin neither answered nor looked at him. But Erlend could feel that she was less angry now than before, when he had mocked at her home. He wondered much at this — but he saw that so it was.

They came to Medalby, and Erlend's tenant came out and would have them into the dwelling-house. But Erlend said 'twere well they should look round the farm-buildings first — and Kristin must come with them. "The farm is hers now — and she understands these things better than I, Stein," he said, laughing. There were some other farmers there, come to act as witnesses — some of them too were Erlend's tenants.

Stein had come to the farm last term-day, and ever since he had been praying that the landlord would come up and see the state of the houses when he took them over, or would send men to act for him. The other farmers bore witness that not one of the houses had been weather-tight, and that those which now were tumble-down had been no better when Stein came. Kristin saw that it was a good farm, but that it had been ill cared for. She could see that this Stein was a hard-working man. Erlend, too, was reasonable and promised him some relief in his rents, till such time as he had got the houses mended.

Then they went into the hall, and found the board set out with good food and strong ale. The farmer's wife begged Kristin to forgive that she had not come out to meet her. She said her husband would not suffer her to go out under open sky till she had been churched after her childbed. Kristin greeted the woman kindly, and had her take her over to the cradle to look at the child. It was these people's first-born: a son twelve days old, and big and sturdy.

Next Erlend and Kristin were led to the high-seat, and all the folks sat down, and ate and drank a good while. Kristin was the one that talked most during the meal; Erlend said little, and the peasants not much; but Kristin thought she could mark that they liked her well.

Then the child awoke, and began first to whimper, and then to shriek so fearfully that the mother had to fetch it and give it the breast to stop its cries. Kristin looked more than once across at the two, and when the boy was full-fed and quiet, she took him from the woman and laid him on her arm.

"Look, husband!" she said. "Is not this a fair and lusty knave?"

"Doubtless it is," said her husband, not looking towards them.

Kristin sat holding the child a little before she gave it back to the mother.

"I will send over a gift for this little son of yours, Arndis," she said, "for that he is the first child I have held in my arms since I came hither north of Dovre."

Hot and defiant, with a little smile, she looked at her husband and then along the row of peasants on the bench. There was the least little twitching at the corners of the mouths of one or two of them; but immediately they stared before them, stiff with solemnity. Then stood up a very old fellow who had drunk well already. He took the ladle out of the ale-bowl, laid it on the table and lifted the heavy vessel aloft:

"Then will we pledge you, mistress, on a wish: that the next child you hold in your arms may be the new master of Husaby!"

Kristin rose up and took the heavy bowl. First she held it out to her husband. Erlend but touched it with his lips, but Kristin drank deep and long.

"Thanks for that good wish, Jon o' the Woods," she said, nodding to him, her face shining and gleeful. Then she passed on the bowl.

Erlend sat there darkly flushed and, Kristin could see, in great wrath. She herself felt naught now but an unthinking need to laugh and be glad. Some time after, Erlend gave the sign for breaking up, and they set out on their homeward way.

They had ridden a good way in silence, when Erlend broke out of a sudden:

"Think you it was needful to let our very peasants know you were with child when you were wedded? You may stake your soul that 'twill be no time now ere the tale about us two is all over every parish by the Trondheim Fjord. . . ."

Kristin made no answer at first. She looked straight forth over her horse's head, and her face grew so white that Erlend was afraid.

"As long as I live, I shall not forget," she said at last, without looking at him, "that this was your first greeting to your young son that is beneath my girdle."

"Kristin!" said Erlend beseechingly. "My Kristin," he implored, when she answered not, nor looked at him. "Kristin!"

"Sir?" she said in cold, measured tones, without turning her head.

Erlend swore furiously, set spurs to his horse and dashed forward along the road. But, a little after, he came riding back to meet her.

"Now had you vexed me so sorely," he said, "that *I* was nigh riding off and leaving *you*."

"And if you had," answered Kristin quietly, "it might have been that you had had long and long to wait ere I came after you to Husaby."

"How you talk!" said the man despairingly.

Again they rode for a space without speaking. In a while they came to a place where a bridle-path led off over a ridge. Erlend said to his wife:

"I had meant that we should ride home by this way over the hill — 'tis a little farther, but I had a mind to take you up here with me some time."

Kristin nodded listlessly.

In a little, Erlend said it would be better they should go on foot. He tied their horses to a tree.

"Gunnulf and I had a fort on the hill-top here," he said. "I would like well to see if any of our castle is left — "

He took her hand. She let him hold it, but walked with her eyes cast down, looking to her footing. It was not long before they reached the top. Over the rime-covered woods in the gorge of the little stream they saw Husaby on the hill-side right over against them, lying wide-stretched and brave, with its stone church and the many great houses, wide plough-lands around it and dark pine-clad ridges behind.

"Mother," said Erlend in a low voice, "she would come with us up here — often. But always she sat gazing south, up towards the Dovrefjeld. I trow she longed both early and late to be gone from Husaby. Or sometimes she would turn to the north and look towards the hill-glen where you see the far-off blue — the hills beyond the fjord. Never did she look across at Husaby."

His voice was soft and beseeching. But Kristin neither spoke nor looked at him. Soon he went off and began kicking the frozen heather:

"No, I can see there's naught left here of Gunnulf's and my stronghold. True enough, 'tis many a long day since we played about here, Gunnulf and I — "

There was no answer. — Right below where they stood lay a little frozen pool — Erlend took up a stone and threw it down on to the ice. The pool was frozen to the bottom, so the stone did but make a small white star on the black mirror. Erlend took another stone and threw harder — then another and yet another, till at last he was showering down stones furiously, bent on splintering the ice to shards. Then he caught sight of his wife's face — she stood there with eyes dark with scorn, smiling disdainfully at his childishness.

Erlend turned sharp round — but at the same moment Kristin grew deadly pale, and her eyelids closed. She stood clutching in the air with her hands, swaying as if about to fall — then caught the trunk of a tree and held to it.

"Kristin — what is it?" he asked fearfully.

She made no answer, but stood as if she were listening for something. Her eyes were far away and strange.

Now she felt it again. Deep down within her she felt as though a fish moved its tail. And again it was as if the whole earth swayed around her, and she grew dizzy and weak, but less now than at first.

"What is amiss with you?" said Erlend once again.

She had waited so for this — hardly daring to acknowledge to herself the anguish of her waiting. She could not speak of it — now, when they had been unfriends this whole day. But then *he* said it:

"Was it the child that grew quick within you?" he asked in a low voice, touching her shoulder.

At that she cast from her all her wrath against him, and clung to her child's father and hid her face in his breast.

Soon after, they went down again to the place where their horses were tied. The short day was nearly done; behind them in the southwest the sun went down behind the tree-tops, a blurred red ball in the frost-haze.

Erlend tried his wife's saddle-girths and buckles with care before he lifted her up to the saddle. Then he went and untied his own horse. He felt under his belt for his gloves, which he had stuck in there, and found but one. He began to look about on the hill-side.

Kristin could not forbear saying:

"'Tis of no use seeking here for your glove, Erlend."

"You might as lief have told me, if you saw me lose it — though you were never so wroth with me," he said. The gloves were those Kristin had sewn for him and given him with her betrothal-gifts.

"It fell from your belt when you struck me," said Kristin very low, and with downcast eyes.

Erlend stood by his horse's shoulder, with his hand on the saddle-bow. He looked abashed and unhappy; but of a sudden he burst out laughing:

"Never had I dreamed, Kristin — in those days when I

was wooing you, running around beseeching my kinsmen to speak for me, and making me so supple-jointed and so humble to win you — that you could ever be such a troll!"

Then Kristin, too, laughed:

"No — for if you had, doubtless you had given up that quest long before — and doubtless that had been best for you."

Erlend took a couple of strides across to her and laid his hand upon her knee:

"Jesus help us, Kristin — when have you ever heard tell of me that I did the thing that was best for me — ?"

He laid his head down on her lap and looked with shining eyes up in his wife's face. Flushed and happy, Kristin bent her head and tried to hide from him her smile and her eyes.

He took her horse by the bit and let his own follow after them; so he led her till they were come down from the hill-side. Every time he looked at her, he laughed; and she turned away her head from him to hide that she was laughing too.

"Now," he said gaily when they were down again on the road, "now will we ride home to Husaby, my Kristin, and be as happy as two thieves!"

2

On Yule Eve it blew and rained in torrents. 'Twas no fit weather for sleighing, so Kristin had to stay at home when Erlend and the house-folk rode off to midnight mass at Birgsi church.

She stood in the doorway of the hall and looked after them. The fir-root torches they bore shone red on the murky old house-walls, and were mirrored in the watery glaze of the courtyard. The wind took the flames and blew them flat out sidewise. Kristin stood till the noise of their going died away in the night.

Within, in the hall, tapers burned on the board. It was littered with the leavings of the supper — slabs of porridge in platters, half-eaten bread-slices and fishbones in puddles of spilt ale. The serving-maids who were to stay at home had lain down already in their resting-places on the floor-straw. Kristin was alone on the manor with them and one old man that they called Aan. He had served at Husaby since the days of Erlend's grandfather; he lived now in a little hut down by the lake, but often came up to the manor in the day-time, and went pottering about, doing, as he thought, a deal of work. Aan had fallen asleep at the board to-night, and Erlend and Ulf had borne him off to a corner, laughing, and laid him there, covered with a rug.

By now, the floor would be strewn thick with rushes at home at Jörundgaard; for all the house-folk slept in the hall together the holy nights of Yule-tide. Ere they set forth to the church, it was their use to clear away the broken food of the fast supper, and her mother and the wenches set out the board as fairly as they could, with butter and cheeses, piles of thin, light-brown bread-slices, shining white bacon and the thickest of smoked knuckles of mutton. The silver flagons and mead-horns stood shining on the board; and her father had himself put the ale-cask up on the bench.

Kristin turned her chair round to the hearth — she would look no longer at the sluttish table. One of the girls was snoring — the sound was horrible to hear.

'Twas one of the things she could not like in Erlend — at home in his own house he ate in such a slovenly fashion, raked about in the dishes for tit-bits, and would scarce so much as wash his hands before he went to meat. And then he would let his dogs get up on his lap and snatch at bits of food while people were eating. So 'twas only what might have been looked for that the serving-folk had no manners at the board. . . .

At home she had been taught to eat daintily — and slowly. For 'twas not seemly, her mother said, that the folk of the house should sit waiting while the servants ate — and those who swinked and toiled must be given time to eat well and be filled.

"Gunna," Kristin called softly to the great yellow bitch that lay with a whole litter of whelps up against the stone border of the hearth. She was so snappish that Erlend had called her after the ill-tempered old lady of the house at Raasvold.

"Poor old barebones!" whispered Kristin, patting the beast, as it came and laid its head on her knee. She was sharp as a scythe along the backbone, and her dugs almost swept the floor. The whelps were eating their mother quite up. "So, so, my poor old barebones!"

Kristin laid her head back against the chair and looked up at the sooty rafters. She was weary. . . .

Oh, no — no easy time had she had, these months she had been at Husaby. She had had some talk with Erlend in the evening of the day they had been at Medalby; and had seen that he believed she was bitter against him because he had brought this upon her.

"I mind me well," he said in a whisper, "the day in the spring when we went in the woods north of the church. I mind well you prayed me to let you be — "

Kristin was glad because he said this. For at other times she had often wondered to see how many things Erlend seemed to have forgotten. But now he said:

"Yet had I not believed of you, Kristin, that you could go about thus, bearing a hidden grudge against me, and yet seeming kind and joyous as ever. For you must have known long since how things were with you. I had thought you were as clear and open as the sun in heaven — "

"Ah, Erlend," she said sadly, "you should know best of any in this world that I have followed secret ways and been false to them that trusted me most." But she was fain that he should understand. "I know not if you remember now, my dearest, that, long before that, your deeds towards me had been such as none would call fair. And yet God and Mary Virgin know that I bore you no grudge, nor loved you any less — "

Erlend's face grew tender.

"So thought I," he said low. "But this too I trow you know — through all those years I strove to set up again what I had broken down. I took comfort in the thought that things would go in the end so that I could reward you for being so long-suffering and so true."

Then she had asked him:

"You have heard of my grandfather's brother and the Lady Bengta, who fled together from Sweden against the will of her kin. God punished them by giving them no

child. Have you never been afraid, in these last years, that
He would punish us in like wise — ?"

And she had said to him, softly and trembling:

"You may well believe, my Erlend, that small joy was
mine last summer, when first I grew ware of this. And yet
methought — methought if you should die and leave me
before we were wedded, I had liefer you left me with a
child of yours than all alone. And I thought, if *I* should
die in bearing you a child — 'twould be better than that
you should have no true-born son to mount into the high-
seat in your place when you have to part from this earthly
home."

Erlend answered hotly:

"For me, I would deem my son all too dear bought if
he should cost you your life. Speak not so, Kristin. . . .
So dear to me Husaby is not," he said in a while. "And
least of all since I have been sure that Orm can never
inherit after me — "

"Care you more for *her* son than for mine?" asked
Kristin then.

"*Your* son — " Erlend laughed a little. "Of him, see you,
I know but this, that he comes hither a half-year or so too
soon. Orm I have loved for twelve years — "

A while after, Kristin asked:

"These children of yours — you miss them sometimes?"

"Aye," said the man. "Before, I would often go inland to
Österdal, where they are, to see them."

"You could go there this Advent," said Kristin in a low
voice.

"Would it not mislike you if I went?" asked Erlend
eagerly.

Kristin had answered that she would think it but right.
Then he had asked whether 'twould be against her liking
if he brought the children home with him for Yule. "Soon
or late, look you, you must see them." And again she had
answered that this seemed to her but right.

While Erlend was gone to Österdal, Kristin had worked
hard making things ready for Yule. It irked her much to
go about now amidst these strange henchmen and serving-
women — she had much ado to force herself to dress and
undress in the presence of the two maids, who Erlend had
said were to sleep near her in the hall. She had to remind

herself that she could never have borne to lie alone in that great house — where another before her had slept with Erlend.

The serving-women of the manor were no better than one might have looked they should be. Such peasants as took good heed of their daughters had had no mind to send them to service in a house where the master lived in open adultery with a wedded woman, and had set her to rule his house. The maids were idle and unused to obey a mistress. But some of them soon began to like the new order that Kristin brought in, and that she took a hand herself in their tasks. They grew full of talk and cheer, since the mistress hearkened to them and answered them kindly and cheerily. And Kristin showed the house-folk daily a calm and gentle face. She rebuked none harshly, but if any serving-maid should gainsay her bidding, the mistress seemed to think the girl knew not what to do, and quietly showed her how she would have the work done. It was thus Kristin had seen her father take things with new serving-men who grumbled — and no man at Jörundgaard had ever offered to gainsay Lavrans a second time.

Thus they might get through this winter well enough. Later she must contrive to get rid of the women that she could not come to like, or that she could not bring into shape.

One piece of work there was that 'twas beyond her to take in hand, except when she was free from the eyes of these strangers. But in the mornings when she sat alone in the hall, she sewed on clothes for her child — swaddling-clothes of soft wadmal, bands of bought stuff, red and green, and white linen for the christening-gown. While she sat sewing on her seam, her mind was tossed about between fear and trust in the holy friends of mankind she had prayed to intercede for her. True, the child lived and moved within her now, so that she had no rest night or day. But she had heard of children born with a belly where their face should have been, or with heads turned backwards, or toes where the heels should be. And she could see before her eyes Svein, who was bluish-red over half his face, because his mother had taken fright at a fire. . . .

Then she would throw down her seam, go and bend the knee before the picture of the Virgin Mary, and say seven Aves. Brother Edvin had said that God's Mother was filled with exceeding great joy each time she heard the Angel's greeting, even if it were in the mouth of the vilest sinner. And 'twas the words *Dominus tecum* that most rejoiced Mary's heart; therefore must she ever say them three times over.

It helped her always for a while.

One evening when she was sitting at the table with the house-folk, one of the women, a young maid that helped in the indoor work, had said:

"Methinks, mistress, 'twere better we should begin sewing swaddling-bands and baby-clothes now, before we set up this web you speak of — "

Kristin made as though she had not heard, and went on speaking of the dyeing of the web. Then the girl began again:

"But maybe you brought baby-clothes with you from home?"

Kristin smiled a little and turned again towards the others. When, a little while after, she glanced round, the wench was sitting, fiery-red in the face, peeping fearfully across at her mistress. Kristin smiled again and began speaking to Ulf across the table. Then of a sudden the girl burst out crying. Kristin laughed a little, and the girl wept more and more, till she was snivelling and sobbing.

"Nay, Frida — let us have no more of this," said Kristin at last, quietly. "You came hither as a grown serving-maid; try now not to behave like a baby girl."

The maid whimpered — she had not meant to be saucy — Kristin must not be wroth.

"No," said Kristin, still smiling. "Eat your supper now and weep no more. We have none of us more wit than God hath been pleased to grant us."

Frida jumped up and ran out, sobbing bitterly.

Afterwards, when Ulf Haldorssön stood talking with Kristin of the work to be done next day, he said with a laugh:

"Erlend should have betrothed him to you ten years

agone, Kristin. Then had things stood better with him now in every wise."

"Think you so?" said she, smiling as before. "In those days I was but nine winters old. Think you Erlend was the man to go around waiting long years for a child-bride?"

Ulf laughed and went out.

But that night Kristin lay awake, weeping tears of loneliness and shame.

Then Erlend had come home, the week before Yule, with Orm, his son, riding at his father's side. A stab of pain went through Kristin's heart when he led the boy up to her and bade him greet his stepmother.

He was a most comely child. 'Twas thus she had thought *he* might look, the son she was to bear. Sometimes when she dared to be glad, to trust that her child would be born sound and shapely, to dream ahead of the boy that should grow up by her knee, it was thus he looked in her dream — so like his father.

He was, maybe, somewhat small of his age, and slight, but well-shaped, fine-limbed and fair of face, dark of skin and hair, but with great blue eyes and red, soft mouth. He greeted his stepmother in seemly wise, but his face was hard and cold. Kristin had had no chance to speak much with the boy. But she felt his eyes upon her, wherever she went, and she felt as though she grew yet more heavy and awkward of body and gait when she knew that the lad was gazing at her.

She saw not that Erlend spoke much with his son, but she could see that of the two it was the boy that held back. Kristin spoke to her husband of Orm, saying that he was a comely lad and seemed of a good wit. His daughter Erlend had not brought with him; he deemed Margret was too small to make such a long journey in winter-time. She was fairer still than her brother, he said proudly when Kristin asked of the little maid — and much quicker of wit; she could turn her foster-father and mother round her little finger. She had gold-yellow, curling locks and brown eyes.

Then must she be like her mother, thought Kristin. In spite of herself, jealousy gnawed her heart. Did Erlend love this daughter of his as her father had loved her? His voice had been so soft and warm when he spoke of Margret.

Kristin stood up now and went to the outer doorway. It was so dark without and so heavy with rain that neither moon nor stars were seen. She thought, though, it must be nigh on midnight now. She brought in the lantern from the outer room and lighted it. Then she threw her cloak around her and went out into the rain.

"In Jesu name," she whispered, crossing herself thrice, as she stepped out into the night.

At the top of the courtyard stood the priest's house. It was empty now. Though Erlend had long since been freed from the ban, no priest had yet come to dwell at Husaby; now and then one of the chaplains from Orkedal would come over and say mass; but the new priest appointed to the church was in foreign parts with Master Gunnulf; it seemed they had been school-friends. They had been looked for home the last summer — but now Erlend said they could scarce be there before the late spring. Gunnulf had had the lung-sickness in his youth — he would scarce travel in the winter-time.

Kristin let herself into the cold, empty house, and found the church-key. Then she paused awhile. The ground was a slippery glaze — there was pitch-darkness, and wind, and rain. 'Twas a parlous thing for her, as she was now, to go out at night-time, and most of all on Yule Eve, when all evil spirits are in the air. But she could not give it up — she must come into the church.

"In the name of God the Almighty I go forth here," she whispered out into the storm. Lighting herself with the lantern, she set her feet with care where grass-tufts and stones showed above the ice-crust. In the dark the road to the church seemed long; but at last she stood on the threshold-stone.

Inside, it was bitter cold, much colder than outside in the rain. Kristin went forward to the chancel and knelt down before the crucifix, which gleamed dimly in the darkness above her.

When she had said her prayers and risen up, she stood still a little. It was as though she had looked that some-

thing should befall her. But there happened nothing. She
was cold and afraid in the dark, desolate church.

She crept up to the altar and turned her light on the
pictures. They were old, harsh and ugly. The altar-table
was of naked stone — altar-cloths, books and vessels she
knew were locked away in a chest.

In the nave was a bench running along the wall. Kristin
went down and sat on it, placing her lantern on the floor.
Her cloak was wet, and her feet were wet and cold. She
tried to draw up one foot under her, but it hurt her to sit
so. So she wrapped her cloak well about her, and strove
to gather and fix her thoughts on this one thing, that now
was come again the holy midnight hour, when Christ had
Himself born of Mary Virgin in Bethlehem.

*Verbum caro factum est et habitavit in nobis.**

She remembered Sira Eirik's deep, clear voice. And
Audun, the old deacon, that was never to be aught but
deacon. And their church at home where she had stood
by her mother's side and heard the Christmas mass. Every
single year she had heard it. She tried to remember more
of the holy words, but she could think of naught but the
church and all the well-known faces. Farthest in front, on
the men's side, stood her father and gazed with far-off
eyes into the blaze of light from the choir.

'Twas so unbelievable that their church was no more. It
was burnt down. She burst into tears at the thought. And
here was she sitting all alone this night, when all Christian
folk were gathered together in joy and gladness in God's
house. But 'twas like this was as it should be — that she
was barred out to-night from the rejoicings for the birth
of God's Son by a pure and stainless maid. — Her father
and mother were surely at Sundbu this Christmas. But
there would be no mass to-night in the chapel there; she
knew on Yule Eve the Sundbu folk ever rode over to mass
in the head church at Ladalm.

It was the first time, as far back as she could remember,
that she had not been at Christ's Mass. She must have been
quite small the first time her father and mother took her
with them. For she could remember that they had stuffed
her into a sheepskin bag with the wool inside, and her

* St. John's Gospel, i. 14.

father had borne her in his arms. It was a night of fearful
cold, and they rode through a forest — the light of the
fir-root torches gleamed on snow-laden pines. Her father's
face was purple-red in the glare, and the furred rim of
his hood was snowy-white with rime. Now and again he
bent his head a little and bit the tip of her nose, asking
her if she felt the bite; then called laughing over his
shoulder to her mother that Kristin's nose was not frozen
off yet. It must have been while they still dwelt at Skog —
belike when she was three winters old. In those days her
father and mother were quite young folk. She remembered now her mother's voice that night — high and glad
and full of laughter as she called out to her husband and
asked about the child. Aye, her mother's voice had been
young and fresh then. . . .

— Bethlehem — it betokens in Norse the place of heavy
bread. For there was given to men the bread that nour-
isheth unto life ever-lasting. . . .

'Twas at the day mass that Sira Eirik stepped up into
the lectern and set forth the evangel in the people's own
tongue.

Between the masses the folks sat in the guest-shed northward of the church. They had drink with them and the
cups went round. Betweenwhiles the men would go out
to the stalls and see to the horses. But on vigil nights
in the summer-time the congregation sat out on the
church-green, and between the services the young folks
danced.

— And the blessed maid, Mary, herself wrapped her son
in the swaddling-clothes. And she laid him in the manger
from which oxen and asses were wont to eat. . . .

Kristin pressed her hands strongly against her sides.

Little son, sweetest son, son of mine, God will have
mercy on us for His blessed Mother's sake. Blessed Mary,
thou brightest star of the sea, thou dawn of life eternal,
who didst bring forth the sun of all the world — help us!
Little child, what ails thee to-night, that thou art so unquiet — canst thou feel, even beneath my heart, that I
am so bitter cold — ?"

On Childermas Day last Yule-tide Sira Eirik had set
forth the gospel concerning the innocent children whom
the cruel soldiers slaughtered in their mothers' arms. But

so it was, he said, that God had chosen out these young children to enter into the hall of heaven before all the other blood-witnesses. And this was for a sign that the Kingdom of Heaven is of such as these. And He took a little lad and set him in the midst of them. Except ye make yourselves over in the likeness of these, dear brothers and sisters, ye cannot enter into the halls of the heavenly kingdom. And let this be for a comfort for everyone, man or woman, that mourneth the death of a young child. . . . At that Kristin had seen her father's and mother's eyes meet across the church; and she looked away quickly, for she knew that in this she had no part. . . .

This had been last year. The first Yule-tide after Ulvhild's death. Oh — but not *my* child! Jesus, Mary! Let me keep my son!

Her father had been loath to go for the St. Stephen's riding last year — but all the men had begged and prayed him till at last he joined them. The ride set out from the church-green at home and galloped down to the rivers-meet by Loptsgaard; there they met the men from Ottadal. She remembered her father dashing past on his golden-chestnut stallion — he stood in his stirrups and leaned forward along his horse's neck, whooping and cheering on the beast, the whole ride thundering after.

But last year he had been home early, and he came quite sober. Other years the men were used to come home late that day, and beyond measure drunken. For they had to call in at all the farmyards by the way and drink the healths brought out to them, to Christ and to St. Stephen, who was the first to see the Star in the East as he was riding King Herod's colts to water them in Jordan river. The horses, too, were given ale to drink that day, to make them wild and fiery. On Stephen's day it was ever so that the farmers must be busy with horse-games even till the time of evensong — scarce could the men be got to think of aught or speak of aught but horses. . . .

She could remember one Yule-tide when they had had the great common drinking-feast at Jörundgaard, and her father had promised a priest that was among the guests that he should have a young chestnut colt, a son of Gulds-veinen, if he could catch it and back it, as it ran, loose and bare-backed, in the courtyard.

'Twas a long time since — before the mishap to Ulvhild. Their mother stood before the house-door with the little sister on her arm, and Kristin stood holding to her skirt — a little afraid.

The priest ran after the horse, seized the halter, leapt from the ground so that his long gown flowed out on all sides; but had to loose the rearing fiery beast again. "So — coltie, coltie — heia, coltie, heia, sonny!" he sang out, hopping and dancing like a billy-goat. Her father and an old farmer that was there stood helpless, holding each other up, their features all drawn awry with laughter and strong drink.

Either the priest must have earned and won the colt, or Lavrans must have given it him unearned, for Kristin remembered that he was riding it when he left the manor. Then were they all sober enough; as he mounted, Lavrans held the stirrup for him with great reverence, and he blessed them with three fingers as he said farewell. 'Twas like he had been a priest of some dignity. . . .

Aye, her home was often right merry at Yule-tide. There was the coming of the guisers, too. Her father tossed her up on to his back, and she felt his coat all icy and his hair wet. To clear their heads before they went to Vespers, the men went off to the well and poured icy water over each other. They laughed when the women scolded about this. Her father took her little cold hands and pressed them against his forehead, which was red and burning-hot still. This was out in the courtyard, in the evening — a young white sliver of moon hung over the mountain ridge in the sea-green sky. When he was bearing her into the hall, he hit her head by mischance against the door-lintel, so that a great bump rose up on her forehead. Afterwards she sat on his lap at the board. He held the hilt of his dagger against the bump, and fed her with tit-bits, and let her drink mead from his beaker. And, sitting there, she felt no fear of the noisy guisers that were ramping about the hall.

— O father, O father — my kind, dear father!

Sobbing aloud, Kristin hid her face in her hands. Oh, if her father knew how things were with her this Christmas Eve!

* * *

As she went back to the hall, she saw sparks flying up above the kitchen-house roof. The maids were getting ready food for the churchgoers.

It was dark in the hall. The candles on the table were burnt out, and the hearth was almost black. Kristin laid on more wood, and blew up the embers. Then she saw that Orm was sitting in her chair. He rose up as soon as she caught sight of him.

"Orm!" said Kristin. "Went you not to the mass with your father and the others?"

Orm swallowed once or twice:

"He must have forgot to wake me, methinks. He bade me lie down awhile on the south bed. He said he would wake me — "

" 'Twas pity, Orm," said Kristin.

The boy made no answer. In a little while he said:

"I thought you were gone with them after all — I woke up all alone in the hall."

"I was over awhile in the church," said Kristin.

"Dare you go out on Yule night, then?" asked the lad. "Know you not that the Asgards-ride might have come by and taken you — ?"

"I trow 'tis not only the evil spirits that are abroad this night," she answered. "On Yule night they say all spirits — I knew a monk once, that is now dead — I trust well he stands before God's face, for there was naught in him but good. He told me once — heard you ever of the beasts in their stalls, how they spoke together on Christmas night? They could talk Latin in those days. So the cock crowed: *Christus natus est* — nay, I remember not the whole. But the other beasts asked: Where? and the goat bleated: *Betlem, Betlem* — and the sheep said: *Eamus, eamus* —"

Orm smiled scornfully.

"Think you I am such a babe you can comfort me with nurses' tales — ? Why offer you not to take me in your lap and give me suck — ?"

"Methinks I said it most to comfort my own self, Orm," said Kristin, quietly. "I too had been fain to go with them to the mass."

She felt now that she could not bear to look at the dirty, littered table any longer. She went over, swept all the leavings on to a platter, and set it on the floor for the dogs. Then she took out the mop of sedge-grass from beneath the bench and wiped the table-top dry with it.

"Will you come with me to the west-storehouse, Orm, to fetch bread and salt meat?" asked Kristin. "Then will we set out the table for Christmas morning."

"Why bid you not your serving-women to do all this?" asked the boy.

" 'Twas taught me at home in my father's and mother's house," said the young wife, "that at Yule-tide none should ask another for aught, but each should strive to do the most. He was most blest that most could serve the others throughout the holy days."

"Yet you ask me," said Orm.

" 'Tis another thing to ask you — that are the son of the house."

Orm took the lantern, and they went together across the courtyard. In the storehouse Kristin filled two great platters with Christmas fare. She took, too, a bundle of great tallow candles. While they were about this, the boy said:

"I trow 'tis farmer's fashion that you spoke of but now. For I have heard he is but a wadmal-farmer, Lavrans Björgulfsön."

"Of whom have you heard this?" asked Kristin.

"Of mother," said Orm. "Many a time I heard her say to father, when we lived at Husaby before, that he might see now, not even a grey-clad farmer would give his daughter to him in marriage."

"A pleasant home was Husaby in those days," said Kristin, shortly.

The boy made no answer. His mouth quivered a little.

Kristin and Orm bore the laden platters back to the hall, and she set the table for the meal. But some things were still lacking, to be fetched from the storehouse. Orm took a platter and said, a little bashfully:

"I will go for you, Kristin; 'tis so slippery in the yard."

She stood outside the door, and waited till he came back.

Afterward they sat them down by the hearth — she in the armchair and the boy on a joint-stool near her. In a while Orm Erlendssön said in a low voice:

"Tell me something more, while we sit waiting here, stepmother."

"Tell you — ?" asked Kristin in the same tone.

"Aye — a story or the like — something fitting for Christmas night," said the boy, shyly.

Kristin leaned back in her chair, grasping in her thin hands the carven beasts' heads at the arm-ends.

"That monk I named but now, he had been in England, too. And he used to tell that there is a place there where grown thornbushes that bloom with white blossoms each Christmas night. St. Joseph of Arimathea came to land in that country-side, when he fled before the heathen, and there he thrust his staff into the earth, and it took root and blossomed— he was the first that brought the Christian faith to Bretland. Glastonborg that place is called — I mind me now. Brother Edvin had seen those bushes himself. . . . 'Twas there in Glastonborg that he was buried, along with his Queen, that King Arthur that you will have heard tell of — he that was one of the Seven Champions of Christendom.

"They say in England that Christ's Cross was made of alder-wood. But we at home used to burn ash in the holy days; for 'twas ash-wood he made up the fire with, St. Joseph, Christ's stepfather, when he was to light a fire to comfort Mary Virgin and the new-born Son of God. Father heard that too of Brother Edvin — "

"But there's little ash grows here north of Dovre," said the boy. "They used it up in the old times for spear-shafts, you know. I know not of one other ash on all the lands of Husaby but the one that stands by the eastern yard-gate, and that one father cannot cut, for the Brownie of the Yard dwells under it. — But, Kristin, I wot they have the Holy Rood itself at Romaborg; surely they can find out if 'tis true that it is made of alder-wood — "

"Aye," said Kristin, "I know not if it be true. For you know 'tis said that the Cross was made of a shoot from the Tree of Life, that Seth was given grace to fetch from the Garden of Eden and bear home to Adam before he died — "

"Aye," said Orm. "But tell it to me."

A while after, Kristin said to the boy:

" 'Twere well you lay down now, kinsman, and slept awhile. It will be long yet till the church-folk are back."

Orm stood up.

"We have not pledged each other yet as kin, Kristin Lavransdatter."

He went and brought over a drinking-horn from the table, drank to his stepmother, and handed her the horn.

She felt as though an icy stream ran down her back. She could not but remember that hour when Orm's mother would have drunk with her. And the child in her womb moved unquietly. How is it with him to-night? thought the mother. It was as though the unborn babe felt all that she felt, was cold when she was cold, shrank in fear when she was afraid. But since 'tis so, I must not be so weak, thought Kristin. She took the horn and drank to her stepson.

As she gave it back to Orm, she passed her hand lightly over his black mane of hair. No, she thought, to thee I shall be no hard step-dame, be sure — thou fair, fair young son of Erlend's. . . .

She had fallen asleep in her chair when Erlend came in and flung his frozen mittens on the table.

"Are you come back already?" asked Kristin, wondering. "I deemed you would have stayed for the morning mass."

"Oh, two masses will serve my turn for a long time," said Erlend. The cloak that Kristin took from him was heavy with ice. "Aye, now 'tis clear again, and freezing hard — "

" 'Twas pity you should have forgot to wake Orm," said his wife.

"Was he vexed about it?" asked the father. " 'Twas not that I forgot, either," he said in a low voice. "But he was sleeping so sound that I thought — You may be sure the good folk gaped at me enough in church, for that I came there without you — I had no mind, on the top of that, to go forward with the boy at my side."

Kristin said naught; but the words hurt her. She could not think this well done of Erlend.

3

THEY saw not much of outside folk at Husaby that Yuletide. Erlend would not go abroad to any place where he was bidden, but stayed at home on the manor, in no pleasant mood.

The thing was that this misadventure galled him more nearly than his wife could know. He had boasted not a little of his betrothed ever since his kinsmen had gone to Jörundgaard and won her father's consent. It was the last thing he had wished, that any should believe he held her or her kin of less account than his own kindred. No — all men should know that he held himself honoured and uplifted again to worship by Lavrans Björgulfsön's betrothing his daughter to him. Now would all folks say he could scarce have held the maid much better than a peasant's child, since he had dared to do her father such despite as to sleep with his daughter before she was given him in wedlock. At his wedding Erlend had pressed his bride's parents strongly to come to Husaby the next summer and see how things were with him there. Not alone was he fain to show them it was no mean condition he had brought their daughter to; but he had been glad, too, at the thought of going about and showing himself in the company of these comely and stately new kinsfolk, for he knew that Lavrans and Ragnfrid could hold their own with the foremost, wherever they might come. And he had deemed, since the time he was at Jörundgaard when the church burned down, that, in spite of all, Lavrans liked him none so ill. But now there was small reason to think that the next meeting between him and his wife's kin would bring joy to either part.

It vexed Kristin that Erlend vented his ill-humour so often upon Orm. The boy had no playfellows of his own age, and so it came that he was often troublesome and in the way. He did his share of mischief, too. One day he had taken his father's French cross-bow without leave, and had broken something in the lock. Erlend was in great wrath; he struck Orm a box on the ear, and swore that the boy should never more touch a bow at Husaby.

" 'Twas not Orm's fault," said Kristin without turning. She was sitting sewing with her back to the two. "The spring was out of gear when he took the bow, and he tried to put it right. You should not be so unfair as to deny a great boy like your son the use of one from among all the

bows you have in this house. Rather give him one of the
bows that are up in the armoury."

"You can give him a bow yourself, if you have a mind,"
said Erlend, wrathfully.

"That will I gladly," answered Kristin as before. "I will
speak of it to Ulf, next time he goes in to the city."

"You must go and thank your kind stepmother, Orm,"
said Erlend, in a voice of anger and scorn.

Orm did as he was bid, and then flew out of the room as
swiftly as he might. Erlend stood still awhile.

" 'Twas most to vex me you did this, Kristin," he said.

"Aye, I know I am a troll. You have told me that already," she replied.

"But mind you too, my sweet," said Erlend sorrowfully,
"that I spoke not in earnest when I said that word?"

Kristin made no answer, nor looked up from her seam.
Soon he went out, and when he was gone she sat there
weeping. She had come to care for Orm, and she deemed
that Erlend was often unjust to his son. But then, too, her
husband's silence and unjoyous looks were tormenting her
so that she lay weeping half the night; and then her head
would ache all the day after. Her hands were grown so
thin now that she had to thrust on some small silver rings
she had had since childhood above her betrothal ring and
her wedding-ring, to keep them from falling off while she
slept.

The Sunday before the beginning of the fast, late in the
afternoon, Sir Baard Peterssön with his daughter, the
widow, and Sir Munan Baardssön with his lady came to
Husaby as unlooked-for guests. Erlend and Kristin went
out into the courtyard to bid the strangers welcome.

The moment Sir Munan set eyes on Kristin, he clapped
his hand on Erlend's shoulder:

"I see well, kinsman, you have known how to care so
for your wife that she hath thriven in your house. You
are nowise so thin and peaked, now, Kristin, as you were
at your wedding — and far fresher of hue are you too,"
said he, laughing, for Kristin had flushed red as a berry.

Erlend made no answer. Sir Baard's face was clouded;
but the two ladies seemed neither to hear nor see aught;
they greeted their hosts seemly and quietly.

Kristin had ale and mead brought forth to them by the hearth while they waited for the meal. Munan Baardssön talked without cease. He had letters with him for Erlend from the Duchess — she had asked what was become of him and his bride; and was the maid he had wedded now the same that he would have carried off to Sweden? 'Twas the devil's own journey in midwinter that he had made — up through the dales and by ship to Nidaros. But he journeyed on the King's errand, and it booted not to murmur. He had looked in on his mother at Haugen, and he brought her greetings.

"Were you at Jörundgaard?" asked Kristin, in a low voice.

No; for he had come to know they were gone from home to the grave-ale at Blakarsarv. There had been a grievous mischance. The mistress of the house, Tora, Ragnfrid's cousin — she had fallen down from the storehouse balcony and broken her back — and 'twas her husband coming against her unawares that had pushed her over — it was one of those old storehouses where there was no right balcony, but only a few boards laid on the beam-ends of the upper story. He heard they had had to bind Rolf and watch him night and day since the mishap befell — to keep him from laying hands on himself.

The listeners sat very still, shuddering. Kristin knew but little of these kinsfolk, but they had been at her wedding. Suddenly she felt strange and faint — there was a blackness before her eyes. Munan, who sat over against her, leapt up and came to her. When he stood by her with his arm around her shoulders, he looked kind — Kristin thought, 'twas maybe not so strange that this cousin of Erlend's was dear to him.

"I knew him, Rolf, when we were young," he said now. "Folk were used to pity Tora Guttormsdatter — they said he was wild and hard-hearted. Yet one can see now that he held her dear. Aye, aye — many a man blusters and talks big about how glad he would be to be quit of his wedlock, but most men know well that a wife is the worst loss they can have — "

Baard Peterssön rose suddenly and went over to the bench by the wall.

"Beshrew my mouth," said Sir Munan softly. "To think I can never remember to watch my tongue — "

Kristin understood not what was amiss. The dizziness had passed now, but she had a feeling of discomfort — they seemed all so strange. She was glad when the serving-folk brought in the meal.

Munan looked at the table and rubbed his hands:

"Sure I was that we should do well to look in on you, Kristin, before we settle down to munching Lenten fare. Where have you gotten savoury dishes like these from in such a little while? A man might go nigh to think you had learned witchcraft of my mother. But I see well you are quick to bring forth all things that a housewife should gladden her husband withal."

They sat down to the table. Velvet cushions had been laid for the guests on the wall-bench on either side of the high-seat. The serving-folk sat on the outer bench, Ulf Haldorssön in the midst, over against his master.

Kristin talked a little, quietly, with the strange ladies, striving to hide how ill at ease she was. Time and again Munan Baardssön broke in with would-be playful words, ever harping on Kristin's state. She made as though she did not hear.

Munan was a man fatter than the common. His small well-formed ears were sunk right into the flesh of his thick red neck, and his belly got in his way when sitting down to table.

"Aye, often have I wondered about that matter of the resurrection of the body," he said; "whether I shall be raised up with all this blubber I have laid on around my bones, when that day comes. *You* will soon enough be slim-waisted again now, Kristin — but with me 'tis no such easy matter. You will scarce believe it, but my belt was no wider than Erlend's there when I was twenty winters old — "

"Be still now, Munan," Erlend begged, in a low voice. "You are plaguing Kristin — "

"I shall be so, since so you say," the other took him up. "You are a proud man now, I dare swear — sitting at your own table with your wedded wife by you in the high-seat. Aye, and God that's over all knows that 'twas none too soon either — you are old enough, boy! Surely I will hold my tongue, since you bid me. But never did any tell you when you should speak or be silent, in days gone by when *you* sat at *my* table. Often and long were you my guest, and I deem not you marked at any time that you were not welcome.

"But much I wonder whether indeed it likes Kristin so ill that I jest a little with her — what say you, fair wife of my kinsman? — you were not wont to be so startlish in days gone by. I have known Erlend since he was as high as my knee, and methinks I can say I have wished the boy well all his days. Mettlesome and manful are you, Erlend, with a sword in your hand, either a-horseback or a-shipboard. But I will pray to St. Olav to cleave me in two halves with his axe the day I see you stand up on those long legs of yours, look man or woman straight in the face, and answer for the mischief you have wrought in your light-mindedness. No, dear kinsman of mine, then do you hang your head like a bird in the snare, and wait upon God and your kindred to help you out of the pinch. Aye, and so clear-witted a woman are you, Kristin, that I trow you know this — and methinks you may well have need to laugh a little now; for I wager you have seen enough this winter past of shamefaced looks and sorrow and repentance — "

Kristin sat with face darkly flushed. Her hands were shaking, and she dared not look towards Erlend. Anger seethed within her — here sat the strange ladies and Orm and her serving-folk. So these were the courtly ways of Erlend's rich kinsmen. . . .

Then said Sir Baard, in a low voice that only those who sat next him were meant to hear:

"I see not that 'tis aught to jest about — that Erlend should have behaved him thus before his wedding. I pledged my word for you, Erlend, with Lavrans Björgulfsön."

"Aye, devil knows 'twas unwisely done of you, fosterfather," said Erlend, loud and hotly. "And I marvel how you could be so foolish. For you — I trow you know me too — "

But now was there no checking Munan any longer:

"Aye, but now will I say why this seems to me a merry jest. Mind you what answer you made me, Baard, when I came to you and said we must needs help Erlend to make this marriage — nay, now I *will* tell of it; Erlend shall know what you believed of me — thus and thus it stands between them, said I, and if he wins not Kristin Lavransdatter to wife, God and Mary Virgin alone know what mad doings we shall next hear of. Then it was you asked me, was I so fain to have him wedded to the maid he had betrayed because I thought belike she was barren, since after so long she had yet shown no sign? But I trow you know me, you others — you know me for a trusty kinsman to my kin — " Quite overcome, he burst out weeping. "God and all holy men be my witness — never have I coveted your goods, kinsman — and then, to be sure, there is Gunnulf besides between me and Husaby. And I answered you, Baard, well you know it — the first son that Kristin bore, I would give him my gold-mounted dagger with the ivory sheath — and you can have it now!" he shouted though his tears, throwing the costly weapon along the table to her. "If it be not a son this time, 'tis like there will come one next year — "

Tears of shame and wrath rolled down on Kristin's cheeks. She had a hard struggle not to give way altogether. But the two stranger ladies sat eating as calmly as if they were well used to such scenes. And Erlend whispered to her to take the dagger: "Or Munan will keep on with this all night."

"Aye, and I deny not, Kristin," went on Munan, "that I am not so ill pleased your father should be made to see he was too rash when he answered for your mind. So haughty was Lavrans — we were not good enough for him, forsooth — and you were all too fine and pure to suffer a man like Erlend for your bedfellow. He spoke as if he deemed you could not bear to do aught of nights but sing in a nunnery choir. I said to him: 'Dear Lavrans,' I said, 'your daughter is a fair and fresh and sprightly young maid, and the winter nights are long and cold in this our land — '"

Kristin drew the linen of her coif across her face. She sobbed aloud, and would have risen, but Erlend drew her down again into her seat.

"Be still, " he said vehemently. "Pay no heed to Munan
— surely you can see he is raving drunk — ?"

She felt that Lady Katrin and Lady Vilborg deemed that
'twas poor-spirited of her not to be better mistress of
herself. But she could not stay her weeping.

Baard Peterssön said in fury:

"Hold your rotten tongue. A swine have you been all
your days — yet might you well leave a sick woman in
peace from your filthy talk — "

"Said you 'swine' — ? Aye, true it is, I have more bastards than you. But one thing have I never done — nor
Erlend either — bought another man to be called our
child's father in our place — "

"Munan!" shouted Erlend, springing up. "Now call I
for peace in my hall!"

"Oh, call for peace in your tail! — *Our* children call
father him that got them — in swinish living, as you call
it!" Munan smote the board so that cups and platters leapt
in the air. "*Our* sons dwell not as serving-men in the house
of their kindred. But here sits son of yours at the board
with you, and he sits on the varlets' bench. Now should I
deem that the worst shame of all — "

Baard sprang up and drove a flagon into the other's face.
The two men grappled, half upsetting the table-top, so
that food and vessels went sliding down into the laps of
those on the outer bench.

Kristin sat deadly white, with mouth half opened. Once
she glanced across at Ulf — the man was laughing aloud,
with a coarse, evil laughter. Then, taking hold of the table-
top, he heaved it into place and thrust it against the two
struggling men.

Erlend leapt on to the table. Kneeling on it amid the
litter of the feast, he caught Munan round the arms below
his shoulders and dragged him bodily up beside himself —
his own face purpling with the strain. Munan kicked out
at the old man, drawing blood from his mouth, but the
next moment Erlend had flung him clean across the table
on to the open floor. He himself leapt after him — and
stood panting like a bellows.

Munan scrambled to his feet and rushed at Erlend, who
slipped clear of his grasp once or twice, then suddenly
leapt upon him and held him tightly grappled with his

long sinewy legs and arms. Erlend was lithe as a cat, but Munan, solid and heavy, kept his footing and would not be thrown. They swayed struggling about the hall, while the serving-women shrieked and screamed, and none of the men moved a hand to part them.

Then Lady Katrin, fat, heavy and slow-moving as ever, arose from her seat and stepped over the table as calmly as though she were mounting her storehouse-steps.

"Have done, now," she said in her thick, dragging voice. "Loose him, Erlend! This was ill done of you, husband — to speak thus to an old man and a near kinsman — "

The men obeyed her. Munan stood meekly and let his wife stanch the blood from his nose with the hem of her coif. She bade him go to bed, and he followed obediently when she led him over to the southern box-bed. His wife and one of his men pulled the clothes off him, rolled him into the bed, and shut the bed-doors on him.

Erlend had gone over to the table. He leant across it beside Ulf, who had not stirred from his place.

"Foster-father!" he said in an unhappy voice. He seemed quite to have forgotten his wife. Sir Baard sat rocking with his head, the tears trickling down his cheeks.

"There had been no need, either, for Ulf to serve," he brought out, through the weeping that made him gasp and sob. "You could have had the farm when Haldor died — you know well 'twas my intent you should."

"The farm you gave to Haldor was none so brave — you bought a husband for your wife's maid good cheap," said Ulf. "He cleared and tilled and bettered it — me-thought, for one thing, it was but reason that my brothers should have it after their father. And besides, little was I minded to sit down and be a farmer — and least of all up on yonder hill-side, gaping down into the Hestnes court-yard — meseemed I could hear every day up there the voices of Paal and Vilborg, cursing that you had given all too rich a gift to your bastard son — "

"I proffered you help, Ulf," said Baard, still weeping, "when you were bent on faring forth with Erlend. I told you all the truth of this matter, as soon as you were of age to understand. I prayed you to turn to your father — "

"I call him my father who fostered me when I was a child. And that man's name was Haldor. He was good to mother and to me. He taught me to ride a horse and to handle a sword — as a churl doth his cudgel, I mind me Paal once said."

Ulf hurled from him the knife he had in his hand, so that it flew ringing across the table. He rose and picked it up again, wiped it on the back of his thigh, and stuck it in its sheath. Then he turned to Erlend:

"Make an end now with this feast of yours, and send your people to bed! See you not your wife is unused as yet to the fair fashions of our kindred in their feasting?"

And with that he was gone from the hall.

Sir Baard looked after him — he seemed of a sudden wretchedly old and feeble, as he sat there, huddled together among the velvet cushions. His daughter, Vilborg, and one of his men helped him to his feet and led him out.

Kristin sat alone on the high-seat, weeping and weeping. When Erlend tried to take hold on her, she struck his hand aside vehemently. She swayed about on her feet once or twice when she walked across the floor; but she answered curtly: "No," when her husband asked if she were sick.

She liked not these shut-up box-beds. At home the beds were only curtained off from the hall by hangings, so that the air inside was less hot and stifling. And to-night it was worse in there than ever — for at best she could scarce draw breath. The hard lump she felt pressing right up under her breast-bone she thought must be the child's head—she fancied that it lay with its little head bored in amongst the roots of her heart — it choked her breathing as Erlend had done in old days, when he pressed his dark-haired head against her breast. But to-night there was no sweetness in the thought. . . .

"Will you never make an end of weeping?" asked her husband, trying to pass his arm beneath her shoulders.

He was quite sober. He could bear much liquor, and for the most part he drank but little. Kristin was thinking — never in the world could aught like this have befallen in her home. Never had she heard folk there revile each other, or rake up in their talk things that were best left unnamed. Often as she had seen her father reeling in drunkenness, and the hall full of drunken guests, not once had it befallen even then that he had not been fit to keep order in his house — peace and goodwill had ever ruled, even till the folks dropped off the benches to the floor, and fell asleep together in joy and harmony.

"Dearest one, take not this so hardly," Erlend begged.

"And Sir Baard," she broke out through her tears. "Fie on such doings — he that talked to my father as if he bore a message from God himself — aye, Munan told me of it at our betrothal-feast — "

Erlend answered softly:

"Well enough I know it, Kristin, that I have cause to cast down my eyes before your father. He is a good man — but my foster-father is no worse than he. Inga — Paal and Vilborg's mother — she lay crippled and sick for six years before she died. 'Twas before I came to Hestnes, but I have heard all the story, and never has a husband cherished a sick wife more truly and lovingly. But 'twas at that time Ulf was born — "

"All the more the shame, then — with his sick wife's maid — "

"You show you so childish sometimes, a man can scarce talk with you," said Erlend in despair. "God help us, Kristin, you will be twenty come next spring — and more winters than one are gone by since you must needs be accounted a grown woman — "

"Aye, 'tis true *you* have the right to scorn me for it — "

Erlend groaned aloud:

"You know yourself that I meant it not so. — But you have lived there at Jörundgaard and hearkened to Lavrans — and for all he is a bold man and a manful, he talks oft-times as if he had been a monk and not a whole grown man — "

"Heard you ever of any monk that had six children?" she said angrily.

"I have heard of one, Skurda-Grim, that had seven," said Erlend, desperately. "The Abbot of Holm that was — Nay, Kristin, Kristin, weep not so, in God's name! Methinks you have lost your wits — "

Munan was passing meek the next morning. "I could never have thought you would take my drunken pratings so much to heart, Kristin, girl," he said gravely, patting her cheek. "Else had I kept a better watch on my tongue, be sure."

He spoke to Erlend of Orm, saying it must be irksome for Kristin now to see this boy about. 'Twould be best to send him out of the way at this time — he offered to take the boy for a while. Erlend liked the proffer well — and Orm was glad to go with Munan. But Kristin missed the child much — she had come to hold her stepson dear.

Again now in the evenings she was left alone with Erlend, and he was no great company for her. He sat by the hearth, said a word now and then, or took a draught from the ale-bowl, and played a little with his dogs. Then he would go and stretch himself on the bench — then go to bed — would ask once or twice if she should not go to rest soon, and then fall asleep.

Kristin sat and sewed. Her breaths came short and so heavy they could be heard. But there was not long to wait now. She could scarce remember, it seemed, what 'twas like to be free and supple in the waist, to be able to tie her shoe without pain and labour.

Now that Erlend slept, she need not even try to keep back the tears. There was no sound in the hall, save when a crumbling brand would drop on the hearth, or a dog would move in its sleep. Sometimes she would wonder — what had they spoken of in the days that were gone — Erlend and she? Like enough they had not spoken much — they had other pastime in their short, stolen trysts. . . .

This was the time of year when her mother and the serving-maids were wont to sit of evenings in the weaving-house. And her father and the men too would come in and sit down by the women with their own tasks—mending leather gear and farm tools, and carving in wood. The little house was filled full of folk, and talk ran on quietly and easily amongst them. When one had gone over to get him a drink from the ale-tub, he asked ever, before he hung up the ladle again, if any other had a mind to drink — 'twas a firm, fixed rule.

Then would there be someone who could say forth a snatch of some saga — of champions in the old age that had fought with mound-dwellers and giantesses of the hills. Or her father would tell them, as he sat at his wood-carving, tales of knighthood, such as he had heard read aloud in his Lord's hall, when he was a page to Duke Haakon in his youth. Fair outlandish names — King Osantrix, Sir Titurel the knight — and Sisibe, Guniver, Gloriana and Isood were the Queens' names . . . . But other evenings they would tell cock-and-bull stories and merry tales, till the menfolk guffawed with laughter, and her mother and the maids shook their heads and tittered.

Ulvhild and Astrid would sing. Her mother had the sweetest voice of all, but it took much praying before they could bring her to sing to them. Her father was not so backward — and he could play so tunably on his harp.

Then Ulvhild would lay from her wheel and spindle and press her hands to her back.

"Is your back weary now, little Ulvhild?" her father would ask, and would take her up in his lap. Someone would bring the draughtboard, and father and Ulvhild would play till bedtime came. She remembered her little sister's yellow locks flowing down over her father's greenish-brown wadmal sleeve. He held up the weak little back so tenderly with the circle of his arm.

Father's long slender hands, with a heavy gold ring on each little finger. The rings had been his mother's. The one with the red stone, her bridal-ring, he had said that Kristin should have after him. But the one on his right hand, with a stone that was half blue and half white like the device of his shield, that had Sir Björgulf had made for his wife when she went with child of Lavrans — she was to be given it when she had borne him a son. Three nights had Kristin Sigurdsdatter worn the ring; then she tied it round her boy's neck; and Lavrans said he would take it with him to his grave.

Oh, what would her father say when he heard this of her? When 'twas noised abroad all over the countryside at home, and he could not but know, wherever he might fare, to church, to Thing, or to meetings, that all men were laughing at him behind his back, that he had let him be so fooled. At Jörundgaard they had decked out a wanton with the Sundbu bridal crown above the flowing hair of maidenhood —

"Folk say of me, I know well, that I cannot rule my children." She remembered her father's look when he said it — he would fain have been sad and stern of face, but his eyes were merry. She had done amiss in some little matter — spoken to him, unspoken to, before strangers, or the like. "Aye, Kristin, sooth it is you go not much in fear of your father." Then a laugh broke out, and she laughed too. "Aye, but 'tis an ill thing, Kristin." And neither of them knew what it was that was so ill a thing — that she stood not in seemly awe of him; or that he could in no wise keep a sad brow when he had to chide her.

It was as though the unbearable dread that there should be somewhat amiss with the child grew fainter and farther off as Kristin's pains and bodily trouble grew. She tried to send her thoughts forward — in a month — she would have had her boy for a while then already. But she could not make it seem true. She could only long and long for her home.

Once Erlend had asked if she would he should send for her mother. But she said: No — she deemed not her mother was fit to journey so far in winter-time. Now she repented this. And she repented that she had said: No, to Tordis of Laugarbru, who had been so fain to come north with her and help her through her first winter as the mistress of a house. But she had thought shame to have Tordis by. Tordis had been Ragnfrid's maid at her home at Sundbu, and had followed her mistress to Skog and back again to the Dale. When she married, Lavrans had made her husband steward at Jörundgaard, since Ragnfrid could not bear to be parted from her dear hand-maiden. Kristin had no mind to have with her any woman from her home.

But now it seemed to her a fearful thing that she should not have a single known face to look upon when her time came to lie in the straw. She was afraid — she knew so little of the bringing to bed of women. Her mother had never spoken to her of it, and would never have young maids with her when she helped women in their labour — it would but frighten the young things, she said. But Kristin knew it must sometimes be fearful — she remembered the time Ulvhild was born. That, though, Ragnfrid had said was because she had forgot herself and crept under a fence-rail — her other children she had borne easily. But Kristin remembered now that she herself had been thoughtless, and passed under a rope on the ship. . . .

Yet this did not always bring heavy labour — she had heard her mother and other women talk of such things. Ragnfrid had the name in their country-side of the best midwife far around, and never would she deny her help, not if it was to a beggar woman or the poorest man's daughter that had fallen in trouble, nor if the weather was such that three men must go with her on ski and take turns at bearing her on their backs. . . .

But surely, it came to Kristin in a flash, 'twas not to be believed that a woman of such skill in these things as her mother had, should not have seen what was amiss with her last summer, when she was so ailing. But then — sure it was that her mother would come, even though they had not sent to call her. Ragnfrid would never suffer that her daughter should go through that struggle in a stranger's arms. Her mother would come — surely she was even now on her way thither. . . . Oh, and then she could pray her mother's forgiveness for all that she had sinned against her — her own mother would hold her up, she would kneel at her mother's knee when she bore her child. Mother comes, mother comes. . . . Kristin sobbed behind her hands from a lightened heart. O mother — forgive me, mother.

The thought that her mother was on her way up to her grew so fixed in Kristin that one day she deemed she could feel within herself: Mother will come to-day. And on in the morning she took her cloak about her and went out to meet her on the road that leads from Gauldal over to Skaun. None marked her as she left the manor.

Erlend had had timber driven down from the woods for the mending and bettering of the houses, so that the road was well trodden; but 'twas heavy going for her none the less — she lost her breath, her heart beat hard, and sharp pain came in her sides — it felt as though the overstretched flesh would break asunder when she had walked a little while. And most of the way was through thick forest. She was afraid indeed — but there had been no word of wolves in the country-side this winter. And surely God would guard her that went forth to meet her mother, fall at her feet, and pray her for forgiveness — and she could not but go on and on.

She came by and by to a little lake, where there lay

some small farmsteads. Where the road led on to the ice, she sat down upon a log — and sometimes sitting, sometimes walking to keep warm, stayed waiting there many hours. But at last she must needs turn home again.

The day after, she strayed out again by the same road. But as she crossed the yard of one of the small farms by the lake, the woman of the house came running after her.

"In God's name, mistress, what is this you do?"

No sooner had she spoke than Kristin herself grew so afraid that she could not move from the spot — trembling, with eyes wild with fear, she gazed at the peasant women.

"Through the woods — think but if the wolf got scent of you! And other ill things too might well come upon you — how can you bear you so witlessly?"

The goodwife threw her arms around the young lady of the manor to hold her up — and looked into her thin face, all yellowish-white and flecked with brown:

"You must come into our house and rest awhile — and then we will take you home — someone from here," she said, as she led Kristin away with her.

It was a little and a poor house, and within all was in much disorder, for there were many small children playing on the floor. The mother sent them out to the kitchen-house, took her guest's cloak, led her to a seat on the bench, and drew off her snowy shoes. Then she wrapped a sheepskin round her feet.

For all Kristin prayed the other not to put herself about, her hostess was not to be hindered from serving her with food and with ale from the Yule-cask. And she was thinking the while — a rare rule must they keep at Husaby! She herself was but a poor man's wife; little help had they had on their farm, and often none at all; but never would Öistein suffer her to go alone without the farm-yard fence when she was with child — nay, if she but went across to the byre after dark had fallen, someone must ever keep an eye on her. But the richest lady in all the country-side might stray out and run the risk of the most dreadful death, and not a Christian soul to take care of her — though the serving-folk at Husaby were tumbling over each other and doing naught. 'Twas like, then, the folks said sooth that said Erlend Nikulaussön was weary of his marriage already, and cared not for his wife. . . .

But she chatted away to Kristin all the time, and forced her to eat and drink. And Kristin was much ashamed — but she had such a stomach to her meat as she had not felt — not since the last spring; the kind woman's food tasted so good. And the woman laughed and said 'twas like great folks' womankind were made no otherwise than poor. 'Twas often so that when a body could not bear to look at food at home, one would be right greedy for strangers' fare even if 'twere coarse and poor.

Her name was Audfinna Andunsdatter, and she was from Updal, she said. When she marked that it cheered her guest, she took to telling of her home and her country. And before Kristin was aware, *her* tongue too was loosed — and she was talking of *her* home and her parents and her own country-side. Audfinna saw well that the young wife's heart was near breaking with homesickness — so she tempted and beguiled Kristin into going on. And Kristin, hot and dizzy with the strong ale, went on talking till she was laughing and crying in the same breath. All that she had tried in vain to sob away from her heart in the lonesome evening hours at Husaby, seemed to melt now little by little as she told her tale to this kind peasant wife.

It was quite dark now above the smoke-vent, but Audfinna would have Kristin stay till Öistein or their sons came home from the wood and could take her home. Kristin grew silent and drowsy, but she sat smiling, with shining eyes — so happy she had not felt since she came to Husaby.

Suddenly the door was flung open, and a man shouted in to ask if they had seen aught of the mistress — then caught sight of her sitting there and rushed out again. A minute after, Erlend's long shape came stooping low through the doorway. He set from him the axe he bore, and staggered back against the wall — he had to prop himself with hands thrust behind him, and he could not speak.

"You have been afraid for your lady?" asked Audfinna, going over to him.

"Aye — I take no shame to say it." He passed his hand up under his hair. "So frighted has man scarce ever been, I trow, as I have been this night. When I heard she had gone off into the woods —"

Audfinna told how it was Kristin had come thither. Erlend took the woman's hand.

"Never will I forget what I owe you and your husband for this," he said.

Then he went across to where his wife sat, and, standing beside her, laid a hand upon her neck. He spoke not a word to her, but stood still thus as long as they were in the house.

Now came crowding in henchmen from Husaby and men from the nearest farms. All looked as though they needed a heartening draught, so Audfinna bore round the ale-bowl before they set forth again.

The men went off on ski across the fields, but Erlend had given his to one of his followers; he walked down the hill holding Kristin inside his cloak. It was quite dark now, and the stars shone bright.

Then came a sound from the woods behind them — a long-drawn howl that mounted higher and higher in the night. It was wolves — and there were many. Erlend stopped short, shivering, loosed his hold of her, and Kristin knew that he crossed himself, while he gripped the axe in his other hand. "Had you now been — oh, no — " He crushed her to him so fiercely that she moaned with pain.

The ski-runners in the fields turned sharp about, and toiled back to the pair as fast as they could climb. Then they flung the ski over their shoulders, and made a close ring around her with their spears and axes. The wolves followed them all the way to Husaby — so near that now and then they could see a glimpse of them through the darkness.

When they came into the lighted hall, many of the men's faces showed grey and white. One said: "This was the grimmest — " and straightway fell a-vomiting into the hearth-fire. The frightened maids brought their mistress to her bed. Eat she could not. But now that the sick, awful dread was overpast, it yet seemed to her comforting after a fashion to see that all had been so afrighted for her sake.

When they were left alone in the hall, Erlend came across and sat himself down on the edge of her bed.

"Why did you this?" he whispered. And when she made no answer, he said, yet lower:

"Is it such grief to you that you have come into my house — ?"

It was a little while before she understood what he meant:

"Jesus, Maria! How can you think such a thought?"

"What had you in mind that time you said — when we had been at Medalby, when I would have ridden from you — that I might have had long to wait ere you came after me to Husaby?" he asked in the same low tone.

"Oh, I spoke but in wrath," said Kristin, bashfully, in a low voice. And she told him now what it was that had taken her out these days. Erlend sat very still and listened.

"Much do I wonder when the day will come that my house of Husaby will seem home to you," said he, bending over her in the dark.

"Oh — in not much more than a week now, maybe," whispered Kristin, with a wavering laugh. When he laid his face down against hers, she threw her arms round his neck and gave back his kiss eagerly.

" 'Tis the first time you have laid your arms about my neck of your own accord since I struck you," said Erlend in a low voice. "You are slow to forgive, my Kristin — "

It came into her mind that this was the first time since the night when he had learned she was with child, that she had had courage to offer him a caress unasked.

But after this day Erlend showed such kindness towards her that Kristin repented each hour she had felt anger against him.

4

GREGORY's *Mass** came and went by. Kristin had believed so surely that her time must come then, at the latest. But now it would soon be Mary's Mass in Lent.† and still she was about on her feet.

Erlend was forced to go to Nidaros for the mid-fast Thing; he said he would surely be home Monday night,

---

* 13th February.
† 25th March.

but now it was Wednesday morning and he was not yet come. Kristin sat in the hall, scarce knowing what to be about — it was as though she had no power to begin upon any work.

The sunlight streamed in through the smoke-vent — she felt that without it must be like spring to-day. She rose up and threw a cloak about her.

One of the maids had told her that folk said if a woman went beyond her due time, a good way was for her to let the horse she rode at her bridal eat corn from her lap. Kristin stood a little while in the hall-door — in the blinding sunlight the yard lay all brown, but glittering rills of water ran in bright frozen runnels through the horse-dung and litter. The skies were spread bright and silky-blue above the old houses — on the two figure-heads fixed to the beams of the east storehouse, the traces of their old-time gilding shone out to-day in the clear air. Water dribbled and ran from the roofs, and the smoke whirled and danced in little mild puffs of wind.

She went to the stables and in, and filled the lap of her skirt with oats from the corn-bin. The stable smell, and the sound of the horses stirring in there in the dark, did her good. But some of the folks were in the stable; and she was ashamed to do what she had come to do.

She went out and threw the grain to the hens that were scratching around and sunning themselves in the yard. Her thoughts far off, she looked at Tore, the stable-man, currying and brushing down the grey gelding — it was fast shedding its winter coat. Now and then she shut her eyes, and turned her face, faded and pale with the indoor air, up to the sunshine.

So she was standing when three men rode into the court-yard. The foremost of them was a young priest whom she did not know. As soon as he was aware of her, he leapt from his horse and came straight up to her with outstretched hand.

"You had not meant me this great honour, I trow — that you, the lady of the house, should come forth to welcome me," said he, smiling. "But I must thank you for it, none the less. For I wot well you must be my brother's wife, Kristin Lavransdatter?"

"Then must you be Master Gunnulf, my brother-in-law," she answered, flushing red. "Well met, sir! And welcome home to Husaby!"

"Thanks for a fair welcome," said the priest; and he stooped and kissed her cheek, after the fashion she knew was used in foreign lands, when kinsfolk met. "Happy be your coming hither, Erlend's wife!"

Ulf Haldorssön came out, and bade a groom take the strangers' horses. Gunnulf greeted Ulf right heartily:

"Are you here, kinsman? — I had looked to hear that you were a wedded man and a householder by now."

"Nay, no wedding for me, till I must choose between a wife and the gallows," said Ulf, with a laugh; and the priest laughed too. "I have pledged me to the devil to live unwed as firmly as you have promised it to God."

"Aye, then should you be scatheless whichever way you turn you, Ulf," answered Master Gunnulf, laughing. "For you will do well the day you break your promise to yonder man; and yet 'tis said also that a man should keep his word, were it to the fiend himself. . . . Is Erlend not at home?" he asked in wonder. He proffered Kristin his hand, as they turned to go into the hall.

To hide her bashfulness, Kristin moved about among the serving-women, and saw to the spreading of the board. She bade Erlend's learned brother sit in the high-seat, but, when she would not sit there with him, he moved down to the bench beside her.

Now she was sitting at his side, she saw that Master Gunnulf must be shorter than Erlend by half a head at the least — but he seemed to bulk larger. He was stronger-built and more thickset in body and limbs, and his broad shoulders were quite straight — Erlend's slouched a little. He was clad in dark raiment, most seemly for a priest, but the long cassock, reaching to his feet, and upward almost to the band of his linen shirt, was fastened with buttons of enamel, and from his woven belt his eating-gear hung in a silver sheath.

She looked up into the priest's countenance. He had a round strong head and a round but thin face, with broad low forehead, cheek-bones a little large, and a fine rounded chin. The nose was straight and the ears small and comely, but his mouth was wide and thin-lipped, and the upper lip came forward never so little and overshadowed the

lower lip's little splash of red. Only his hair was like
Erlend's — the close-cropped ring round the priest's ton-
sure was black, with a dry, sooty gleam, and looked as
though 'twere as silky-soft as Erlend's mane. For the rest
he was not unlike his cousin Munan Baardssön — she could
see now it might be true that Munan had been comely in
his youth. Nay, 'twas Aashild, his mother's sister, that he
favoured — now she saw that he had the same eyes as
Lady Aashild — amber-yellow eyes, shining under narrow straight black brows.

At first Kristin was a little shy of this brother-in-law
of hers that had laid up such store of learning at the great
schools in Paris and Valland.* But little by little she forgot
her bashfulness. It was so easy to talk to Master Gunnulf.
It seemed not as though he talked of himself — far less
that he was fain to flount his learning. But when she had
time to bethink herself a little, she found he had told her
so much that Kristin thought she never had known before
how great a world there was outside Norway. She forgot
herself and all her affairs, as she sat looking up in the
priest's round large-boned face, with its subtile sprightly
smile. He had laid one leg over the knee of the other
under his cassock, and sat with his white sinewy hands
clasped round his ankle.

When, late in the afternoon, he joined her in the hall, he
asked if they should play draughts. Kristin could but answer that she thought not there was any draught-board in
the house.

"Is there not?" asked the priest in wonder. He went over
to Ulf:

"Know you, Ulf, what Erlend has done with mother's
gilded draught-set? — The things for pastime that she left
behind her here — surely he has not let any other have
them?"

"They are in a chest above in the armoury," said Ulf.
" 'Twas in his mind, methinks, that they should not come
into the hands of others — that were on the manor hereto-
fore," he added low. "Would you have me fetch the chest,
Gunnulf?"

---

* A general name for the Latin countries.

"Aye — Erlend cannot, surely, have aught against it now," said the priest.

A little after, the two came back bearing a great carven chest. The key was in it, and Gunnulf opened it. On top lay a cithern and another stringed instrument whose like Kristin had never seen before. Gunnulf called it a salterion — he let his fingers stray over the strings, but it was untuned. There were rolls of ribands, reels of silk, broidered gloves, and silken hoods, and three books with clasps. At length the priest found the draught-board; the squares were in white and gold, and the men were of narwhal ivory, white and golden.

Kristin had to own now to her brother-in-law that she was slow-witted at the draughts and had no great skill of stringed music. But the books she was eager to look into.

"Aye — belike you have learnt to read in books, Kristin?" the priest asked; and now she could answer, a little proudly, that so much she had indeed learned while she was yet a child. And in the cloister she had been praised for her skill in reading and writing.

The priest stood over her with a smile on his face while she turned the leaves of the books. One was a knightly saga of Tristan and Isolde, and the other held histories of holy men — she opened it at St. Martin's saga. The third book was in Latin, and was in a passing fair script with great capital letters painted in many hues.

"This one belonged to our ancestor, Bishop Nikulaus," said Gunnulf.

Kristin read, half aloud:

> *Averte faciam tuam a peccatis meis —*
> *et omnes iniquitates meas dele.*
> *Cor mundum crea in me, Deus —*
> *et spiritum rectum innova in visceribus meis.*
> *Ne projicias me a facie tua —*
> *et Spiritum Sanctum tuum ne auferas a me.**

"Understand you this?" asked Gunnulf, and Kristin nodded and said she understood a little. She knew enough

---

* Psalm li. 9–11.

of the words' meaning to be strangely moved that her eyes should fall on them just now. Her face quivered a little, and she could not keep back her tears. Then Gunnulf took the psaltery on his lap, and said he would try if he could not mend it.

While they sat thus, they heard the trampling of horses in the court-yard — and straightway after Erlend burst into the hall, beaming with gladness — he had heard who it was that was come. The brothers stood with their hands on each other's shoulders, Erlend asking questions and not waiting for the answers. Gunnulf had been in Nidaros two days, so it was pure chance that they had not met there.

" 'Tis strange too," said Erlend. "Methought that all the priesthood of Christ's Church would have gone forth in procession to meet you, when you came home — so wise and stuffed with learning as you must be now — "

"And know you so surely that they did not?" asked his brother, laughing. "You come not over nigh to Christ's Church when you are in the city, I have heard tell."

"True, boy — I draw not nigh to my Lord Archbishop when I can steer clear of him — he hath singed my hide for me once already." Erlend laughed unrepentantly. "How like you your brother-in-law, my sweet? — I see you have made friends with Kristin already, brother — she cares not much for our other kindred. . . ."

It was not till they were sitting down to the supper-board that Erlend marked that he still had his fur cap and his cloak on and the sword at his belt.

It was the merriest evening Kristin had yet had at Husaby. Erlend forced his brother to sit with her in the high-seat; he himself carved for him and filled his cup. The first time he drank to Gunnulf, he kneeled down on one knee, and made as though to kiss his brother's hand.

"All hail, Lord! We must use us, Kristin, to show the Archbishop all seemly honour — nay, for Archbishop you surely will be one day, Gunnulf!"

It was late before the house-folk left the hall, but the two brothers and Kristin sat on over their drink. Erlend had set himself on the table facing his brother.

"Aye, this gear I had thought on at our bridal," said he, pointing to his mother's chest, "and thought that Kristin should have it. But 'tis so easy for me to forget; and

you, brother — you forget nothing. But the ring my
mother left hath found its way on to a fair hand, methinks?" He laid Kristin's hand on his knee and turned
round her betrothal-ring.

Gunnulf nodded. He laid the psaltery in Erland's lap:

"Sing now, brother; you were wont to sing so sweetly
and play so well in old days — "

" 'Tis many years since," said Erland, more gravely.
Then he ran his fingers over the strings:

> Good King Olav, Harald's son,
> Rode in the thick woods' shade;
> Found in the earth a footprint small,
> — Here be tidings great!
>
> Out spoke he then, Finn Arnessön,
> — Rode of the meiny foremost:
> "Fair would show such a little foot,
> All in scarlet hosen . . . ."

Erland smiled as he sang, and Kristin looked up at the
priest a little timidly — not knowing but he might mislike this ditty of St. Olav and Alvhild. But Gunnulf sat
smiling — yet she felt sure, of a sudden, that 'twas not at
the song, but at Erland.

"Kristin need not sing to-night. I trow you are short of
breath now, my dearest," said Erland, stroking her cheek.
"But now 'tis your turn — " He gave the psaltery to his
brother.

One could tell from the priest's playing and singing that
he had been well schooled:

> The King rode northward into the hills —
>
> He heard a dove that made her moan,
> Lamenting that her mate was gone:
>
> > *Lully, lulley! lully lulley!*
> > *The falcon hath borne my mate away!*
>
> After the hawk he is fain to ride;
> It flies through the wild hills far and wide.
>
> It led him up, it led him down,
> It led him into an orchard brown.
>
> In that orchard there was an hall
> That was hangèd with purple and pall.

There lieth a fair knight in his blood —
He is the Lord so brave and good.

At his bed's head there standeth a stone,
Corpus Domini written thereon.

Lully, lulley! lully lulley!
The falcon hath borne my mate away! *

"Where learned you that song?" asked Erlend.

"Oh — some boys sang it outside the hostel where I lodged in Kanterborg," said Gunnulf. "And methought I would try to turn it into our Norse tongue. But it goes not so well in Norse." He sat playing snatches of the tune on the strings.

"Well, brother — 'tis long past midnight. Like enough, Kristin needs to come to her bed now — are you weary, my wife?"

Kristin looked up at the men in fear; she was very pale:

"I know not. . . . Methinks 'twere best now I should not lie in the bed in here — "

"Are you sick?" they both asked, bending over her.

"I know not," she said as before. She pressed her hands behind her hips. "My back feels so strange — "

Erlend sprang up and went towards the door. Gunnulf followed him:

" 'Tis an ill chance that you brought them not here before this, the ladies that are to help her," he said. "Is it come much before she looked for it — ?"

Erlend flushed a burning red.

"Kristin deemed she would need none other than her own maids — they have borne children themselves, some of them — " He tried to laugh.

"Are you beside yourself?" Gunnulf gazed at him. "Hath not every cottar's wife skilled women and neighbours' wives † to help her when she is brought to bed — and shall your wife creep off and hide herself in a hole, like a tib-cat kittening? Nay, brother — so much of a man I would have you be as to fetch the foremost ladies of the country-side to Kristin — "

---

* See Note 3.
† See Note 4.

Erlend bent his face, flushed with shame:

"You say truly, brother. I will ride myself to Raasvold — I must send men to the other manors. And do you bide here with Kristin!"

"Are you going forth?" asked Kristin fearfully, when she saw Erlend put on his riding-cloak.

He came across and threw his arms around her.

"I go to fetch the best women in the country-side for you, my Kristin. Gunnulf will stay by you, while the maids make ready for you in the little hall," he said, kissing her.

"Could you not send one to Audfinna Andunsdatter?" she begged. "But not before daylight — I would not have her waked from sleep for my sake — she has so much on her hands, I know —"

Gunnulf asked his brother who Audfinna was.

"It seems not to me over-seemly," said the priest. "The wife of one of your tenants —"

"Kristin shall have it as she will," said Erlend. And as the priest went out with him and he stood waiting for his horse, he told the other how Kristin had come to know the farmer's wife. Gunnulf bit his lip and stood deep in thought.

There was noise and commotion now throughout the manor; men rode away into the night, and serving-women came running in to ask how it fared with their mistress. Kristin said there was not much amiss as yet, but they must make all things ready for her in the little hall. She would send word when she would be brought in there.

Then she was left alone again with the priest. She strove to speak evenly and cheerily with him as before.

"*You* are not afraid," he said with a little smile.

"Nay, but I *am* afraid!" She looked up into his eyes — her own were dark and frightened. "Know you, brother-in-law — were they born here at Husaby, Erlend's other children?"

"No," said the priest, quickly. "The boy was born at Hunehals, and the little maid up in Strind — on a farm he once owned there. — Is it," he asked in a little, "that it has troubled you to remember that this other woman lived here with Erlend before?"

"Aye," said Kristin.

" 'Tis hard for you to judge justly of Erlend's doings in this matter of Eline," said the priest gravely. "It was no easy thing for Erlend to rule himself — never has it been easy for Erland to know what right was. For, ever since we were little children, so has it been, that whatever Erlend did, mother thought it was well done, and father that it was ill. Aye, he has told you, doubtless, so much of our mother, that you know of all this — "

"For all I can remember, he has named her but twice or thrice," said Kristin. "But I have seen well enough that he loved her."

Gunnulf said softly:

"Surely there has never been such love between a mother and her son. Mother was much younger than my father. Then there befell this mischance of her sister, Aashild — Baard, our father's brother, died, and 'twas said — aye, doubtless you know of that? Father believed the worst, and he said to mother — Erlend flung his knife at father once, when he was yet a boy — he flew at father's throat more than once for mother's sake, when he was half-grown. . . .

"When mother fell sick, he parted him from Eline Ormsdatter. Mother fell sick with sores and scabs on her flesh, and father said 'twas leprosy. He sent her from him — would have forced her to dwell as a commoner* with the Sisters at the spital. Then Erlend fetched mother away and bore her with him to Oslo — they went, on the way, to Aashild, who is skilled in leechcraft, and she, and the King's French leech too, said that she was no leper. King Haakon welcomed Erlend kindly then, and bade him try the virtue of the holy Erik Valdemarssön's grave — the King's mother's father. Many had there found healing for skin-sicknesses.

"Erlend set forth for Denmark with mother, but she died aboard his ship, south of Stad. When Erlend came home with her — aye, you must bear in mind that father was stricken in years, and Erlend had been an unruly son all his days — when Erlend came to Nidaros with mother's body — father was in our town dwelling then, and he

---

* See Note 13 to *The Bridal Wreath (The Garland)*.

would not take Erlend in — before he saw whether the
boy had taken the sickness, he said. Erlend took horse and
rode off, and rested not till he came to the farm where
Eline was with her son. And after that he held fast to her,
in despite of all, despite that he himself was weary of her;
and so it came about that he brought her to Husaby and
set her to rule his house, when he was once master here.
She had this hold on him, that she said if he failed her
after this, he were worthy to be smitten down himself
with leprosy. — But now 'tis time, I trow, for your women
to see to you, Kristin — " he said, looking down at her
young face, grown grey and stiff with horror and torment. But when he would have gone to the door, she cried
aloud after him:

"No, no, go not from me — "

" 'Twill be all the sooner over," the priest said to comfort her, "since you are so sick already."

" 'Tis not that!" She gripped his arm hard. "Gunnulf — " It seemed to him he had never seen such terror
in a human face.

"Kristin — remember — you must remember, this is no
worse for you than for other women!"

"Yes. Yes." She pressed her face down on the priest's
arm. "For now I know that Eline and her children should
be sitting in my place. For he had pledged her his faith
and wedded troth, ere ever I came to be his paramour — "

"Know you that?" said Gunnulf calmly. "Erlend knew
no better himself then. But you know that that word of
his he could not keep — never had the Archbishop given
his leave that they two should wed. Think not it can be
that your marriage holds not good. You are Erlend's true
wife — "

"Oh, I had thrown away all right to tread the earth
long before he wed me. And 'twas yet worse than I knew
— oh, would I might die, and this child never be born — I
dare not see what 'tis I have borne within me — "

"God forgive you, Kristin — you know not what you
say! Would you wish that your child die unborn and unchristened — ?"

"Aye, for, whatever befall, what I bear beneath my heart
must be the devil's. It cannot be saved. Oh, had I but drunk
the draught Eline proffered me — it had mayhap been

some atonement for all we had sinned, Erlend and I. — Then had this child never been gotten — oh, Gunnulf, all the time have I known it — that when I should see the thing I had nourished within me, then would I know full well it had been better for me to drink the draught of leprosy that she proffered me, than to drive her to death, to whom Erlend had first bound himself — "

"Kristin," said the priest, "you speak you know not what. 'Twas not you that drove that hapless woman to her death. Erlend *could* not keep the word he had pledged her when he was young and knew little of law or of right. Never could he have lived with her but in sin. And she had let herself be led astray by another too, and Erlend would have wed her to that other when he heard it. 'Twas not your doing that she took her life — "

"Would you know how it came to pass that she took her life?" Kristin was so hopeless now that she spoke quite calmly. "We were at Haugen together, Erlend and I, and she came thither. She had a horn with her, she would have had me drink with her — 'twas for Erlend she had meant it, I can see now, but when she found me there with him, she would have had me — I knew that there was treachery — I saw that she drank no drop herself when she set her lips to the horn. But I would as lief have drunk — I cared not whether I lived or died, since I had come to know he had had her with him at Husaby all the time. Then came Erlend in — he threatened her with his knife, to make her drink first. She begged and prayed, and he would have let her go. Then the devil took hold on me — I took the horn. — 'One of us two, your paramours,' said I — I egged Erlend on — 'you cannot keep us both,' said I. So it was that she slew herself with Erlend's knife — but Björn and Aashild found a device to hide how it had come about — "

"So, Moster* Aashild was of this counsel!" said Gunnulf grimly. "I understand — she had beguiled you into Erlend's hands — "

"No," said Kristin vehemently. "Lady Aashild prayed us — she prayed Erlend and she prayed me, in such wise that I know not how I could hold out against her — that we should deal honourably, so far as that yet might be —

---

* Moster = mother's sister.

fall at my father's feet and pray him to forgive us our misdeeds. But I dared not. I made pretence that I was fearful lest father should slay Erlend — oh, though I knew well father would never have harmed a man who yielded himself and his cause into his hands. I made pretence that I feared to bring on him such sorrow that he could never hold up his head again. Oh, but I have shown since, I feared not so much to bring my father sorrow — You cannot believe, Gunnulf, how good a man my father is — none could know, that knows not my father, how good he has been to me all my days. Ever has father loved me so. 'Twas that I could not bear he should know I had borne me so shamelessly, that while he deemed I was sitting among the Sisters at Oslo, learning all that was good and right — aye, for I bore the novices' weed while I was lying with Erlend in barns and in lofts down in the city — "

She looked up at Gunnulf. His face was white and hard as stone.

"See you now why I am afraid? She that took him to her, when he came tainted with leprosy — "

"Would *you* not have done it?" asked the priest, in a still voice.

"Yes, yes, yes." A shadow of the wild smile of former days flew over the woman's ravaged face.

"And, besides, Erlend was not tainted," said Gunnulf. "None but father ever believed that mother died of leprosy."

"But surely *I* must be as a leper in the sight of God," said Kristin. She laid her face down on the priest's arm, to which she was clinging. "Such as I am now, tainted with all sin — "

"My sister," said the priest softly, laying his other hand on her linen coif, "so sinful sure you cannot be, you young child, that you have forgotten that as sure as God can cleanse a man in the flesh from leprosy, so surely can He cleanse your soul from sin — "

"Oh, I know not," she sobbed, her face still hidden in his arm. "I know not — and I repent not either, Gunnulf. Frighted am I, but yet — frighted was I when I stood before the church-door with Erlend and the priest joined us together — frighted was I when I went with him in to the bride's mass — with golden crown on flowing hair, for I dared not speak of my shame to my father — with all my sins unatoned, aye, I dared not confess the truth to my own parish priest. But when I went about here at Husaby in the winter, and saw myself grow fouler with each day that passed — then was I yet more afraid, because Erlend was not towards me as he once had been — I thought of the time when he came to me in my bower at Skog of nights — "

"Kristin — " The priest tried to lift up her face. "You dare not think of such things now! Think but that God sees now your sorrow and your repentance. Turn you to the merciful maid, Mary, that hath compassion on all that are sorrowful — "

"But understand you not? — I have driven another to cast away her own life — "

"Kristin," said the priest sternly, "dare you think in your wicked pride that sin of yours can be so great that God's loving-kindness is not greater?"

He stroked and stroked her linen hood.

"Mind you not, my sister, how it was when the devil would have tempted St. Martin? Did not the fiend ask if St. Martin dared believe fully his own word when he promised all the sinners he shrove God's mercy? But the Bishop answered: 'To thee also I dare to promise God's forgiveness, in the hour that thou prayest for it — wouldst thou but cast away thy pride and believe that His love is greater than thy hate — ' "

Gunnulf stood for a little space, still patting the weeping woman's head. And he thought the while — his mouth white-lipped and hard-set — was it *so* that Erlend had dealt with his young bride!

Audfinna Andundsdatter was the first woman to come. She found the lying-in woman in the little hall; Gunnulf sat by her, and a couple of maids were busy in the room.

Audfinna greeted the priest with reverence, but Kristin rose up and went toward her with outstretched hand:

"Have thanks, Audfinna, for your coming — I know 'tis no light thing for your folks at home to do without you — ?"

Gunnulf had looked searchingly at the woman. Now he too rose:

" 'Twas bravely done of you to come so quick; there is need that my brother's wife should have one with her she can trust — she is strange to this country-side, young and unaccustomed — "

"Jesus, she is white as her coif!" whispered Audfinna. "Think you, sir, I might give her a little sleeping-draught? — methinks she had need to win some rest before it comes on her more sorely."

She set about making ready busily but quietly; felt the couch that the serving-women had made up on the floor, and bade them bring more cushions and more straw. Next she placed small stone pots with herbs in them against the fire. Thereafter she set about loosing all bands and knots in Kristin's dress, and last of all drew out all the pins from the sick woman's hair.

"Never did I see fairer," said she, when the whole flood of gold-brown silky locks rolled down around the white visage. She could not forbear to laugh a little: "Methinks it can scarce have lost much either of strength or brightness, even if so be that you bore it uncovered a little longer than was right — "

She got Kristin softly bedded among the cushions on the floor and covered her well with rugs:

"Drink this now; then will you not feel the pains so sorely; and try to get a little sleep between-times."

It was time for Gunnulf to go. He went across and bent over Kristin.

"You will pray for me, Gunnulf," she asked, beseechingly.

"I will pray for you even till I see you with your child upon your arm — and after too," he said, and laid her hand back again under the coverlid.

Kristin lay and dozed. She felt almost well. The shooting pains across her loins came and went and came again — but it was so unlike all she had felt before, that each time they had passed she wondered almost if it were not but her fancy. After the torment and horror of the early morning hours she felt as though she were already happily over the worst dread and anguish. Audfinna went about so softly, hanging up child's clothing, rugs and furs to warm at the hearth — and stirring her pots a little, so that a spicy smell stole out into the room. At length Kristin fell half asleep between the fits of pain, and dreamed she was at home in the brewhouse at Jörundgaard, and was helping her mother with the dyeing of a great web of cloth — doubtless 'twas the steam from the ash-bark and nettles.

Then came the lady-midwives, one after another — ladies from the manors of their parish and of Birgsi. Audfinna drew back among the serving-women. And when evening was drawing on, Kristin felt the pains grew sore. The ladies said she should walk about the room as long as she could bear to do so. It was torture to her — the room was chock-full of women now, and she must pace about like a mare put up for sale. Between-whiles, too, she must let the strange ladies press and feel round about her body with their hands; and then they would talk together. At length Lady Gunna of Raasvold, who was to have the ordering of all things in the room, said that now she might lie down upon the floor. The lady divided the women in two parts, one to sleep while the other waked and watched: "Aye, 'twill not be quickly over, this — but scream all you will, Kristin, when you feel the pains sore — take no heed of the sleepers. We are all here for naught but to help you, poor child!" she said gently and kindly, patting the girl's cheek.

But Kristin lay gnawing her lips and crushing the edges of the coverlid in her sweat-bathed hands. It was stiflingly hot — but they said that so it should be. After every fit of pain the sweat poured from her.

Between-times she would lie thinking of the food for all these women. She was so fain they should deem that she had her house in good order. She had bidden Torbjörg, the cook, pour butter-milk into the water the fresh fish were boiled in. If only Gunnulf would not deem it a breach of the fast. Sira Eirik had said 'twas no breach, for buttermilk is not milk food, and, besides, the fish-broth is thrown away. The dried fish that Erlend had gotten for the house last autumn, they must nowise be let touch — spoiled and full of maggots as it was.

Mary, Blessed Lady of mine — will it be long, think you, till you will help me? — oh — now 'tis so hard, so hard — so hard —

She must try to hold out a little longer yet, before she gave way and screamed. . . .

Audfinna sat over by the hearth, tending the pots of hot water. Kristin wished so that she dared pray her to come to her and hold her hand. She knew not what she would have given now for the clasp of a friendly hand that she knew. But she was ashamed to ask it. . . .

All through the next forenoon a sort of bewildered stillness lay over Husaby. It was the eve of Mary's Mass, and all the work of the place should have been out of hand by the hour of nones; but the men were bemused and cast down, and the scared serving-wenches scrambled through the house-work in slovenly wise. The house-folk had grown fond of their young mistress — and things were going none too well with her, 'twas said.

Erlend stood out in the courtyard talking with his smith. He tried to keep his thoughts on what the man was saying. Then Lady Gunna came towards him swiftly:

"We can come no way with your wife, Erlend — now have we tried all shifts. You must come down — maybe 'twill help if she be set in your lap. Go in and put on you a short coat — but be hasty; she is hard bested, the poor young thing."

Erlend had grown red as blood. He remembered, he had heard — if a woman could no otherwise be delivered of a child she had conceived in secret, 'twas said it might help if she were set on the father's knee.

Kristin lay on the floor under some rugs; two women sat by her. As Erlend came in, he saw that she shrank together, bored her head into the lap of one of the women, and rolled it about here and there — but not a moan came from her.

When the fit was over, she looked up with wild, terrified eyes; the brown, cracked lips gasped open. Every trace of youth and comeliness was gone from the swollen, red, flaming face — even the hair was tangled up into a dirty mat, with bits of straw and wool from the sheepskins. She looked at Erlend as if she knew him not at first. But when she understood why the women had sent for him, she shook her head vehemently:

" 'Tis not our use, where I come from — that men should be by when a woman bears a child — "

"They use it sometimes here north of Dovre," said Erlend softly. "If it might shorten the pain for you a little, my Kristin, you must suffer it — "

"Oh — !" As he knelt beside her, she threw her arms about his waist and crushed herself against him. Crouched together and shaking, she fought through the fit without a cry.

"Can I speak two words with my husband alone?" she said, swiftly, breathlessly, when it was past. The women drew back.

"Was it when she was in labour that you promised her what she said — that you would wed her when she was a widow — that night when Orm was born?" whispered Kristin.

Erlend gasped for breath, as though he had been struck a blow above the heart. Then he shook his head vehemently:

"I was at the castle that night — 'twas my troop that had the watch. 'Twas when I came home to our lodging in the morning and they laid the boy in my arms — Have you been lying here thinking of this, Kristin?"

"Aye — " Again she clung tight to him, while a wave of pain swept over her. Erlend dried away the sweat that poured down over her face.

"Now you know this," he asked when she was still again, "would you not I should bide with you as Lady Gunna says?"

But Kristin shook her head again. And at last the women were forced to let Erlend go.

But with that it seemed as if her strength to hold out were broken. She shrieked aloud in wild terror of the pangs she felt coming, and wailed out prayers for help. Yet when the women talked of fetching the husband in again, she screamed out: no — she had rather be tortured to death —

Gunnulf and the clerk that was with him went to the church to hold evensong. Every soul on the manor who was not with the lying-in woman went with them. But Erlend stole out of the church before the service was over, and went southward toward the houses.

In the west over the hill-tops on the farther side of the

Dale the sky was yellowish red — dusk was beginning to fall in the clear mild spring evening. A star came forth here and there, white in the light-hued sky. A little flake of mist was drifting over the wood down by the lake — there were bare patches where the fields faced sunwards, and the smell of mould and melting snow was in the air.

The little hall lay westmost of the houses, out towards where the ground dipped to the valley. Erlend went over to it and stood awhile behind its wall. The timbers were still warm with the sun, when he leaned against them. Oh, her cries — ! He had heard a heifer once bellowing in the grip of a bear — it was up at their sæter, when he was a half-grown lad. Arnbjörn, the cowherd, and himself had run south through the woods. He remembered the shaggy mass that stood up and turned into a bear with hot red open maw. Arnbjörn's spear broke off in the bear's paws — then the man snatched Erlend's from him, for he stood palsied with horror. The heifer lay there still living, but udder and thighs were eaten away —

Kristin mine — oh, Kristin mine — ! Lord, for Thy blessed Mother's sake, have mercy —

He fled back to the church.

The maids came into the hall with the supper — the board was not set up, but they put down the food by the hearth. The men took bread and fish for themselves over to the benches, and sat in their places silent; they ate a little, but none seemed to have a stomach to his food. No one came to take away the dishes after the meal, and none of the men got up to go to rest. They sat on, gazing into the hearth-fire, and spoke not to each other.

Erlend had hidden himself in the corner by the bed — he could not bear that any should see his face.

Master Gunnulf had lit a little hand-lamp and set it on an arm of the high-seat. He sat himself on the bench below it with a book in his hands — and sat there, his lip just moving, soundlessly and without cease.

Once Ulf Haldorssön rose up, went over to the hearth and took a slice of soft bread, then searched a little among the sticks of firewood and picked out one. Then he went down the hall, to the corner near the entrance-door, where old Aan sat. The two busied themselves with the bread, hidden behind Ulf's cloak; and Aan cut and carved at the stick. The other men glanced across at them now and then. In a while Ulf and Aan stood up and left the hall.

Gunnulf looked after the two, but said no word. He went on again with his prayers.

Once a young lad fell off the bench in his sleep and rolled out on the floor. He rose up and looked about him bewildered. Then he sighed a little and sat down again on the bench.

Ulf Haldorssön and Aan came in again quietly and went to the places where they had sat before. The men looked over at them, but none spoke.

Of a sudden Erlend sprang up. He went across the hall to his house-folk. His face was grey as clay, and his eyes hollow.

"Is there none of you that knows a way?" said he. "You, Aan?" he whispered.

"It availed not," Ulf whispered back.

"Methinks 'tis written that she shall not have this child," said Aan, wiping his nose, "and then neither runes nor offerings avail. 'Tis pity of you, Erlend — to lose this kind young wife so soon — "

"Oh, speak not as though she were dead already," said Erlend, broken and despairing. He went back to his corner and threw himself down with his head within the bed-end.

Once a man went out and came in again. "The moon is up," he said. " 'Twill soon be morning."

A little while after, Lady Gunna came into the hall. She sank down on the beggars' bench by the door — her grey hair bunched out on all sides; her head-dress had slipped back on her shoulders.

The men rose — drew near to her slowly.

"One of you must come down there and hold her," she said, weeping. "We cannot, any more. You must go to her, Gunnulf — none can tell how it may end — "

Gunnulf stood up and thrust the prayer-book into the pouch at his belt.

"*You* must come too, Erlend," said the Lady.

* * *

The rough, hoarse crying met him in the doorway —
Erlend stopped, trembling. He caught a glimpse of Kristin's distorted unknowable face in the midst of a group of
weeping women — she was on her knees, they holding her
up.

Near by the door some serving-women had flung themselves down with their faces hidden on the benches; they
were praying aloud, unceasingly. He threw himself down
beside them and hid his head in his arms. Shriek after
shriek came from her, and each time it was as though an
icy pang of unbelieving horror went through him. This
thing *could* not be. . . .

Once he plucked up heart and looked across. Gunnulf
was sitting now in front of her on a stool, holding her
under the arms. Lady Gunna knelt by her side, and had
her arms round Kristin's waist, but Kristin was struggling,
in deadly terror, to thrust the other away.

"Oh, no — oh, no — loose me — I cannot bear — God,
God, help me — "

"God will help you soon now, Kristin," said the priest
each time. A woman stood by, holding a basin of water,
and after every spasm she took a wet cloth and wiped the
sick woman's face — the sweat from under her hair-roots
and the slime from between her lips.

Then her head fell forward between Gunnulf's arms,
and she slept for a moment — but the torments dragged
her out of her sleep again, at once. And the priest went
on saying:

"Now, Kristin, will you soon be helped — "

None thought any more what time of the night it might
be. But through the smoke-vent already the dawn grinned
greyly down.

Then, after a long frantic shriek of anguish, there came
a sudden utter stillness. Erlend heard the women bustling
about — he would have looked up; but he heard someone
weeping aloud, and he shrank down again — he dared not
know —

Then Kristin screamed again — a high wild scream of
lamentation, unlike the mad inhuman animal cries that had
gone before. Erlend started up.

Gunnulf stood bending over, holding Kristin, who still

knelt. She was looking in deadly horror at something
Lady Gunna was holding in a sheepskin — a raw, dark-red
mass, like naught but the entrails of a slaughtered beast.

The priest drew her close to him:

"Kristin mine — you have borne as fine and fair a son
as ever mother had need to thank God for — and he
breathes!" he said vehemently to the weeping women.
"He breathes — God will not be so cruel as not to hear
us — "

Even while the priest was speaking, it came to pass.
Through the mother's weary, bewildered head there
flitted, half remembered, the vision of a bud she had once
seen in the convent garden — something from out of
which broke red crinkled silken petals — and spread themselves out into a flower.

The shapeless lump of flesh moved — sounds came from
it — it stretched itself out and turned into a quite small
wine-red child in human likeness — it had arms and legs
and hands and feet, with full-formed fingers and toes on
them — it struggled and wheezed a little. . . .

"So little, so little, so little he is — " she cried aloud in
a thin hoarse voice, and sank down, helpless between
laughter and weeping. The women round about burst into
laughter and dried their tears, and Gunnulf passed her
over into their arms.

"Roll him in a trough, that he may scream the better,"
said the priest, following the women who bore the new-
born boy away to the hearth-place.

When Kristin waked from her long swoon, she was
lying in her bed. Someone had taken off the dreadful
sweat-drenched clothes, and a blessed sense of warmth
and healing was streaming into her body — they had laid
small bags of hot nettle-porridge upon her, and packed
her in with heated rugs and skins.

One bade her hush when she would have spoken. There
was a great stillness in the room. And through the stillness
came a voice that she could scarce call to mind:

" — Nikulaus, in the name of the Father, the Son and
the Holy Ghost — "

There was a trickling of water.

Kristin rose a little on her elbow and looked out. Out

there by the hearth stood a priest in white vestments, and Ulf Haldorssön lifted a red, sprawling naked child up out of the great brazen cauldron, gave it to the godmother, and took from her the lighted taper.

She had her child — it was he that was shrieking now so as almost to drown the priest's words. But she was so weary — she cared but little, and only wished to sleep —

Then she heard Erlend's voice, saying hastily and in fear:

"His head — his head is so strange."

" 'Tis swollen up," said a woman calmly. "That is nothing strange — he has had to fight hard, this lad, for his life."

Kristin cried out something. It was as though she grew awake right into her inmost heart — this was her son, and he had striven for his life even as had she.

Gunnulf turned quickly, laughing — caught the little bundle of swaddling-clothes from Lady Gunna's lap and bore it over to the bed. He laid the boy in his mother's arms. Sick with tenderness and joy, she rubbed her face against the little glimpse of a red silky-soft face within the linen cloths.

She looked up at Erlend. She knew that once before she had seen him with a grey, ravaged face like this — she could not remember when, she was so strange and dizzy in the head — but she knew that it was well that she need not remember. And it was good to see him stand thus by his brother — the priest had laid a hand upon his shoulder. A sense of measureless peace and safety came upon her, as she looked up at the tall man in alb and stole; the round, lean face under the black ring of hair was so strong, but his smile was comely and kind.

Erlend drove his dagger deep into the timber wall-post behind the mother and child.

" 'Tis needless now," said the priest, laughing, "for the boy is baptised."

Kristin came to think of somewhat Brother Edvin had once said. A new-christened child, he said, was as holy as the holy angels in heaven. 'Twas washed clean from the sins of its parents, and as yet it had done no sin itself. Timidly and warily she kissed the little face.

Lady Gunna came over to them. She was worn out and

weary, and wroth with the father, who had not had wit enough to say a word of thanks to the lady helpers. And the priest had taken the child from her and borne it to the mother — she should have done that, for she had delivered the woman, and, besides that, she was godmother to the boy.

"You have not greeted your son yet, Erlend, or taken him in your arms," she said angrily.

Erlend lifted the babe out of its mother's arms, and laid his face against it for a moment.

"I doubt I shall scarce come to like you from my heart, Naakkve, till I have forgotten that you tormented your mother so cruelly," he said, and laid the boy down again by Kristin.

"Aye, blame *him* for it, do," said the old lady wrathfully. Master Gunnulf laughed, and then Lady Gunna laughed too. She would have taken the child and laid it in the cradle, but Kristin begged hard to keep him with her a little longer. Soon after, she fell asleep, with her son close in by her — knew dimly that Erlend touched her, warily, as if he feared to hurt her by a touch, and then slept again.

5

THE TENTH day after the child's birth, Master Gunnulf said to his brother, when they were alone in the hall in the morning:

"Methinks 'tis full time now, Erlend, that you send word to your wife's kin of how things stand with her."

"I see not that there is such haste," answered Erlend. "They will scarce be overjoyful at Jörundgaard when they hear that there is a son in our house already."

"Can you believe," said Gunnulf, "that Kristin's mother knew not in the autumn that her daughter was ailing? And if she knew, then must she now be going in fear — "

Erlend made no answer.

But a little later in the day, as Gunnulf sat in the little hall talking with Kristin, Erlend came in. He had a skin-cap on his head, a short and thick outer coat of wadmal, long breeches and shaggy boots. He bent over his wife and patted her cheek:

"Tell me, Kristin mine — would you have me bear any greeting from you to Jörundgaard? — for now am I bound southward to tell them of our son."

Kristin flushed deeply — she looked both affrighted and glad.

" 'Tis no more than your father has a right to crave of me," said Erlend gravely, "that I come myself with these tidings."

Kristin lay still a little.

"Tell them at home," she said softly, "that I have longed every day, since I left my home, to fall at my father's and mother's feet and pray them for forgiveness."

Soon after, Erlend left her. Kristin did not think to ask him how he was journeying. But Gunnulf went with his brother out into the courtyard. Outside the hall-door stood Erlend's ski and a spear-headed staff.

"You go on ski?" said Gunnulf. "Who goes with you?"

"None," said Erlend, laughing. "You should know best, Gunnulf, that 'tis no easy thing for any to bear me company on ski."

"Methinks 'tis folly and rashness," said the priest. "There are many wolves in the Höiland woods this year, they say — "

Erlend only laughed and began to fasten on his ski.

"I shall be up by the Gjeitskar sæters, I trow, before 'tis dark. There is long light already. By evening the third day I should be at Jörundgaard — "

" 'Tis ill going from Gjeitskar on to the beaten road — and there are bad fog-pockets to pass through. And you know that those sæters are ill places in winter-time."

"You can give me your flint and steel," said the other, still laughing, "lest by chance I should have to cast away my own — at some elf-woman if she should crave such *kurteisi* of me as beseems not a wedded man. Come, brother; now am I to do as you would have me — betake me to Kristin's father and bid him crave such amends from me as he deems fair and right — so far you can sure let me guide myself, as to choose how I shall journey."

With this Master Gunnulf was forced to be content. But he warned the house-folk strictly to keep it hidden from Kristin that Erlend had gone forth alone.

* * *

The southern sky stretched pale-yellow over the deepening blue of the mountain snowfields, the evening that Erlend rushed down past the Sil churchyard, the snow-crust hissing and crunching under his ski. High up rode a half-moon, shining misty-white in the evening twilight.

At Jörundgaard dark smoke was whirling up from the vent-holes towards the pale clear sky. The strokes of an axe rang out, measured and cold through the stillness.

From the gateway a pack of farm-dogs rushed out barking at the new-comer. Inside the courtyard a flock of shaggy goats were picking their way about, dark in the clear dusk — they were tugging at a heap of pine-branches in the midst of the yard. Three little children in thick winter clothes ran about amongst them.

The homely peace of this place took hold on Erlend strangely. He stood uneasily and waited for Lavrans, who came out to meet the stranger — he had been standing down by the woodshed talking to a man who was splitting fence-staves. He stopped short when he saw it was his son-in-law — thrust the spear he bore in his hand hard down into the snow.

"Is it *you*?" he asked in a low voice. "Alone — ? Is there — is aught — ? How is it that you come in this wise?" he added in a moment.

"Thus it is." Erlend pulled himself together and looked his father-in-law in the face. "Methought that less I could not do than come myself to bear you these tidings: Kristin bore a son on Mary's Mass in the morning. — Aye, she does well now," he added quickly.

Lavrans stood still awhile. He set his teeth hard in his under lip — his chin shook and quivered a little.

"These were tidings!" he said at last.

Little Ramborg had come up, and stood at her father's side. She looked up, her face glowing red.

"Be still," said Lavrans harshly, though the little maid had not said a word, but only blushed. "Stand not here — begone — "

He said no more. Erlend stood, bent forward, leaning on the staff clenched in his left hand. He looked down at the snow. His right hand was thrust into his breast. Lavrans pointed:

"Are you hurt?"

"A little," said Erlend. "I came over some bare rocks last night in the dark."

Lavrans took hold of the wrist and felt it warily: "No bone is broken, methinks," he said. — "You must tell her mother yourself — " He went off towards the hall as Ragnfrid came into the yard. She looked in wonderment after her husband — then she knew Erlend and went swiftly towards him.

She hearkened without a word, while Erlend for the second time had to bring forth his tidings. But her eyes shone with moisture when he said at the end:

"Methought that you had maybe seen somewhat before she left you last autumn — and that you might be fearful now for her — "

" 'Twas kindly done of you, Erlend," said she, in a voice that shook, "to think of this. True it is that I have been afraid for her every day since you took her from us."

Lavrans came back:

"Here is fox's fat — I see your cheek is frozen, son-in-law. You must bide awhile in the outer room while Ragnfrid dresses it with this and gets you thawed — how stands it with your feet? — you must take off your boots that we may see — "

When the house-folk came in to supper, Lavrans told them the tidings, and bade that strong ale be brought in for them to make merry on. But there was no right merriment over the ale-drinking — the master himself sat there with a cup of water. He prayed Erlend to forgive this — 'twas a vow he had made while yet a boy, to drink naught but water in fast-time. And so the folks sat somewhat soberly, and the talk went but tardily over the good ale. The children would come round to Lavrans now and then — he put an arm round them when they came close to his knee, but he answered absently when they spoke to him. Ramborg answered short and sharply when Erlend tried to jest with her — bent, it seemed, on showing that she misliked this brother-in-law of hers. She was in her eighth winter now; lively and comely, but bearing no likeness to her sisters.

Erlend asked who the other children might be. Lavrans answered that the boy was Haavard Trondssön, the

youngest of the Sundbu children. 'Twas dull for him there
amid his grown-up brothers and sisters; and last Yule he
had set his heart on coming home with Ragnfrid, his
father's sister. The little maid was Helga Rolfsdatter of
Blakarsarv — there had been naught for it but for the
kinsfolk to take the children away home with them after
the grave-ale there — 'twere pity they should see their
father as he was now. For Ramborg it was a happy thing
to have this foster-brother and sister. "We begin to grow
old now, Ragnfrid and I," said Lavrans, "and this little one
is more frolicsome and fond of play than Kristin was" —
he stroked his daughter's curly hair.

Erlend went and sat by his mother-in-law, and she asked
him of Kristin's childbed. He saw that Lavrans was listening to what they said. But soon he stood up, crossed the
room, and put on hat and cloak. He had a mind to go over
to the parsonage, he said — he would pray Sira Eirik to
come and drink with them.

Lavrans went by the well-trodden path over the fields to
Romundgaard. The moon was dipping behind the hills
now — but thousands of stars glittered above the white
mountains. — He hoped that the priest might be at home
— he could bear no longer to sit alone with the others.

But when he was come in between the fences near the
farm, he saw a little taper coming towards him. Old
Audun was bearing it — when he marked that there was
someone on his path, he rang his little silver bell. Lavrans
Björgulfsön threw himself on his knees in the snow-drift
by the path.

Audun went by bearing the taper, and the bell that still
tinkled gently. Behind him came Sira Eirik a-horseback.
He lifted the pyx high in his hands when he came by the
kneeling man — looked not to right or left, but rode
calmly past, while Lavrans bowed himself down and
stretched his two hands up in greeting towards his
Saviour.

— 'Twas Einar Hnufa's son, the man that was with the
priest — so 'twas drawing to an end with the old man
now — ! Aye, aye. Lavrans said the prayers for the dying
before he rose from the snow and went homewards. Even

so, this meeting with God in the night had strengthened and comforted him much.

When they had gone to rest, he asked his wife:

"Knew you aught of this — that 'twas *so* with Kristin?"

"Did you not know?" said Ragnfrid.

"No," answered her husband, so shortly that she understood it must none the less have been in his thought at times.

" 'Tis true I was afraid one while this last summer," said the mother, haltingly. "I saw that she had no joy in her food. But as time wore on, I deemed I must have been deceived. She seemed so joyous all the time we were making ready for the wedding — "

"Aye, for *that* she had good reason," said the father somewhat grimly. "But that she said naught to you — you that are her mother — "

"Aye, you can remember that, now she has done amiss," answered Ragnfrid bitterly. "You know full well, never has Kristin been used to turn to me — "

Lavrans said no more. In a little while he gently bade his wife sleep well, and lay quiet beside her. He felt that sleep would scarce come to him for a long time yet.

Kristin — Kristin — his little maid —

— Never had he touched with a single word on what Ragnfrid had confessed to him that night of the bridal. And she could not with reason think that he had let her feel he thought of it. He had not changed in his bearing towards her — rather had he striven to show her yet more friendliness and love. But 'twas not the first time this winter that he had marked this bitterness in Ragnfrid, or seen her search for some hidden offence in innocent words of his. He understood it not, and he knew of no remedy — he must let it be.

Our Father which art in heaven — He prayed for Kristin and her child. Then he prayed for his wife and himself. Last of all he prayed for strength to bear with Erlend Nikulaussön in patience of spirit, for so long as he needs must have his son-in-law dwelling on this, his manor.

Lavrans would not have his daughter's husband set forth for home till it was seen what turn things would take with

his wrist. And he would not hear of Erlend's going back
alone.

"Kristin would be joyful if you bore me company,"
said Erlend, one day.

Lavrans was silent awhile. Then he brought forward
many lets and hindrances. Ragnfrid would like little to be
left alone here on the manor. And should he fare so far
north, he could scarce hope to be back for the spring
sowings. But the end was that he set out with Erlend. He
took no man with him — he would come back by ship to
Raumsdal; he could hire horses to bring him thence down
the Dale — he had acquaintance everywhere along that
road.

They spoke not much together on their way, but they
journeyed in good accord. It tasked Lavrans' strength to
keep up with Erlend, for he would not own that the other
went too fast for him. But Erlend soon marked this, and
suited his pace to his father-in-law's. He gave himself great
pains to please his wife's father — he had this quiet compliant way with him when he wished to win the friendship
of any.

The third night they took shelter in a stone hut. They
had had foul weather, with mist, but Erlend seemed to
find the way as surely as ever. Lavrans had marked that
Erlend had a marvellous sure eye for all signs and marks
in the air and on the earth, and for the nature and ways
of beasts — and he ever knew where he was. All that he
himself, used as he was to the hills, had taught himself by
looking and marking and remembering, the other seemed
to know blindly. Erlend laughed himself about it — he
did but feel it all within him.

They found the hut in the pitch-darkness, at the very
hour Erlend had foretold. Lavrans thought to himself of
a night like this when he had made a bed for himself in the
snow but a bow-shot from his own horse-camp. The snow
was drifted high around the hut, so that they had to break
in through the smoke-hole. Erlend covered over the opening with a horse-hide that was lying in the hut, and fastened it with slats of wood which he pressed in under the
rafters. He scraped away with a ski the snow that had

sifted in, and managed to make up a fire in the fire-place with the frozen wood that lay there. He drew out three or four ptarmigan from under the bench — he had hid them away there on his way south — plastered them round with clay from the floor by the fire-place where it had thawed, and threw the lumps into the glowing embers.

Lavrans lay on the earthen bench, where Erlend had made a couch for him as well as he could with their wallets and cloaks.

" 'Tis the fashion the soldiers use with stolen fowls, Erlend," said he, laughing.

"Aye, for I learned one thing and another when I was serving the Count," said Erlend, in the same tone.

He was as brisk now and full of life as he had been quiet and almost sluggish at most times his father-in-law had seen him. He began to tell tales, as he sat on the ground in front of the other, of the years when he had served Count Jacob of Halland. He had been troop-leader in the Castle, and he had cruised about with three small ships to guard the coast. Erlend's eyes were like a child's now — he did not brag, he but let his tongue run on. Lavrans lay looking down at him. . . .

He had prayed God to grant him patience with this husband of his daughter — and now he was well-nigh angry with himself that he liked Erlend so much better than he had a mind to. He remembered, too, that that night when their church burned down he had liked his son-in-law well. 'Twas not in that long carcass of his that Erlend lacked manhood. A stab of pain went through the father's heart — 'twas pity of Erlend, he might have been fit for better things than beguiling of women. But, as things were, little more had come of all the rest of him than boyish pranks. Had the times been such that a chieftain could have taken this man in hand and used him — but as the world now was, when each man must trust to his own judgment in many things — and a man in Erlend's place had in his hands not his own welfare only, but so many others' — And this was Kristin's husband. . . .

Erlend looked up at his father-in-law. He too grew grave. Then he said:

"One thing would I pray you, Lavrans — before we come home to my house — that you would say to me what you must needs have on your heart."

Lavrans was silent.

"You know well," said Erlend, as before, "that I will willingly submit me to you in any fashion you may wish, and make such amends as you deem may be a fitting punishment for me."

Lavrans looked down into the young man's face — then he smiled strangely:

"That might be a hard matter, Erlend — for me to say and for you to do. — But at the least you must give fitting gifts to the church at Sundbu and to the priests whom you two fooled along with other folk," he said vehemently. "I will speak of it no more. You cannot blame it on your youth either. More honour had it been to you, Erlend, had you confessed and made submission before I made your bridal — "

"Aye," said Erlend. "But I knew not then that things stood so that the wrong I had done you must come to light."

Lavrans sat up.

"Knew you not, when you were wedded, that Kristin — ?"

"No," said Erlend, with a crestfallen look. "We had been wedded nigh on two months before I knew it."

Lavrans looked at him with some wonder, but he said naught. Then Erlend went on, haltingly and weakly:

"Glad am I that you came with me, father-in-law. Kristin has been all this winter so heavy of mood — she has scarce cared to speak a word to me. Many a time has it seemed to me as though she found but little happiness either at Husaby or with me."

Lavrans answered somewhat coldly:

" 'Tis the same, I trow, with all young wives. Now she is well once more, doubtless you will soon be as good friends again as you were before," he added, smiling a little mockingly.

But Erlend sat gazing into the heap of embers. It came over him so surely, of a sudden — though truly he had felt it ever since he first saw the little red baby face against Kristin's white shoulder — things would never-more be between them as once they had been.

When her father came into the little hall where Kristin lay, she sat up in bed and stretched out towards him. She threw her arms around him, and wept and wept so sorely that Lavrans was afraid.

She had been up for a time, but then she came to know that Erlend had gone south over the hills alone, and when time dragged on and he returned not, she grew so fearful that fever came on her and she must needs lie abed again.

It was plain to see that she was weak still — weeping came on her for never so little things. — The new chap-lain, Sira Eiliv Serkssön, had come to the manor while Erlend was away. He had taken it on him to go now and then to sit by the lady of the house and read to her — but she wept at the least thing so causelessly that he scarce knew what he dared let her hear.

One day when her father sat with her, Kristin had set her heart on swaddling the child herself, so that he might see rightly how fair and well-shaped the boy was. As he lay sprawling amidst his swaddling-clothes on the coverlid in front of his mother:

"What is yonder mark he has on his breast?" asked Lavrans.

Right over his heart the boy had some small blood-red spots — it looked as though a bloody hand had touched him there. Kristin had herself been troubled when first she saw this mark. But she had tried to comfort herself with the reason she now gave:

" 'Tis but a fire-mark, belike — I caught at my breast when I saw the church burning."

Her father started. Aye — true it was that he knew not how long — or how much — she had kept hidden. And he could not understand how it had been possible for her — his own child — from him. . . .

"Methinks you have no right liking for my son," Kris-tin said to her father many times; and Lavrans laughed and said: nay, that he like him right well. He had brought rich gifts, too, to lay both in the cradle and in the lying-in woman's bed. But Kristin deemed that on one truly thought enough of her son — Erlend least of all. "Look on him, Father," she begged; "see you, he is laughing — did you ever see so fair a babe as Naakkve, Father?"

She asked this, time and again. Once Lavrans said, as though in thought:

"Haavard, your brother — our second son — was a passing fair child."

In a little Kristin asked, in a weak voice:

" 'Twas he that lived the longest of my brothers?"

"Aye, he lived to be two winters old. . . . Nay, now, my Kristin, you must not weep again," he prayed her gently.

Neither Lavrans nor Gunnulf Nikulaussön liked the boy to be called Naakkve; he was christened Nikulaus. Erlend would have it that 'twas the same name; but Gunnulf said: no; sagas told of men called Naakkve back in heathen times. But naught could bring Erlend to use the name his father had borne; and Kristin ever called the boy the name by which Erlend had first greeted their son.

So to Kristin's mind there was but one at Husaby besides herself who fully understood how noble and hopeful a child Naakkve was. This was the new priest, Sira Eiliv. — In this matter his judgment was scarce less sound than the mother's own.

Sira Eiliv was a short spare-limbed man with a little round belly, and this gave him a somewhat laughable look. His presence was unremarkable — folks who had spoken with him more than once still found it hard to know the priest again, so common was his face. His hair and skin were of the same hue — like reddish-yellow sand — and his round watery-blue eyes were flat with his head. In his bearing he was quiet and retiring; but Master Gunnulf said that Sira Eiliv was so learned that he, too, could have passed through all the degrees, had he but had more forwardness. But much more than by learning was he adorned with purity of life, humility and devout love towards Christ and His Church.

He was of low kindred, and though he was but little older than Gunnulf Nikulaussön, he seemed already not far from an old man. Gunnulf had known him since they went to school together at Nidaros, and he spoke always with great love of Eiliv Serkssön. Erlend deemed 'twas no great matter of a priest they had gotten for Husaby, but Kristin soon looked on him with trust and love.

Kristin still lived down in the little hall with the child, even after she had been churched. 'Twas a heavy day for Kristin that — Sira Eiliv led her within the church-door, but he dared not give her the Lord's body. She had confessed herself to him, but for the sin she had committed, as partner in the guilt of another's unblessed death, she must seek absolution from the Archbishop. That morning when Gunnulf had sat with her in her soul's agony, he had strictly charged her that, as soon as she was free from danger of bodily death, she must haste to seek healing for her soul. So soon as health and strength enough were hers, she must fulfill her vow to St. Olav. Now that he by his intercession had saved her son and brought him alive and whole to the light and to cleansing baptism, she must walk barefoot to his grave and lay down upon it the golden garland of maidenhood, which she had guarded so ill and borne so wrongfully. And Gunnulf counselled her to prepare herself for this pilgrimage by solitary life, prayers, reading and meditation, also by fasting, but this with due measure, for the sake of the child at her breast.

The evening of her churching, when she was sitting sorrowful, Gunnulf had come to her and given her a rosary. He said that in foreign lands 'twas not only cloister-folk and priests that used such bead-rolls for a help in their pious exercises. This rosary was a most fair one; the beads were of a sort of yellow wood that came from India, and smelt so sweet and delicate that they were well fitted to bring to mind that which a good prayer should be — the heart's sacrifice and yearning for help to live righteously before God. Some among the beads were of amber and gold, and the cross was of a fair enamel.

This spring Erlend Nikulaussön busied himself much with setting his estate in order. This year all fences were mended and gates set up in due time, the ploughing and spring-sowing well and early gotten out of hand, and Erlend bought some right good horned cattle. He had had to slaughter many at the new year, and 'twas no great loss, so many old and wretched beasts as there had been in his herd. He got together folk for tar-burning and birch-bark-peeling, and the houses of the manor were timbered up and their roofs repaired. There had not been such order at Husaby, folks said, since old Sir Nikulaus was in his full strength. Aye, and 'twas known, too, that the master sought counsel and help from his wife's father. With him and with his brother, the priest, Erlend went about and visited friends and kinsfolk in the country round when he had leisure from his work. But now he went around in seemly wise, with a few brisk and likely serving-men. In former days Erlend had used to ride about with a whole troop of unruly hotheads. And so the talk of the country-side, which so long had seethed with wrath over Erlend Nikulaussön's shameless evil life and shiftless and ruinous husbandry at Husaby, died away now into good-humoured jesting. Folks smiled and said that the young housewife Erlend had gotten had brought much to pass in six months.

A while before Botolph's Mass, Lavrans Björgulfsön set forth for Nidaros in Master Gunnulf's company. He was to be the priest's guest for some days, while he sought out the shrine of St. Olav and the other churches in the city, before he journeyed south to his home again. He parted from his daughter and her husband in all love and kindness.

6

KRISTIN was to set out on her walk to Nidaros three days after the mass of the Seljemen* — later on in the month the city would be full of bustle and commotion making ready for the Olav's Mass† festival; and earlier the Arch-bishop would not be in the town.

The evening before, Master Gunnulf had come to Husaby, and early in the morning he went with Sira Eiliv to the church to sing Matins. The grass was as a grey fur coverlid with the heavy dew, as Kristin walked to the church; but the sunlight was golden on the woods that

_____
* 8th July.
† 29th July.

topped the ridge, and the cuckoo called from the hill-side
— it looked as though she would have fair weather for her
pilgrimage.

There was none in the church save Erlend and his wife,
and the two priests in the lighted choir. Erlend looked
across at Kristin's naked feet. Ice-cold must it be for her,
standing on the stone floor. She was to walk the twenty
miles with no other company than their prayers. He strove
to lift up his heart to God, so as he had not striven for
many a year.

She was clad in an ashen-grey kirtle, and had a rope
about her waist. Underneath, he knew, she wore a shift of
sackcloth. A tightly bound wadmal cloth hid her hair.

When they stepped from the church out into the morning sunshine, a maid met them with the child. Kristin set
herself down on some logs of wood. With her back to her
husband she sat and let the boy suck his fill, that he might
be full-fed when she set out. Erlend stood, unmoving, a
little space away from her — he was white and cold in
the face with the strain.

The priests came out a little later — they had taken off
their vestments in the sacristy. They stopped by Kristin.
Then Sira Eiliv went on down towards the manor, but
Gunnulf stayed and helped her to get the child securely
bound on her back. In a bag that hung from her neck she
had the golden garland, money and a little bread and salt.
She took her staff in her hand, bowed deeply before the
priest, and began to walk quietly northwards up the path
that led to the woods.

Erlend was left standing — deadly white of face. Of a
sudden he began to run. North of the church were some
small hillocks covered with scanty grass and close-cropped
juniper and birch — goats were used to graze there. Erlend ran up them — from there he could see her yet a little
way on — till she was swallowed up in the woods.

Gunnulf walked slowly up after his brother. The priest
looked tall and dark in the bright morning light. He, too,
was very pale.

Erlend stood with mouth half open, the tears running
down over his white cheeks. Of a sudden he fell on his
knees — then threw himself forward headlong on the short

grass, and lay, sobbing, sobbing, and tearing at the heather with his long brown fingers.

Gunnulf stood motionless. He looked down at the weeping man — then out towards the woods where the woman had vanished.

Erlend lifted his head a little.

"Gunnulf, was it needful that you should lay this upon her? — Was it needful?" he asked again. "Could not *you* have absolved her?"

The other made no answer; and he went on again:

"Had not I confessed and done penance?" He sat up. "I bought *her* thirty days' masses and vigils and a yearly mass on her death-day for ever and a grave in hallowed ground — I confessed the sin to Bishop Helge, and made pilgrimage to the Holy Blood at Schwerin — could not all this help a little for Kristin — ?"

"If you have done this," said the priest quietly, "laid before God a contrite heart and won His full forgiveness — you sure must know that the marks left by your sin here on earth you must yet strive, year in, year out, to wipe away. What you brought on her that is now your wife, when you first dragged her down into unclean living and after into manslaughter — that cannot you amend for her, but only God. Pray that He may hold His hand over her on this journey, where you cannot bear her company and guard her. And forget not, brother, so long as you two live, that you saw your wife go forth from your house in this wise — by reason of your sins, more than of her own."

Erlend said in a little:

"I had sworn by God and my Christian faith, before I took her honour, that I would never have another to wife; and she promised that never would she have another to husband, so long as we lived on the earth. You have said yourself, Gunnulf, that they that vow thus are bound in wedlock before God; any of them who thereafter wedded another would be living in adultery in His eyes. But if so it be, then was it not unclean living, I trow, for Kristin to be mine — "

"The sin was not in that you lived with her," said the priest in a while, "had that been possible without your transgressing other laws — but you had led this child away into sinful revolt against all whom God had set over her — and at last you brought blood-guiltiness upon her. I told you this also, that time when we spoke of the matter: therefore hath the Church ordained laws concerning marriage, that banns be published forth to the world, and that we priests shall not wed man and maid against the will of the kindred." He sat down, clasped one knee with his hands, and gazed out over the country-side bright with summer, with the little lake gleaming blue in the valley-bottom. "You must have known it yourself, Erlend — a thicket of briers and thorns and nettles had you sowed around you — how could you draw a young maid in to your side and she not be torn and wounded and bleeding — "

"You stood my friend more than once, brother, in the days when it was Eline and I," said Erlend, low. "I have ever been thankful for it — "

"I scarce deem that I had done it," said Gunnulf in a voice that shook, "could I have believed of you that you had the heart to deal as you have done with a fine and pure young maid — a child in years beside you."

Erlend made no answer. Gunnulf asked in a low voice:

"That time in Oslo — thought you never of how it would have gone with Kristin, if she had been found with child — while dwelling in a cloister of nuns? — and was the betrothed of another — her father a proud man, jealous of his honour — all her kindred high-born folk, unused to suffer shame — "

"You may well believe I thought of it — " Erlend had turned his head aside. "I had Munan's word to stand by her — I told her, too, of this — "

"Munan! Could you find in your heart to talk to a man like Munan of Kristin's honour?"

"He is not what you think him," said Erlend, shortly.

"And how of his wife, our kinswoman, Lady Katrin? For 'twas not your meaning, I trow, that he should carry her to one of those other places of his where he keeps his paramours — ?"

Erlend smote the ground with his fist till the knuckles grew bloody:

"Aye, 'tis the devil's own business for a man, when his wife goes to shrift with his brother!"

"She hath not confessed her to me," said the priest. "And I am not her parish priest. She made her moan to me in her bitter agony and fear — and I strove to help her and to give her such counsel and such comfort as seemed best."

"Well," Erlend threw back his head and looked at his brother, "I know it myself — I should not have done it — have had her come to me in Brynhild's house —"

The priest sat speechless a moment.

"In Brynhild Fluga's — ?"

"Aye; told she not of that, when she told the rest — ?"

"'Twould be hard enough, methinks, for Kristin to tell such things of her own husband in confession," said the priest in a while. "I think that she would rather die than tell them any other place." He sat awhile, and then said with a harsh vehemency:

"If so it were, Erlend, that you deemed you were her husband before God, he that should protect and guard her — then do your doings seem yet worse to me. You tempted her astray in groves and hay-lofts, you led her over a harlot's threshold — and, last of all, you brought her to Björn Gunnarssön and Lady Aashild. . . ."

"You shall not speak so of Moster Aashild," said Erlend softly.

"You have said yourself, before now, that you believed her guilty of our uncle's death — she and this man Björn —"

"I care naught for that," said Erlend vehemently. "Moster Aashild is dear to me —"

"Aye, I understood as much," said the priest. His mouth twisted into a little crooked scornful smile. "Since you grudged it not to her that she should meet Lavrans Björgulfsön after you had borne away his daughter. 'Twould seem, indeed, Erlend, that you deem your friendship can scarce be bought too dear —"

"Jesus!" Erlend hid his face in his hands. But the priest went on:

"Had you seen your wife's anguish of soul, as she shuddered with horror of her sins, unshriven and helpless — and she sat there and was to bear your child, and death stood at her door — so young a child herself, and so unhappy —"

"I know, I know!" Erlend trembled. "I know that she lay thinking of it, in her torment. For Jesu sake, Gunnulf, be still now — I am your brother after all!"

But the priest went on without mercy:

" Had I been a man like you and not a priest — and had I led so young and good a maid astray — I had freed me from the other — God help me, rather would I have done as Moster Aashild did with her husband and burned in hell for it world without end, than I would have borne that she should suffer such things as you have brought down on the head of your innocent love — "

Erlend sat awhile, trembling.

"You call yourself a priest," he said in a low voice. "Are you *so* good a priest that you have never sinned — with a woman?"

Gunnulf looked not at his brother. A wave of red flooded his face:

"You have no right to ask such things — but yet I will answer you. He knows that died for us upon the Cross what bitter need I have of His mercy. But I say to you, Erlend — had He on all the earth's round not one single servant that was pure and unstained by sin, and were there not in His holy Church one single priest more faithful and worthy than I, wretched traitor to my Lord that I am — yet is it the Lord's laws and commandments that are taught in it. Never can His Word be polluted by the mouth of an unclean priest, it will not burn and consume our lips — but this, maybe, you cannot understand. Yet this you know as well as I and every other filthy thrall of Satan whom He hath bought with His blood — God's law cannot be shaken nor His honour diminished. As surely as His sun is alike mighty, whether it shine upon the barren sea and waste grey mountain or upon these fair and fertile lands — "

Erlend had hidden his face in his hands. He sat long silent, and when he spoke his voice was dry and hard:

"Priest or no priest — since you are not so absolute in saintly living — understand you not — ? Could you do to a woman who had slept in your arms — borne you two children — could you do to her as Aashild did to her husband?"

The priest was silent a little. Then he said, a little mockingly:

"You were not wont to judge Moster Aashild so hardly — "

"It cannot be the same for a man, I trow, as for a woman. I mind me the last time they were here at Husaby, and Sir Björn was with them. We were sitting by the hearth, mother and Aashild, and Sir Björn was playing his harp and singing to them — I was standing by his knee. Then Baard called her — he was in bed, and he would have her go to rest at once — he used words so shameless and immodest — Moster Aashild rose, and Sir Björn too; he left the hall, but first they looked at each other — Aye, afterwards I thought, when I grew old enough to understand — it may well be it is true — I had begged leave to light Sir Björn across to the house he was to sleep in, but I dared not, and I dared not lie in the hall. I ran out and laid me down by the men in the servants' house. By Jesus, Gunnulf — for a man it can never be as it was for Aashild that night —

"No, Gunnulf — kill a woman that — except I took her with another — "

Yet had he done that very thing. But *this* Gunnulf could not say to his brother. So he asked coldly:

"Was it not true, then, either, that Eline had been untrue to you?"

"Untrue?" Erlend turned on his brother, suddenly afire. "Mean you that I should have laid it to her charge that she had given herself to Gissur — after I had made plain to her again and yet again that betwixt us all must be at an end?"

Gunnulf bowed his head.

"No, like enough you are right," he said wearily, in a low voice. But under this little breath of approval Erlend flamed up — he threw back his head and looked at the priest.

"You are so tender of Kristin, Gunnulf. Strange how you have hung over her all the spring — almost more than is seemly for a brother and a priest. Almost it might seem you grudged her to me — Were it not that things were with her as they were when you first saw her, folks might deem — "

Gunnulf looked at him. Beside himself under his brother's

look, Erlend sprang up — and Gunnulf, too, rose. He did not withdraw his look, and Erlend struck out at him with his clenched fist. The priest caught and gripped his wrist. Erlend tried to close with his brother, but Gunnulf stood immovable, holding him off.

Erlend grew quiet straightway. "I should have remembered you were a priest," he said low.

"You see that on that score you have no need to repent," said Gunnulf, with a little smile. — Erlend stood chafing his wrist.

"Aye, you ever had the devil's own strength in your hands — "

"This is like the days when we were boys." Gunnulf's voice had grown strangely soft and mild. "I have thought of it often all these years I have been from home — the times when we were boys. Often were we unfriends then, but never did it last long, Erlend."

"But now, Gunnulf," said the other sorrowfully, "it can nevermore be as when we were boys."

"No," answered the priest quietly, " 'tis like it cannot — "

They stood still a long time. At last Gunnulf said:

"I leave you now, Erlend. I will go now down to Eiliv, and bid him farewell, and then I shall set out. Aye, 'tis to the priest in Orkedal that I go; I shall not come to Nidaros while *she* is there." He smiled a little.

"Gunnulf! I meant it not so — go not from me like this — "

Gunnulf stood still. He drew one or two deep breaths, then said:

"One thing I would have you know of me, Erlend — since you know that I have knowledge of all this concerning you. — Sit down."

The priest sat himself down as before. Erlend stretched himself on the ground before him, and, hand under chin, looked up into his brother's strangely stiff and strained face. Then he smiled a little:

"What is it, Gunnulf — would you make confession to me?"

"Aye," said his brother, softly. Yet then he sat a long time silent. Erlend saw his lips move once, and he clenched his folded hands tightly round his knee.

"What is it?" A smile flickered over Erlend's face. "Sure it cannot be so — that some fair lady, far off in southern lands — "

"No," said the priest. His voice grew rough and hoarse. " 'Tis not of love —

"Know you, Erlend, how it came about that I was vowed to be a priest?"

"Aye. When our brothers died, and they feared they would lose us too — "

"No," said Gunnulf. "Munan they thought was well again, and Gaute had not taken the sickness — 'twas not till the winter after that he died. But you were lying, choking to death, and mother vowed me to the service of St. Olav if he would save your life — "

"Who has told you this?" asked Erlend in a while.

"Ingrid, my foster-mother."

"Aye, 'tis true, indeed, that *I* had been a strange gift to give St. Olav," said Erlend, laughing a little. "He had been ill-served with me. — But you have said yourself, Gunnulf, you were well content that you had been called to the priesthood from childhood up — "

"Aye," said the priest. "But 'twas not always so. I mind me well the day you rode from Husaby with Munan Baardssön, to fare to the King, our kinsman, and become his man. Your horse danced under you; your new arms glittered and shone. *I* was never to bear arms. — Fair were you, my brother — you were but sixteen winters old; and already I had seen, for many a day past, that you were well loved of dames and damsels — "

"That glory endured not long," said Erlend. "I learned to cut my nails straight across, swear by Jesus at each second word, and use my dagger to ward with while I struck with my sword. Then was I sent north, and met *her* — and was hunted from the bodyguard with shame, and my father shut his door against me — "

"And you fled forth from the land with a fair lady," said Gunnulf, quietly as before. "And we heard at home that you were Captain of Count Jacob's castle."

"Aye, that was no such great matter as it sounded here at home," said Erlend, laughing.

"Father and you were not friends — me he thought not enough of to care to be unfriends with me. Mother held me dear, I know — but how little she deemed me worth, weighed against you — that marked I best when you had fled from the land. You, brother, were the only one that loved me truly. And God knows that you were my dearest friend on earth. But in those days when I was young and witless, times have been when I deemed you had been given all too much more than I. This it was that I would have told you, Erlend."

Erlend lay face downwards on the hill-side.

"Leave us not, Gunnulf," he begged.

"Yes," said the priest, "I go. Too much have we said now one to the other. God and Mary Virgin grant that we may meet again in a better hour. Farewell, Erlend — "

"Farewell," said Erlend, not looking up.

When Gunnulf, some hours after, came forth from the priest's house, ready for his journey, he saw a man riding over the fields towards the southern woods. He had a bow slung across his back, and three hounds ran by his horse's side. It was Erlend.

Meanwhile Kristin was walking swiftly on her way, by the forest path that led over the hills. The sun stood high in the heavens now, and the pine-tops shone against the summer sky, but within the woods was still the cool and freshness of morning. The air was full of the balmy scent of pine-needles and peaty soil, and of the twinflower that sprinkled all the knolls with its small pink bells; and the grassy pathway was damp and soft and comforting to her feet. Kristin walked, saying over her prayers, and now and then she looked up at the little white fair-weather clouds that swam in the blue above the tree-tops.

All the time she could not but think of Brother Edvin. So had he walked and walked, year in, year out, from early spring to deepest winter. Over the mountain paths, under black scaurs and white snow-fields. He had rested him at the sæters, drunk of the beck, and eaten of the bread that sæter-girls and horse-herds bore out to him — then had he bidden farewell and called down God's peace and benison on fold and cattle. Down through the sighing woods of the hill-sides he had passed, down into the Dale; tall and stoop-shouldered, with head bent forward, he had wandered along the high-road past the well-tilled farms and the dwelling-places — and everywhere, wheresoever he passed, his lo in intercessions for all men left, as it were, fair weather in his track.

She met not a living thing, save now and then a few cattle — there were sæters on these hill-tops. But the path was well-trodden, and over the marshy grounds were cordwood bridgeways. Kristin was unafraid — she felt as though the monk walked invisibly by her side. Brother Edvin, if in truth thou art a holy saint, if thou standest before the face of God, pray for me now!

Lord Jesus Christ, holy Mary, St. Olav — She longed to reach the goal of her pilgrimage — she longed to cast from her the burden of the hidden sins of years, the weight of masses and offices that she had filched unlawfully while yet unshriven and unrepentant — she longed to be free and cleansed, yet more keenly than she had yearned to be delivered of her burden in the spring before the boy was born. . . .

He slept so soundly and securely on his mother's back. He waked not till she had passed through the woods and come down to the Snow-bird farms, and could look out over Budvik and the Saltnes arm of the fjord. She sat down in a pasture off the path, slung the bundle with the babe in it round into her lap, and opened the breast of her kirtle. It was sweet to feel him against her breast; sweet to be able to sit awhile; a blessed sweetness through all her body to feel the stone-hard, milk-swollen breasts grow soft as he sucked.

The country-side lay still below her, baking in the sunshine — with green meadows and bright cornfields amid dark woods. Here and there a little smoke went up above the house-roofs. In some places the hay-harvest had begun.

She had leave to cross from Saltnes-sand over to Steine by boat. Once across, she had come to a quite unknown country. The road she must follow across the Bynes led up for a while among farms, then she came into woods again, but here she never had to go for long out of sight of human dwellings. She was passing weary. But she thought of her parents — had they not walked barefoot all the way from Jörundgaard in Sil, over Dovre and down to Nidaros, bearing Ulvhild between them on a litter? She must not think how heavy Naakkve was upon her back.

— Worse was the dreadful itching in her head from the thick sweat-drenched wadmal hood. And round her waist, where the rope pressed her garments against her body, her shift had gnawed at the flesh till there must be raw places on it.

Other wayfarers began to pass her on the road. Now and again folk would ride by her, going one way or the other. She overtook a peasant cart bound for the city with wares — the heavy solid wheels bumped and jolted over roots and stones, creaking and squeaking. Two men were dragging along a beast for the slaughter-house. They looked a little at the young pilgrim woman, for her comeliness' sake — but such wayfarers were a common sight in this country-side. At one place some men were busy putting up a house a little off the road; they called out to her, and an oldish man came running up and proffered her a drink of ale. Kristin curtsied, drank, and gave thanks in words such as poor folks were wont to say to her when she gave alms.

Soon after she must needs rest again. She found a little green slope near by the road, with a stream running by it. Kristin laid her child in the grass; he wakened and began to shriek piteously, so that she hastened with a wandering mind through the prayers she should have said. Then she took Naakkve in her lap and unwound his swaddling-clothes. He had fouled his cloths, and she had but little with her to put in their stead; so she washed out the cloths and laid them on a warm smooth rock to dry in the sun. The outer clothes she wrapped loosely round the boy. He liked this well, for he could kick and sprawl now, as he drank from his mother's breast. Kristin looked with joy at his fine rosy-white limbs, and pressed down one of his hands between her breasts as she suckled him.

Two horsemen rode by at a sharp trot. Kristin glanced up — 'twas a master and his man; but suddenly the master reined up his horse, sprang from the saddle, and came back on foot to where she sat. It was Simon Andressön.

"Maybe you had liefer I should not greet you?" he asked. He stood holding his horse and looking down at her. He was clad as for a journey, in a sleeveless leather jerkin over a light-blue linen coat — a silk cap was on his head, and his face was red and shone with sweat. " 'Tis

strange to see you — but maybe you have no mind to
talk with me — ?"

"Surely you must know — how is it with you, Simon?"
Kristin drew up her bare feet under her skirt, and tried to
take the child away from her breast. But the boy screamed
and gulped and groped about so that she was forced to
lay him to it again. She gathered her kirtle over the breast
as well as she could, and sat with eyes downcast.

"Is it yours?" asked Simon, pointing to the child. "Nay,
that was a foolish question!" he said, laughing. " 'Tis a son,
I warrant? Erlend Nikulaussön has fortune with him." He
had bound his horse to a tree, and now he sat down on a
stone a little way from Kristin. He brought his sword
forward between his knees, and sat with his hands on the
hilt, turning up the earth with the scabbard-end.

"I looked not to meet you here, north of Dovre, Simon,"
said Kristin, that she might say something.

"No," said Simon. "I have had no errands in this part
of the land before."

Kristin called to mind that she had heard somewhat — at
her home-coming feast — of the youngest son of Arne
Gjavvaldssön of Ranheim being to wed Andres Darre's
youngest daughter. — Had he been at Ranheim, she asked.

"You know of it?" said Simon. "Aye, it has got about
already in this country-side, I can well believe."

"It is so, then," said Kristin, "that Gjavvald is to wed
Sigrid?"

Simon looked up sharply, pressing his lips together:

"I see you know *not* of it yet."

"I have not set foot outside Husaby courtyard all the
winter," said Kristin. "And few folks have I seen. I heard
there was talk of this wedding — "

"Aye — as well that you hear it from my lips, I trow —
it must needs get abroad up here." He sat silent a little.
"Gjavvald died three days before winter-night* — he fell
with his horse and broke his back. Mind you, just before
you come to Dyfrin, where the road runs east of the river,
and the ground falls sheer — oh, no, you would scarce
remember. We were on our way to their betrothal-ale;

---

* Winter-night = 14th October, the beginning of the winter
half-year.

Arne and his sons had come by sea to Oslo — " Simon stopped short.

"Maybe she loved Gjavvald — Sigrid — and had been joyful that she was to wed him?" asked Kristin, shyly and fearfully.

"Aye," said Simon. "And she bore a son to him — at Apostles' Mass last spring — "

"Oh, Simon!"

Sigrid Andresdatter, with the brown curls about her little round face. When she laughed there came deep pits in her cheeks. Dimples and little childish white teeth — Simon had these too. Kristin remembered that when she was in her less gentle moods towards her betrothed, this had seemed to her unmanly — most of all after she had come to know Erlend. They were much alike, Sigrid and Simon; but that she was so plump and laughing made *her* but the fairer. She was fourteen winters old then. — Such joyful laughter as Sigrid's Kristin had never heard. Simon was ever teasing and jesting with his youngest sister — Kristin had felt that he held her dearest of them all.

"You know that father was fondest of Sigrid," said Simon. "So he was fain that she and Gjavvald should see if they could like one another, before he clinched the bargain with Arne. And they did — methought a little more than seemed good — they must ever be flinging at each other when they met and glancing and laughing — this was last summer at Dyfrin. But they were so young — none could have thought of this. And Astrid — you know she was betrothed when you and I — Aye, she said naught against it; and Torgrim, you know, has great wealth, and is kind, too, after a fashion — but nothing, nor no one, can please him, and he thinks ever he has all the hurts and ailments that folks ever heard tell of. So we were glad, all of us, that Sigrid was so joyful in the match that had been made for her. . . .

"And when we brought Gjavvald home like this — Halfrid, my wife, managed things so that she went home with us to Mandvik. And then, afterward, it came to light that Gjavvald had not left her — alone — "

They were silent awhile. Then Kristin said, softly:

"This has been no joyous journey for you, then, Simon?"

"Oh, no." Then he laughed a little. "But soon I shall be

used now to riding on woeful errands, Kristin. And for this one, you see, I was the properest — father had not the strength, and 'tis with me at Mandvik that Sigrid and the boy are. Now have we ordered things so that he will take his father's place among the Ranheim kindred, and I could see on them all, I trow, that unwelcome he will not be, the poor little lad, when presently he is sent thither — "

"And your sister," asked Kristin, catching her breath, "where will she be?"

Simon looked down at the ground.

"Father will have her home to Dyfrin now," he said in a low voice.

"Simon! Oh, have you the heart to let this be — ?"

"You sure can see," he answered, without looking up, "how great a gain it is for the boy to be taken into his father's kindred from the very first. Halfrid and I would gladly have them both with us. No sister could have been more faithful and loving towards another than Halfrid was to Sigrid. None of her kindred have been hard to her — believe not that of us. Not even father — though this thing has made of him a broken man. But see you not? — unrighteous had it been if any of us had set ourselves against that which gives the innocent boy his father's name and heritage."

Kristin's babe let go the breast. The mother drew the clothes quickly over her bosom, and pressed the little one to her, trembling. He gulped contentedly a couple of times, and slobbered over himself and his mother's hands.

Simon glanced over at the two, and said with a sort of smile:

"You had better fortune, Kristin, than befell my sister."

"Aye — surely it must seem to you no righteous fate," said Kristin softly, "that I am called wife and my son is true-born. For had I been left with a fatherless bastard, I had been rightly served — "

"I had deemed that the worst tidings I could hear," said Simon. "I wish you naught but good, Kristin," he said in a lower voice.

Soon after, he asked her of the road. Northward he had journeyed by ship from Tunsberg, he said, "I must ride on now and see to it that I overtake my man — "

"Is it Finn that you have with you?" asked Kristin.

"No, Finn is wedded now; he is with me no longer. Do you remember him still?" asked Simon, with a little gladness in his voice.

"Is he a fair child, Sigrid's boy?" asked Kristin, looking at Naakkve.

"I hear them say so. To me one babe in arms looks much like all others," answered Simon.

"Then I trow you have no children yourself," said Kristin, and could not forbear to smile.

"No," he answered shortly. And thereupon he bade farewell and rode away.

When Kristin set out again, she had the child on her back no longer. She bore it in her arms, pressing its face into the hollow of her neck. She could think of naught else but Sigrid Andresdatter.

But *her* father could never have done it. Lavrans Björgulfsön ride out to beg for his daughter's base-born child part and lot with its father's kindred! — never could he have done it. And never, never would he have had the heart to take her babe from her — tear a little being out of his mother's arms, tear him from her breast, with her milk still on his innocent lips. My Naakkve, no, *he* would never have had the heart — if 'twas righteous ten times over, my father would never have done it. . . .

But she could not drive a picture from her mind. A troop of riders vanishing northward through the gorge, where the Dale grows narrow and the mountains crowd together, black with pine-woods. Cold gusts come from the river that rushes thundering over the great rocks, ice-green, foaming, with here and there black pools. He that should fall over there would be hurled from rock to rock and crushed straightway — Jesus, Maria —

Then she saw the fields at home at Jörundgaard of a clear summer night — saw herself running down the path to the little green clearing in the alder-brake by the river — where they were used to wash clothes. The river ran with a changeless harsh roar among the great stones of its shelving bed — Lord Christ, I cannot do aught else —

Oh, but father had never had the heart to do it. Not if it

were never so right. When I prayed, prayed on my bare
bended knees: Father, you must not take my child from
me —

Kristin stood on Feginsbrekka and saw the city lying
below her in the golden sunlight. Beyond the river's broad
shining curves lay brown houses with green turfed roofs,
dark domes of leaves in the gardens, light-hued stone
houses with pointed gables, churches that heaved up black
shingled backs, and churches with dully gleaming leaden
roofs. But above the green land, above the fair city, rose
Christ's Church,* so mighty, so gloriously shining, 'twas
as though all things else lay prostrate at its feet. With the
evening sun blazing full upon its breast and on the shining
glass of its windows, with towers and giddy spires and
golden vanes, it stood pointing up into the bright summer heavens.

Around lay a country-side green with summer, bearing
worshipful great manors on its slopes. Outside again the
fjord stretched wide and bright, with shadows drifting
upon it from the great summer clouds that rose over the
shining blue hills beyond. The cloister-holm, low among
plashing wavelets, lay like a green garland, white-flowered
with its stone houses. So many ships' masts out in the road-
stead, so many fairest houses —

Quite overcome, sobbing, the young woman flung herself down before the cross by the wayside, where thousands of pilgrims had lain before her, thanking God for
that helping hands were stretched out towards human
souls on their journey through this fair and perilous world.

The bells were ringing to Vespers in churches and
cloisters when Kristin came into Christ's churchyard. She
dared to glance for a moment up at the church's west
front — then, blinded, she cast down her eyes.

Human beings had never compassed this work of their
own strength — God's spirit had worked in holy Öistein,
and the builders of this house that came after him. Thy
Kingdom come, Thy will be done on earth as it is in
heaven — now she understood the words. A reflection of

---

* See Note 5.

the glory of God's kingdom witnessed in these stones that His will was all that was fair. Kristin trembled. Aye, well might God turn in wrath from all that was foul — from sin and shame and uncleanness.

In the galleries of the heavenly dwelling stood holy men and women, so fair that she dared not look upon them. Lovely unwithering tendrils of eternity wound silently upwards — broke into leaf on tower and spire, and blossomed in stone monstrances. Over the midmost door hung Christ upon His cross, Mary and John the Evangelist stood by His side, and they were white as though fashioned out of snow, and gold glinted on the white.

Three times she walked around the church, praying. The mighty wall-masses with bewildering riches of pillars and arches and windows, the glimpses of the huge slopes of the roof, the tower, the gold of the spire far up in the skyey spaces — Kristin sank under her load of sin.

She trembled when she kissed the hewn stone of the portal. In a lightning flash she saw the dark carven wood round the church-door at home — that she had kissed with childish lips after her father and mother. . . .

She sprinkled holy water over the child and herself — and thought of the time her father used to do this, when she was small. With the child pressed tight in her arms, she went forward up the church.

She went as through a forest — the columns were furrowed like ancient trees, and in through the forest flowed the light, many-hued and clear as song, from the pictured windows. High up above her, beasts and men sported among the stone leafage, and angels played — and yet far, dizzily far higher, the vaulting soared, lifting the church towards God. In a hall that lay to one side, worship was being held at an altar. Kristin sank down on her knees by a pillar. The singing cut into her like a too strong light. Now she saw how low she lay in the dust. . . .

Pater noster. Credo in unum Deum. Ave Maria, gratia plena. She had learnt her prayers by saying them after her father and mother before she understood a word — longer since than she could remember. Lord Jesus Christ! Was there ever woman so sinful as she — ?

High under the triumphal arch, uplifted over the people, hung Christ the Crucified. The stainless Virgin that was His mother stood gazing in deathly anguish up at her innocent Son, suffering a death of torment like an evildoer.

And here knelt she, with the fruit of sin in her arms. She pressed the child to her — he was fresh as an apple, red and white as a rose — he was awake now, and lay looking up at her with his clear sweet eyes. . . .

Conceived in sin. Borne under her hard evil heart. Drawn from her sin-polluted body, so fair, so whole, so unspeakably lovely and fresh and pure. The undeserved mercy broke her heart asunder; she knelt, crushed with penitence, and the weeping welled up out of her soul as blood flows from a death-wound.

Naakkve, Naakkve, child of mine — God visiteth the sins of the fathers upon the children. Knew I not that? — Ah, yes, I knew it. But I had no mercy in my heart for the innocent life that might be wakened in my womb — to be accursed and condemned to torment for my sin —

Repented I my sin, when I bore you within me, my beloved, beloved son? Oh, no, 'twas not repentance. — My heart was hard with anger and evil thoughts in the hour when I first felt thee move, so little and so defenceless. — *Magnificat anima mea Dominum. Et exaltavit spiritus meus in Deo salutari meo* — thus she sang, the gentle Queen of all women, when she was chosen out to bear Him that was to die for our sins. — I called not to mind Him that had power to take away the burden of my sin and my child's sin — oh, no, 'twas not repentance, I but feigned me lowly and wretched, and begged and begged that the commands of righteousness be broken; for that I could not bear it if God upheld His law and chastened me according to the Word that I had known all my days —

Oh, aye, now she knew it. She had thought that God was such an one as her own father, that St. Olav was as her father. She had thought all the time, deep in her heart, that when her punishment grew to be heavier than she could bear, then would she meet, not with justice, but with compassion. . . .

She was weeping so that she could not rise when the people stood up during the worship. She lay on, crouched in a heap over her child. Near her there were other folk that did not rise — two well-dressed peasant women with a young boy between them.

She looked up towards the choir. Behind the golden grated doors St. Olav's shrine gleamed in the darkness, towering high behind the altar. An icy-cold shuddering ran through her. There lay his holy body awaiting the day of resurrection. Then would the lid fly open, and he would arise. With his axe in hand he would stride down this mighty church; and up from the paven floor, up from the earth outside, up from every graveyard in Norway's land would the dead yellow skeletons arise. — They would be clothed upon with flesh and would muster themselves round their King. They that had striven to tread in his blood-marked footsteps, and they that had only sought him that he might help them with the burdens of sin and sorrow and sickness that here in life they had bound on themselves and their children. Now they throng around their lord and pray him to lay their needs before God. Lord, hearken to my prayer for this folk, which I held so dear that I would rather suffer outlawry and need and hatred and death, than that man or maid should grow up in Norway and not know that Thou diedst to save all sinners. Lord, Thou who didst bid us go out and make all the peoples Thy disciples — with my blood did I, Olav Haraldssön, write Thy Evangel in the Norse tongue, for these my poor freedmen. . . .

Kristin shut her eyes, sick and dizzy. The King's countenance was before her — his flaming eyes saw to the bottom of her soul — she trembled under St. Olav's glance.

Understand you now, Kristin, that you need help?

Aye, Lord King, now I understand it. Sore need is mine that thou support me, that I may not turn me from God again. Be with me, thou Chief of His people, when I bear forth my prayers, and pray thou that I be granted mercy. Holy Olav, pray for me!

Cor mundam crea in me, Deus, et spiritum rectum innova in visceribus meis.

Ne projicias me a facie tua —

Libera me de sanguinibus, Deus, Deus salutis meæ —

The worship was at an end. Folks were leaving the church. The two peasant women, who had knelt near

Kristin, rose up. But the boy between them did not rise; he began to move over the pavement by pressing the knuckles of his clenched hands on the flags and jerking himself along like a young unfledged crow. He had tiny legs, twisted up close under his body. The women walked so as to hide him as well as they could with their garments.

When they were out of sight, Kristin threw herself down and kissed the church-floor where they had passed by her.

Somewhat doubtful and at a loss, she stood by the entrance to the choir, when a young priest came out of the grated door. He stopped before the red-eyed young woman, and Kristin told him her errand, as well as she might. At first he scarce understood. She brought out the golden garland and held it toward him.

"Oh, are you Kristin Lavransdatter, Erlend's wife, of Husaby — ?" He looked at her with a little wonder; her face was all swollen with weeping. "Aye, aye, your brother-in-law, Master Gunnulf, spoke of this, aye — "

He led her out into the sacristy, took the wreath, and unwrapped the linen cloth from around it and looked at it; then he smiled a little.

"Aye, you will understand, I trow — there must be witnesses and such-like by — you cannot give away such a precious thing, mistress, as though 'twere a piece of buttered bread — but I can take it in charge, meanwhile — 'tis like you would not choose to bear it about with you in the city. — Oh, pray Canon Arne to be so good as to come hither," he said to a church-servant.

"Your husband should be with you too, I trow, rightly. But it may be Gunnulf has some letter from him. . . .

"You are to be brought before the Archbishop himself — was it not so? Else 'tis Hauk Tomassön who is Pœnitentiarius — I know not whether Gunnulf has talked with Lord Eiliv — but you must come hither to matins tomorrow, and you can ask for me after lauds! I am called Paal Aslakssön. Him" — he pointed to the child — "you must leave in the hostel. You are to sleep in the Sisters' hostel at Bakke, I mind me your brother-in-law said."

Another priest came in, and the two spoke together awhile. The first opened a little locker in the wall, took out a pair of scales, and weighed the wreath, while the other wrote in a book. Then they laid the wreath away in the press and locked it.

Canon Paal was about to lead her out — but he asked first if she wished he should lift her son up to St. Olav's shrine.

He took the boy with the sure, somewhat careless deftness of a priest used to holding babies at baptism. Kristin went with him into the church, and he asked if she, too, would not fain kiss the shrine.

I dare not, thought Kristin; but she followed the priest up the steps to the high platform whereon the shrine was set. There came, as it were, a great blinding white light before her eyes when she approached her lips to the golden tabernacle.

The priest looked at her a moment, fearing she might fall down in a swoon. But she rose again to her feet. Then he let the child's forehead touch the sacred shrine.

Canon Paal went with her to the church-door and asked if she were sure she could find her way to the ferry. Then he bade her good-night — he had spoken all the time in smooth, dry fashion, like any other mannerly young courtier.

— It had begun to rain a little, and a breathing of sweet scents came balmily from the gardens and from the street, which was fresh and green like a country courtyard, save for the strips worn bare by the coming and going of people and carts.

Kristin sheltered the boy from the rain as best she could; he was so heavy now, so heavy that her arms were numb and dead from bearing him. And he whimpered and cried unceasingly — like enough he was hungry again.

His mother was deadly tired — from the day-long walking, and from all the weeping and the vehemency of her emotion in the church. She was cold — and the rain grew heavier; the leaves of the trees glanced and shivered under the spattering drops. She threaded her way through the lanes and came out on an open place, where she could see ahead to where the river ran broad and grey, its surface pitted like a sieve by the falling drops.

There was no ferry-boat. Kristin spoke to two men who

had taken shelter under the floor of a warehouse that stood on piles at the water's edge. They said she must go out to the landing-place — the nuns had a house there, and there was a ferry-man.

Kristin dragged herself up again across the open place, footsore and wet and weary. She came to a little grey stone church — behind it lay some houses inside a fence. Naakkve was shrieking wildly, so that she could not go into the church. But the sound of singing came to her through the unglazed window-openings, and she knew the antiphon: *Lætare Regina Cæli* — Rejoice, thou Queen of Heaven — for He whom thou wert chosen to bear — is arisen even as He said. Alleluja!

It was the song the Minorites sang after complin. Brother Edvin had taught her this hymn to the Lord's Mother when she watched by his bed those nights when he lay in mortal sickness at home with them at Jörundgaard. — She stole into the churchyard, and, standing by the wall with her child upon her arm, she said it softly over to herself.

— Nothing, Kristin, that you could do could turn your father's heart from you. 'Tis therefore you must give him no more cause for sorrow. . . .

— Even as Thy nail-pierced hands were outstretched upon the Cross, O dear Lord of heaven — howsoever far away a soul had strayed from the straight path, yet were the nail-pierced hands stretched out towards it yearningly. Naught was needful save this one thing: that the sinful souls should turn them to those open arms, freely, as a child goes to its father, not like thralls hunted home to their cruel master. Now she understood how hateful sin was. Again there came that pain in her breast; as though her heart would break asunder in penitence and in shame at her unworthiness of the mercy shown her. . . .

Close in to the church-wall there was a little shelter from the rain. She sat down upon a grave-stone and began to still the child's hunger. Now and then she bent forward and kissed the little down-covered head.

She must have fallen asleep. Someone touched her shoulder. A monk in orders and an old lay brother with a sexton's spade in his hand stood before her. The barefoot friar asked if she sought lodging for the night.

It flashed across her — she would far rather abide here to-night with the Minorites, Brother Edvin's brethren. And it was so far to Bakke — and she was sinking with weariness. Then the monk bade the lay brother bring this woman to the women's hostel — "and give her a little calamus-wash for her feet — she is footsore, I see."

It was close and dark in the women's hostel — it stood without the fence, out by the lane. The lay brother brought her water to wash with and a little food, and she sat by the hearth and tried to quiet the child. 'Twas plain that Naakkve felt it in his food that his mother was worn out and had fasted that day; he wept and whimpered between-times while sucking at her exhausted breasts. Kristin took mouthfuls of the milk the lay brother had brought her; she tried to spirt it from her mouth into the child's — but the little knave clamoured loudly against this new fashion of taking in food, and the old man laughed and shook his head. She must drink it herself, he said, then the boy would soon get the good of it. . . .

At last he went away. Kristin crept up into one of the uppermost box-beds, close under the roof. From it she could reach up to open a trap in the roof — and there was need, for there was a sickening smell in the hostel. Kristin opened the trap — the rain-washed air of the bright cool summer night streamed in about her. She sat in the short bed with her head and neck propped against the wall-timbers — there were so few pillows in the bed. The boy slept on her lap. She had meant to shut the trap again in a while, but she fell asleep unawares.

Far on in the night she awoke. The summer moon shone in, pale and honey-yellow, across the child and her, and lit up the wall over against them. And she was ware of a figure in the midst of the stream of moonlight, hovering between floor and roof-tree.

He was clad in an ashen-grey frock — he was tall and stoop-backed. Now he turned his old, old furrowed face towards her. It was Brother Edvin. He smiled, and his smile was unspeakably tender — a little roguishly merry, just as when he lived on earth.

Kristin wondered not at all. Humble, happy, full of hope and trust, she looked at him, awaiting that which he might say or do.

The monk laughed and held up an old, heavy fur mitten towards her — then he hung it on the moonbeam and left is hanging there. And then he smiled still more, nodded to her, and melted away.

THE MISTRESS OF HUSABY

PART TWO

HUSABY

I

ON a day early in the new year, there came to Husaby some unlooked-for guests. They were Lavrans Björgulfsön and old Smid Gudleikssön from Dovre, and with them were two gentlemen whom Kristin knew not. And Erlend wondered much to see his father-in-law come in their company — they were Sir Erling Vidkunssön of Giske and Haftor Graut of Godöy — he had not deemed that Lavrans knew these men. But Sir Erling made things clear, telling how they had met together at Nes of Raumsdal, where he had sat with Lavrans and Smid on the Court of Six * that had at last set at rest the dispute between Sir Jon Haukssön's distant heirs. Lavrans and he had fallen into talk of Erlend, and the thought came to Sir Erling that, since he had an errand to Nidaros, he would like well to wait upon the folks at Husaby, if Lavrans would join him and sail north with him. Smid Gudleikssön said, laughing, that he had all but bidden himself to come along with them:

"For I was fain to see our Kristin again — the fairest rose of the Norddal. And then, methought my kinswoman, Ragnfrid, would be beholden to me if I kept an eye on her husband, to mark what weighty counsels he may be hatching with wise and mighty men like these. Aye, your father has other gear to guide this winter, my Kristin, than making the rounds of the manors with us, drinking out Yule till the Fast comes in. Now have we sat at home on our lands in peace and quiet all these years, and looked, each man of us, to his own affairs. But now would Lavrans have all us King's-men of the Dale ride in a troop to Oslo at hardest midwinter — now are we to counsel the great lords of the King's council in the King's behoof — they are ruling things so ill for the poor boy in his nonage, says Lavrans — " †

---

* See Note 6.
† See Note 7.

Sir Erling looked somewhat ill at ease. Erlend raised his eyebrows.

"Are you of these counsels, father-in-law — the calling of the great meeting of King's-men?"

"No, no," said Lavrans. "I but ride to the meeting like the other King's-men in the Dale, since we have been summoned thither — "

But Smid Gudleikssön took up his tale again: 'Twas Lavrans that had talked him over — and Herstein of Kruke, and Trond Gjesling and Guttorm Sneis and others that had had no mind to go. . . .

"Nay but — is it not your wont in these parts to bid stranger folks step into the house?" asked Lavrans. "Let us try now whether Kristin brews as good ale as her mother!" . . .

Erlend looked doubtfully, and Kristin wondered greatly.

"What is this, father?" she asked a little later, when he was with her in the little hall, whither she had brought the child to be out of the strangers' way.

Lavrans sat dancing his grandson on his knee. Naakkve was ten months old now, a great child and comely. Already, at Yule-tide, he had been put into short coats and hose.

"Never before, father, have I known you put in your word in matters such as this," went on his daughter. "I have ever heard you say that for the welfare of the land in peace and war and of his subjects 'twas best the King should take order, and the men whom he called to his side. Erlend says that this venture is the work of the nobles in the south — they would set aside Lady Ingebjörg and the men her father gave her for councillors — and seize again for themselves such power as they had when King Haakon and his brother were children. But you yourself were used to say that the kingdom suffered great scathe from their rule — "

Lavrans whispered to her to send away the child's nurse. When they were alone, he asked:

"Whence came these tiding to Erlend — was it from Munan?"

Kristin told him that Orm had brought with him a letter from Sir Munan when he came home in the autumn. She did not tell that she herself had read it to Erlend — he had no great skill in making out writing. But in the letter Munan had made bitter complaint that every man in Norway that bore arms on his shield deemed now that he understood the governing of the realm better than the men who had stood at King Haakon's side when he lived; and they held that they knew better what was for the young King's welfare than the high-born lady, his own mother. He had warned Erlend that should there be signs that the Norwegian nobles had a mind to copy what the Swedes had done at Skara in the summer — hatch plots against Lady Ingebjörg and her old and tried councillors — the lady's kinsmen must hold them ready, and Erlend should come south and meet Munan at Hamar.

"Said he not, too," said Lavrans, pushing his finger in under Naakkve's fat chin, "that I was one of the men that set their faces against the unlawful call to arms that Munan brought with him up the Dale — in our King's name?"

"You!" said Kristin. "Met *you* Munan Baardssön in the autumn?"

"I did so," answered Lavrans. "And we agreed not overwell together."

"Spoke you of me?" asked Kristin, quickly.

"No, little Kristin," said her father, laughing a little. "I cannot call to mind that at that meeting your name came up between us. — Know you if it is so that your husband has a mind to fare south and seek out Munan Baardssön?"

"I believe it," said Kristin. "Sira Eiliv drew up a letter for Erlend not long since — and he spoke of having soon to journey south."

Lavrans sat silent a little, looking at the child groping with its fingers about his dagger-hilt and trying to bite the rock-crystal set in it.

"Is it true that they would take away the rule of the realm from Lady Ingebjörg?" asked Kristin.

"The lady is as old as you, or thereabouts," answered her father, still smiling a little. "None would take from the King's mother the honour or the power whereto she is born. But the Archbishop, and certain of the friends and kinsmen of our King that is gone, have called a meeting to take counsel how the lady's power and honour and the good of the people of the land can best be guarded."

Kristin said low:

"I can see well, father, that you are not come to Husaby this time only to see Naakkve and me."

"Not only for that," said Lavrans. Then he laughed: "And I can see, my daughter, that this likes you but little!"

He laid one of his hands over her face and stroked it up and down. 'Twas so he had been used to do, ever since she was a little girl, whenever he had been scolding or teasing her.

Meantime Sir Erling and Erlend sat up in the armoury — so was called the great storehouse that lay north-east of the courtyard, near by the main gate. It was high as a tower, having three stories; in the uppermost was a chamber with loopholed walls, and there were kept all the arms that were not in daily use on the manor. King Skule had built this house.

Sir Erling and Erlend had fur cloaks on, for it was bitter cold in the room. The guest walked about, looking at the many fair weapons and suits of armour that Erlend had inherited from his mother's father, Gaute Erlendssön.

Erling Vidkunssön was a somewhat short man, slightly built, though a little plump, but he bore himself lightly and with grace. Fair of face he was not, though his features were well formed — but his hair was of a light reddish hue, and his eyelashes and brows were white — the eyes themselves, too, were of the palest blue. That Sir Erling was deemed, none the less, to be a well-looking man may have been because all knew he was the richest knight in Norway. But 'twas true he had a rarely winning, quiet way with him. He was of excellent understanding, well-taught and rich in knowledge, and since he never strove to show off his learning, but always was found ready to hearken to others, he had gotten him the name of one of the wisest men in the land. He was much of the same age as Erlend Nikulaussön, and they were of kin to each other, though far off, through their kinship to the Strovreim house. They had known each other long, but there had been no close friendship between the two men.

Erlend sat on a great chest, talking of the ship he had built him last summer; 'twas a thirty-two-oar ship, and he deemed that she would prove a rarely swift sailer and easy steerer. He had had two shipwrights down from the Nordland, and had himself overseen the work along with them.

"Ships are among the few things whereof I know a little, Erling," said he; "and you shall see, 'twill be a fair sight to watch *Margygren* cleave the sea-surges — "

"*Margygren* * — 'tis a fearsome, heathenish name you have given your ship, kinsman," said Sir Erling, laughing a little. "You mean to sail south in her, then?"

"You are as holy as my wife, I see — she too says 'tis a heathenish name. Aye, she likes not the ship either; but she is inland-bred — she cannot endure the sea."

"Aye, she seems right holy and fine and gracious, your lady," said Sir Erling, courteously. "As one might look she should be, seeing the kindred she comes of."

"Aye — " Erlend laughed a little. "There goes by no day when she hears not mass. And Sira Eiliv, our priest whom you saw, reads aloud for us from godly books — 'tis what he likes best, next to ale and dainty dishes, to read aloud. And poor folk come hither to Kristin for counsel and help — they would be fain to kiss the hem of her garment, I well believe — my men I scarce know again any more. She is likest one of the ladies of whom there is record in the holy sagas that King Haakon forced us to sit and hear the priest read aloud — mind you? — when we were pages together. Much is changed at Husaby since last you were my guest, Erling. — I marvel, indeed, that you would come to me as things were then," he said, a little after.

"You spoke of the time when we were pages together," said Erling Vidkunssön, with a smile that became him well. "We were friends then, were we not? In those days all of us, Erlend, deemed that you would go far in this land of ours — "

But Erlend only laughed: "Aye, so did I deem too."

"Can you not sail southward along with me, Erlend?" asked Sir Erling.

---

* *Margygren* = the Sea Ogress.

"My purpose is to journey by land," answered the other.

"A toilsome journey for you — over the mountains in mid-winter," said Sir Erling. " 'Twould be pleasant if you would bear Haftor and me company."

"I have given my promise to certain other folks to go with them," answered Erlend.

"Doubtless you go along with your father-in-law — aye — that is but reason."

"No, not so — I know so little of these men from the Dale he is to ride with." Erlend sat silent a little. "No, I have promised to look in on Munan at Stange," he brought out hurriedly.

"You can spare yourself the pains," answered the other. "Munan is gone down to his estates in Hising, and it may well be that 'twill be long ere he come north again. Is it long since you heard from him?"

" 'Twas at Michaelmastide — he wrote to me from Ringabu."

"Aye — but you know what befell in the Dale this last autumn?" asked Erling. "You know it not? Sure you must know that he rode round himself to the Wardens, all about Mjös and up the Dale, with letters, bearing that the farmers should bring forth full levies of horses and supplies — one horse to every six farmers — and the nobles and freeholders should send horses, but might stay at home themselves? *Have* you not heard of this? And that the men of the nothern Dale denied to furnish this levy when Munan came with Eirik Topp to the Thing at Vaage? Moreover, 'twas Lavrans Björgulfsön that led the opposers — he challenged Eirik, if there were aught outstanding from the lawful levies, to gather it in in lawful wise; but he called it high-handed oppression on the people to crave war-taxes from the farmers only to help a Danish man to wage war on the Danish King. And, should our King call for service from his King's-men, said Lavrans, he should find them ready enough to come to tryst with good weapons and horses and men-at-arms — but *he* sent not from Jörundgaard so much as a he-goat in a hempen halter, except the King craved he should ride it himself to the muster. Nay, now, know you not this? Smid Gudleikssön says that Lavrans had promised his farmers he would pay their levy-fines * for them, if need should be — "

Erlend sat in wonderment:

"Did Lavrans so? Never have I heard before that my wife's father meddled in aught but what might touch his own estates, or his friends' — "

" 'Tis not often he does so, like enough," said Sir Erling. "But so much I could see, when I was in at Nes with him, that when Lavrans Björgulfsön speaks his mind in a matter, he lacks not followers in plenty — for he speaks not except he know the business so well that his word can hardly be overthrown. Now, touching these supplies, 'tis said he has changed letters with his kin in Sweden — as you know, Lady Ramborg, his father's mother, and Sir Erngisle's grandfather were cousins, so that his kinsfolk there are many and worshipful. Quiet a man as he is, your father-in-law, he has no little power in the country-side where folks know him — though he use it not often."

"Aye, then do I understand why you seek his company, Erling," said Erlend, laughing. "I marvelled that you were grown such hot friends."

"Marvel you at that?" answered Erling, unmoved. "A strange man must be he that would not be fain to call Lavrans of Jörundgaard his friend. You, kinsman, would serve your turn better by hearkening to him than to Munan."

"Munan has been like an elder brother to me ever since the day I first went forth from my home," said Erlend, a little hotly. "Never did he fail me when the pinch came. And if, now, a pinch be come for him — "

"Munan will be safe enough," said Erling Vidkunssön, calmly as before. "The letters he bore around were sealed with the Great Seal of Norway — unlawfully, but that touches not him. True, there is more — that which he was privy to, and attested with his seal, when he was witness to the Lady Eufemia's betrothal — but that can scarce be brought to light without touching one whom we would not — Truth to tell, Erlend — I trow Munan can save his own skin without your help — but you may harm yourself — "

---

* See Note 8.

" 'Tis the Lady Ingebjörg you aim at, I see well," said Erlend. "I have promised our kinswoman to serve her both here and in foreign lands — "

"And even so have I promised," answered Erling. "And I mean to keep that promise — and so, I trow, does every Norseman that served and loved our lord and kinsman, King Haakon. And the best service that can be shown her is to part her from those councillors who counsel so young a lady to her own hurt and her son's."

"Believe you," asked Erlend in a low voice, "that you can compass *that?*"

"Aye," said Erling Vidkunssön firmly. "I believe it. And I trow all believe it who hearken not to" — he shrugged his shoulders — "evil-minded — and loose — talk. And that should we, the lady's kinsmen, be the last of all to do."

A serving-maid lifted the trap-door in the floor, and asked if it would suit them now that the mistress have the supper borne into the hall. . . .

While the house-folk were still at the board, the talk ever and anon kept glancing towards these weighty matters that were in question. Kristin marked that both her father and Sir Erling tried to hinder this; they turned the talk with news of wedding-bargains and deaths, strife among heirs, and dealings in farms. She was uneasy, though she scarce herself knew why. They had some errand of weight to Erlend, she could see. And though she would not own it to herself, she knew her husband so well now as to feel that, headstrong as he was, he might mayhap easily enough be turned aside by one who had a firm hand in a soft glove, as the saying went.

After the meal, the men moved over to the hearth and sat there drinking. Kristin settled down on the bench, took her broidery frame into her lap, and began plaiting the fringes. Soon after, Haftor Graut came across to her, laid a cushion on the floor, and sat on it at her feet. He had found Erlend's cithern, and he held it on his knee and sat thrumming on it and prattling. Haftor was a quite young man with yellow ringleted hair; most comely of face, but freckled beyond measure. Kristin soon marked that he was a most heedless talker. He had but just wedded richly; but he had grown weary at home on his estates, he said; 'twas therefore that he was going now to the meeting of the King's-men.

"But 'tis no more than reason that Erlend Nikulaussön bide at home," he said, and laid his head back on Kristin's lap. She drew a little aside, laughed, and said, if she mistook not, her husband too was minded to journey south — "whatever the cause may be" — she said, with an air of innocence. "There is so much unrest in the land these days; 'tis no easy thing for a simple woman to judge of such things."

"Yet 'tis a woman's simplicity that is most to blame for it," answered Haftor, laughing, and shifting nearer again. "Aye, so at least say Erling and Lavrans Björgulfsön — I would be fain to know what they mean by it. What think you, Mistress Kristin? Lady Ingebjörg is a good, simple woman — maybe she is sitting now, even as you are, plaiting silk with her snow-white fingers, and thinking: hard-hearted would it be to deny her departed husband's trusty vassal a little matter of help towards bettering his fortunes — "

Erlend came over and sat by his wife; Haftor had to move a little to make room.

" 'Tis such trumpery tales the dames hatch out in their lodging when their husbands are fools enough to take them with them to the assemblies — "

"Folks say, where I am from," said Haftor, "that there is no smoke without fire — "

"Aye, we have that byword too," said Lavrans — he and Erling had joined them — "yet was I cheated, Haftor, last winter — when I would have lit my lantern with a piece of fresh horse-dung." He sat himself down on the edge of the table. Straightway Sir Erling fetched Lavrans' beaker across and proffered it to him with a bow, then sat down near him on the bench.

" 'Tis not possible, Haftor," said Erlend, "that you can know, away north in Haalogaland, what Lady Ingebjörg and her councillors know of the purposes and undertakings of the Danes. I know not if you were not short-sighted when you set you up against the King's call for help. Sir Knut — eye, we may as lief call him by his name, 'tis he we all have in our thoughts — he seemed not to me

to be the man to let himself be caught dozing on his perch.
You folk dwell too far away from the great cauldrons to
smell what is cooking in them. And, better timely ware
than after yare, say I — "

"Aye," said Sir Erling. "Almost a man might say, they
cook for us now in our neighbour's manor — soon will we
Norsemen be most like folk living on a pittance from their
heirs; they send us in at the door the porridge they have
cooked in Sweden — eat it, if you would have meat! I
deem that 'twas an error of our Lord, King Haakon —
that, as it were, he moved the kitchen to an outskirt of
the farm, when he made Oslo the first city of the land.
Before, it lay midway of the farmplace, if we may keep
to this way of speech — Björgvin * or Nidaros — but now
there is none to rule in these parts but the Archbishop and
chapter. — Aye, what say you, Erlend — you that are a
Trönder, with all your goods and all your lordship in
Trondheim here — "

"Aye, God's blood, Erling — if 'tis that you would be
at — carry home the pot and hang it up at the right hearth-
stone once again — "

"Aye," said Haftor. "All too long have we here in the
north had to be content with a smell of singeing and a
gulp of cold broth — "

Lavrans broke in:

"Thus it stands, Erlend — I had not taken it on me to be
spokesman for the folk of our country-side at home, but
that I had in my keeping letters from my kinsman, Sir
Erngisle. So I knew none of the men who bear rightful rule
either in the Dane-King's or in our King's realms have any
thought of breaking the peace and friendship between
our lands."

"If you know who bears rule in Denmark now, father-
in-law, I wot you know more than do most other men,"
said Erlend.

"One thing I know. There is *one* man whom none would
see bear rule, neither here nor in Sweden, nor in Denmark. And that was the drift of the Swedes' doings at
Skara last summer, and that is the drift of the meeting we

* Bergen.

would now hold at Oslo — to make clear to all who have not yet understood it, that on this all prudent men are at one."

By now they had all drunk so much that they were grown somewhat loud-voiced — all but old Smid Gud-leiksön, who sat nodding in his chair by the hearth. Erlend cried out:

"Aye, you folks are so prudent that the devil himself cannot trick you! 'Tis no marvel you should be afeared of Knut Porse. You cannot understand him, you good gentry — he is not the man to be content to sit mumchance watching the days slip by and the grass grow as God wills it. Fain would I be to meet that knight again; I knew him when I was in Halland. And naught would I have against it if I stood in Knut Porse's shoes."

"So much dared not *I* have said where my wife could hear me," said Haftor Graut.

But Erling Vidkunssön, too, was now well on in drink. He tried to keep a hold on his courtly ways, but they slipped from his grasp:

"You!" said he, bursting into a great laughter. "You, kinsman! — Nay, Erlend!" He slapped the other on the shoulder and laughed and laughed.

"Nay, Erlend," said Lavrans, bluntly; "there needs more for that than beguiling of fair ladies. Were there no more in Knut Porse than that he can play the fox in the goose-pen, I trow we gentlefolk of Norway were all too slothful to turn out of our houses to hunt him off — even were the goose our King's mother. But whomsoever Sir Knut can beguile to play the fool for his sake, he himself plays no tricks that have not a meaning in them. *He* has his goal, and be sure that he takes not his eyes from it — "

There was a pause in the talk. Then Erlend said, his eyes glittering:

"Then I would that Sir Knut were a Norseman."

The others sat silent a little. Sir Erling took a draught from his goblet, then said:

"God forbid — had we such a man among us here in Norway, I fear me there would quickly be an end of peace in the land."

"Peace in the land!" said Erlend scornfully.

"Aye, peace in the land," answered Erling Vidkunssön. "Bear in mind, Erlend — 'tis not we of the knighthood alone that own this land and live in it. To you 'twould mayhap seem sport if there should arise here a man greedy of adventure and of power like Knut Porse. So it was in bygone days that when a man raised revolt in this our land, 'twas ever easy for him to find followers from among the nobles. Either they would gain the upper hand and win titles and fiefs, or their kinsmen won and they were given grace for life and goods — aye, the record tells who lost their lives, but the more part saved their skins, whether things went so or so — the more part of *our* fathers, mark you. But the mass of the common farmers and the towns-men, Erlend — the working-folk that many a time had their dues wrung from them twice over in one year, and might be joyful, moreover, each time a troop had passed through the country-side without burning their farms and slaughtering their cattle — the common folk that had to suffer such unbearable burdens and oppressions — *they*, I trow, thank God and St. Olav for old King Haakon and King Magnus and his sons, who strengthened the laws and made the peace sure." . . .

"Aye, I well believe that you believe 'tis so." — Erlend threw back his head. Lavrans sat looking at the younger man — he was wide awake enough now, was Erlend. His dark vehement countenance was flushed red, his throat seemed to swell into an arch in the slender brown neck. Then Lavrans looked at his daughter. Kristin had let her work fall on her lap and was following the men's talk intently.

"Are you so sure that the peasants and the common folk think thus and praise so much the new order of things? 'Tis true they had hard times often — in the old days when kings and rival kings warred with each other through all the land. I know that they remember still the time when they must often take to the hills with cattle and wives and children, while their farms went up in flames all down the valleys. I have heard them speak of it. But I know that they remember somewhat else as well — that their own fathers were in the armies; not we alone played for power, Erling, the peasants' sons played too — and time and again

they won our lands from us. 'Tis not when law rules in a
land that the son of a Skidan * trull, that knows not his
own father's name, wins a Baron's widow with her lands
and goods, as did Reidar Darre — you deemed a son of his
house a good enough match for your daughter, Lavrans, and
now hath he wedded your lady's niece, Erling! But now
law and justice rule; and — I know not how it comes, but
I know that 'tis so — the peasants' land comes into our
hands more and more, and that *with* the law — the more
law and right rule, the quicker it goes, the quicker their
power in the kingdom's affairs and their own slips from
their hands. And, Erling, the common farmers know it
too! Oh, no! be not too sure, good sirs, that the common
folk have no yearnings back to that time when they might
lose their farms by fire and by rapine — but they might
also win by arms more than they can win by law."

Lavrans nodded:

"There is some truth in what Erlend says," he said in a
low voice.

But Erling Vidkunssön rose:

"I can well believe it — that the common folk remember
better the few men that rose from nothing and came to
might and mastery in the time of the sword, than the num-
berless men that went under in black poverty and misery.
And yet none were harder masters to the small folks than
these first — I trow 'tis of them the byword was first
made: none more unkind than kin. If a man be not born to
mastery, he is ever a hard master — but if he have grown
up in childhood amid serving-men and serving-women, 'tis
far easier for him to understand that without the common
folk we are in many a wise as helpless as children all our
days, and that, not only for the love of God, but quite as
much for our own sake, we should serve them on our part
with our knowledge and guard them with our knighthood.
Never yet has a kingdom stood except there were in it
great men with the strength and the will to use their
power to make sure the lesser folk's rights — "

"You could preach against my brother for a wager,
Erling," said Erlend, laughing. "But my belief is that these

---

* The modern Skien.

stubborn Trönders liked us great folk better in the old days when we led their sons to battle and foray, let our blood flow out over the deck-planks mixed with theirs, and hewed rings in sunder and shared the booty with our house-carls. — Aye, you see, Kristin, sometimes I sleep with one ear open when Sira Eiliv reads from his great books."

"Goods that are won by unright come not down to the third heir," said Lavrans Björgulfsön. "Have you not heard this, Erlend?"

"Surely have I heard it!" Erlend laughed aloud. "But seen it I have not — "

Erling Vidkunssön said:

"So it is, Erlend, that few are born to be masters, but all are born to serve; the right lord is the servant of his servants — "

Erlend clasped his hands together behind his neck and stretched himself, smiling:

"Thereon have I never thought. And I deem not that my tenants have any service to thank me for. And yet, strange though it be, I trow they like me well." — He rubbed his cheek against Kristin's black kitten, which had sprung up on his shoulder, and now, arching its back, was walking, purring, around his neck. "But my wife here — she is the most serviceable of ladies to all — though in faith you have no cause to believe it — for our cans and flagons are empty, Kristin mine!"

Orm, who had sat silent, following the men's talk, got up at once and went out.

"The lady grew so weary she fell asleep," said Haftor smiling, "and the fault is yours — 'twere liker you had left her in peace to speak with me that have the wit to know how to talk to ladies."

"Aye, this talk has run on too long for you, I fear, mistress," began Sir Erling, in excuse; but Kristin answered with a smile:

"So indeed it is, sir, that I have not understood all that has been said here to-night; but I bear it well in mind, and I shall have good time to think over it hereafter." . . .

Orm came back with some maids bearing in more liquor. He went round filling the cups. Lavrans looked sorrowfully at the comely boy. He had tried to have some speech with Orm Erlendssön, but the boy was of few words, though his bearing was mannerly and courteous.

One of the maids whispered to Kristin that Naakkve was awake down in the little hall and was screaming terribly. On this the lady of the house bade her guests good-night, and left the hall with her maids.

The men turned to the ale-cup again. Sir Erling and Lavrans changed glances now and again, and at last the knight said:

"There is a matter, Erlend, that I had a mind to speak of with you. 'Tis sure that a levy of ships will be called out from the lands around the fjord here and from Möre — folks in the north dread that the Russians will come again in strength in the summer and that without help they will not be strong enough to guard the land. This Russian feud is the first profit we have drawn from our sharing a King with Sweden — 'twould not be just that the Haaloga-landers should be left to enjoy it all alone. Now, it so falls out that Arne Gjavvaldssön is too old and sickly — and there has been talk of naming you as chief of all the ships from this side of the fjord. How would that like you — ?"

Erlend smote one fist into the palm of the other hand — his whole face shone: "How would it like me — !"

" 'Twill scarce be possible to raise any great force," said Erling, as though to sober him. "But we thought that, if you will, 'twere well you should set things in train with the wardens round about. — You know all this country well — and 'twas said among the Lords of the Council that mayhap you were the man best fitted to make somewhat of this matter. There are they that still remember that you won no small honour when you were warden of the Halland coast under Count Jacob — I mind myself that I heard him say to King Haakon he had been unwise to deal so harshly with a likely young fellow; he said there was the stuff in you for a trusty servant to your King — "

Erlend snapped his fingers. "Nay, now, never tell me *you* are to be our King, Erling Vidkunssön! Is this the plot you are hatching," he said, laughing loudly, "to make Erling King?"

Erling said, testily:

"Nay, Erlend, see you not that I speak now in all sadness — "

"God help us — were you jesting before, then? Methought you had spoken in sadness enough all the night. — Aye, aye, let us speak in sadness, then, kinsman — tell me all concerning this affair."

Kristin lay sleeping with the child by her breast when Erlend came down to the little hall. He lit a fir-root brand at the embers on the hearth, and looked at the two for a little while by its light.

So fair she was — and a fair child was he, too, their son. She was always so sleepy in the evenings now, Kristin — the moment she had laid her down and drawn her child close to her, both of them fell asleep. Erlend laughed a little and threw the stick back on to the hearth. Slowly he drew off his garments.

Northward in the spring with *Margygren* and three or four warships. Haftor Graut with three ships from Haalogaland — but Haftor was new and untried in such work, him 'twas like he could manage as seemed good to him. Aye, he saw well he could have things as he would, up there; for this Haftor looked not as though he were a coward or half-hearted. Erlend stretched himself out and smiled in the dark. He had thought to raise a crew for *Margygren* from Möre, outside the fjord. But both this country-side and Birgsi swarmed with stout bold young fellows — the finest choice of men was his. . . .

Not much more than a year since he was wedded. Childbearing, penance and fasting, and now the boy first and the boy last, both day and night. Yet — she was the same young sweet Kristin — when he could get her to forget the prating priests and her greeding suckling for a little while. . . .

He kissed her shoulder, but she marked it not. Poor child, let her sleep — he had so much to think on to-night. Erlend turned him away from her and gazed out into the room at the little glowing spark on the hearth. Aye, maybe he should get up and cover up the embers — but he had no mind to. . . .

Memories of his youth came in shreds and snatches. A quivering ship's-stem standing as 'twere waiting a second for the oncoming wave — and the sea that came washing inboard. The mighty clamour of storm and waves. The whole vessel groaned in the press of the seas — the mast-head cut its wild curve on the flying clouds. 'Twas some-where off the Halland coast. — Erlend lay overwhelmed — feeling the tears fill his eyes. He had not known, him-self, how these years of idleness had irked him.

The next morning Lavrans Björgulfsön and Sir Erling stood at the upper end of the courtyard, looking at some of Erlend's horses that ran loose without the fence.

" — Methinks," said Lavrans, "should Erlend appear in this assembly, such rank and birth are his — since he is kinsman to the King and the King's mother — that he must needs come forth among the foremost. Now I know not, Sir Erling, whether you deem you can be sure that his judgment in these affairs will not rather lead him to the other side. If Ivar Ogmundssön should try a countermove — Erlend has near ties, too, with the men who will follow Sir Ivar — "

"I have little thought that Sir Ivar will do aught," said Erling Vidkunssön. "And Munan" — his lip curled a little — "*he* will be wise enough, I trow, to stay away — he knows that else it might well be made clear to all men how much or how little Munan Baardssön counts for." They both laughed. "And there is this to be thought on — aye, you know yet better than I, Lavrans Lagmandssön,* you who have kith and kin there, that the Swedish lords are loath to count our knighthood as the equals of theirs. It might seem needful, then, that we should let no man be lacking from among those we have richest and of highest birth — we can ill afford to let a man like Erlend have leave to sit at home, dallying with his wife and tending his farms — tend he them well or ill," he added, as he saw Lavrans' look.

Lavrans smiled slightly.

"But should you deem it unwise to press Erlend to come with us, I will let it be."

"Methinks, dear sir," said Lavrans, "that Erlend would do better service here, in his own country. As you said

_____
* See *The Bridal Wreath (The Garland)*, Note 13.

yourself — we look for no goodwill towards this levy in the parishes south of Namdalseid — whose folk deem they have naught to fear from the Russians. It might be that Erlend would be the very man to change folks' thoughts of the matter in some measure — "

"He hath such a cursed loose tongue," Sir Erling burst out.

Lavrans answered with a little smile:

"Maybe his kind of talk will be understood by many better than — more clear-headed people's speech — " They looked at each other again, and both laughed. "Howsoever that may be — he could sure do greater hurt should he go to the meeting and talk too loudly there."

"Aye, if so be that you cannot stay him."

"That can I at no rate do, when once he meets with birds of the flock he has been wont to fly with — my son-in-law and I are too unlike."

Erlend came up to them:

"Have you had so much profit of the mass that you have no need of breakfast?"

"I heard not aught of breakfast — I am hollow as a wolf — and thirsty — " Lavrans caressed a dirty white horse he had been handling as he stood there. "The man that tends your farm-horses, son-in-law, I would drive off the place before I sat down to table, were he *my* man."

"I dare not, for fear of Kristin," said Erlend. "One of her maids is with child by him — "

"Nay, but count you that so great a deed in this country-side," said Lavrans, raising his eyebrows a little, "that you deem, because of it, you cannot be lacking him — ?"

"No, but see you," said Erlend, laughing, "Kristin and the priest will have them wed — and they will have me put him in the way of earning a living for himself and her. The girl would not, and her guardian would not, and Tore himself is none too willing — but they will not let me turn him out; she fears he would fly the parish. Then, too, there is Ulf Haldorssön to oversee him — when Ulf is at home — "

Erling Vidkunssön went to meet Smid Gudleikssön. Lavrans said to his son-in-law:

"Methinks Kristin looks somewhat pale in the day-time — "

"Aye," said Erlend, eagerly. "Can you not speak to her, father-in-law? — that boy of hers is sucking the very marrow out of her. I trow she will keep him at the breast till the third fast comes round, like any cottar woman — "

"Aye, she loves her son much," said Lavrans, with a little smile.

"Aye," Erlend shook his head. "They will sit for three hours — she and Sira Eiliv — and talk of it, should a little rash show on him here or there; and for every tooth he cuts, it seems to them a great miracle has come to pass. I have never known aught else but that children were apt to cut teeth; and methinks it had been a greater marvel if our Naakkve had had none."

2

THE YEAR after, one evening at the end of the Yule-tide festival, Kristin Lavransdatter and Orm Erlendssön came, quite unlooked-for guests, to Master Gunnulf's house in the city.

It had blown and sleeted the whole day since before noon, and now in the late evening a heavy snowstorm had come on. The two were thick with snow when they came into the room where the priest sat at the supper-board with his household.

Gunnulf asked, in some fear, whether there was aught amiss at home on the manor. But Kristin shook her head. Erlend was away from home, at a feast in Gelmin, she answered when the priest questioned; but she had been so tired, she could not go with him.

The priest thought how she had now ridden all the way into the city — her horse and Orm's were quite worn out, the last of the way they had scarce had strength to struggle through the snow-drifts. Gunnulf sent the two women of his household with Kristin — they were to get dry garments for her. They were his foster-mother and her sister — other women there were none in the priest's house. He himself cared for his brother's son. Meantime Orm was speaking:

"Kristin is sick, I trow. I said it to father, but he grew angry — "

She had been quite unlike herself in these last days, said the boy. He knew not what it was. He could not remember whether 'twas she or himself that had first thought of coming hither — oh, yes, 'twas she that had first spoken of how much she longed to come in to Christ's Church, and he had said that if that were so, he would go with her. So this morning, straightway after his father had ridden off, Kristin had said that now they would go. Orm had let her have her way, though he saw the weather was threatening — but he liked not the look in her eyes.

Gunnulf thought that neither did he like it, when Kristin now came in. Sorely thin she looked in Ingrid's black habit; her face was wan as bast, and the eyes were sunken, with blue-black rings beneath them — their glance was dark and strange.

It was more than three months since he had seen her — when he was at Husaby for the chistening-feast. She had looked well then, as she lay in state in her bed, and she had said that she felt strong — it had been an easy delivery. So he had spoken against it when Ragnfrid Ivarsdatter and Erlend would have the child given to a foster-mother — while Kristin wept and begged that she might nurse Björgulf herself — the second son was called after Lavrans' father.

The priest asked, therefore, first after Björgulf — for he knew too that Kristin had not liked the nurse they gave the child to. But she said that he was thriving well and Frida was fond of him, and tended him much better than any could have deemed she would. And Nikulaus? asked the uncle; was he as bonny as ever? A little smile came to the mother's face. Naakkve grew comelier and comelier every day. No, he spoke not much; but else he was ahead of his age in every wise, and so big — no one would believe he was in his second winter — Lady Gunna said so too.

Then she sat lost again. Master Gunnulf looked at the two, his brother's wife and his brother's son, sitting one on each side of him. They looked so weary and sorrowful, he grew sad at heart to see them.

Orm, indeed, seemed ever heavy of mood. The boy was fifteen years old now; he would have been the comeliest of youths, had he not looked so weak and ailing. He was well-nigh as tall as his father, but his form was all too slender and narrow of shoulder. In face, too, he favoured Erlend, but his eyes were a much darker blue, and the mouth under its first short black down was yet smaller and softer than his father's, and ever, when it was closed, a sad little furrow showed at its corners. Even Orm's narrow brown neck under his black curled hair looked strangely unhappy as he sat, stooping a little over his food.

Kristin had not sat at the board with her brother-in-law before in his own hall. The year before, she had been with Erlend in the city at the spring-tide Thing,* and then they lodged in this house, which Gunnulf had from his father — but the priest himself had lived then in the house of the Crossed Friars, as vicar for one of the canons. Now was Master Gunnulf parish priest of Steine; but he kept a chaplain in his charge, and himself oversaw the work of engrossing books for the churches of the Archbishopric, while the precentor, Eirik Finssön, was sick. Thus he lived for the time in his own house.

The hall was unlike the rooms Kristin was used to see. 'Twas a timber house, but in the midst of the eastern gable-wall Gunnulf had had built a great stone fire-place, such as he had seen in the southlands; a log fire burned between cast-iron dogs. The table stood along one of the long walls, and by the wall over against it were benches with writing-desks; before a picture of Mary Virgin burned a lamp of yellow-metal, and near by stood frames filled with books.

Strange seemed the room to her and strange her brother-in-law, now she saw him sitting here at his board with his household, clerks and serving-men with a strange half-priestly look. There were some poor folk too — old men and a young boy with film-like thin red eyelids clinging tightly over the empty sockets. On the women's bench with the two old women-servants sat a girl with a two years' child in her lap; she swallowed down the meat hungrily, and stuffed the child's mouth till its cheeks seemed like to burst.

So it was that all the priests of Christ's Church fed the poor at eventide. But Kristin had heard that to Gunnulf

---

* See Note 9.

Nikulaussön came fewer beggars than to the other priests, although — or because — he made them sit with him in his own hall and welcomed every beggar man as an honoured guest. They were given food from his own dish and drink from the priest's own casks. Therefore they came hither when they felt the need of a meat-meal — but else they would liefer go to the other priests, where they were given porridge and small ale in the kitchen.

And as soon as the scribe had said grace after meat, the poor folk made ready to depart. Gunnulf talked kindly to each one, asked if they would not lodge here for the night, or whether there were aught else they desired; but only the blind boy stayed. The girl with the child in special the priest asked to stay, and not take the little one out into the night; but she muttered an excuse and hastened out. Gunnulf bade a serving-man see to it that blind Arnstein was given ale and a good bed in the guest-house. Then he rose and cast a hooded cloak about him: "You are weary, I trow, Orm and Kristin, and would go to rest. Audhild will see to you — belike you will be asleep when I come from the church."

But Kristin begged that she might go with him. " 'Tis for that I am come hither," she said, fixing desperate eyes on Gunnulf. He bade Ingrid lend her a dry cloak, and she and Orm went with the little company that followed him from the house.

The ringing of the bells sounded as though they were right above their heads in the black night sky — the church was not many paces off. They dragged heavily through deep wet new-fallen snow. The weather was still now — now and then single flakelets of snow still floated down, shining faintly in the dark.

In deadly weariness Kristin tried to lean against the pillar she stood by; but the stone chilled her through. She stood in the dark church and looked up towards the lights of the choir. She could not see Gunnulf up there. But she knew he sat there amidst the priests, with his taper by his book — no, after all she was sure she could not speak to him. . . .

To-night it seemed as though there could be no help found for her anywhere. Sira Eiliv at home reproved her,

that she took her everyday sins so hardly — he said 'twas
the lure of spiritual pride: let her but be diligent in
prayer and good works, and she would have no time to
brood so much over such things. "The devil is no such
fool as not to see then that your soul must needs escape
him at last, and he will no longer care to tempt you so
much — "

Oh, no! 'Twas like the devil was none so sure that he
must lose her soul — But when she had lain here before,
crushed by sorrow for her sins, for the hardness of her
heart, her unclean life and her soul's blindness — then had
she felt that the sainted King had taken her in beneath the
sheltering hem of his cloak. She had felt the clasp of his
strong warm hand on hers, he had pointed out to her the
light that is the source of all strength and holiness. St. Olav
had turned her eyes toward Christ upon His Cross — see,
Kristin, God's loving-kindness. — Aye, she had begun to
understand God's long-suffering love. — But since then
she had turned her again from the light, and shut her heart
against it, and now was there naught else in her soul but
disquiet and wrath and fear.

Wretched, wretched was she. She had seen it for herself: such a woman as she was had need of hard trials ere
she could be healed of her unloving spirit. And yet so
rebellious was she that it seemed her heart must break
under the trials that had been laid upon her. They were
but small trials — but they were so many — and she so
rebellious.

— She had a glimpse of her stepson's tall slender form
over on the men's side —

She *could* not help herself. Orm she loved like a child of
her own; but 'twas not possible for her to grow to love
Margret. She had striven and striven, and tried to force
herself to like the child, ever since the day last winter
when Ulf Haldorssön had brought her home to Husaby.
She deemed herself it was a fearful thing — could she feel
such misliking and wrath towards a little maid of nine
years old! And well she knew that in part it was that the
child was so strangely like her mother — she understood
not Erlend, he showed naught but pride that his little
brown-eyed daughter was so fair; never did the child
seem to wake discomforting memories in her father. 'Twas

as though Erlend had forgotten all that concerned these children's mother. — But it was not *only* because Margret favoured her mother that Kristin misliked her. Margret would not endure to be taught by any; she was haughty and harsh to the serving-folk; untruthful was she too, and, with her father, a flatterer. She loved him not, as did Orm — it was ever to gain something that she clung to Erlend with kisses and caresses. And Erlend poured out gifts upon her and humoured all the little maid's whims. Orm liked not his sister either, she had seen that. . . .

It was pain to her to feel herself so hard and cruel that she could not look on at Margret's doings without anger and harsh judgment. But 'twas much more pain to see and hear the endless bickerings between Erlend and his eldest son. The pain was the keener because she saw that Erlend in his inmost heart had a boundless love for the boy — and he was unjust and rough towards Orm because he was at his wit's end to know what he should do with his son or how he could make his future sure. He had given his bastard children lands and goods — but it seemed unthinkable that Orm could ever make a farmer. And Erlend grew desperate when he saw how weak and sinewless Orm was — he would call his son rotten; strive fiercely to harden him, practise him, hour after hour, in use of heavy weapons that 'twas impossible the boy could wield; force him to drink himself sick in the evenings, and bring him back half dead from perilous and toilsome huntings. Beneath all this Kristin could see the anguish of Erlend's heart — he was wild with sorrow often, she saw, that this fine-grained and comely son of his was fit but for one place — and from that his birth barred him out. And thus had Kristin come to understand how little patience Erlend had when he must fear for one that was dear to him or feel pity for him.

She saw that Orm understood it too. And she saw that the youth's heart was torn between love of his father and pride in him — and scorn for the man's unreason when Erlend made his child suffer because troubles lay before them for which he and not the boy was to blame. But Orm had drawn close to his young step-mother — it was as though with her he breathed freely and felt lighter of mood. When he was alone with her, he would jest and laugh by times — in his still fashion. But Erlend liked not this — 'twas as though he misdoubted that these two might sit in judgment on his doings.

Oh, no; 'twas no easy lot for Erlend — no marvel that he was sore and hasty in all that touched the two children. And yet —

She winced with pain still when she thought on it.

They had had the manor full of guests the week before. Now when Margret came home, Erlend had had the loft at the lower end of the hall, over the outer room and the closet, set in order for a sleeping-room — it should be her maiden bower, he said — and there she slept, with the maid whom her father had set to tend and serve her; Frida slept there too, with Björgulf. But now, in this press of Yuletide guests, Kristin had made this loft a sleeping-room for the young men, and the two maids and the babe must sleep in the serving-women's house. But since she had thought Erlend would maybe not like that she should send Margret to sleep with the serving-folk, she had made up a bed for her on one of the benches in the hall, where the ladies and maidens were sleeping. Margret was ever loath to rise in the mornings; that morning Kristin had waked her many times, but she had turned over again, and still lay sleeping after all the others were up. Kristin must needs have the hall cleared and made ready — the guests must have their morning meal — and at length she quite lost patience. She pulled the down pillows from under Margret's head, and took away the coverings from above her. But when she saw the child lying there naked on the fur rug, she threw over her the cloak from her own shoulders. It was a piece of plain undyed wadmal — she wore it only when she went to and from the kitchen and storehouses to see to the service of the food.

Erlend came in at that moment — he had slept in one of the storehouse lofts with some other men, for Lady Gunna lay with Kristin in the great bed. When he saw what was towards he had fallen into a fury. He had grasped Kristin by the arm so hard that the marks of his fingers still showed on her flesh.

"Think you it is fit that this daughter of mine lie in straw and wadmal? Margit is mine, see you, even though she be not yours — what is not too good for your own

children is none too good for her. But since you have held
up the innocent little maid to scorn before these women,
you must even make it good again before their eyes —
spread over Margit again what you took off her — !"

True it was that Erlend had drunk heavily the night
before, and when that was so he was ever fretful next
morning. And he might well have thought that there was
talk among the women when they saw Eline's children.
And he was thin-skinned and sore in all that touched the
esteem they were held in. — And yet —

She had tried to speak of it with Sira Eiliv. But here he
could not help her. Gunnulf had said the sins she had confessed and done penance for, before Eiliv Serkssön came
to be her father confessor, she need not name to him unless
she saw that he must know them in order that he might
judge and counsel in her concerns. So there was much that
she had never told him, though she felt herself that thus
she had come to seem in Sira Eiliv's eyes a better woman
than she was. But yet 'twas so comforting for her to have
the friendship of this good and pure-hearted man. Erlend
mocked — but she had so much comfort from Sira Eiliv.
With him she could talk as much as she would of her
children; all the little things that she wearied Erlend to
death with, the priest was ready to talk over with her.
He had a way with little children, and good skill in their
little troubles and sicknesses. Erlend laughed at her when
she went herself to the kitchen and cooked dainty dishes
to send to the priest's house — for Sira Eiliv was fond of
good food and drink, and it pleased her to busy herself
with such things and try her hand with what she had
learnt from her mother or seen in the cloister. Erlend cared
not what he ate, if one but let him have flesh-meat at all
times, when 'twas not a fast. But Sira Eiliv came and talked
and thanked her and praised her skill when she had sent
him a spitful of young ptarmigans in wrappings of fine
bacon, or a dish of reindeer's tongues in French wine and
honey. And he counselled her with her garden, and got her
cuttings from Tautra, where his brother was a monk, and
from Olav's Cloister, where the prior was his good friend.
And then he read to her, and he could tell her so many
fair things of the life out in the great world.

But for the very reason that he was so good and so

simple-hearted a man, 'twas often hard to speak to him of
the evil she saw in her own heart. When she confessed to
him how wroth she had been with Erlend for his behaviour
in this matter of Margret, he had enjoined on her her duty
to bear with her husband. But he seemed to deem that
'twas Erlend alone that had offended in speaking so un-
justly to his wife — and that in the hearing of strangers.
And Kristin, indeed, thought the like. But in her inmost
heart she felt that she shared in the guilt; she could not
make it clear, but it troubled her heart sorely.

Kristin looked up at the shrine, gleaming faintly golden
up in the twilight behind the high altar. She had looked
so surely that when she stood here once more, something
should again befall her — some deliverance of her spirit.
Again would a living spring well up in her heart and
wash away all the unrest and fear and bitterness and doubt
that filled it.

But to-night there was none that had patience with her.
Have you not learnt it once already, Kristin — to bring
your self-righteousness forth into the light of God's righteousness, to hold up your heathenish and selfish desire in
the light of love? 'Tis that you *will* not learn it Kristin. . . .

But when last she knelt there, she had had Naakkve in
her arms. His little mouth against her breast sent such
warmth into her heart that it grew even as soft wax, easy
for the heavenly love to refashion. And still she *had*
Naakkve — he ran about the hall at home, so fair and so
sweet that her breast ached if she but thought of him. His
soft curly hair began to darken now — he would be black-
haired like his father. And he was so bursting with life
and naughtiness. — She made beasts for him out of old bits
of fur rug, and he threw them away and ran after them,
racing the young hounds. And 'twould often end with the
bear falling into the hearth-fire, and burning up with
much smoke and an evil smell, while Naakkve stood howl-
ing and jumping and stamping, and then buried his head
in her lap — all his adventures ended there as yet. The
maids fought for his favour, the men caught hold of him
and threw him up to the roof, whenever they came into
the hall. Did the boy see Ulf Haldorssön, he ran at once
and clung to the man's leg — Ulf had taken him out into
the farm-place now and then. Erlend snapped his fingers

to his son and set him on his shoulder for a moment — but his father was the one at Husaby that paid least heed to the boy. Though he *was* fond of Naakkve, Erlend *was* glad that he had two true-born sons now.

The mother's heart turned in her:

Björgulf they had taken from her. He whimpered now when she would have held him; and Frida put him straightway to her breast — the foster-mother watched over the boy jealously. But the new child she would never give away from her. They had said, her mother and Erlend, that she must spare her strength now, and so they took her new-born son and gave him to another woman. 'Twas as though she had felt a sort of vengeful joy at the thought that all that had come of this was that now she looked to have yet a third child before Björgulf was full eleven months old.

She dared not speak of it with Sira Eiliv, he would maybe only think that she was vexed that already she must go through all this again. But it was not that. . . .

From her pilgrimage she had come home with a deep awe in her soul — nevermore should this mådness overmaster her. All summer through she had sat alone with her child down in the old hall, weighed in her mind the Archbishop's words and Gunnulf's sayings, been vigilant in prayer and penance, diligent in working to restore the neglected manor and to win her house-folk by kindness and thought for their welfare, eager to help and serve all around her so far as her hands and her power might reach. There had sunk over her a calm delicious peace. She upheld herself with thoughts of her father, with prayers to the holy men and women Sira Eiliv read of, and meditation on their courage and steadfastness. And with a heart tender with happiness and thankfulness she remembered Brother Edvin, as he had appeared to her in the moonlight that night. She had understood full well his message, when he smiled so gently and hung his mitten on the beam of moonlight. Had she but faith enough, she could grow to be a good woman.

When the first year of marriage was at an end, she had to remove back to her husband. She comforted herself, when she felt herself unsure — the Archbishop himself had enjoined upon her that in the life in common with her husband she should show her new heart. And indeed she strove with an eager care for his welfare and his honour. Erlend himself had said it: "It has come to pass, after all, Kristin — you have brought back honour to Husaby." Folks showed her so much kindness and respect — all seemed willing to forget that she had begun her wedded life a little over-hastily. Where housewives came together, she was taken into council; they praised her ordering of her home; she was fetched to be brideswoman and to be helper at births on the great manors, none made her feel that she was young and unskilled, and a new-comer to the country. The serving-folk sat on in the hall through the evening, even as at Jörundgaard — all had somewhat to ask their mistress of. It came over her, in a glow of joy, how kind folks were to her and that Erlend was proud of her. . . .

Then Erlend had gone to work to take order for the ship-levy from the harbours south of the fjord. He journeyed about the country, riding or sailing, and was busy with folk who came to him, and letters that were to be sent off. He was so young and glad and comely — the sluggish unjoyous air she had often seen upon him in former days seemed blown clean away from the man. He glittered new-wakened like the morning. He had little time to spare for her now — but she grew dizzy and wild again when he came near her with his smiling face and his venture-loving eyes.

She had laughed with him over the letter that had come from Munan Baardssön. The knight had not been at the meeting of the King's-men himself, but he scoffed at the whole affair, and most of all at Erling Vidkunssön's being made Regent. It seemed, said Munan, his first task was to give himself new titles — he would have folk call him High Steward now, 'twas said. Munan wrote of her father, too:

"The hill-wolf from Sil crept under a rock and sat mum. That is to say, your father-in-law took lodging with the priests of St. Laurentius Church and raised not his honeyed voice in the parleys. There in his keeping had he letters under the seals of Sir Erngisle and Sir Karl Turessön; if they be not yet worn to shreds, then must the parchment be tougher than the devil's shoe-soles. This also you must know: that Lavrans gave eight marks, pure silver, to Nonneseter. Like enough the man hath got it in his head that Kristin had not so wearisome a life there as of rights she should — "

She had, indeed, felt a pang of shame and pain, but yet she could not help laughing with Erlend. The winter and spring had passed over her in a whirl of mirth and gladness. Now and then a storm on Orm's account — Erlend knew not if he should take the boy north with him. It ended with an outburst at Easter-tide — that night Erlend wept in her arms; he dared not take his son aboard with him; he feared that Orm must needs come short in war. She had comforted him and herself — and the lad — maybe the boy would grow stronger as he grew in years.

The day she rode down with Erlend to the haven at Birgsi, she could not feel aught of fear or sorrow at his going. She was drunken, as it were, with him and his overflowing gladness.

She had not known herself then that already she was with child again. She had thought when she felt sick, 'twas but with the bustle over Erlend's going, with the unquiet times and drinking-feasts at home, and that she was worn out with suckling Naakkve. When she felt new life quicken within her, she had been — She had so looked forward to the winter, to going about in city and country-side with her fair and gallant husband — while yet she, too, was fair and young. She was so sure that in his Russian warfare Erlend would show he was fit for other things than wasting his name and fame and goods. She had thought to wean the boy in autumn — 'twas troublesome to take him and his nurse with her wherever she might go. — And now — no, she had not been glad, and she had told this to Sira Eiliv. Then had the priest rebuked her most sharply for her unloving and worldly spirit. And the whole summer she had passed striving to be glad and to thank God for the new child she was to bear, and for the good tidings that came to her of Erlend's worthy deeds in the north.

Then he came home just before Michaelmas. She had seen well that *he* was not overglad when he saw what was at hand. And that night he said it:

"Methought that when once you were mine — 'twould

be like drinking Yule-tide every day. But it looks as though most of the time would be long fasts."

Each time she remembered it, a wave of blood flooded her face, as hotly as that evening when she had turned her from him, darkly flushed and tearless. Erlend had tried to make amends with love and kindness. But she could not forget it. The fire in her that not all her tears of penitence had had power to quench, nor her anguish for her sin to choke — 'twas as though Erlend had trod it out with his foot when he said those words.

Late at night, the service over, they sat before the chimney-place in Gunnulf's hall, he and Kristin and Orm. A flagon of wine, with some small goblets, stood on the edge of the fire-place. Master Gunnulf had asked more than once if his guests would not rest now. But Kristin begged that she might sit on.

"Mind you, brother-in-law," she asked, "I told you once that our priest at home counselled me to give myself to a cloister, should father not consent that Erlend and I should wed."

Gunnulf glanced swiftly across at Orm. But Kristin said with a little sick smile:

"Think you this grown youth knows not that I am a weak and sinful woman?"

Master Gunnulf answered softly:

"Had you a call to the life of the cloister, then, Kristin?"

"God might sure have opened my eyes, when once I was come into His service."

"Mayhap He deemed your eyes had need to be opened to understand that you should be in his Service wheresoever you may be. Husband, children, house-folk at Husaby have surely need that a trusty and patient servingwoman of God go about among them and care for their welfare. . . .

"Surely that maid weds best who chooses Christ to her bridegroom and gives herself not into the power of any sinful man. But a child that hath already sinned — "

" 'I would have had you come to God, wearing your garland,' " whispered Kristin. "So said he to me, Brother Edvin Rikardssön, of whom I have often told you. Think you the same — ?"

Gunnulf Nikulaussön nodded:

" — Though many a woman has raised herself up out of a life of sin with such strength and steadfastness that we now may safely pray for her intercession. But this befell more often in the days gone by, when she was threatened with torture and the stake and red-hot pincers, if she avowed herself a Christian. I have oft-times thought, Kristin, that 'twas easier then to tear oneself free from the bonds of sin, when it could be done thus by might and main at a single wrench. Even though mankind is so corrupted — yet does courage still dwell by nature in many a breast — and 'tis courage that oftest drives on a soul to seek out God. And so 'tis like the tortures spurred on as many souls to steadfastness as they terrified to apostasy. But a young wildered child that is torn from the lusts of the flesh ere yet she has learned to understand what they bring upon her soul, is brought into sisterhood with holy maids who have given themselves up to watch and pray for them that sleep without in the world. . . .

"Would it might soon be summer!" he said of a sudden, rising up. The two others looked up at him in wonder.

"Aye — it came into my mind, the cuckoo's call from the hill-sides at Husaby of a morning. Always we heard it first from the eastward in the hills behind the houses, and then an answer came from far off in the woods round By — it sounded so sweet across the lake in the morning stillness. Think you not 'tis fair at Husaby, Kristin?"

"East-cuckoo is grief-cuckoo," said Orm Erlendssön in a low voice. "Methinks Husaby is the fairest place in all the world."

The priest laid his hands on his nephew's narrow shoulders for a moment:

"So thought I too, kinsman. 'Twas *my* father's home, too. The youngest son stands no nearer the heritage than do you, my Orm."

"When father lived with my mother, you were nearest heir," said the youth, softly as before.

"We cannot help it, I and my children, Orm," said Kristin, sorrowfully.

"I trow you will have marked, too, that I bear you no grudge," he answered quietly.

" 'Tis such a wide, open country-side," said Kristin in a

little. "One sees so far around from Husaby — and the heaven is so — so wide. Where I am from, it lies like a roof right down upon the hill-tops. The Dale lies low in shelter, so hollow and green and fresh. The world is of so fit a size — not too great and not too narrow." She sighed and moved her hands in her lap.

"His home was there, the man your father would have wed you to?" asked the priest, and the woman nodded.

"Have you ever repented that you took him not?" he asked again, and she shook her head.

Gunnulf crossed the room and took a book from the case. He sat down again by the fire with it, opened its clasps, and turned over the leaves. He did not read from it, but sat with the book open in his lap.

"When Adam and his wife had defied the will of God, then felt they in their own flesh a power that defied their wills. God had created them, man and woman, young and fair, for that they should live in wedlock, and bring forth co-heirs with themselves of His bounteous gifts, the loveliness of the Garden of Paradise, the fruit of the tree of life and bliss everlasting. They needed not to feel shame of their bodies, for, as long as they obeyed God, their whole body and all their limbs were in the power of their will, even as are hand and foot."

Flushing red as blood, Kristin pressed her hands crosswise under her breast. The priest bent towards her a little; she felt his strong yellow eyes on her bowed visage.

"Eve made spoil of that which belonged to God, and her husband received it, when she gave him that which by right was the possession of their Father and Creator. They would fain have made themselves His like — now they marked that they were like unto Him first in this: as they had betrayed His lordship in the great world, so now was their lordship over the little world, the soul's house of flesh, betrayed. As they had played false to the Lord their God, so now would the body play false to its lord, the soul.

"Then did these bodies seem to them so ugly and hateful that they made them garments to hide them. First but a short apron of fig-leaves. But as they came to know more and more the inwardness of their own fleshly nature, they drew the garments up higher and higher, over the place of their hearts and over their backs so unwilling to bow. Until these last days when men clothe themselves in steel to the outermost joints of fingers and toes, and hide their faces behind the bars of the helm — so are strife and treachery spread abroad in the world."

"Help me, Gunnulf," prayed Kristin. She was white to the lips. "I — I know not my own will."

"Say then: 'Thy will be done,'" answered the priest softly. "You know that His will is that you should open your heart to His love, and that then must you love Him again with all your soul's might."

Kristin turned suddenly towards her brother-in-law:

"You know not — you — how dearly I loved Erlend. And my children — "

"My sister — all other love is but as an image of heaven in the water-puddles of a muddy road. Bemired must you needs be if you will dabble in them. But if you bear ever in mind that 'tis a mirroring of the light from yonder other home, then will you rejoice in its fairness, and will take good heed not to destroy it by stirring up the mud beneath — "

"Aye. But you, Gunnulf, are a priest — you have vowed to God to fly these — lures — "

"That have you too, Kristin — when you promised to forsake the devil and all his works. The devil's work is that which begins in sweet desire, and ends in them that work it stinging and biting each other like toad and asp. 'Twas that Eve learnt — that when she would have given her husband and her offspring that which was God's possession, then brought she them naught but outlawry and blood-guiltiness and death, that came into the world when brother slew brother on that first small field where thorns and thistles grew on the stone-heaps between the little plots — "

"Aye. But *you* are a priest," she said, as before. "You have not to strive every day to agree with another in your will" — she burst into tears — "to be patient — "

The priest said with a little smile:

"In that matter there is strife between soul and body in each mother's son. Therefore are bride-mass and wedlock appointed, that man and woman shall find help to live, wedded pair and parents and children and house-

mates, as trusty and helpful travelling companions on the
journey towards the home of peace."

Kristin said low:

"Methinks it must be easier to watch and pray for them
that sleep without in the world, than to strive against one's
own sins — "

"It is so," said the priest sharply. "But think you, Kristin, there has lived one holy man that hath not had to
defend *himself* against the enemy, at the same time as he
strove to guard the lambs against the wolf — ?"

Kristin said, low and timidly:

"I had thought — they that move from holy place to
holy place and have command of all prayers and words of
power — "

Gunnulf leaned forward, mended the fire, and remained
sitting with his elbows on his knees:

" 'Twas six years ago, about this time of year, that we
came to Rome, Eiliv and I and two Scots priests that we
had come to know in Avignon. We had gone afoot the
whole way. . . .

"We came to the city just before the Fast. At that time
the folk in the south-lands hold great feastings and banquets — they call it *carnevale.* Then does wine, both red
and white, run like rivers in the tavern-houses, and folks
dance at nights without doors, and have torches and bon-
fires in the open places. It is spring in Italia then, and the
flowers are blooming in meadows and gardens, and the
women deck themselves with these, and throw roses and
violets to the passers-by in the streets — they sit up in the
windows, and they have carpets of silk and gold brocade
hanging from the sills down over the wallstones. For all
the houses are of stone down there, and the knights have
their castles and strong places in the midst of the cities.
Belike there is no town-law or market-peace in that city —
for they and their serving-men fight in the streets till the
blood runs down —

"There stood such a castle in the street where we dwelt,
and the lord who ruled there was named Ermes Malavolti.
It shadowed all the narrow lane where we lodged, and our
chamber was as dark and cold as a dungeon in a stone
fortress. Often when we went out, we must needs press
close to the wall while he rode by, with silver bells on

his garments and a whole troop of armed followers, while filth and rottenness splashed up from under the horses' hoofs — for in that land the people do but cast all filth and sweepings without doors. The streets are cold and dark and strait as rock-clifts — little like the green roadways of our towns. In those streets they hold races when the time of *carnevale* comes — send out wild Arab horses to race against each other — "

The priest paused a little, then he went on:

"This Sir Ermes had a kinswoman dwelling in his house. Isota was her name, and she might well have been Isold the Fair herself. Her skin and hair were of the hue of honey, but her eyes, I trow, were black, I saw her time and again at a window —

"But outside the city the land is more waste than the most desolate uplands in our land, where nothing haunts but wolf and reindeer, and the eagle screams. Yet are there towns and castles in the mountains round about, and out on the green uplands a man can see everywhere marks of folk who must have dwelt there in days gone by; and great flocks of sheep and herds of white oxen are grazing. Herdsmen armed with long spears follow them about a-horseback, and these are perilous folk for wayfarers to meet, for they murder and rob them and cast the bodies into holes in the earth —

"But out upon these green plains lie the churches of pilgrimage."

Master Gunnulf was silent for a while:

"Maybe that land seems so unspeakably desolate because the city lies in the midst of it, she that was queen over all the heathen world and was chosen to be the bride of Christ. For now the watchmen have forsaken the city, and in the whirl of feasting and riot the town seems like a forsaken spouse. Ribalds have set up their abode in the castle, where the husband is not at home, and they have lured the lady on to wanton with them in their lust and strife and bloodshed.* . . .

"But under the earth are treasures, dearer than all the treasures the sun shines upon. There are the graves of the holy martyrs, hewn out of the living rock beneath — and

---

* See Note 10.

there are so many that a man grows dizzy but to think of it. When one calls to mind how many they are, the torture-witnesses that here have suffered death for the cause of Christ, one might deem that each grain of dust that whirls up from the hoofs of the ribalds' horses must be holy and worthy of worship." . . .

The priest drew a thin chain out from under his clothing, and opened the little silver cross that hung from it. Within was something black that looked like tinder, and a little green bone.

"Once we had been down in these passages all the day, and we had said our prayers in caves and oratories, where the first disciples of St. Peter and St. Paul had met together to hold mass. Then the monks, who owned the church where we went down, gave us these holy relics. 'Tis a little piece of such a sponge as the holy maids used to wipe up the blood of the martyrs withal, that it might not be lost; and a knuckle of a finger-joint of a holy man whose name God alone knows. Then did we four promise one another that we would every day call upon this holy one whose honour is unknown to men, and we took this nameless martyr to witness that we should never forget how quite unworthy we were to be rewarded of God and honoured of men, and ever remember that naught in the world is worthy to be desired save only His mercy." . . .

Kristin kissed the cross reverently and gave it to Orm, who did the like. Then Gunnulf said suddenly:

"I will give you this holy relic, kinsman."

Orm knelt on one knee and kissed his uncle's hand. Gunnulf hung the cross around the youth's neck.

"Would you not be fain to see these places, Orm?"

The boy's face lighted up in a smile:

"Yes — and I know now that some day I shall come thither."

"Have you never had a mind to be a priest?" asked Gunnulf.

"Yes," answered the boy. "When father has cursed these weak arms of mine. But I know not if he would like that I should be a priest. And then there is that you wot of," he said softly.

"For your birth a dispensation could be had, I trow,"

said the priest quietly. "Maybe, Orm, some day we might journey southward together, you and I — "

"Tell us more, uncle," prayed Orm softly.

"Aye, that will I," Gunnulf clasped the arms of the chair with his hands and looked into the fire.

"While I wandered there, seeing naught else but memorials of the blood-witnesses, and remembered the unbearable torments they had endured for Jesu name's sake, there came upon me a sore temptation. I thought of how the Lord had hung nailed to the Cross those six hours. But His witnesses had been tormented with unutterable tortures for many days — women saw their children tortured to death before their eyes, young tender maids had their flesh torn from their bones with iron combs, young boys were driven on the claws of wild beasts and the horns of mad oxen. . . . Then seemed it to me as though many of these had borne more than Christ Himself. . . .

"I brooded over this till I thought that my heart and my brain must break in sunder. But at last the light I had begged and prayed for came to me. And I understood that as these had suffered, so ought we all to have strength to suffer. Who would be so foolish that he would not willingly endure pains and torment, when this was the path that led to a faithful and steadfast bridegroom, who waiteth with arms stretched out and breast bloody and burning with love?

"For He loved mankind. And therefore did He die, as the bridegroom who hath gone forth to save his bride from the hands of robbers. And they bind him and torment him unto death, while he sees his dearest love sit feasting with his slayers, jesting with them and mocking his torments and his faithful love — "

Gunnulf Nikulaussön buried his face in his hands:

"Then did I understand that this mighty love upholdeth all things in the world — even the fires of hell. For if God would, He could take the soul by force — we should be strengthless motes in His hand. But He loves us as the bridegroom loves his bride, who will not force her, but if she yield not to him willingly, must suffer that she flee him and shun him. But I have thought, too, that may-hap no soul can yet be lost to all eternity. For every soul must desire this love, methinks, but it seems so dear a purchase to give up all other delights for its sake. But when the fire hath burnt away all stiff-necked and rebellious will, then at last shall the will to God, were it no greater in a man than a single nail in a whole house, remain in the soul unconsumed, as the iron nail is left in the ashes of a house burned down — "

"Gunnulf" — Kristin half rose — "I am afraid — "

Gunnulf looked up, with white face and flaming eyes:

"I too was afraid. For I understood that this torment of God's love can have no end so long as man and maid are born upon this earth and He must be fearful that He may lose their souls — so long as He daily and hourly giveth His body and His blood on a thousand altars — and there are men who scorn the offering. . . .

"And I was afraid to think of myself, that unclean had served at His altar, said mass with unclean lips and lifted Him with unclean hands — and methought I was even as the man who had brought his beloved to a house of shame and betrayed her — "

He caught Kristin in his arms as she sank down, and he and Orm bore the swooning woman over to the bed.

In a little she opened her eyes — sat upright and covered her face with her hands. She burst into a wild, wailing weeping:

"I cannot, I cannot — Gunnulf, when you speak, I see that I can never — "

Gunnulf took her hand; but she turned her face away from the man's wild pale visage:

"Kristin, never can you content you with any lesser love than the love that is between God and the soul —

"Kristin, look about you — see what the world is. You who have borne two children — have you never thought on this: that every child that is born is baptised in blood, and that every human being born into this world draws in with his first breath the scent of blood? Think you not that you, who are a mother, should fix all your intent on this: that your sons fall not back to that first baptismal pact with the world, but hold fast to the other pact they made with God when made clean by the waters of the font — ?"

She sobbed and sobbed.

"I am afraid of you," she said again. "Gunnulf, when

you speak thus, I see well that never can I find my way onward to peace."

"God will find you," said the priest softly. "Be still, and fly not from Him who hath sought after you before you were conceived in your mother's womb."

He sat awhile there by the bedside. Then he asked quietly and calmly if he might wake Ingrid and ask the woman to come and help her to undress. Kristin shook her head.

Then he made the sign of the cross over her thrice, and, bidding Orm good-night, went into the closet where he slept.

Orm and Kristin drew off their clothes. The boy seemed sunk in fathomless thoughts. When Kristin had lain down, he came across to her. He looked at her grey-flecked face and asked if she would have him sit by her till she slept.

"Oh, aye — oh, no, Orm; you must be weary, you that are young. The night must be far spent — "

Orm stood yet a little while.

"Seems it not strange to you?" he said of a sudden; "father and my Uncle Gunnulf — unlike as they are to each other — yet they are like, too, after a fashion — "

Kristin lay a little, thinking:

"Aye, maybe so — they are unlike other men."

Soon after, she fell asleep and Orm went over to the other bed. He stripped and crept in. There was a linen sheet below, and linen covers on the pillows. The boy stretched himself out at ease on the smooth cool couch. His heart beat, thrilled with the thought of these new adventures whereto his uncle's words had shown the way. Prayers, fasts, all the observances he had practised because he had been taught to — all grew new of a sudden — weapons in a goodly warfare that he longed for. Maybe he would be a monk — or a priest — if he could win dispensation for his birth in adultery.

Gunnulf's couch was a wooden bench with a sheet of skin spread over a little straw, and a single small pillow — so that he was forced to lie stretched straight out. The priest took off his frock, lay down in his underclothes, and drew the thin coverlid of wadmal over him. The little lighted wick, twined round an iron rod, he left burning.

His own words had left him crushed with unrest and fear. He was faint with yearning for that time gone by — would he never again find that bridal gladness of heart that had filled his whole being that spring in Rome?

It was when he had come back to Norway that strong disquietude took hold upon him.

There was so much to disquiet him. There were his riches. The great inheritance from his fathers — and the rich benefice. There was the path that he could see lying before him. His place in the Cathedral chapter — he knew that it was meant for him. If he forsook not all his possessions, to go into the cloister of the Preaching Friars, take the vows, and bow himself under the rule — this was the life he desired — with but half a heart.

— And then when he was old enough and hardened to the fight — There were men in Norway's realm that lived and died in utter heathendom, or led astray by the false doctrines that the Russians put forth under the name of Christendom. The Lapps and the other half-wild folks that he could never cease to think on — was it not as though God had wakened in him this longing to fare forth to their land, bringing the Word and the Light — ?

— But he thrust the thought of this mission from him, on the plea that he must obey the Archbishop. Lord Eiliv counselled him against it. Lord Eiliv had hearkened to him and spoken with him, showing him clearly that he spoke as to the son of his old friend, Sir Nikulaus of Husaby: "But you can never keep you within measure, you that come of the line of old Skogheim-Gaute's daughters, whether 'tis good or ill you have set your heart upon." The Lapp-folks' salvation he had much at heart himself, he said — but they had no need of a teacher that wrote and spoke Latin as well as his mother-tongue, and was learned in the Law no less than in Arithmetica and Algorismus. Sure it was that he had been given learning that he might use it. "And to my mind 'tis uncertain whether you have been granted the gift of speaking to the poor and simple peoples up north."

Ah, but in that sweet spring his learning had seemed to him no more to be held in reverence than the learning every little maid gets from her mother — to spin and brew and bake and milk — the teaching every child needs that it may do its work in the world.

Gunnulf stood up. On one end-wall of the closet hung a great crucifix, and in front of it was a great flat stone upon the floor.

He knelt down upon the stone and stretched his arms out side-wise. He had used his body to endure this posture, so that he could kneel thus by the hour, still as stone. With eyes fastened on the crucifix he awaited the comfort that came to him when he could lose himself wholly in contemplation of the Cross.

But the first thought that came into his mind now was this: had he the will to part with this crucifix? St. Franciscus and his brethren had crosses that they joined together themselves out of branches of trees. He should give away this fair Rood — to the church at Husaby he might give it. Peasants, children and women that came thither to mass — they might well be strengthened by feeding their eyes with such a visible picture of the Saviour's lovely mildness in His Passion. Simple souls like Kristin — For himself it should not be needful.

Night after night he had knelt with close-shut senses and limbs benumbed, till he saw the vision. The hill with the three crosses against the sky. Yonder cross in the midst, which was destined to bear the Lord of earth and heaven, trembled and shook, it bent like a tree before the storm, affrighted that it should bear that all too precious burden, the sacrifice for the sins of all the world. The Lord of the Tents of Storm held it in, as the knight curbs his charger, the Chief of the Castle of Heaven it bore to battle. Then was made manifest the wonder that was the key to deeper and ever deeper wonders. The blood that ran down the Cross for the remission of all sins and the boot of all sorrows, that was the visible miracle. By this first wonder the soul's eye could be opened to behold the yet darker mysteries — God that descended unto earth, and became the Son of a Virgin and Brother to mankind, that harried hell, and stormed with his booty of souls set free up to the blinding sea of light, wherefrom the world hath issued and whereby the world is upholden. And towards those bottomless and eternal deeps of light his thoughts were drawn up, and there they passed into the light and vanished, as a flight of birds passes away into the glory of the evening sky.

Gunnulf did not move until the church-bell rang for matins. All was still when he passed through the hall — they slept, Kristin and Orm.

Out in the pitch-dark yard the priest tarried a little. But none of his house-folk came to go with him into the church. He required not that they should go to worship more than twice in the day's round; but Ingrid, his foster-mother, well-nigh always bore him company to matins. This morning it seemed she, too, slept. Aye, she had been late up the night before —

All that day the three kinsfolk spoke but little together, and of naught but small matters. Gunnulf looked weary, but he talked jestingly of this and that. "Foolish were we yester-evening — we sat there as sorrowful as three father-less children," he said once; and told of some of the many merry little haps that befell here in Nidaros — with the pilgrims and such folk — which the priests jested over among themselves. An old man from Herjedal had had errands here for all the folks of his parish, and got the prayers all mingled together — things would have looked but ill in the parish, it came into his head after, if St. Olav had taken him at his word!

Late in the evening Erlend came in, dripping wet — he had sailed in to the city and it was blowing hard again. He was raging, and fell upon Orm at once with furious words. Gunnulf listened awhile in silence; then:

"When you speak so to Orm, Erlend, you are like our father — as he used to be when he spoke to you —"

Erlend went silent at once. Then he flung round:

"Well I wot, so witlessly did I never bear myself when I was a boy — make off from the manor, a sick woman and a whelp of a boy, in a snowstorm! Else 'tis not much to brag of, Orm's manfulness, but you see that at least he fears not his father!"

"You feared not your father either," said his brother, smiling.

Orm stood up before his father, saying naught and striving to seem careless.

"Aye, you can go," said Erlend. And then: "I am nigh sick of the whole affair at Husaby. But one thing I know — this summer shall Orm fare with me northward — then I trow we will lick this pet lamb of Kristin's into shape. — He is not a bungler, either," he went on eagerly to his brother. "He shoots with a sure aim — and a coward he is not — but ever is he cross-grained and mopish, and 'tis as though he had no marrow to his bones — "

"Nay, if you berate your son often as you did but now, 'tis no marvel if he mope," said the priest.

Erlend changed his tone, laughing as he said:

"For that matter, I had often to suffer worse things from father — and God knows that I moped not much for that. But let that be — now I am come hither, let us even be merry for Yule, since Yule it is. Where is Kristin? — What was it she had to speak of to you now, again?"

"I believe not that she had aught to speak of with me," said the priest. "She had set her heart on hearing mass here at Yule."

"She might well make shift with those she hears at home, methinks," said Erlend. "But 'tis pity of her — as things are going, all youth is being worn away from her." He struck one hand against the other. "I see not how the Lord can think we have need of a new son every year — "

Gunnulf looked up at his brother:

"What — ! Nay, I know not what our Lord may deem that you two need. But what Kristin needs most, I trow, is that you be kind to her now — "

"Aye, like enough she does," said Erlend in a low voice.

The next morning Erlend went to the day-mass with his wife. They were bound for St. Gregory's Church — Erlend heard mass there always when he was in the city. The two walked alone; and down the street, where the snow lay swept up into drifts, heavy and wet, Erlend led his wife by the hand fairly and courteously. He had not said a word to her of her flight from home, and he had been friendly to Orm after the first storm.

Kristin walked quiet and pale, with head a little bent;

the long black fur cloak with the silver clasps seemed to weigh down her slight thin body.

"Would you that I should ride with you homeward — and let Orm sail with the boat?" said the man, to say somewhat. "Maybe you would scarce care to cross the fjord?"

"No — you know that I like not to go on boats — "

It was calm now, and the weather mild — every moment a load of heavy, wet snow would fall from the trees. The skies hung low and dark-grey above the white town. There was a greenish-grey watery tinge on the snow, and the houses' timber walls and the fences and tree-trunks showed black in the damp air. Never, Kristin thought, had she seen the world look so cold and wan and faded. . . .

3

KRISTIN sat with Gaute on her lap, gazing from the little hill north of the manor. The evening was so fair. The lake lay below her bright and still, mirroring the hills and the farms of By, and the golden clouds in the sky. It had rained earlier in the day, and the smell of leaves and earth rose strongly up. The grass on the meadows below must be knee-deep already, and the fields were hidden with the spears of corn.

Sounds came from far to-night. Now the pipes and drums and fiddles struck up again, down on the green at Vinjar — up here the sounds came so sweetly to the ear.

The cuckoo would fall into long spells of silence, and then send forth its call once and again from far off in the southern woods. And in all the groves round about the manor the birds whistled and sang — but in scattered notes and softly, for the sun was still high.

The home cattle came tinkling and lowing out of the pasture above the courtyard gate.

"See, see — my Gaute will get his milk soon," she babbled to the child, lifting him up. The boy, as was his wont, laid his heavy head down on his mother's shoulder. Now and then he clung closer to her — Kristin took it for a sign that he *did* take in somewhat of her petting and her prattle.

She walked down to the houses. Before the door of the hall, Naakkve and Björgulf were running about, and trying to coax back the cat, which had fled from them up on to the roof. But in a moment they set to work again with the broken dagger they owned in common, digging deeper the hole they had made in the earthen floor of the outer room.

Dagrun came into the hall with goat's milk in a wooden pail, and the mistress gave Gaute ladleful after ladleful of the warm drink. The boy grunted angrily when the serving-woman spoke to him, and struck out at her and hid himself against his mother's breast when the woman would have taken hold of him.

"Yet methinks he comes on better," said the byre-woman.

Kristin lifted the little face in her hand — it was yellow-white like tallow, and the eyes were always weary. Gaute had a great, heavy head and thin, strengthless limbs. He would be two years old a week after Lawrence Mass, but as yet he could not set foot to ground, and he had but five teeth and could not speak a word.

Sira Eiliv said 'twas not rickets; for neither the alb nor the altar-books had availed. High and low, wherever he came, the priest sought for some remedy for this sickness that was come upon Gaute. She knew he remembered the child in all his prayers. But to her he could but say that she must bow in patience before the will of God. And she must give him plenty of warm goat's milk to drink. . . .

Poor, poor little boy of hers! Kristin hugged and kissed him when the woman was gone. So fair, so fair he was! She thought she saw he favoured her father's kindred — his eyes were dark grey and his hair flaxen white, thick, and soft as silk.

Now he began to whimper again. Kristin got up and walked about the room with him. Little and thin as he was, he grew heavy after a while — but Gaute would be content nowhere but in his mother's arms. So she bore him up and down the murky hall, crooning to him as she walked.

Someone rode into the courtyard, and Ulf Haldorssön's voice echoed loudly from the house-walls. Kristin came out to the doorway of the outer room, with the child in her arms.

"You must unsaddle your own horse to-night, Ulf, I fear — they are at the dancing — all the men. Shame it is you should be troubled, but you must forgive it — "

Ulf muttered testily as he unsaddled his horse. Naakkve and Björgulf pressed about him the while, and begged for a ride on the horse up to the garden-close.

"Nay, you must bide here with Gaute, my Naakkve — play with your brother and let him not cry while I am away in the kitchen — "

The boy looked glum. But in a moment he was down on all-fours, butting and lowing at the little one, whom Kristin had set down on a cushion at the door. The mother bent down and stroked Naakkve's hair. He was so good to his little brothers.

When Kristin came back to the hall bearing the great platter on her arms, Ulf Haldorssön sat on the bench playing with the children. Gaute was happy with Ulf, if his mother were but out of sight — but now he whimpered at once and reached out after her. Kristin set down the platter and took Gaute in her arms.

Ulf blew the froth off the newly-drawn ale, drank, and began eating from the small dishes on the platter.

"Are they out, all your maids, to-night?"

Kristin said:

"There are both fiddles and drums, and pipes too — a troupe of gleemen — come over from Orkdal from the bridal there. You may well believe, when they heard of it — they are but young girls after all — "

"You let them gad and frisk about their fill, Kristin. A man might think you were afraid 'twould be hard to find a wet-nurse here this autumn — "

Without thinking, Kristin smoothed her kirtle down over her slender waist. She had flushed darkly red at the man's words. Ulf laughed short and harshly:

"But if you will go about ever dragging Gaute with you, 'twill go with you as it went last year, I trow. . . .

Come hither to your foster-father, boy, and you shall eat out of one dish with me." . . .

Kristin made no answer. She set her three small sons in a row on the bench by the other wall, and fetched a bowl of milk-porridge, and, for herself, a little stool. There she sat and fed them, though Naakkve and Björgulf grumbled — they would have had spoons to eat with themselves. One was four years old now, and the other nigh three.

"Where is Erlend?" asked Ulf.

"Margret had a mind to go to the dancing, so he went with her."

" 'Tis well at least he hath wit enough to watch his own maid," said Ulf.

Again the wife said nothing. She undressed the children and put them to bed, Gaute in the cradle and the two others in her own bed. Erlend had resigned himself to have them there, since she had grown well of her great sickness the year before.

When Ulf was done, he stretched himself out on the bench. Kristin dragged the block-chair over to the cradle, fetched a basket of woollen yarns, and began winding balls of wool for her weaving, while she gently and softly rocked the cradle with her foot.

"Will you not go to rest?" she asked once without turning her head. "You must be weary, Ulf?"

The man rose, put some fuel on the fire, and came across to his mistress. He sat down on the bench over against her. Kristin saw he was not so worn out with hard living as he was wont to be when he had been some days in Nidaros.

"You ask not even what the tidings are from the city, Kristin," he said, leaning forward, elbows on knees, and gazing on her.

Her heart began to beat with fear — she understood from the man's looks and bearing that there must again be tidings that were not good. But she answered with a calm and gentle smile:

"You must tell me, Ulf; have you heard aught?"

"Oh, aye — " But first he brought his wallet and took out from it the things he had fetched for her from the town. Kristin thanked him.

"I see you have heard news in the city?" she asked a little after.

Ulf looked at his young mistress — then he turned his eyes to the pale child, asleep in the cradle.

"Does he always sweat so much in the head?" he asked, low, touching the hair gently where it was dark with damp. "Kristin, when you were wed to Erlend — the letter that was drawn out concerning the settlement of your goods — stood it not so that you should deal yourself at your will with the lands of his extra-gift and your morning-gift?"

Kristin's heart beat harder, but she spoke calmly:

"Aye, and so it is, Ulf, that Erlend has ever asked my counsel and sought my consent in all dealings with these lands. Speak you of the parcels of farms in Verdalen that he has sold to Vigleik of Lyng?"

"Aye," said Ulf. "Now has he bought Hugrekken * from Vigleik, so he will keep up two ships now, it seems. . . . And what are you to have, Kristin?"

"Erlend's part of Skjervastad, half a hide † in Ulfkel-stad, and what he owns of Aarhammar," said she. "You sure did not believe that Erlend had sold those lands without my will and without making good their worth to me?"

"Hm." Ulf sat silent a little. "Yet will your incomings be less, Kristin. Skjervastad — 'twas there Erlend got hay last winter and released the rent to the farmer for three years — "

" 'Twas not Erlend's fault that we got no dry hay last year — I know, Ulf, you did all you could — but with all the misery there was here last summer — "

"Of Aarhammar he sold more than half to the Sisters of Rein convent, that time when he made ready to flee the land with you" — Ulf laughed a little — "or pledged it — 'tis the same thing with Erlend. Free from the King's levy — the whole of that is on Audun's shoulders, who holds the farm that now is to be called yours!"

"Can he not rent the land that is come under the convent?" asked Kristin.

"The Sisters' tenant on the neighbour farm hath rented

---

* Hugrekken = *The Valiant.*
† See Note 11.

it," said Ulf. " 'Tis a hard and an unsure task for tenants to make ends meet when the farms are split up as Erlend is busy doing."

Kristin was silent. She knew well enough it was true.

" 'Tis ever quick work with Erlend," spoke out Ulf again. "His goods wane as swiftly as his household waxes."

As the woman made no answer, he spoke again:

"You will soon have many children, Kristin Lavrans-datter."

"Yet none that I can spare," she answered in a voice that shook a little.

"Be not so afraid for Gaute — he will grow strong, I warrant him," said Ulf, in a low voice.

"It must be as God will — but 'tis long waiting."

He heard the hidden suffering in the mother's voice — a strange look of helplessness seemed to come upon the dark, heavy man.

"It avails so little, Kristin — much have you brought to pass here at Husaby, but if Erlend is to take the sea again with two ships — I believe not overmuch in peace coming in the north, and your husband is so little crafty that he knows not how to turn to his profit what he has won in these two years. Ill years have they been — and all through them you have been a sick woman. Should things go on in this wise, 'twill break your courage at last, young wife that you are. I have helped you all I could on the manor here — but this other thing — Erlend's unwisdom — "

"Aye," she broke in, "that God knows you have — you have been the staunchest kinsman to us, friend Ulf, and never can I thank you enough or repay you — "

Ulf stood up, lit a candle at the hearth, set it on the taper-holder on the table, and stood there with his back turned to his mistress. Kristin had let her hands sink into her lap — now she began again winding wool and rocking the cradle.

"Can you not send word home to your folks," asked Ulf softly, "so that Lavrans might come up too this autumn when your mother comes to you?"

"I had not meant to trouble my mother this autumn. She begins to grow old — and it befalls so often that I must lie in the straw that I can scarce ask her to come to me every time — " She forced a smile.

"But do it this time," answered Ulf. "And pray your father to bear her company — so that you may ask his counsel in these matters — "

"In this matter I will not ask my father's counsel," she said quietly and firmly.

"But Gunnulf, then?" asked Ulf in a little. "Can you not speak with him?"

" 'Twould not be seemly to trouble him with such things now," said Kristin as before.

"Mean you because he has withdrawn him into a cloister?" Ulf laughed mockingly. "Never did I mark that monks knew less of guiding goods and gear than other folks.

"If so be you will not take counsel of any, Kristin, then must you speak yourself with Erlend," he said, when she made no answer. "Think on your sons, Kristin!"

Kristin sat long in silence.

"You that are so good to our children, Ulf," she said at last, "methinks 'twere liker you should wed and have your own folk to care for than that you should go on here — plaguing yourself with Erlend's — and my — troubles."

Ulf turned towards the woman. He stood with his hands grasping the table-edge behind him and looked at Kristin Lavransdatter. Still was she as straight and slim and fair as ever, as she sat there. Her dress was of dark home-dyed woollen stuff, but the linen coif that lay around her still, pale face was fine and soft. The belt from which hung her bunch of keys was set with little silver roses. On her breast glittered two chains bearing crosses; the great one with gilded links hung down well-nigh to her waist — 'twas it she had had of her father. Above lay the thin silver chain with the little cross that Orm had prayed them to give to his stepmother and say that she should wear it always.

As yet she had arisen from each child-bearing fair as ever — only a little stiller, with the weight on her young shoulders a little heavier. A little thinner in the cheeks, the eyes a little darker and more sad under the broad white forehead, the mouth a little less full and red. But 'twas like her comeliness would be worn away ere she was many years older, if things went on as they were going. . . .

"Think you not, Ulf, it were happier for you if you

settled down on your own farm?" she went on again. "You have bought twenty marks' worth more of Skjoldvirkstad land, Erlend said — you own nigh half the farm now. And Isak has but the one child — Aase is both comely and kindly, and a notable woman — and she seems to like you—"

"Yet will I not have her, if I must marry her," the man said gruffly, with a harsh laugh. "And Aase Isaksdatter is too good for—" His voice changed. "I never knew other father than a foster-father, Kristin — and I trow 'tis my lot that I shall have no children either but foster-children."

"Nay, I will pray Mary Virgin that you may have better fortune, kinsman."

"I am not so young, either. Five-and-thirty winters, Kristin," he laughed. "There wants not much but that I might be your father—"

"Nay, but then must you have sinned full early," answered Kristin, striving to speak lightly and laughingly.

"Would you not go to bed now?" asked Ulf soon after.

"Yes, in a little — but *you* must be weary, Ulf — you should go to rest."

The man bade her good-night quietly and went out.

Kristin took the candle from the board, and looked in by its light at the two boys sleeping in the box-bed. There was no matter on Björgulf's eyelashes — God be thanked for that. There had been fair weather for a while past. As soon as the wind came a little keen, or the weather was so rough that the children must play within by the hearth-fire, his eyes grew sore. She stood long looking at the two. Then she bent over Gaute in the cradle.

They had been as fresh and healthy as little birds, all her three young sons — until the sickness came to the country-side last summer. Folk called it the scarlet fever — it carried off children from the homes all around the fjord, so that 'twas a piteous thing to see or hear tell of. She had been granted grace to keep all hers — all her own. . . .

For five days and nights had she sat by the southern bed, where they lay, all three, with red spots over all their skin, and sick eyes that shunned the light — the little bodies fiery-hot. She sat with her hand under the coverlid patting the soles of Björgulf's feet, and sang and sang, till her slender voice was sunken to a hoarse whisper:

> Shoe, shoe, guardsman's steed —
> How can we shoe him best at need?
> Iron shoes are fitting for the guardsman's steed.

> Shoe, shoe, Earlie's steed —
> How can we shoe him best at need?
> Silver shoes are fitting for the Earlie's steed.

> Shoe, shoe, King's own steed —
> How can we shoe him best at need?
> Gold shoes are fitting for the King's own steed.

Björgulf was the least sick and the most restless. If she stopped singing but a moment, he would cast the coverlid off him at once. Gaute was only ten months old — he was so deathly sick, she deemed he could not live. He lay at her breast, wrapped in rugs and furs, and had no strength to suck. She held him in one arm and patted Björgulf's foot-soles with the other hand.

From time to time, when it chanced that all three slept for a little space, she would lay herself down on the front of the bed beside them, fully clothed. Erlend came and went, looking helplessly at his three sons. He tried to sing for them, but they cared naught for their father's mellow voice — 'twas their mother they would have sing, though she had no singing-voice.

The serving-women hovered round and would have had their mistress spare herself; the men came and asked tidings; Orm tried to make sport for his little brothers. Margret Erlend had sent away to Österdal, by Kristin's counsel; but Orm was set upon staying — besides, he was grown up now. Sira Eiliv sat by the children's bed when he was not out visiting his sick. Much toil and sorrow stripped the priest of all the fat he had laid on at Husaby — it went hard with him to see so many fair young children die. And some grown folks died too.

The sixth evening all the children were so much better that Kristin promised her husband that to-night she would take her clothes off and go to bed — Erlend proffered to watch along with the maids and call her if there were need. But at the supper-board she saw that Orm's face was a fiery red — and his eyes shining with fever. He said 'twas nothing — but suddenly he started from the table and out. When Erlend and Kristin followed, they found him vomiting in the courtyard.

Erlend threw his arms about him:

"Orm — my son — are you sick — ?"

"My head is aching so," moaned the boy, and let it sink heavily on his father's shoulder.

And so that night they sat watching Orm. For the most part he lay muttering in brain-sickness — he would shriek aloud and fight the air with his long arms — it seemed as though he had ugly visions. What he said they could not understand.

And in the morning Kristin broke down. It proved that she had been with child again; now she miscarried, and afterward she lay sunk in a drowse, as though half dead, and then fell into a high fever. Orm had lain in his grave more than two weeks before she knew of her stepson's death.

She was so weak then that she scarce had strength to feel sorrow. She was so bloodless and faint that naught could come home to her keenly — it seemed to her that it was well with her now, as she lay there but half alive. There had been a dreadful time, when the women hardly dared touch her or do what was needed for cleanliness — but it all seemed part of her fevered wanderings. Now it was good to lie and be tended. Round her bed hung so many sweet-scented wreaths of mountain flowers to keep the flies away — folk had sent them down from the sæters, and they smelt so sweet, most of all when there was rain in the air. Erlend brought their children in to her one day — she saw that they were wasted by the sickness, and that Gaute knew her not again, but even that did not hurt her yet. She only felt that Erlend seemed ever to be by her.

He went to mass every day, and he knelt long praying by Orm's grave. The churchyard was by the parish church at Vinjar, but some of the little children of the house had been given burial in the chapel of ease at Husaby — Erlend's two brothers and a little daughter of Munan Bishopssön. Kristin had often felt pity for these little

ones, lying all alone under the stone flags. Now had Orm
Erlendssön found his last resting-place amongst these
children.

It was while the others still feared for Kristin's life that
the companies of beggars, which made into Nidaros as
Olav's Wake drew nigh, came through the parish. 'Twas
mostly the same mumpers, men and women, that came
thither each year — the pilgrims were always open-handed
to them, for it was held that the intercession of the poor
availed much. And they had grown used to come round
through Skaun in these years of Kristin's rule at Husaby,
for they knew that there they would be given lodging for
the night, food in plenty, and alms before they passed on
their way. This time the serving-folk would have turned
them away, since the mistress lay sick. But when Erlend,
who had been away in the north the last two summers,
heard that his wife had been wont to deal so lovingly
with the beggars, he bade that they should be given lodging and entertainment even as they had had them of her.
And in the morning he went himself among the beggars,
helped to pour their liquor and bear round food for
them, and gave them the almspennies himself, while he
meekly begged them for their prayers for his wife. Many
of the beggars wept when they heard that the gentle young
woman lay at death's door.

Sira Eiliv had told her all this when she grew better. It
was not till nigh on Yule-tide that she was strong enough
to take up her keys again herself.

Erlend had sent word to her parents as soon as she fell
sick, but then they were gone south to the wedding at
Skog. Later they came to Husaby; she was better then,
but so weary she was not fit to talk much with them. She
was best pleased to have none but Erlend by her bedside.

Weak and chilly and bloodless, she crept for shelter into
his health and strength. The old fire in her blood was
gone, so utterly gone that she could not call to mind any
more what it was like to love in such wise; but with it was
gone the unrest and bitterness of the last years. It seemed
to her that she was happy now — even though grief for
Orm lay heavy on them both, and though Erlend knew
not how afraid she was for little Gaute, yet was she happy

with him now. She had understood how sorely he had
feared to lose her. . . .

'Twas a nice and a hard matter, then, to have to speak
to him now — to touch on things that might break the
peace and the content that were between them.

She stood without before the door of the hall in the
shining summer night when the house-folk came back
from the dancing. Margret hung on her father's arm. She
was clad and adorned more fittingly for a bridal feast
than for a dance on the church-green, where all kind of
folk come together. But the stepmother had quite given
over making or meddling in the maid's upbringing. Erlend
must guide his own daughter as he would.

They were thirsty, Erlend and Margret, and Kristin
went to fetch ale for them. The girl sat awhile prattling
— she and her stepmother were good friends, now that
Kristin no longer tried to teach her. Erlend laughed at all
his daughter's chatter of the dancing. But at last Margret
and her maid went up to their loft to go to rest.

The man went on wandering up and down the hall —
stretched himself and yawned, but said he was not weary.
He ran his fingers through his long black hair:

"There was not time for it, when we came from the
bath-house — because of this dancing — I trow you must
set to and cut my hair, Kristin — I cannot go about in
this wise in the holy-days — "

Kristin made answer that it was dark — but Erlend
laughed and pointed up at the smoke-vent — 'twas daylight again already. So she lit the candle again, bade him
sit down, and spread a cloth over his shoulders. While she
clipped, he shifted about ticklishly, and laughed when the
scissors came near his neck.

She gathered the shorn hair carefully together and burnt
it in the hearth-fire, and shook the cloth, too, over the
fire. Then she combed Erlend's hair down smooth from
the crown, and snipped with the scissors here and there
where the edge was not quite even.

Erlend caught her hands as she stood behind him, held
them together round his throat, and looked up at her with
smiling face thrown backward.

"You are tired," he said then, letting go her hands and rising with a little sigh.

Erlend sailed to Björgvin when midsummer was but just past. He complained much because this time again his wife was unfit to bear him company — she smiled wearily; howsoever things had been, she said, she had not been able to leave Gaute.

Thus it came that Kristin was alone at Husaby again this summer. At least it was well that this year she looked not that the child should come till Matthews Mass; * 'twas doubly hard for both herself and for the ladies who came to tend her when it came at the busy harvest season.

She wondered whether things would go on thus ever. Times were not now as they had been when she was a girl. The Danish war she had heard her father tell of, and she remembered when he was from home on the war against Duke Eirik. 'Twas from that he had brought back the great scars on his body. But, all the same, at home in the Dales they seemed, as it were, so far from war — thither it would come nevermore — so, she felt, did all men think. Most of her memories were of peace, of her father dwelling at home, guiding his possessions, caring and taking thought for all of them.

Now all the time there was unrest — all men spoke of contention and warlike levies and the government of the realm. In Kristin's mind it all went together with the picture of the sea and the coast, as she had seen them that single time when first she came hither to the north. Along the coast they came sailing, men that had their heads full of counsels and plans and counter-plans and deliberations, spiritual lords and laymen. Among these did Erlend belong, by his high birth and his riches. But she felt that he stood but half within their circle.

She pondered much on why it was that he stood thus, half without. What were his fellows' real thoughts of him?

When he was but the man she loved, she had never asked such questions. She had seen, indeed, that he was sudden and vehement, unthinking and ever apt to bear him unwisely. But then she had found excuses for all —

---

* 21st September.

had never troubled to think what his humour might bring
down upon them both. When once they two had got leave
to wed, all would be changed — so she had comforted herself. Sometimes she dimly felt that 'twas not till the hour
when she knew they two had given life to a child that she
had begun to think — what manner of man was Erlend, he
whom folk called light-minded and unwise, a man in
whom none could trust. . . .

She had trusted in him. She remembered Brynhild's
loft; she remembered how the bond between him and that
other had been cut asunder in the end. She remembered
his dealings, after she was his lawfully betrothed bride.
But he had held fast to her in despite of all rebuffs and
abasements; and she had seen that, now too, he would not
lose her for all the gold on earth. . . .

She could not but think of Haftor of Godöy. Ever,
when they had met, he had been dangling about her with
toying gallantry; but she had never troubled her for this.
She deemed it was but his fashion of jesting. She could
scarce believe aught else even now; she had liked the
comely gamesome young man — aye, she liked him still.
But that anyone could deem such things to be but a jest
— no, that she could not understand.

She had met Haftor Graut again at the royal banquets at
Nidaros, and he hung about her there too, after his wont.
One evening he got her to go with him into a loft-room,
and she lay down with him on a made-up bed that was
there. At home in the Dales she could never have thought
of doing aught of the kind — there 'twas no custom at the
feasts for men and women to steal aside, thus alone, two
and two together. But here all were used to it, and it
seemed not that any found it unseemly — 'twas said it was
a fashion of the knights and ladies in foreign lands. When
first they came in, Lady Elin, Sir Erling's wife, lay on the
other bed with a Swedish knight; she could hear that they
were talking of the King's ear-ache — The Swede looked
pleased when Lady Elin made a motion to get up and go
back to the hall.

When she understood that Haftor meant in sober sadness what he prayed her for as they lay there talking, she
had been so amazed that she seemed unable to be either
afraid or greatly angered. Were they not, both of them,

wedded, and had not both children by their wedded spouses? She felt she never could have fully believed before that such things happened. Even after all she herself had done and gone through — no; this she must have believed could not befall. Laughing and gay and coaxing had Haftor been with her — she could not bring herself to say he had tried to lure her astray; for that he had not been earnest enough. And yet it seemed he would have had her do the deadliest sin. . . .

He stepped down from the bed the moment she bade him begone — he had grown meek enough, but he seemed more amazed than beshamed. And he asked in sheer unbelief — did she truly dream that married folk were never unfaithful — ? Sure, she must know that few men could swear they had kept no paramours. Women, maybe, were somewhat better, but truly —

"Believed you then, too, when you were a young maid, all the priests preach of sin and suchlike?" he asked. "But then I understand not, Kristin Lavransdatter, how it could come to pass that Erlend had his will with you."

He had looked up into her face — and her eyes must have spoken, though not for much gold would she have talked with Haftor of this. For 'twas in a high singing voice of wonderment that he said:

"The like of this I had deemed was but a thing they told of — in songs and ballads — "

She had told no one of this; not even Erlend. He liked Haftor well. And truly it was fearful that there could be any folk as light-minded as Haftor Graut — but 'twas as though she could not feel that it touched her at all. Nor had he ever tried to be too free with her since — he but sat and stared when they met, his sea-blue eyes wide with wonderment.

No, if Erlend were light-minded, at least it was not in that way. And, she thought, *was* he so unwise? She saw that folk startled at things he said, and afterward laid their heads together over them. There was often much right and reason in what Erlend Nikulaussön put forth. 'Twas but that he never saw what the other great folk never lost from sight — the cautious heedfulness with which they kept watch on each other. Trickery, Erlend called it, and laughed his reckless laugh, which nettled folks somewhat but disarmed them in the long run. They too, would laugh then, and slap him on the shoulder, saying he might be sharp-witted enough, but short-witted he was for sure.

Then would he undo the work of his own words with wanton and malapert jesting. And folk would suffer much of that from Erlend. Dimly his wife felt — and was humbled by the feeling — why all men bore with his unbridled tongue. Erlend would flinch and give way the moment he met a man who stood firmly by his own judgment — even if, to Erlend, that judgment seemed folly, he would yet give up his own, whatever the matter might be, but would cover his retreat with fleering talk about the man. And folk were well content to know that Erlend had this timorousness of mind — reckless as he was of his own welfare, hungry for adventure, desperately in love with every peril that could be met by force of arms. After all, they felt, they need not be too much disquieted by Erlend Nikulaussön.

The year before, when the winter was well-nigh gone, the High Steward had been in Nidaros, and he had brought the little King with him. Kristin had been in the city for the great banquet in the palace. Still and stately she had sat in her silken coif, bearing her red bridal dress with all her richest adornments, amidst the most high-born ladies of the court. With watchful eyes she had followed her husband's doings among the men, watched and listened and pondered — even as she watched and listened and pondered wherever she went with Erlend, or whenever she marked that folk spoke of him.

One thing and another she had understood. Sir Erling Vidkunssön was willing to stake all on upholding the rule of Norway northward toward the Icy Sea, on guarding and securing Haalogaland. But the Council and the Knighthood were against him, and were loath to agree to any undertaking great enough to serve this purpose. The Archbishop himself and the priesthood of the archdiocese were not unwilling to stand by him with money help — this she knew from Gunnulf — but else all churchmen throughout the land were set against him, even though 'twas a war against God's enemies, heretics and heathens. And the great laymen worked against the High Steward, here in Trondheim at least. They had grown used to pay-ing small regard to the words of the law-books and the rights of the Crown, and it liked them but little that Sir Erling stood so sharply in these matters for the spirit of his kinsman, King Haakon of happy memory. But it was not on this account that Erlend would not let himself be used, as she now understood the High Steward had meant to use her husband. With Erlend it was but that the other's grave and stately bearing wearied him — and he avenged himself by scoffing a little at his powerful kinsman.

Kristin thought she understood Erlend's footing with Sir Erling now. One thing was that the knight had had a sort of kindness for Erlend from their youth up; and then he had doubtless thought that, could he win over the high-born and valiant master of Husaby, who had gained some skill too in the craft of war from his service with Count Jacob — who at least knew more of war than most of his fellows, that had done naught but sit at home — he might thereby serve both his own plans and Erlend's welfare. But it had not fallen out as he planned.

Two summers had Erlend kept the sea till late autumn, wallowing in the seas that wash the long northern coast, and hunting the robbers' barks with the four small ships that followed his banner. He had come in for fresh meat to a new Norse settlement far north in Tana, just as the Karelians were hard at work plundering it — and, with the handful of men he had with him ashore, he had caught eighteen of the robbers and hung them to the roof-tree of the half-burnt barn. He had cut to pieces a band of Russians that was flying to the hills, and had burnt some enemy ships amid the outer skerries and destroyed their crews. The fame of his swiftness and daring had spread wide in the north; his Trönder and men of Möre loved their leader for his hardihood's sake and for his will to share all toil and all hardness with his men. He made friends both among the small folk and among the young sons of the great manors north in Haalogaland, where before the people had well-nigh grown used to thinking they must guard their coasts unaided.

Yet could not Erlend be of service to the High Steward in his plans for a great northern crusade. True, the folk of Trondheim bragged of his deeds against the Russians — if the talk turned on them, they let no one forget he was of their country-side. Aye, 'twas proven clearly enough that there was plenty of the good old mettle in the young fellows round about the fjord here. But what Erlend of Husaby said and what he did were not things that counted with full-grown and prudent men.

She saw that Erlend was still reckoned as one of the young men — though he was a year older than the High Steward. She understood that it suited many folk well that such he should be held to be, so that his words and his deeds could be belittled as being but the deeds and counsels of a hot-headed young man. Thus was he liked, spoiled and bragged of — but not accounted as a man come to man's estate. And she saw how willing he seemed to fall in with this and be what his fellows would have him be.

He spoke up for the Russian war; he spoke of the Swedes who owned half our King, and yet would not reckon the Norse gentry and knights as nobles, the equals of their own. Or had the like ever been heard of in any land, he asked, as long as the world had stood, that anyone had craved war-levies from noblemen in other wise than that they should ride their own horses and bear their own shields to the field? — Kristin knew that this was much what her father had said at the Thing in Vaage some years back, and he had pressed it on Erlend when his son-in-law had been loath to part company with Munan Baardssön and his counsels. No, said Erlend now — and he named his father-in-law's powerful kinsmen in Sweden — he knew well enough what account these Swedish gentry made of us. If we show not what we can do, we shall soon be fit but to be reckoned as pensioners of the Swedes. . . .

Aye, folk would say, there was somewhat in all this. But then they would talk again of the High Steward. Sir Erling had his own pot to boil in the north there; the Karelians had burned Bjarkö over his steward's head one year, and harried his farmers. And then Erlend changed his note and grew merry over the knight: Erling Vidkunssön thought not on his own concerns, of that he was sure. He was so noble and fine and stately a knight — no more worshipful man could they have found to be the cornerstone of their affairs. By God's Cross, Erling was as worshipful and as venerable as the bravest golden capital letter in the Book of the Laws. Folk laughed, and bore in mind not so much Erlend's praise of the High Steward's honour as that he had likened him to a gilded letter in a book.

No, they took not Erlend in earnest — not even now, when he was honoured after a fashion. But in those days when, young and headstrong and desperate, he had lived in whoredom with a woman, and would not put her away in despite of King's command and Church's ban — then they had taken him sadly enough, turning their backs on him in furious wrath over his godless and shameless life. Now was all this forgotten and forgiven — and Kristin understood that there was something of thankfulness for this in her husband's willingness to yield, and be what folk would have him to be — he had suffered bitterly, she knew, in the days when he lived an outcast from among his fellows here in his home.

There was but one thing — she must needs think of her father when he forgave some good-for-naught his rent or his debt — with the slightest shrug of his shoulders. 'Twas our Christian duty to bear with them that could not play a man's part. Was it thus that Erlend had gained forgiveness for his sins of youth — ?

But Erlend *had* paid for those deeds of his when he lived with Eline. He had answered for his sin till the day when he had met *her* and she had followed him, nothing loath, into new sins. Was it she, then, who — ?

No. She grew afraid now of her own thoughts.

And she tried to shut out from her mind all care for things wherein she could take no hand. She would only think of those matters in which she could do some good by her carefulness. All the rest she must leave in God's hand. God had helped her in all things wherein her own toil and pains could avail. The home-farm at Husaby had now been worked up again into a good farm as of yore — in despite of the bad years. Three healthy comely sons had He vouchsafed her to bear — every year had He granted her life anew when she must face death in child-bearing; He had let her arise in full health after each childbed. All her sweet little ones had she been given grace to keep last year when the sickness bore away so many fair little children in the country round. And Gaute — Gaute *would* grow strong, that she believed full surely.

Doubtless it must be as Erlend said — he must needs spend freely as he did and have all things costly about him. Else could he never play his part amongst his peers or win his way to such rights and rewards under the King as were his due by birth. She must believe that he understood such things better than she.

'Twas witless to think things could have been better with him in any wise in those days when he lived as in bonds of sin with that other — and with herself. Glimpse after glimpse came before her eyes of his face as it was in those days, ravaged with sorrow, drawn with passion. No, no, 'twas well as now it was. He was but somewhat too careless and unthinking.

Erlend came home at Michaelmas-tide. He had hoped to find Kristin in bed; but she was still up and about, and she came to meet him a little way. She was piteously heavy-footed this time — but she bore Gaute on her arm, as ever; the two bigger boys ran before her.

Erlend leapt from his horse and set the two boys up on it. Then he took the little one from his wife and would have borne him. Kristin's white, worn face lighted up when Gaute showed no fear of his father — it must sure be that he knew him again. She asked not aught of her husband's doings; she talked only of the four little teeth Gaute had gotten. He had been so sick when he cut them.

Then the boy burst out screaming — he had scratched his cheek on his father's neck-brooch. He fought to go back to his mother again, and she would take him, in despite of all that Erlend could say.

It was not till the evening, when they sat in the hall and the children slept, that Kristin asked her husband of his sojourn in Björgvin — as though it were a thing she had but now remembered.

Erlend stole a glance at his wife. Poor love — she looked so wretched. So first he brought forth odds and ends of news. Erling had prayed him to greet her and give her this — 'twas a bronze dagger, green and eaten up with copper rust. They had found it in a stone-heap out at Giske — they said such things would be good to lay in the cradle, if 'twere rickets that ailed Gaute.

Kristin wrapped the cloth about the dagger again, rose toilsomely from her chair, and went across to the cradle. She put the little bundle in amongst all the other things that already lay there under the coverings — a flint axe found in the earth, some beaver-grease, a little cross of mezereon, heirloom silver, a fire-steel, roots of purples and finger-fern.

"Lie down now, my Kristin," he begged lovingly. He came over and drew off her shoes and hose — and he told his tidings the while.

Haakon Ogmundssön was come back, and peace with the Russians and Karelians was made and sealed. He himself would have to journey north again now, this autumn. For 'twas nowise sure that things would calm down so quickly, and there was need that Vargöy * should be held by a man who knew the land and people. Aye, he would have full power as the King's Governor there — the fortress needed strengthening so that the King's peace might be upheld in the lands within the new boundary marks.

Erlend looked in suspense up at his wife's face. She seemed a little affrighted — but she asked not many questions, and it was clear that she understood not much of the full meaning of his tidings. He saw how weary she was — so he spoke no more of these things, but stayed by her awhile, sitting on the bed's edge.

He knew himself what he had undertaken. Erlend laughed quietly to himself as he lingered over his undressing. 'Twould be no sitting with silver-belted belly, giving ale-feasts to friends and kinsmen, and trimming your nails fine and even, while you sent your sheriffs and lieutenants hither and thither on your errands — after the fashion of the King's Governors in the castles down here in the south. For the castle of Vargöy — 'twas a stronghold of another kidney.

Lapps, Russians, Karelians, and the mixed spawn of all the races — troll-pack, wizards, heathen hounds, the foul fiend's own pet lambs — had to be taught to pay their

---

* The modern Vardö.

dues again to the Norse commissaries, and to leave in
peace the Norse homes lying scattered, with as far between them as from here out to Möre maybe. Peace —
maybe the land up there would be at peace some time — in
his time 'twould be but the peace there is while the devil
is at mass. And then there would be his own dare-devils to
keep in check. As they would be towards spring, when
they began to grow brain-sick with the dark and the
storms and the cold and the hellish noise of the sea — and
meal and butter and drink began to run low, and they
fought about their womenfolk, and life on the island was
more than flesh and blood could bear. He had seen somewhat of it when he was there with Gissur Galle as a
young lad; ho, ho — 'twould be no bed of roses.

Ingolf Peit, the man there now, was a good man enough.
But Erling was right. A man from among the knighthood
must take things up there in hand — till this was done,
none would understand that 'twas the Norse King's firm
intent to uphold his rule over the land. Ho, ho — in that
land would he be stuck like a needle in a blanket. The
nearest Norse parish down at Malang, the devil knows
how far.

Ingolf was a worthy fellow — when he had someone
over him. He would give Ingolf the command of *Hugrekken*. *Margygren* was the finest ship of them all, he had
proved that now. Erlend laughed softly and happily. He
had said it to Kristin so often — that was a henchwoman
she must needs suffer him to cleave to.

He was waked by the noise of a child crying in the dark.
Over in the bed by the other wall he heard Kristin moving
and speaking coaxingly in a low voice — it was Björgulf
that was crying. Sometimes the boy would wake in the
night and could not open his eyes for the matter on the
lashes — and then the mother would wet them with her
tongue. It had ever seemed to him ugly to look on.

Kristin lulled the child softly. The thin small tones of
her voice irked him.

Erlend remembered what he had dreamed. He was
walking somewhere on a rocky strand — it was ebb-tide,
and he leapt from stone to stone. The sea lay pale and
bright, licking at the tangle far outside — 'twas like a still,

clouded summer night, no sun. Against the silvery light
at the fjord-mouth he saw his ship lying at anchor, black
and slender, rocking gently, gently on the swell. There
was an unearthly sweet smell of sea and sea-weed. . . .

His heart within him grew sick with longing. Now, in
the darkness of the night, lying here in the guest bed with
the long-drawn tones of the nurses' lullaby chafing at his
ears — now he felt how great his longing was. Away from
his home and the children that the house overflowed with,
away from talk of husbandry and housefolk and tenants
and young ones — and from heart-heaviness for her who
was ever sick and ailing, and whom he must for ever
pity. . . .

Erlend pressed his clenched hands over his heart. 'Twas
as though it had ceased beating and did but lie shaking
with fear in his breast. He longed to leave her! When he
thought on what she was to go through, weak and
strengthless as she now was — it might come at any moment, he knew — 'twas as though he strangled with fear.
Should he lose Kristin — he saw not how he could endure
to live without her. But neither could he endure to live
with her — not now; he must needs come away from it
all, and take breath again — 'twas as though *his* life were
at stake too.

Jesus, my Saviour — oh, what sort of man was he! Now,
to-night, he saw it clearly — Kristin, his sweet, his dearest
love — true, deep-hearted joy he had never known with
her, save in those days when he was leading her astray in
sin.

And he had believed so surely that the day when he won
Kristin to have and to hold her before God and man —
that day all evil would be wiped away from his life so
wholly that he would forget it had ever been.

He must be such an one that he could not suffer aught
that was truly good and pure near him. For Kristin — aye,
since she was escaped from the sin and uncleanness he had
led her into, she had been as an angel from God's heaven.
Mild and trusty, gentle, diligent, worthy of honour. She
had brought honour to Husaby once more. She was become again what she had been on yonder summer night
when the pure young maiden soul nestled in under his
cloak out there in the cloister garden and he had thought,

as he felt the slender young body against his side — the devil himself could not find in his heart to hurt this child or cause her sorrow. . . .

The tears ran down over Erlend's face.

— Then belike it was true, what they had told him, the priests, that sin ate up a man's soul like rust — for no rest, no peace was his, here with his own sweet love — he but longed to be gone from her and all that was hers. . . .

He had wept himself into a half-slumber, when he marked that she was up, and walking about the room lulling and crooning to the child.

Erlend leapt out of bed, stumbling in the dark over some children's shoes on the floor, came to his wife, and took Gaute from her. The boy shrieked aloud and Kristin said plaintively:

"I had almost gotten him to sleep."

The father shook the screaming child, gave him some slaps behind — and as the boy shrieked still louder bade him hush in such a harsh voice that Gaute suddenly stopped in terror. No such thing had ever befallen him before in his life. . . .

"Now, for God's sake, use any wits that you have left, Kristin." His vehemency seemed to strip him of all strength as he stood there in the pitch-dark room, naked, shivering and half awake, with a sobbing child in his arms. "An end of this there must be, I tell you — what have you nursemaids for? — the young ones must sleep with them. You cannot go on thus."

"Can you not suffer me to have my children with me in the time that is left to me?" answered his wife in a low, wailing voice.

Erlend *would* not understand what she meant.

"In the time that is left, what you need is *rest*. Lay you down now, Kristin," he begged, more mildly.

He took Gaute with him to his own bed — lulled the child a little and groped in the darkness till he found his belt on the bed-step. The small silver scales it was set with chinked and tinkled as the boy played with it.

"The dagger is not in it?" asked Kristin fearfully from her bed; and Gaute set up a fresh howl when he heard his mother's voice. Erlend hushed him again and tinkled the belt — and at last the child gave way and grew quiet.

Poor miserable little soul, maybe one should scarce wish
he might grow up — 'twas unsure if Gaute had all his wits.

Oh, no, oh, no — most blessed maid Mary — he meant
it not — he wished not that his own little son should lose
his life. No, no — Erlend took the child close within his
arms and laid his face down on the warm downy hair.

Their fair sons — But he grew so weary of hearing of
them early and late; of stumbling over them wherever he
went at home here. That three small young ones could be
in all places at once on a great manor like this passed his
understanding. But he remembered his burning wrath with
Eline because she had troubled herself little about their
children. An unjust man he must surely be — for he was
vexed now because he never saw Kristin anywhere without children hanging about her.

Never had he known, when he took his true-born sons
in his embrace, the like of what he had felt the first time
they laid Orm in his arms. Oh, Orm, Orm, my son — He
had been so weary of Eline even then — sickened with her
self-will and her rages and her ungovernable love. He had
seen that she was too old for him. And he had begun to
understand what this madness was like to cost him. But he
had thought: give her up he could not — since she had
lost all for his sake. The boy's birth had given him, he
thought, a cause the more to bear with the mother. He had
been so young when he became Orm's father that he had
not fully understood what the child's standing would be —
with a mother that was another's wedded wife.

Weeping came over him again, and he drew Gaute
closer to him. Orm — none of his children had he loved
as he loved that boy; he missed him so, and he repented so
bitterly every hard and hasty word he had said to him. It
could not be that Orm had known how his father loved
him. It had all come from his bitterness and despair, as he
came to see clearly that never could Orm be counted for
his true-born son, never could he bear his father's arms.
And from jealousy, too, as he saw his son draw closer to
his stepmother than to him; and this, too, that Kristin's
even, gentle kindness to the youth seemed to him like a
reproach.

And then came the days he could not endure to remember. Orm lay in the loft-room in the dead-straw, and the

women came and told him they thought not that Kristin could live through her sickness. They dug Orm's grave over in the chapel, and asked if Kristin were to lie there or were to be taken in to St. Gregory's and buried where his father and mother lay.

Oh, but — and at this he held his breath in fear. Behind him lay all his life, filled with memories he fled from, because he could not endure to think of them. Now, to-night, he saw it — He could forget, after a sort, in the daily fellowship with his kind. But he could not so guard himself that it rose not up in some hour such as this — and then 'twas as though an evil spell had robbed him of all courage.

Those days at Haugen — at most times he had well-nigh managed to forget them. He had not been at Haugen since yonder night he and Björn had driven away from it; and he had not seen Björn and Aashild since his wedding-day. It was Sir Björn he had been afraid to meet. And now — He thought of what Munan had told him — 'twas said they walked there; Haugen was so felly haunted that the houses stood empty; none would live there now, not if they were given the farm free.

Björn Gunnarssön had had a kind of hardihood that Erlend knew he could never attain to. He had been steady of hand when he stabbed his wife — right in the heart, Munan said.

'Twould be two years next winter since Björn and Lady Aashild died. No smoke had been seen from the houses at Haugen for a week; and at last some men plucked up heart and went thither. Sir Björn lay in the bed with his throat cut across; he held his wife's body in his arms. Before the bed, on the floor, lay his bloody dagger.

None had doubted how this had come to pass. . . . Yet did Munan Baardssön and his brother so order things that the two were buried in hallowed ground. — 'Twas put about that it might well have been robbers; though the chest with Björn and Aashild's goods was untouched. The bodies were untouched by rats or mice — the truth was such vermin were not to be found at Haugen — and folk took this for a sure sign of the lady's skill in witch-craft.

Munan Baardssön was fearfully shaken by his mother's

end. He had set forth straightway on pilgrimage to St.
James of Compostella.

Erlend remembered the morning after the night his own
mother died. They lay at anchor inside Moldöy Sound,
but the white fog was so thick that 'twas but in short moments now and then they caught a glimpse of the cliff-wall
they lay close under. Yet did it give back a muffled echo
of the hollow sounds as the boat rowed landwards with
the priest. He stood in the fore-part and watched them
row away from the ship. All things he came near were
wet with the fog; the wet stood in beads on his hair and
his clothes, and the stranger priest and his acolyte sat in
the boat's bow crouched with updrawn shoulders over
the sacred elements in their lap. They looked like hawks
in rainy weather. The oar-strokes and the creaking of the
rowlocks and the echoes from the cliff sounded on faintly
long after the boat was blotted out in the fog.

Then he too had vowed a pilgrimage. He had had but
one thought then — that he must see again his mother's
sweet and lovely face as it had been of old — with the soft
smooth skin of palest brown. Now she lay dead below
there, with face ravaged by the fearful sores, that cracked
and oozed small clear drops of moisture when she had
tried to smile to him. . . .

Was it his fault that his father had met his return in
such a fashion? Or that he had turned him then to one
who was outcast like himself — ? And after that he had
thrust all thought of pilgrimage from his mind, and had
not troubled to think of his mother any more. Ill as
things had gone with her on earth 'twas like that now she
was come where there was peace — and but little peace
had fallen to *his* lot after he had sought Eline again. . . .

Peace — but once in his life, it seemed now, had he
known it — that night when he sat behind the stone wall
out towards the woods by Hofvin, and held Kristin in
his lap, sleeping her soft, secure, unbroken, childlike sleep.
Not for long had he been able to refrain him from breaking that calm. And 'twas not peace that he had found with
her since — that he found with her now. Though he saw
that all others in his home found peace with his young
wife.

And now his one longing was to be gone to strife

again. He longed wildly for that outermost barren rock, for the sea thundering round the northern forelands, for the endless coast, and the mighty fjords where all manner of snares and pitfalls might await him, for the folks whose tongues he knew but by bits and scraps, for their sorceries and fickleness and slippery wiles, for war and the sea, and the song of his men's weapons and his own —

He fell asleep at last, but wakened again — what was it he had just been dreaming? Aye, black Lapp girls — something half forgotten that had befallen when he was in the north with Gissur — a wild night when they had all been crazed with drink.

And here lay he with his little sick son in his arms and dreamed such dreams. — He grew so frighted of himself that he dared not try to sleep any more. And he could not endure to lie awake. Aye, truly he must be an unhappy wretch. — Stiff with dread, he lay unmoving and felt the heart tolling in his breast, while he longed for the dawn to release him.

He talked Kristin over into keeping her bed the next day; for he felt he could not bear to see her go dragging herself about the house — in such wretchedness. He sat by her and played with her hand. She had had the comeliest arms — slim, but so round that the fine small bones in the slender joints were not seen. Now they stood out like knots on the gaunt arms whose skin on the underside was more blue than white.

Without, it blew, and rained till the water came streaming from the hill-sides. Once, well on in the day, as he came down from the armoury, he heard Gaute screaming somewhere in the courtyard. In the narrow passage between two houses he found his three small sons, sitting in the midst of the runnels splashing from the roofs. Naakkve held the little one tight, while Björgulf tried to force a living earthworm into his mouth — he had his hand quite full of writhing pink worms.

The boys stood with injured looks when their father seized and scolded them. 'Twas old Aan, they said, that had told them of it — Gaute would get his teeth without pain or trouble if they could but get him to take a bite or two of living earthworms.

All three were dripping wet from top to toe. Erlend roared out for the children's nurses — they came rushing, one from the wright's shop and one from the stable. Their master cursed them heartily, then thrust Gaute under his arm like a sucking-pig, and drove the others before him into the hall.

Soon after, the three were sitting dry and happy in their blue holy-day kirtles on the step before their mother's bed. Their father had drawn up a stool for himself, and he chattered and romped, and, laughing, hugged the young ones to him, to deaden in his own mind the memory of last night's fear. But the mother smiled happily to see Erlend playing with their children. Erlend kept a Lapland witch, he said; she was two hundred winters old, and dried up till she was no bigger than *that*. He kept her in a skin bag in the great chest that stood in his ship-house. Food? Aye — she got food — every Yule night the thigh of a Christian man — she got through a whole year on that. And if they were not good and quiet and ceased not plaguing their mother, that was so sick, they should go into the skin bag too. . . .

"Mother is to have our little sister — that is why she is sick," said Naakkve, proud of having the clue to the riddle. Erlend pulled the boy by the ears on to his knee:

"Aye — and when she is born, this sister of yours, I will have my old Lapp hag throw a spell over you three, and you shall turn into white bears and root about in the wild woods; but my daughter shall inherit all my goods and gear."

The children shrieked, and clambered up to their mother in her bed — Gaute understood not what was amiss, but he shrieked and crawled up to keep his brothers company. Kristin chid her husband — such jesting was too uncanny. But Naakkve tumbled out again — in a rapture of laughter and fear he rushed at his father, hung on to his belt, and snapped at Erlend's hands, with mingled shrieks and shouts of joy.

Erlend did not get the daughter he would so fain have had this time either. Kristin bore him two great and comely sons, but they had well-nigh cost her life.

Erlend had them baptised, one after Ivar Gjesling and the other after King Skule. Skule's name had not been kept up among their kindred — Lady Ragnrid had said that her father was an ill-omened man, and it was best therefore that his name be let drop. But Erlend swore that none of his sons bore a prouder name than this, his youngest.

The autumn was so far spent that Erlend must needs set forth for the north as soon as Kristin was through the worst of the danger. And he thought in his heart 'twas as well he should be gone before she came upon her feet again. Five sons in five years — 'twas enough for any man; and he was loath to have cause to dread that she might die in childbed while he was tied up there at Vargöy.

He saw that Kristin, too, thought somewhat of the kind. She murmured no longer that he was to leave her alone. She had taken each child as it came, as a precious gift of God, and the troubles it brought as things she must bear without repining. But this time it had gone so fearfully hard with her that Erlend saw 'twas as though all heart had been wrung out of her. She lay there, her face yellow as clay, and looked on the two small bundles of swaddling-clothes by her side, and her eyes were not so happy as when she had gazed first on the others.

Erlend went through the whole journey north in his thoughts as he sat beside her. A hard voyage 'twould be, belike, so late in autumn — and strange to come up there into the long night. But he yearned to be gone, unspeakably. This last terror for his wife had broken down all resistance in his soul — will-lessly he gave himself up to his longing to flee away from home.

4

ERLEND NIKULAUSSÖN held the post of Captain of the Vargöy stronghold and keeper of the Northern Marches for well-nigh two years. In all that time he came not further south than to Bjarköy, and there but once, when he and Sir Erling Vidkunssön had made tryst there. The second summer Erlend was in the north, Heming Alfssön died at

last, and Erlend was made Warden * of Orkdöla County
in his stead. Haftor Graut went north to take his place at
Vargöy.

Erlend was a glad man when he sailed for the south,
some days after Mary's Mass in autumn. It was the cure
for his honour that he had wished for all these years — to
be given the Wardenship his father once held. Not that
this had been a goal he had ever wittingly worked to reach.
But it had ever seemed to him that 'twas this he needed,
so that he might come into the place where he rightfully
belonged — both in his own and in his fellows' eyes. Now
'twas no matter if men still deemed him to be somewhat
unlike the other, the home-keeping nobles — there was no
disgrace in the unlikeness any more.

And he longed to be home. Things had been more
peaceful in Finmarken than he had looked for. Even the
first winter had worn on him — he sat there idle in the
castle, and could do naught at that season towards the
mending and bettering of the works. They had been put
in good order seventeen years before, but now were quite
fallen in ruin.

Then came the spring and summer, with life and bustle
— meetings here and there in the fjords with the Norse
and half-Norse tax-gatherers and the spokesmen from the
tribes of the uplands. Erlend roved the seas and fjords
with his two ships and amused himself royally. On the island the houses were mended, and the works strengthened.
But the next year there was but little doing.

Haftor would see to it, doubtless, that the quiet did not
last long. Erlend laughed. They had sailed together eastward well-nigh as far as Trianema, and there had Haftor
taken a Russian Lapp woman and had brought her back
with him. Erlend had talked to him gravely: he must
remember, 'twas above all needful that the heathen should
understand always that we were the masters — and to that
end, seeing one had but a handful of men, one must bear
one so as not to stir up trouble needlessly. No making or
meddling should the Lapps fight and slay each other; that
pleasure one must let them enjoy in peace. But be ever
ready to pounce like a hawk on Russians and Kolbjags and

---

* See Note 12.

whatever else the pack might call themselves. And leave the womenfolk in peace — for one thing, they were witches, every one — and, for another, there were enough to proffer themselves. — But the Godöy lad must steer his course as he would; he would learn in time. Haftor was joyful at getting free from his farms and his wife, and now Erlend was fain to come home to his. He had a right blissful longing now for Kristin and Husaby and his own country-side and all his children — for all things at his home where Kristin was.

In Lyngsfjord he heard tidings of a ship with some monks aboard; 'twas said they were Preaching Friars from Nidaros, who were journeying north, bent on planting the true faith amidst the heathen and heretics of the marches.

Erlend felt sure within him that Gunnulf was of the company. And, true enough, three nights later he sat alone with his brother in an earthen hut on a little Norse farm that lay by the strand where they had met.

Erlend was strangely moved. He had heard mass and taken the sacrament with his crew — the only time since he came here to the north, save that once when he had been at Bjarköy. The church at Vargöy was without a priest; a deacon had been left in the fort, and he had striven to keep count of the holy-days for them, but else had there been but scurvy provision for the souls of the Norsemen in these northern lands. They must even comfort themselves with the thought that it was a crusade of a kind they were on, and 'twas like they would not be held to such strict account for their sins.

He sat speaking to Gunnulf of this, and his brother listened with a far-off, strange smile on his wide thin lips. It looked as if he ever sucked in the under-lip a little, as a man may often do when he is thinking hard of some matter, and is nigh to understanding, but has not yet come to full clearness in his thoughts.

The night was far spent already. All other folks on the farm were sleeping up in the shed; the brothers knew that they alone were waking. And they were both stirred by the strangeness of their sitting here — they two alone.

The roaring of the sea and of the storm came to them lulled and deadened by the turf-walls. Now and then a

puff of wind would force its way in, blowing up the embers in the fire-place, and flapping the flame of the train-
oil lamp a little. There were no furnishings in the hut; the
brothers sat on the low earthen bench that ran round
three sides of the room, and between them lay Gunnulf's
writing-board, with ink-horn, feather pen and a roll of
parchment. Gunnulf had been writing down one thing and
another his brother had told him of trysting-places and
settled farms, of sailing-marks and weather signs and words
in the Lapps' tongues — just as the things chanced to come
into Erlend's mind. Gunnulf commanded the ship — she
was named the *Sunniva*, for the Preaching Brothers had
chosen St. Sunniva as guardian saint for their mission.

"Aye, if only you come not to the same end as the
Seljemen," * said Erlend, and again Gunnulf smiled a little.

"You tell me I am restless, Gunnulf," Erlend went on.
"What should a man call you, then? First you go wandering about in the south-lands all those years, and no sooner
are you come home but you must needs turn your back
on living and prebends and be off to preach to the devil
and all his imps away north in Velli-aa. You know not
their tongue and they understand not yours. Methinks you
are yet more unstable than I."

"I have neither goods nor kin to answer for," said the
monk. "I have loosed me now from all bonds; but you
have bound you, brother."

"Oh, aye. He is the free man that owns naught."

Gunnulf answered:

"All things that a man owns hold him far more than he
holds them."

"Hm. Nay, by God, 'tis not ever so. Grant that Kristin
holds me — but I have no mind that my lands and my
children should own me."

"Think not só, brother," said Gunnulf, low. "For then
may it easily come to pass that you lose them."

"Nay, no mind have I to grow like to all those other
goodmen — sticking up to their ears in the mud of their
lands," said Erlend, laughing, and again his brother smiled
a little.

---

* For St. Sunniva and the Seljemen, see *The Bridal Wreath*
(The Garland), Note 8.

"Fairer children than Ivar and Skule have I never seen," he spoke. "Methinks 'twas so you must have looked at their age — no marvel that our mother loved you so much."

Each brother rested a hand on the writing-board that lay between them. Even in the faint light of the train-oil lamp it could be seen how unlike these two men's hands were. The monk's, bare of rings and all adornments, white and sinewy, smaller and much more closely set than the other's, looked also much stronger — though Erlend's fist was as hard as horn in the palm, and the bluish-white scar of an arrow-wound furrowed the dark flesh from the wrist up under the sleeve. But the fingers of his narrow brown-tanned hand, dry and knotted at the joints like the twigs of trees, were covered with golden and jewelled rings.

Erlend would fain have taken his brother's hand, but he was ashamed — so he but drank, pulling a wry face over the bad beer.

"She seemed to you to be well and hearty again, Kristin?" Erlend asked in a while.

"Aye, she blossomed like a rose when I was at Husaby in the summer," said the monk, smiling a little. He waited awhile and then said gravely: "One thing I would pray you, brother — to think somewhat more of Kristin's and the children's welfare than till now you have done. And be counselled by her and clinch the bargains that she and Sira Eiliv have agreed for; they wait but for your assent to close them."

"I like not much these plans of hers you speak of," said Erlend haltingly. " — And now too, my standing will be other than it has been — "

"Your lands will be of more worth when you bring your holdings close together," answered the monk. "Methought Kristin's counsels were wise when she told me of the matter."

"I warrant there is scarce a woman in Norway's land that is freer than she to guide things as she will," said Erlend.

"In the end 'tis you that guide things," answered the monk. "And you — you guide Kristin, too, as you will," he added in a low voice.

Erlend laughed softly, low in his throat, stretched himself, and yawned. Then of a sudden he said soberly:

"You have guided her, too, at times, my brother. And I marvel if sometimes your counsels have not come nigh to parting our friendship."

"Mean you the friendship that has been betwixt you and your wife, or the friendship between us two brothers?" said the monk slowly.

"Both," said Erlend, as if it was a thought that but now came to him. "So holy there is sure no need for a lay-woman to be," he said more lightly.

"I have counselled her as I deemed to be best. As *is* best," he corrected himself.

Erlend looked at the monk in the Preaching Brothers' coarse grey-white frock, with the black cowl thrown back, so that it lay in thick folds round the neck and over the shoulders. The crown of the head was shaven now so that there was but a narrow fringe of hair about the round, lean, pallid face — but the hair was thick and black as in Gunnulf's earliest youth.

"Aye, you are no brother of mine now, I trow, any more than you are brother to all mankind," said Erlend, and wondered at the deep bitterness in his own voice.

"So is it not — though so it should be."

"So help me God — almost I believe 'tis therefore you would go to dwell among the Lapps," said Erlend.

Gunnulf bent his head. There was a glow in his yellow-brown eyes.

"Therefore it is — in some measure," he said low and quickly.

They spread out the skins and rugs they had brought with them. It was too cold and raw in the hut for them to take off aught, so they bade each other good-night and lay down on the earthen bench, which, to escape the smoke that hung above, was but little raised above the floor.

Erlend lay thinking of the tidings that had come to him from home. 'Twas not much he had heard in these two years — two letters from his wife had come to his hands, but they had been old already when they reached him. Sira Eiliv had written them for her — she could print herself, fair and plain, but she was ever loath to write, since it seemed to her scarce seemly for an unlearned woman.

Doubtless she would be yet holier now they had a new shrine in their neighbour parish, and that sacred to a man whom she had herself known in his life — and now had Gaute found healing for his sickness there, and she herself won her full health again, after being sickly ever since the birth of the twins. Gunnulf had told him that the Preaching Brothers at Hamar had at last been forced to give back Edvin Rikardssön's body to his brethren at Oslo, and these were now having full record made of all things concerning Brother Edvin's life, and the miracles 'twas said he had wrought both in life and after he was dead. It was their intent to send this writing to the Pope, and try to have the monk beatified. Some peasants from Gaudall and Medaldal had gone south and borne witness to wonders that Brother Edvin had wrought in their parishes by his intercession, and by means of a crucifix that he had carved out, and that now was at Medalhus. They had vowed to build a little church on Vatsfjeld, where he had lived some summers as a hermit, and where was a healing spring that owed to him its virtue. So they were given a hand from his body to enshrine in the church.

Kristin had made offering of two silver cups and of the great clasp set with blue stones that had come to her from her mother's mother, Ulvhild Haavardsdatter, and had had Tiedeken Paus, the goldsmith in the city, make of them a silver hand to hold the bones of Brother Edvin's hand and fingers. And she had been at the Vatsfjeld with Sira Eiliv and her children and a great following, when the Archbishop hallowed the church at St. John's Mass tide the year after Erlend had gone northward.

After this Gaute had gained health swiftly, and had learnt to walk and talk — he was now like other children of his age. Erlend stretched himself — sure it was the greatest joy that could have befallen them that Gaute was grown whole and well. He would give some land to that church. Gaute was fair, Gunnulf said, and comely of face like his mother. Pity that he had not been a little maid — then should he have been called Magnhild. Aye — he was fain now to see all these fair sons of his too. . . .

Gunnulf Nikulaussön lay thinking of the spring day, three years back, when he had ridden up toward Husaby.

On the road he met a man from the manor — the mistress was not at home, he said — she was with a sick woman.

He rode along a narrow grassy path between old stick-fences; there were young leaf-trees covering the steep clay banks, both above him, and down towards the river, that ran below in the bottom, loud with the spring freshets. He rode towards the sun, and the tender green leaves glanced like golden flames on the twigs, but farther in the wood the shade lay cool and deep already on the grassy sward.

He rode on till he caught a glimpse of the lake, lying below him and mirroring darkly the farther shore, with the heavens all blue and the picture of the great summer clouds ruffled and broken by the current ripples. Deep down below the bridle-path lay a little farm on the green, flower-sprinkled slopes. A group of white-coifed house-wives stood out in the courtyard — but Kristin was not amongst them.

A little farther on he saw her horse; it was loose in the close along with some others. The path dipped down in front of him, into a hollow filled with green shadow, and where it wound up over the next billow of the clay banks she was standing by the fence under the leaves listening to the birds' song. He saw her slender black-clad shape bent over the fence in towards the wood; only the coif and an arm showed white. He reined in his horse and rode on towards her at a foot-pace; but when he came near he saw that 'twas the trunk of an old birch tree that stood there.

The next evening, when his serving-folk sailed him in to the city, the priest himself was at the helm. He felt his heart firm and, as it were, new-born in his breast. Nothing now could shake his purpose.

He knew then that what had held him back, had kept him in the world, was the unquenchable longing he had borne within him from his boyhood up — the longing to win the love of men. That he might be beloved he had been generous, mild and mirthful with small folk; he had let his light of learning shine, but with all modesty and humility, among the priests in the city, so that they might like him; he had been compliant with Lord Eiliv Kortin, since the Archbishop had been a friend of his father's, and he knew how Lord Eiliv liked those around him to behave. He had been kindly and gentle with Orm, to win a little of the boy's love away from his fitful father. And he had been stern and unsparing with Kristin, because he knew that she had needed to meet with somewhat that did not give way when she grasped at it for support; something that led her not astray when she came forward, ready to follow.

But that evening he had understood — he had sought to win her trust in himself far more than to strengthen her trust in God. . . .

Erlend had found the word to-night. Not my brother more than all men's brother. That was the way he must go, before his brotherly love could profit *any*.

Two weeks later he had parted all his goods between his kin and the Church, and taken on the habit of a professed Preaching Brother. And last spring, when all souls were deeply shaken by the fearful calamity that had fallen on the land — the lightning had struck Christ's church in Nidaros and half consumed St. Olav's house — Gunnulf had won the Archbishop's support for his old plan. Along with Brother Olav Jonssön, who was a consecrated priest like himself, and three younger monks, one from Nidaros and two from the Preachers' Convent in Björgvin, he was now journeying northward to bring the light of the Word to the unhappy heathen who lived and died in gross darkness within the boundaries of a Christian land.

Christ, Thou Crucified One, now have I given from me all that could bind me. Myself have I given into Thy hands, if Thou wilt deign with my life to buy Satan's household free. Take me, in such wise that I feel I am Thy thrall, for so shall I also possess Thee. — And so should his heart, maybe, one day sing and shout in his breast as it had sung and exulted when he walked the green plains by Romaborg, from pilgrims' church to pilgrims' church — "I am my Beloved's and to Him is my desire — "

The brothers lay, each on his bench in the little hut, thinking and thinking until they slept. A live ember on the hearth between them glowed faintly. Their thoughts drew them farther and farther away one from the other. And

the next day the one set forth for the north and the other
southward.

Erlend had promised Haftor Graut to sail round by
Godöy, and take Haftor's sister with him southward. She
was wed to Thorolf Aasulfssön of Lensvik — he, too, was
a kinsman of Erlend's, but far off.

The first morning, when *Margygren* stood out of
Godöy Sound, her sail bellying against the background of
blue mountains in the fine breeze, Erlend stood on the
after-deck and Ulf Haldorssön was at the helm. Lady
Sunniva came up on deck. She had thrown back the hood
of her cloak, and the wind blew the linen of her coif
backwards, uncovering her curly sun-bright yellow hair.
She had the same sea-blue glittering eyes as her brother,
and, like him, she was fair of face, but thickly freckled,
both on her face and her small plump hands.

From the first evening he saw her at Godöy — their
eyes had met, and then they had looked aside, a secret
smile on each face — Erlend had been assured that she
knew him — and he knew her. Sunniva Olavsdatter — he
could take her with his bare hands; and she looked for
him to do it.

Now, as he stood with her hand in his — he had helped
her up — he chanced to look at Ulf's rough, dark face.
Ulf knew it too, he could see. He was strangely abashed
at the man's look. He remembered in a flash all this kinsman and henchman of his had been privy to in his life —
every coil his folly had snared him in from earliest youth
up. Ulf had no need to look so scornfully at him — he
comforted himself — as though he had meant to be more
free with the lady than right and honour would allow. He
was old enough now, and wise enough from much burning
of his fingers, to be let loose in Haalogaland without
tangling him up in witless folly with the wife of another.
He had a wife himself now — he had been true to Kristin
from the first day he had seen her till now — one or two
matters that had befallen away in the north, no reasonable
man would bring into the account. Else had he not once
looked at a woman — in such wise. He knew it himself —
with a Norse woman — and their equal in birth to boot
— no, he would never have an hour's peace of mind if he

was false to Kristin in such wise. — But this voyage southward with her on board — it might well be perilous.

It was some help to him that they met with rough
weather along the coast, so that he had somewhat else to
do than to dally with the lady. At Dynöy they had to take
shelter, and tarry there in harbour some days. And while
they lay there a thing befell which made Lady Sunniva
seem much less alluring.

Erlend, with Ulf and one or two other men, slept in the
same shed where she and her women lay. One morning
he was alone in there and the lady was not yet risen. She
called him to her — said she had lost a finger-ring in the
bed. He had to come and help her to search — she was
creeping about on the bed. They turned towards each
other now and then in their search, and each time they
had that lurking smile in their eyes. — But when she took
hold of him — Aye, *he* had maybe not borne him in over-
seemly wise — time and place were against it — but she
was so bold and shameless that now of a sudden he grew
hard and cold. Red with shame, he looked away from her
face of laughter and wantonness; freed himself with scant
excuse, then went out and sent in the lady's serving-
women to her.

No, devil take it, he was not so young a bird as to be
caught with chaff. 'Twas one thing to beguile — quite
another thing to be beguiled. But he could not but laugh
— here stood he, and he had just fled away from a fair
dame, like yonder Hebrew, Joseph! Aye, strange things
befall both by sea and land.

Nay — Lady Sunniva — Ah, he could not but remember
one — one whom he knew. She had gone to tryst with him
in a house of call for ribald men-at-arms — and she had
come shamefast and worshipful as a young maid of kingly
birth might go to mass. In woods and barns had she met
him — God forgive him, he had forgotten her birth and
her honour; and she had forgotten them for his sake, but
she had not been able to fling them from her. Her blood
rose up and spoke in her, even when she thought not on it.

God bless thee, my Kristin — so help me God, my faith
that I pledged thee in secret and before the altar, that
will I keep or nevermore be called a man. So be it.

He landed Lady Sunniva soon after at Yrjar, where she

had kinsfolk. The best of the matter was that she seemed not too angry either when they parted. He had had no need to hang his head and mope like a monk — they had had much frolic and dalliance on the way. At parting he gave the lady some costly pelts for a cloak, and she promised he should see her in the cloak. They would surely meet now and then. — Poor woman, her husband was sickly and no longer young.

But he was happy that he was coming home to his wife and had naught on his mind that he need hide from her; and he was proud of his well-proved steadfastness. And he was dizzy and mad with longing for Kristin — she was the sweetest and loveliest of roses and lilies after all — and she was his.

Kristin was at the landing-place to meet him when Erlend came in to Birgsi. Fishers had brought word to Vigg that *Margygren* had been seen out at Yrjar. She had her two eldest sons and Margret with her, and at home at Husaby all was making ready for a great banquet to friends and kinsmen for Erlend's welcome home.

She was grown so fair that Erlend caught his breath when he saw her. But 'twas true she was changed. The girlish look that had still come back to her after she had come through each childbed — the tender, frail, nun-like look under the matron's coif — was gone. She was a young, blooming wife and mother. Her cheeks were round and freshly red between the white lappets of her coif; her bosom full and firm for chains and brooches to glitter on. Her thighs were rounder and fuller under the key-belt and the gilded case that held knife and scissors. Yes, yes — she had but grown more fair — she looked not now as though they could blow her away from him to heaven so lightly as before. Even the long narrow hands were grown fuller and more white.

They tarried at Vigg for the night, in the Abbot's house there. And it was a young, rosy and joyful Kristin, mild and beaming with happiness, that went with him this time to the feast at Husaby, when they rode homeward next day.

There were many grave matters she should have spoken of to her husband when he came home. There were a thousand things about their children; misgivings for Margret; and there were her plans for putting the estates on their feet again. But all was swallowed up in the whirl of festivity.

They passed around from one banquet to another, and she bore the Warden company on his progresses. Erlend kept now yet more men at Husaby, for messages and letters were ever passing betwixt him and his sheriffs and deputies. All the time Erlend was joyous and reckless as ever — how should he not be the very man for Warden, he asked, he that had run his head against well-nigh every rule in the law of the land and the Church's law? Hardly learned was well remembered! — The man was of a quick and ready wit, he had been well taught in boyhood, and this now stood him in stead. He used himself to read his own letters, and took an Icelander into service as scribe. Till now he had been wont to set his seal to whatever others read out to him, and was ever loath to look on a line of writing — Kristin had seen much of the fruits of this in these two years, in which she had made acquaintance with all the papers in his muniment-chests.

But now there came on Kristin a recklessness the like of which she had never known. She grew livelier and less still in her mien when she was out among strangers — for she felt herself very fair, and she was healthful and fresh for the first time since she had been wed. And at nights when Erlend and she lay in a strange bed in a loft on one of the great folks' manors or in the hall of a farm, they laughed and whispered and made sport of the folk they had met, and jested over tidings they had heard. Erlend's tongue was more devil-may-care than ever, and folk seemed to like him better than ever before.

She saw it in their own children — they were almost spellbound with delight when their father would now and again take notice of them. Naakkve and Björgulf did naught now but play with bows and spears and axes and such gear. And it might chance, now and then, that their father would stop in crossing the courtyard, look at their games, and put them right: "Not like that, my son — hold it in this wise" — he changed the grip of the little fist and placed the fingers as they should be. When this chanced, they were beside themselves with eagerness.

The two eldest sons were not to be parted. Björgulf
was the biggest and strongest of the children, as tall as
Naakkve, who was three half-years older, and stouter than
he. He had tight-curled raven-black hair; his little face
was broad but comely; the eyes dark blue. One day Erlend
asked their mother somewhat fearfully if she knew that
Björgulf had not good sight in one eye — and that he had
the slightest squint, too. Kristin said she believed not there
was much amiss; 'twas like he would grow out of it.
Things had so fallen out that she had always made least
ado with this child — he had been born when she was
worn out with nursing Naakkve, and Gaute had followed
so close on his heels. He was the strongest of the children,
and, it seemed, the quickest-witted, but he was most silent.
Erlend was fondest of this son.

Though he did not make it clear to himself, he had a
little grudge against Naakkve, because the boy had come
at an untoward time, and because he had to be called after
his grandfather. And Gaute was not as he had looked to
find him. — The boy had a great head, as was but reason,
since for two years 'twas the only part of him that had
grown — now his body and limbs were making up their
growth. His wits were good enough, but he talked right
slowly, for if he spoke fast he began to lisp or stammer,
and then Margret mocked at him. Kristin was most fond
of this boy — though Erlend could see that, in a manner,
the eldest was still dearest to her — but Gaute had been
so ailing, and he favoured her father somewhat, with his
flaxen hair and dark-grey eyes — and he was ever at his
mother's skirts. He was a little lonely, between the two
elder boys, who held together always, and the twins, who
were still so small that they were ever with their foster-
mothers.

Kristin had less time now to care for her children, and
she was forced to do more as other ladies did and let the
serving-women mind them — but the two eldest ran about,
for choice, among the men on the farm. She no longer
brooded over them with the old overwrought tenderness
— but she laughed and played with them more, when she
had time to gather them about her.

At the New Year there came to Husaby a letter under

Lavrans Björgulfsön's seal. It was written with his own hand and had been sent by the priest of Orkedal who had been south — so 'twas two months old. The weightiest news it brought was that he had betrothed Ramborg to Simon Andressön of Formo. The wedding was appointed for the spring, at the time of the Feast of Holy Cross.

Kristin was amazed beyond measure, but Erlend said he had deemed things might go thus — ever since he had heard that Simon Darre was left a widower, and had come to live on his manor in Sil after old Sir Andres Gudmundssön's death.

5

SIMON DARRE had taken it as a thing that was as it should be, when his father had agreed with Lavrans Björgulfsön on his match with Lavrans' daughter. In his kindred it had ever been the custom that all such matters were in the parents' hands. He had been glad when he saw that his bride was so fair and gracious. He had, indeed, never looked for aught else than that he should be good friends with the wife his father chose for him. Kristin and he suited each other well in age and birth and fortune — if Lavrans were of a somewhat higher kin, his father, on his side, was a knight and had been much about King Haakon, while the other had always lived retired on his estates. And Simon had never marked aught else than that wedded folk agreed well together when they were an equal match.

Then came that night in the loft at Finsbrekken — when evil tongues would have undone the innocent young child. From that hour he had known well enough that his betrothed was dearer to him than if he had but loved her as in duty bound. He thought not much on the matter — but he was glad; he saw that the maid was bashful and coy, but he thought not much on that, either. Then came the time in Oslo, when he was forced to think things over — and then the evening in Fluga's loft.

He had come against something here that he had not thought could hap in this world — amongst honourable folk of a good kindred and in these times. Blinded and stunned, he had flung himself free of his ties — though in bearing he had been cool and calm and steady in talking of the matter with his father and hers.

Thus had he departed from the customs of his house; and next he had done another thing unheard of in his kindred: without even taking counsel with his father, he had wooed the rich young widow at Mandvik. He was dazzled when he saw that Lady Halfrid liked him — she was much more rich and high-born than Kristin, being son's daughter to Baron Tore Haakonssön of Tunsberg, and widow of Sir Finn Aslakssön — and she was comely, and had so fine and noble a bearing that, likened with her, all the women he was used to seemed to him but as farmers' wives. In the devil's name, he would show them all that he could win the noblest wife; in riches and all else she bore the bell from this Trönder that Kristin had let herself be smirched by. And a widow — that, too, was well — plain and aboveboard — the devil might trust in maids any more, for him.

He had been made to learn 'twas not such a plain straightforward thing to live in the world as he had deemed it when at home at Dyfrin. There his father had ruled all things, and his judgments were right. True, Simon had been with the body-guard and served as page for a time, and he had gained a little learning from his father's house-priest at home — it might chance now and again that he deemed his father's wisdom a little behind the time. He would venture to gainsay him too, now and then — but it was but as in jest, and it was taken as a jest — a quick-witted lad, Simon, laughed his father and mother, and so said his brothers and sisters, who would never gainsay Sir Andres. But all things were done as his father willed — Simon himself deemed this but reason.

In the years when he was wedded to Halfrid Erlings-datter and dwelt at Mandvik, he learnt each day more thoroughly that life might be more cross-grained and crooked than Sir Andres Gudmundssön had ever dreamt.

That he should not be able to be happy with such a wife as he had won — such a thought could never have come to him. Deep down in his mind lurked a rueful wonder when he looked at his wife as she moved about the house all day long — so comely was she, with her gentle eyes and the mouth that was so sweet when the lips were shut — no woman had he ever seen wear her robes and her adornments with so much grace. And in the black darkness of the night distaste for her wore all youth and freshness out of him — she was sickly, her breath unhealthful, her caresses tortured him. She was so good that it filled him with a desperate shame — but he could not overcome his misliking.

And then 'twas not long after they were wed before he saw that she could never bear him a living, healthful child. He saw that she sorrowed over it herself even more than he — it cut him to the heart when he thought of *her* fate in that matter. One thing and another he had heard — 'twas so with her because Sir Finn had struck and kicked her more than once while she was wed with him, so that she had miscarried. He had been mad with jealousy of his young fair lady. Her kinsmen would have taken her from him, but Halfrid deemed that it behooved a Christian wife to cleve to her wedded husband, were he good or evil.

But should he not have children of her, then must he ever feel, as now he did, that it was *her* lands they dwelt on, *her* riches that he dealt with and controlled. He dealt with them heedfully and wisely. But all through these years there grew up in his mind a longing for Formo, the manor that was Sir Andres' mother's heritage, which it had always been meant he should take over after his father. He came at last to deem that his home was away yonder north in Gudbrandsdal, almost more than in Romerike.

Folks still went on calling his wife Lady Halfrid, as in the time of her first husband, the knight. And this made Simon feel all the more as if he were but her steward at Mandvik.

It chanced one day that they sat alone in the hall, Simon and his wife. One of the serving-women had come into the hall on some errand. Halfrid looked after her as she left.

"I wonder — " said she. "I fear me Jorunn is with child this summer."

Simon sat with a bow in his lap, mending its lock. He changed the tap-bolt, gazed down into the spring-box, and said, without looking up:

"Aye. And the child is mine."

His wife was silent. When at last he looked up at her, she sat sewing, as intent on her work as he had been on his.

Simon was sick at heart. Sickened because he had so affronted his wife, and sickened at his folly in having to do with the girl, and vexed that he had taken the fatherhood on his shoulders. He was in no wise sure himself. Jorunn was a light piece of goods, he knew. In truth, he had never much liked her; she was ugly, but had a sharp tongue in her head, and was merry to talk with; and it had been she who sat up for him ever when he had come home late during the last winter. He had answered over-hastily, fearing that his wife might complain and blame him. 'Twas a clownish fear; he should have known Halfrid would never stoop to such complaints. But now 'twas done — go back from his own word he would not. He must even put up with being held for the father of his serving-woman's child, whether he were so or not.

Halfrid spoke not again of the matter till a year after; then, one day, she asked if he knew that Jorunn was to be wed over at Borg. Simon knew it well enough, for he had given her dowry himself. Where was the child to be? asked his wife. With its mother's parents, where it was now, answered Simon. Then said she:

"Methinks it would be more seemly that your daughter should grow up here in your manor."

"In your manor, mean you?" asked Simon.

A little tremor passed over the lady's face.

"You know well, my husband, that as long as we both live, you are master here at Mandvik," said she.

Simon went and laid his hands on his wife's shoulders:

"If, indeed, Halfrid, you deem you can bear to see the child here in our home, great thanks shall I owe you for your high-heartedness."

He liked it not. He had seen the young one more than once — 'twas not a comely child, and he could not see that it favoured him or any of his folks. Less than ever did he believe that he was the father. And he had been sorely angered when he heard that Jorunn had had the child christened Arngjerd, after his mother, without leave asked of him. But he must let Halfrid have her way.

She fetched the child to Mandvik, found a foster-mother for it, and saw to it herself that the little one lacked naught. If her eyes chanced to fall upon the child, she often took it on her lap and tended it kindly and lovingly. And by little and little, as Simon saw more of it, he grew fond of the little maid — he had a great love of children. Now, too, he thought he could see a likeness in Arngjerd to his father. It was like that Jorunn had been wise enough to be on her good behaviour after the master had gone too far with her. — So it might well be that Arngjerd was his daughter, and what he had done at Halfrid's asking was the best and most honourable way.

When they had been five years wedded, Halfrid bore her husband a son, a full-formed man-child. She was transfigured with joy, but after her delivery she fell so sick that it was soon plain to all that she must die. Yet was she of good cheer, the last time when for a while she was herself. "Now shall you live on here as master, Simon, and hand on Mandvik and all our lands to your children and mine," she said to her husband.

After this the fever mounted so high that she knew no more, and she had not the grief, while yet on earth, of hearing that her boy had died a day before his mother. And in the other home, Simon thought, 'twas most like she felt no sorrow for such things, but was glad that she had their Erling with her.

Simon remembered afterward that, the night the two bodies lay up in the loft-room, he had stood leaning over the fence of a field that lay down by the sea-shore. It was just before St. John's Mass, and the night was so bright that the full moon's light was well-nigh blotted out. The water lay there palely shining, and plashed and gurgled a little on the strand. Simon had hardly slept more than an hour at a time, ever since the night the boy had been born — it seemed to him now very long ago — and he was so weary that he could scarce feel grief.

He was seven-and-twenty years old at this time.

Well on in the summer, when the estate was settled, Simon made over Mandvik to Stig Haaksonssön, Halfrid's uncle's son. He moved to Dyfrin and stayed there the winter.

Old Sir Andres was bedridden, with dropsy and many other ills and aches; he was nearing his end now, and he bemoaned him much — life had not been so plain and simple for him either in these latter days. Things had not gone so with his fair and likely children as he had wished and looked that they should. Simon sat by his father and tried hard to get back to the easy jesting note of old — but the old man bemoaned himself without cease; Helga Saksesdatter, whom Gyrd had wedded, was so fine that she knew not what follies she should hit upon next — Gyrd dared not belch in his own house without asking his wife's leave. And this Torgrim, ever and always in a pother about his belly — never should Torgrim have had daughter of his, had he known that the man was such a poor wretch he could neither live nor die. Astrid could have no joy of her youth or her wealth so long as her husband lived. And here was Sigrid, broken and grieving — smiles and song had quite gone from her, his good child. That she should have had that child — and Simon no children! Sir Andres wept, an old, unhappy, sick man. Gudmund had set himself against every match his father had spoken of for him, and he was so old and useless now he had let the lad run wild altogether. . . .

— But all the ill fortune had begun when Simon and that maid from the Dale set themselves up against their parents. And 'twas Lavrans' blame — for, bold a man as he was among men, he was chicken-hearted with his womenkind. The girl had snivelled and screeched, no doubt — and straightway he gave in and sent for that gilded whore-monger from Trondheim that could not so much as wait till he had been given his bride in wedlock. But if Lavrans had but been master in his own house, he, Andres Darre, would soon have shown that he could put sense in the head of a beardless whelp of a son. Kristin Lavransdatter — *she* bore children a-plenty — a strapping son every eleventh month, he had heard. . . .

"Aye, but that comes dear, father," said Simon, laughing. "The heritage comes to be split up sadly." He took Arngjerd up and set her on his lap — she had just come trotting into the room.

"Aye, 'twill not be through *her* that your heritage will be split up too small — whoever else it be that shall divide it," said Sir Andres testily. He was fond of his grandchild in a fashion, but it angered him that Simon had a base-born child. "Have you thought upon any new match, Simon?"

"Nay, you must let Halfrid grow cold in her grave first, father," said Simon, stroking the child's pale hair. "I shall wed again in good time — but sure there is no such haste — "

He took his bow and his ski and set forth for the woods, where he could breathe more freely. With his dogs he tracked the elk on the snow-crust, and he shot the capercailzie drowsing in the tree-tops. At night he slept in the Dyfrin forest sæter, and felt that 'twas good to be alone.

There was a scraping of ski outside on the hard snow; his dogs flew up barking, and other dogs answered from without. Simon opened the door upon a night blue with moonlight, and his elder brother Gyrd came in, tall and slender and comely and quiet. He looked younger now than Simon, who had ever been somewhat stout, and had grown a deal heavier in his years at Mandvik.

The brothers sat with the food-wallet between them, eating and drinking, and gazing into the hearth-fire.

"You must have seen," said Gyrd, "that Torgrim means to set us all by the ears when father is gone — and he has gotten Gudmund on his side. And Helga. They would fain keep Sigrid out of her full sister's share — "

"I have seen it. Her full share she shall have — you and I, brother, can sure make that good in despite of them."

"The best, mayhap, would be that father should take order in this before he dies," said Gyrd.

"Nay, let father die in peace," said Simon. "You and I between us should be able to guard our sister and see that they strip her not because she has fallen into such mischance."

Thus it came about that on Sir Andres Darre's death his heirs parted in bitter unkindness. Gyrd was the only one to whom Simon said farewell when he set out from home — and he knew that the life Gyrd's wife was leading him in these days was a none too happy one. Sigrid he took with him to Formo — she was to manage his house and he to see to her lands and goods.

He rode into his manor on a grey-blue day of melting snows, when the alder thickets by the Laagen were brown with blossom. When he had alighted and was entering the hall, with Arngjerd on his arm, Sigrid Andresdatter asked:

"Why smiled you, Simon?"

"Smiled — ?"

He had been thinking how far unlike was this home-coming to what he had looked for once — when the day came when he should take up his abode here on the manor of his father's mother's kin. A sister dishonoured, and a bastard child, these were his belongings now.

The first summer he saw but little of the Jörundgaard folks — he took much pains to shun them.

But the Sunday after the second Mary's Mass in the autumn, it chanced that he stood by Lavrans Björgulfsön's side in the church, so that 'twas they two that had to give each other the kiss of peace when Sira Eirik had prayed that the peace of Holy Church might be increased in us. And when he felt the elder man's thin, dry lips against his cheek, and heard him murmur the prayer for peace upon him, he was strangely moved. He saw that Lavrans meant more by it than but to follow a usage of the Church.

He hastened out when the mass was over; but at the standing for the horses he came on Lavrans, who prayed him to come with him to Jörundgaard and dine. Simon answered that his daughter was sick and his sister sitting by her. Lavrans then prayed that God might heal the child, and shook hands for farewell.

One evening some days after, they had been hard at work at Formo getting in the harvest, for the weather looked doubtful. The most of the corn was housed by the evening, when the first shower came down. Simon ran across the courtyard in pouring rain, while a stream of yellow sunlight from between the clouds shone on the hall-house and the mountain wall behind it — and there he was ware of a little maid standing at the door in sun and rain. She had his favourite dog with her — it broke loose and leaped upon the man, dragging after it a woman's woven belt tied to its collar.

He saw that the girl was the child of a good house — she was cloakless and bare-headed, but her wine-red frock was of city-bought cloth, broidered, and made fast on the breast with a silver-gilt brooch. A silk cord held back the ringleted hair, now dark with the wet, from her forehead. She had a lively little face with broad forehead and pointed chin and great shining eyes, and her cheeks were flaming red, as though she had been running hard.

Simon saw who the maid must be, and greeted her by name, Ramborg.

"How comes it that you do me so much honour as to come hither to us?"

" 'Twas the dog," she said, as she went with him into the house out of the rain. It had a trick of running off to Jörundgaard; now had she brought it back. Aye, she knew it was his dog; she had seen it running after him as he rode by.

Simon rebuked her a little for having come hither on foot quite alone; he said he would have horses saddled and take her home himself. But first, to be sure, she must have some food. Ramborg ran across at once to the bed where little Arngjerd lay ailing, and both the child and Sigrid were much pleased with the guest, for Ramborg was quick and lively. She was not like her sisters, Simon thought.

He rode with Ramborg as far as the by-road to the manor, and would then have turned about; but there he met Lavrans, who had just learned that the child was not with her playmates at Laugarbru, and was setting forth with his people to search — he was much alarmed. Simon was made to come in, and when he had once taken his seat up in the hall his shyness fell from him and he felt quickly at home again with Ragnfrid and Lavrans. They sat late over their drink, and as it was now set in foul weather, he was thankful to stay the night.

There were two beds in the hall. Ragnfrid made up one of them fairly for the guest, and now someone asked where Ramborg was to sleep — with her parents, or in another house.

"Nay, for I will be in my own bed," said the child. "Can I not sleep with you, Simon?" she begged.

Her father said their guest must not be plagued with children in his bed; but Ramborg went on clamouring that she *would* sleep along with Simon. At last Lavrans said sternly that she was too old to share a bed with a strange man.

"No, father, that am I not," she said stubbornly. "I am not too big, am I, Simon?"

"You are too little," said Simon, laughing. "Ask me in five years, Ramborg, and be sure I shall not say you nay. But I warrant that then you will have another sort of man, little Ramborg, than a fat and ugly old widower."

It seemed that Lavrans liked not the jest; he told the girl sharply to hold her tongue and go lie down in her parents' bed. But Ramborg cried out once more:

"Now have you asked for me, Simon Darre, and in my father's hearing."

"So have I, indeed," answered Simon, laughing. "But I fear me, Ramborg, he will answer no."

After this day the Jörundgaard folks and those of Formo were much together. Ramborg was over at the neighbour manor whenever a chance served; she played with Arngjerd as if the child had been her doll, ran about with Sigrid helping in the housekeeping, and would sit in Simon's lap when they were in the hall. He fell again into the habit of petting and romping with the maid, as he had been used to do in old days, when she and Ulvhild had been as little sisters to him.

Simon had dwelt two years in the Dale, when Geirmund Hersteinssön of Kruke made suit for Sigrid Andresdatter. The Kruke kindred were of old yeoman stock, but though one and another of the men had served the Kings in the body-guard, they had never won any name beyond their own country-side. Yet was it as good a match as Sigrid could look to make, and she herself was willing to be wed with Geirmund. So her brother closed the bargain, and Simon held his sister's wedding at his house.

One evening just before, while the hurry and bustle of making ready for the feast were at their worst, Simon said in jest that he knew not what would become of his house when Sigrid had left him. Then said Ramborg:

"You must do the best you can for two years, Simon. At fourteen years a maid is fit to wed, and then you can bring me home."

"Nay, *you* will I not have," said Simon, laughing. "I dare not undertake to bridle so wild a maid as you."

" 'Tis the stillest tarns that run the deepest, says my

father!" Ramborg cried. "*I* am the wild kind, I. My sister, she was meek and mild. Have you forgotten Kristin now, Simon Andressön?"

Simon leapt up from the bench, lifted the maid up against his breast, and kissed her throat so hard that the skin was flecked with red. Aghast and amazed at himself, he loosed her, caught up Arngjerd, and threw her up and crushed her to him in the same way to hide his disorder. He went on romping with the two, the little and the half-grown maid, and chasing them about, while they fled from him up on tables and benches; at last he set them up on the cross-beam next the door, and ran out.

— Kristin's name was scarce ever named at Jörundgaard in his hearing.

Ramborg Lavransdatter grew comely as she grew in years. The talk of the country-side grew busy making matches for her. At one time it was Eindride Haakons-sön of the Valders Gjeslings. They were kin in the fourth degree, but Lavrans and Haakon were both so rich they could well afford to send letters to the Pope in Valland and get dispensation. The match would put an end to some of the old suits-at-law that had gone on ever since the old Gjeslings went out with Duke Skule, and King Haakon took the Vaage lands from them and gave them to Sigurd Eldjarn. Ivar Gjesling the Young had won back Sundbu by weddings and exchanges, but these matters had led to endless jars and dissensions. Lavrans laughed at it all himself; the part of the spoil he could claim in his wife's right was not worth the calfskin and wax he had used up in the suit — to say naught of the trouble and the journeys. But seeing he had been in the broil ever since he was a wedded man, he must hold out to the end. . . .

But Eindride Gjesling took another maid to wife, and the Jörundgaard folks seemed not overmuch cast down. They were at the bride-ale, and Ramborg told her friends proudly, when she came home, that four men had come forward to sound Lavrans about her, either for themselves or for kinsmen. Lavrans had answered that he would not make any bargain for his daughter's hand till she was old enough to say a word in the matter herself.

So things went on till the spring of the year when

Ramborg was fourteen winters old. One evening that spring she was in the byre at Formo with Simon, looking at a calf that had been born. It was white with a brown patch, and the patch seemed to Ramborg to be the very shape of a church. Simon sat on the edge of the cornbin, while the girl leaned across his knees, and he pulled at her plaits:

"Then I wager 'tis a token that you will soon ride to Church, a bride, Ramborg."

"Aye, you know well enough my father will not answer no the day you ask for me," she said. "I am so grown up now, I might well wed this year."

Simon was a little taken aback, but he tried to laugh:

"Are you there again with that foolish old jest?"

"You know well that 'tis no jest," said the girl, looking up at him with her great eyes. "I have known it long — that 'tis to you here, at Formo, I would most fain come. Why have you kissed me and set me in your lap many a time and often, if you would not have me?"

"Right fain would I be to have you, my Ramborg. But I had never thought that so fair and young a maid could be meant for me. I am seventeen years older than you — you have not thought, I trow, how you would like to have an old blear-eyed, big-bellied husband when you were a woman in your best years — "

"*These* are my best years," said she, beaming, "and not yet are you so old and tottering, Simon!"

"But ugly I am — soon would you be sick of kissing *me*."

"That have you no cause to believe," she answered, laughing as before, and held up her mouth to be kissed. But he did not kiss her.

"I will not profit by your simpleness, my sweet. Lavrans will take you with him to the south this summer. Should you not have changed your thoughts ere you come back, then will I thank God and Our Lady for better fortune than I had looked for — but bind you I will not, fair Ramborg."

He called his dogs, took his spear and bow, and went up on the hills that same evening. There was much snow in on the uplands still; he struck off to his sæter and got him a pair of ski, then lay out for a week by the tarn south of the Boar Fells and hunted reindeer. But the evening he set off down towards home he grew uneasy and fearful again. 'Twould be like Ramborg if she had spoken of it to her father in spite of all. When he came over the hill-crest by the Jörundgaard sæter, he saw smoke and sparks going up from the roof. He thought maybe 'twas Lavrans himself that was there, and he went up to the huts.

He thought he could see from the other's bearing that he had guessed right. But they sat there talking of the bad summer last year and of when it would likely be best to move up the cattle this year, of the hunting, and of Lavrans' new hawk, which sat on the floor, flapping its wings over the pluck of the birds that were roasting on a spit over the fire. Lavrans had come up but to see to his horse-shelter in Ilmandsdal — some folks from Alvdal that had come by it that day had told him it was fallen down. So passed the most of the evening. At last Simon spoke up:

"I know not if — has Ramborg said aught to you of a matter she and I spoke of one night?"

Lavrans said slowly:

"Methinks it had been well you had spoken to me first, Simon — you might have known what answer you would have had of me. Aye, aye — I understand how it may have chanced that you named it first to the maid — and it shall make no odds. I am glad that things are so that I can bestow the child in a good man's hands."

After this there was not much more to be said, thought Simon. Strange enough, all the same — here sat he who had never dreamed of making too free with an honest maid or a wife, and he was bound in honour to wed one whom he would liefer not have had. But he made one trial:

"Yet neither is it so, Lavrans, that I have gone wooing your daughter behind your back — I thought I was so old that she would not take it for more than brotherly kindness from old days that I spoke so much with her. And if you deem I am too old for her, I should not marvel at it, nor let it part the friendship betwixt us."

"Few men have I met whom I would rather have in a son's stead than you, Simon," said Lavrans. "And I would fain give Ramborg away myself. You know who will be her guardian when I am gone." It was the first time aught had been said between these two of Erlend Nikulaussön. "In many ways my son-in-law is a better man than I took him to be, when first I knew him. But I know not if he is the man to deal wisely with the giving of a young maid in marriage. And I mark well that Ramborg herself is willing."

"So thinks she now," said Simon. "But she is scarce out of her childhood. Therefore have I no wish to press the matter on you now, if you deem it should stand over yet awhile."

"And I," said Lavrans, frowning a little, "have no wish to force my daughter upon you — believe not that."

"*You* must believe," said Simon quickly, "there is no maid in Norway's land that I would rather have than Ramborg. So it is, Lavrans, that I deemed it all too good fortune for me if I should get me so fair and young and good a bride, rich, and come on both sides of the highest kin. And you for father-in-law," he said a little sheepishly.

A slight laugh came from Lavrans:

"Oh, you know well what I think of you. And I know you will so deal with my child and her heritage that we never shall have cause to repent this bargain, her mother and I — "

"That will I, God and all holy men helping me," said Simon.

On that they shook hands. Simon remembered the first time he had clasped hands with Lavrans on such a bargain; and his heart grew little and sore in his breast.

But Ramborg *was* a better match than he could have looked for. There were only the two daughters to divide between them what Lavrans left. And he would be as a son to the man whom, of all men he knew, he had ever honoured and loved the most. And Ramborg was young and fresh and sweet. . . .

And surely by this time he should be a grown man with a grown man's wit. Had he been waiting here thinking that he might wed as a widow her whom he could not win as a maid — after yon other had enjoyed her youth — and a dozen of stepsons of that breed — nay, then he would be rightly served if his brothers had him adjudged incapable and set him aside from managing his own affairs. Erlend

would live to be as old as the hills — such fellows as he
always did. . . .

Aye, so now they were to be brothers-in-law. They had
not seen each other since that night in yonder house at
Oslo. Well, it must be yet less joyful for the other to
remember than for him.

He would be a good husband to Ramborg, without falsehood or guile. Though it might almost be said the child
had beguiled *him* into a snare —

"You are laughing?" asked Lavrans.

"Did I laugh? 'Twas but a thought that came into my
head."

"You must tell me what it was, Simon — so that I may
have a laugh too."

Simon Andressön fastened his little, sharp eyes on the
other.

"I was thinking of — women. I marvel if any woman
regards men's faith and men's laws as we do amongst
ourselves — when she or hers can gain by setting them at
naught. Halfrid, my first wife — aye, this have I never
told to any Christian soul before you, Lavrans Björgulfsön, and to none other will I ever tell it — she was so good
and holy and upright a woman that methinks her like has
scarce ever lived — I have told you how she took the
matter of Arngjerd's birth. But that time when we saw
how things were with Sigrid — aye — she would have had
me hide away my sister and that *she* should feign to be
with child, and should pass off Sigrid's child for hers. For
thus had we had an heir, and the child had been well pro-
vided, and Sigrid could dwell with us and need not be
parted from it. I verily believe she understood not it would
have been treachery against her own kinsmen — "

Lavrans said, after a pause:

"Then you could have kept Mandvik, Simon — "

"Aye." Simon Darre laughed harshly. "And mayhap with
as good a right as many another man to the land he calls
the heritage of his fathers. Since we have naught to trust
to in such matters but the honour of women."

Lavrans slipped the hood over his falcon's head, and
lifted the bird upon his wrist.

"This is strange talk for a man thinking of marriage,"
he said low. There was something like distaste in his voice.

"Of *your* daughters I trow none thinks such things," answered Simon.

Lavrans looked down at the falcon and scratched its feathers with a stick.

"Not of Kristin either?" he asked yet lower.

"No," said Simon firmly. "She behaved not well towards me, but never did I find that she dealt in falsehood. She told me plainly and honestly that she had met a man whom she loved more than me."

"When you gave her up so willingly," asked the father in a low voice, "was it not because you had heard some — some rumours — about her?"

"No," said Simon as before. "I had never heard rumours about Kristin."

It was fixed that the betrothal-ale should be drunk that same summer, and the wedding be held after Easter the next year, when Ramborg would be full fifteen years old.

Kristin had not seen her home since the day she had ridden away from it as a bride — 'twas now eight winters since. Now she came back with a great company — her husband, Margret, five sons, nurse-girls, handmaids, men and pack-horses with baggage. Lavrans rode out to join them, and met them at Dovre. Kristin no longer wept so lightly as in her youth, but when she saw her father come riding to meet her, her eyes filled with tears. She stopped her horse, slipped down from the saddle, and ran to meet him, and when they met she took his hand and kissed it humbly. Lavrans leapt from his horse at once, and lifted his daughter up in his arms. Then he shook hands with Erlend, who had also alighted, and now came to meet his father-in-law with a reverent greeting.

The next day Simon came over to Jörundgaard to greet his new kinsfolk. Gyrd Darre and Geirmund of Kruke were with him, but their wives they had left at Formo. Simon had chosen to hold his wedding at home, so the women there were in a great bustle.

As to the manner of the meeting — Simon and Erlend greeted each other freely and without constraint. Simon was master of himself, and Erlend was so gay and cheerful that Simon thought he must have forgotten where they

had last seen each other. Then Simon gave Kristin his
hand. They were less sure of themselves, and their eyes
met but for an instant.

Kristin thought to herself that he had fallen off greatly.
In youth he had been comely enough, although even then
too thickset and short-necked. His steel-grey eyes had
looked little under the full eyelids; his mouth had been too
small, and the dimples in his round boyish face too large.
But he had had a fresh-hued visage, and a broad milk-
white forehead beneath goodly light-brown curly hair.
The curly hair he had still, as thick and nut-brown as before, but his face was now an even red-brown all over,
wrinkled under the eyes, and with heavy cheeks and a
double chin. His body, too, was grown heavy — and he
had something of a paunch. He looked not now like a
man who would care to lie at night on the edge of a bed
for the sake of whispering with his betrothed maid. Kristin felt pity for her young sister; she was so fresh and
gracious and so childishly joyful that she was to be wed.
The very first day she had shown Kristin the chests filled
with her bridal gear and Simon's betrothal-gifts — and she
had told how she had heard from Sigrid Andresdatter of
a gilded casket that stood in the bridal-loft at Formo; there
were twelve costly linen coifs in it, and they were to be a
gift to her from her husband the first morning. Poor little
soul, how could she understand what marriage was? 'Twas
pity that she knew so little of this young sister — Ramborg had been at Husaby twice, but there she had ever
been sullen and unfriendly — she could not get to like
Erlend, nor yet Margret, who was of her own age.

Simon thought that he had looked — perhaps hoped —
that Kristin should seem somewhat worn, seeing she had had
so many children. But she bloomed with youth and health,
and she bore her as proudly straight as ever, and walked
as graciously, though it might be she trod the earth now a
little more firmly than of old. She was the comeliest
mother, with her five fair little sons around her.

She was clad in a dress of home-made rusty-brown
woollen with dark-blue birds woven into it — he remembered standing about and leaning up against her loom
while she sat weaving on the stuff.

* * *

There was a little trouble when they came to sit down to table in the upper hall. Skule and Ivar began shrieking, for they were bent on sitting between their mother and their foster-mother as they were used to do. Lavrans thought it not seemly that Ramborg should sit below her sister's serving-woman and small children — so he bade his younger daughter sit in the high-seat beside him, since she was so soon to be parted from her home.

The little lads from Husaby were restless, and seemed not to know much of behaviour at the board. The meal had not gone far, when the little fair-haired boy slid down under the table and came up by the wall-bench beside Simon's knee.

"May I look at that strange sheath you have there in your belt, kinsman Simon?" he said; he spoke slowly and gravely. It was the great silver-mounted sheath to hold a spoon and two knives he had caught sight of.

"You may so, kinsman. What is your name, cousin?"

"Gaute Erlendssön is my name, cousin."

He put down the piece of bacon he was holding on the lap of Simon's festal doublet of silver-grey Flemish cloth, drew the knife from the sheath, and looked closely at it. Then he took the knife Simon was eating with, and the spoon, and put them all in their places, so that he could see how it looked when all the things were in the sheath. He was exceeding grave and exceeding greasy on fingers and face. Simon smiled as he looked at the little visage, so comely and so intent.

Soon after, the two eldest also made their way across to the men's bench; and the twins slipped down under the table and began crawling about there under folk's feet — then out and away to the dogs by the fire-place. There was little chance for the grown-up folk to eat in peace. The children's mother and father spoke to them, indeed, and bade them sit down prettily and be quiet; but the children paid no heed; and the parents, on their side, laughed at them the whole time and seemed to think their ill behaviour no great matter — not even when Lavrans, somewhat sharply, bade one of the serving-men take the dogs down into the room below, so that folks might be able to hear themselves speak in the upper hall.

Ramborg left the hall with her betrothed and went with

him through the spring night a little way up between the fences. Gyrd and Geirmund had ridden on ahead, and Simon stopped to say good-night. He had his foot in the stirrup already — when he turned again, took her in his arms, and crushed the slender child to him so tightly that she moaned softly and happily.

"God bless you, my Ramborg — so fine and fair as you are — all too fine and fair for me," he murmured into her tangled curls.

Ramborg stood looking after him as he rode off in the misty moonlight. She rubbed her upper arm — he had grasped it so hard that it hurt her. Dizzy with joy, she thought: in three days more she would be wed to him.

Lavrans stood with Kristin before the children's bed, and watched her tucking the small bodies into their places. The eldest were big boys already, with thin bodies and slender bony limbs; but the two little ones were plump and rosy, with creases in their flesh and dimples at the joints. A fair sight they seemed to him, lying there red and warm, their thick-growing hair damp with sweat, breathing evenly in their sleep. They were healthy, comely children — but never had he seen young ones so ill brought up as these grandsons of his. 'Twas well, indeed, that Simon's sister and brother's wife had not been there that night. But maybe 'twas not for him to talk about breaking in children — Lavrans sighed a little, and made the sign of the cross above the small heads.

So Simon Andressön drank the bride-ale with Ramborg Lavransdatter, and the wedding was in all ways fair and sumptuous. Bride and bridegroom looked joyous, and many deemed that Ramborg was lovelier on her day of honour than her sister had been — not dazzling fair, like Kristin, but far gentler and more glad; all could see in the clear innocent eyes of this bride that she bore the golden crown of the house of Gjesling with full honour this day.

And glad and proud she sat in the arm-chair before her bride-bed with hair bound up when the guests next morning came up to greet the young folk. With laughter and free jesting they looked on while Simon laid the house-wife's head-dress over his young wife's head. Shouts of greeting and the clashing of arms made the rafters ring, as Ramborg rose and took her husband's hand, upright and red-cheeked under the white coif.

It was not so often that two children of great houses of the one parish were wedded — when the kindred was gone through in all its branches, the kinship would most often be found too near. So all accounted this wedding a rare and joyous festival.

6

ONE of the first things Kristin had marked at her old home was that all the old heads of men, that had stood where the verge-boards crossed at the house-gables, were gone now. Instead there had been set up spires with carven birds and foliage-work, and the new storehouse had a gilded weather-vane. The old posts of the high seat in the hearth-room house, too, had been changed for new. The old ones had been carved in the likeness of two men; ugly enough — but 'twas thought they had been there ever since the house was built, and the custom had been to smear them with fat and bathe them with ale at festivals. On the new posts her father had carved out two men with helms and shields marked with the Cross. 'Twas not St. Olav himself, said he, for it seemed to him unmeet that a sinful man should have images of the holy ones in his house, except to pray before them — but they might, he thought, be two warriors of Olav's guard. All the old carvings Lavrans had himself cut up and burnt — the serving-men dared not touch them. It was with some doubt he still let them bear out food to the great stone at Jörund's grave-mound on holy eves — but yet he deemed 'twere sin and shame to deny to the tenant of the mound what he had been used to be given ever since folk had dwelt upon the place. He had died long before Christendom had come to Norway, so it was not his fault that he was a heathen.

Folks liked these new-fangled doings of Lavrans' but little. 'Twas well enough for him, who could afford to buy himself protection in other quarters. What he got seemed, indeed, to have all the virtue needed, for he had the same good fortune in husbandry as before. But there were those who asked whether yonder folk would not avenge themselves when there came a master to the manor who was less pious and not so open-handed towards the Church and all her belongings. And for small folk 'twas cheaper to give the old ones what they were used to have, rather than make foes of them and trust wholly to the priests.

Besides, it was none too sure, folk deemed, how 'twould go with the friendship between Jörundgaard and the parsonage when Sira Eirik should pass away. The priest was grown old and weakly now, so that he had need of a chaplain to help him. He had first spoken to the Bishop of his daughter's son Bentein Jonssön — but Lavrans, too, spoke to the Bishop, who was a friend of his of old. Folks deemed this misjudged. Truly it might well be that the young priest had been too forward with Kristin Lavransdatter that evening and maybe frighted the girl — but none could know that she might not herself have given some cause for the fellow's boldness. It had come out plainly enough since that she was none so coy as she had seemed. But the truth was, Lavrans had ever put too much faith in that daughter of his, adoring her almost as if she had been a sacred thing.

Afterwards there had been coldness for a time between Sira Eirik and Lavrans. But then came this Sira Solmund as chaplain, and he straightway fell at loggerheads with his parish priest over some lands, whether they belonged to the glebe or were Sira Eirik's own. Lavrans knew more than all other men in the parish of all sales of land and the like from the earliest days, and it was on his witness that the case was adjudged. Since then he and Sira Solmund had not been friends; but Sira Eirik and Audun, the old deacon, now lived at Jörundgaard, one might say; for they went thither daily and sat with Lavrans, bemoaned the wrongs and vexations they had to suffer at the hands of the new priest, and were waited on as they had been two bishops.

Kristin had already heard somewhat of all this from Borgar Trondssön of Sundbu; he had wedded a wife from the Trondheim country, and had been a guest at Husaby more than once. Trond Gjesling was dead some years ago; none deemed him much loss, for he had been a cankered shoot of the old tree, churlish, cross-grained and sickly. Lavrans was the only one who had put up with Trond; he pitied his brother-in-law, and yet more Gudrid, his wife. Now they were gone, all their four sons lived together on the manor; they were comely, bold and likely men, so folk deemed it a good exchange. There was close friendship between them and their uncle at Jörundgaard — he rode over to Sundbu a couple of times each year and went a-hunting with them in the West Hills. But Borgar had told Kristin 'twas beyond all reason, the way Lavrans and Ragnfrid tormented themselves now with penances and godliness. "He swills down water as hard as ever on fast-days; but he communes not with the ale-cup, your father, with the good old heartiness that he used," said Borgar. None could understand the man — 'twas not to be believed that Lavrans had any secret sin to atone for; so far as folks knew, he must sure have lived as Christianly as any son of Adam, saving the holy saints.

Deep down in Kristin's heart there stirred a dim surmise why her father strove thus to come near and ever nearer to his God. But she dared not think it clearly out.

She would not own to herself that she saw how changed her father was. 'Twas not that he was so greatly aged; he had kept his shapely form and his upright and gracious bearing. He was greying fast, but folk marked it not much, because his hair had always been so light in hue. And yet — her remembrance was ever haunted by the picture of the young, fair-shining man — the fresh rounding of the cheeks in the long narrow face, the clear red of the skin under the sunburn, the red full mouth with deep-cut corners. Now was the well-rounded muscular body shrunk into naught but bone and sinew, the face brown and sharp, as though carven in wood, the cheeks were flat and lean, the mouth had a knot of muscle at each corner. Aye, but then he was no young man any more — though neither, after all, was he so old.

Quiet, sedate and thoughtful he had always been, and she knew that from childhood up he had followed the commands of Christ with a rare zeal, had loved masses and prayers in the Roman tongue, and ever sought the church as the place where he found his best solace. But all had felt that a full, gentle tide of courage and joy in life flowed through the quiet man's soul. Now 'twas as though something had ebbed away from him.

She had not seen him drunken more than a single time since she came home — it was one evening of the wedding-feast at Formo. He had staggered somewhat then and been thick of speech, but he had not been out of the way merry. She remembered him as he had been in her childhood at the great ale-drinkings in festival seasons and at banquets — laughing his great laughter and slapping his thighs at each jest; offering to fight and wrestle with any man there who had a name for strength; trying horses; leaping about in the dance, and the first to laugh when his feet failed him; strewing around gifts and overflowing with goodwill and loving-kindness toward all mankind. She understood that her father had need of these great outbursts of revelry, in the midst of his steadfast labours, his strenuous fasting and his quiet home life with his own folks, who saw in him their best friend and surest stay.

She felt, too, that if her husband never had this need to drink himself drunken, 'twas because he kept so little check on himself even if he were never so sober, but ever followed his own devices without much pondering over right and wrong or what folk accounted seemly and wise behaviour. Erlend was the most sober man as to strong drink that she had known — he drank to quench his thirst and for the sake of good-fellowship, without caring much about the matter.

But now Lavrans Björgulfsön had lost his good old heartiness over the ale-cup. No longer had he that within him which needed to be given vent in revelry. It had never come into his mind to drown his cares with hard drinking, and it came not now into his mind — to him it had ever seemed that a man should take his joy to the drinking-feast.

With his sorrows he had gone elsewhere. There was a picture which always hung dim and half-remembered in his daughter's mind — her father on that night when the church was burned. He stood beneath the crucifix that he had saved from the flames, bearing the cross and staying himself by it. Without thinking it clearly out, Kristin felt in her heart that it was in part fear for her and her children's future with the man she had chosen, and the feeling of his own helplessness in this matter, that had changed Lavrans.

This knowledge gnawed secretly at her heart. And, as it was, she had come to Jörundgaard weary of the restless winter they had spent, and of her own weakness in making no stand against Erlend's heedlessness. She knew that he was, and would ever be, a spendthrift; that he had no wit in guiding goods and gear — they dwindled under his hands slowly but ceaselessly. One thing and another she had got him to set in order as she and Sira Eiliv had counselled — but she could not evermore be speaking to him of such things; and it was tempting, too, now to give herself up to gladness with him. She was so weary of striving and struggling with all things both without her and in her own soul. And yet was she such an one that careless pleasure, too, made her careworn and fearful.

Here at home she had looked to find again the peace she had felt in her childhood under her father's safe-keeping. But no — she felt so unsafe. Erlend had good incomings from his Wardenship, but he was living now with yet more pomp, with a greater household and the following of a great chief. And he had begun to keep her quite outside all things in his life that touched not their most private life together. She understood that he would not have her heedful eyes watching his doings. With men he would talk willingly enough of all he had seen and gone through in the north — but to her he never named it. And there were other things. He had met Lady Ingebjörg, the King's mother, and Sir Knut Porse more than once in these years; it had never so chanced at these times that she could be with him. Sir Knut was now a Duke in Denmark, and King Haakon's daughter had bound her to him in wedlock. The marriage had waked bitter wrath in many Norsemen's minds; and steps — Kristin understood not what they were — had been taken against the lady. But the Bishop of Björgvin had sent certain chests to Husaby; they were now on board *Margygren*, and the ship lay out at Ness. Erlend had been given letters and was to sail for Denmark later in the summer. He pressed her to go with him — but she set herself against it. She knew that Erlend moved among these great folk as their equal and dear kinsman, and she feared what might come of it; 'twas unsafe with so rash a man as Erlend. But she could not pluck up heart to go with him — never could she get him to hearken to counsel there, and she was loath to adventure herself in company with folks among whom she, a plain housewife, could scarce hold her own. And then there was the terror of the sea — sea-sickness to her was a thing worse than the hardest childbed.

Thus, as she went about in her old home, her heart was tremulous and ill at ease.

One day she had gone with her father down to Skjenne. And she had seen again the precious rarity which the folk of that manor had in their keeping. 'Twas a spur of the purest gold, huge and in shape old-fashioned, with strange chasings. Like every child in the country-side, she knew where it had come from.

'Twas in the first times after St. Olav had christened the Dale that Audhild the Fair of Skjenne was spirited away into the mountain-side. They dragged the church-bell up on to the mountain and rang for the maid — and the third evening she came walking over the pastures, so decked out with gold that she shone like a star. Then the rope broke, the bell rolled down the scree, and Audhild must turn back again into the hillside.

But many years after there came one night twelve warriors to the priest — he was the first priest that ever was in Sil. They had golden helms and silver corselets, and they rode on dark-brown stallions. 'Twas Audhild's sons by the Mountain-king, and they prayed that their mother might be given burial as a Christian woman and a grave in hallowed earth. She had striven to hold fast her faith and to keep the Church's holy days in the mountain, and she had prayed so sorely for this grace. But the priest denied it to her — folk said because of this he himself had found no rest in his grave, but in autumn nights he could be heard walking in the grove north of the church, weeping and bemoaning his harshness. The same night Audhild's sons had appeared at Skjenne, bearing their mother's greeting to her old parents. And in the morning they found the golden spur in the courtyard. And 'twas clear that yonder folk still accounted the men of Skjenne as kin, for they had ever rare good fortune in the mountains.

Lavrans said to his daughter, as they rode homeward in the summer night:

"These Audhildssons said over the Christian prayers they had learnt of their mother. God's name and Jesu name they could not name; but they said the Lord's Prayer, and the Credo in this wise: I believe on yonder Almighty One, I believe on the only-begotten Son, I believe on the most mighty Spirit. And then they said: Hail thou Lady that art the blessed one among women — and blessed is the fruit of thy womb, the Comfort of all the world — "

Kristin looked timidly up in her father's lean weather-beaten face. In the light summer night it looked so ravaged with sorrows and broodings as never before she had seen it.

"This you had never told me before," she said in a low voice.

"Had I not? Oh, no; I may have deemed it might give you heavier thoughts than your years gave you strength to bear. Sira Eirik says it is written in the books of St. Paul the Apostle: not Manhome alone groaneth in travail — "

One day Kristin sat sewing on the topmost step of the stair to the upper hall, when Simon came riding into the courtyard and stopped just below her, but saw her not. Her parents both came out to him. No, Simon would not alight; Ramborg had but bidden him ask, as he was passing by here — the sheep that had been her pet lamb, they would scarce have sent it to the hills; she would fain have it with her now.

Kristin heard her father smite his hand to his head. Aye, Ramborg's sheep — He gave an angry laugh. An ill thing, this — he had hoped she had forgotten it. For he had given his two eldest grandsons a little hatchet each; and the first thing they had used them for was to slaughter Ramborg's sheep.

Simon laughed a little:

"Aye, the Husaby boys — a rascal crew they are — !"

Kristin ran down the stairway, loosing her silver scissors from its belt-chain:

"Give Ramborg this in amends for my sons' killing her sheep — I know she has wanted this ever sincee she was

small. None shall say that my sons — " She had spoken
hotly, and now she fell suddenly silent. She had seen her
father's and mother's faces — they looked on her gravely
and wonderingly.

Simon made no motion to take the scissors; he seemed
somewhat abashed. Then he caught sight of Björgulf and,
riding across to him, leaned over and lifted the boy to the
saddle-bow in front of him: "So — you're a Viking harrying our coasts? Well, now are you my prisoner, and to-
morrow your parents can come over to me and we will
bargain about your ransom — "

Therewith he waved laughingly to the others and rode
off, with the boy struggling and laughing in his arms.
Simon had come to be right good friends with Erlend's
sons; Kristin remembered that he had ever had a way with
children; her little sisters had loved him. It vexed her
strangely that he should be so fond of children and have
such a turn for playing with them, when her husband was
ever so loath to hearken to children's talk.

The day after, though, when they were at Formo, she
could see well that Simon's wife had given him small
thanks for bringing this guest home with him.

"No one could look that Ramborg should greatly care
for children now," said Ragnfrid. "She is scarce out of
childhood yet, herself. She will change, I warrant, as she
grows older."

"Doubtless she will." Simon and his mother-in-law exchanged a glance and the slightest of smiles. Ah, thought
Kristin — aye, 'twas nigh on two months from the wedding now. . . .

In the trouble and unease of mind that Kristin now suffered, she was apt to vent her disquiet on Erlend. He took
this sojourn on the manor of his wife's father contentedly
and happily, like a man with conscience at ease. He was
good friends with Ragnfrid, and made no secret of his
hearty love for his wife's father. Lavrans, too, seemed fond
of his son-in-law. But so sore and watchful was Kristin
now grown that she felt that in Lavrans' kindness for
Erlend there was much of that pitying tenderness that
Lavrans had always had for every living thing that seemed
to him in some measure unfit to stand on its own feet. His

love for his other daughter's husband was not of this kind
— Simon could meet him as a friend and comrade.

Simon and Erlend, too, were good friends when they
met, but they sought not each other's company. Kristin
still felt a secret shrinking from Simon Darre — both by
reason of what he knew of her, and still more because
she knew that from that day at Oslo he had come off with
honour and Erlend with shame. She raged at the thought
that even this Erlend should be able to forget. Thus she
was not ever good-humoured with her husband. And if
Erlend chanced to be in the mood to bear her testiness
good-humouredly and with meekness, it vexed her that he
took her words so little to heart. Another day it might
chance that his temper was short, and then he would grow
hot, and she would answer him coldly and bitterly.

One evening they sat in the hearth-room house at
Jörundgaard. Lavrans still felt most at home in this house,
most of all in rainy weather when the air was heavy, as
to-day, for in the great hall the roof was flat, and the
smoke from the fire-place was a plague; but in the hearth-
room the smoke rose up and hung under the roof-tree
even when they had to shut the vent-hole against the
weather.

Kristin sat by the hearth, sewing, she was moody and
dull. Over against her, Margret was half asleep over her
seam, and yawned now and then; the children were romping noisily in the room. Ragnfrid was at Formo, and the
most of the serving-folk were gone out. Lavrans sat in his
high-seat, and Erlend at the upper end of the outer bench;
they had the chess-board between them, and moved the
pieces in silence and after much pondering. Once when
Ivar and Skule seemed set on pulling a puppy into two
pieces betwixt them, Lavrans rose and took the shrieking
little beast from them; he said nothing, but sat down again
to the game holding the puppy in his lap.

Kristin went over to them and stood with a hand on her
husband's shoulder, watching the game. Erlend had much
less skill as a chess-player than his father-in-law, and most
often lost when they played in the evenings; but he took
such things with careless good-humour. This evening he
played yet worse than was his wont. Kristin chid him for
it once or twice — not too mildly and sweetly. Then Lav-

rans said at last, somewhat testily:

"How should Erlend keep his thoughts on the game while you stand thus disturbing him? What would you here, Kristin? You have never had any skill of these games."

"No; I trow you folks think I have no understanding of aught — "

"Of one thing I see you have no understanding," said her father sharply, "and that is how it beseems a wife to speak to her husband. Better were it you should go and keep your young ones in bounds — the din they make is worse than the Wild Hunt."

Kristin went and set up her children in a row on the bench and sat down beside them.

"Be still now, my sons," she said. "Your grandfather likes not that you should romp and play in here."

Lavrans looked at his daughter, but said no word. Soon after, the foster-mothers came in, and Kristin, the maids and Margret went out with the children to put them to bed. Erlend said, when he was alone with his father-in-law:

"I could have wished, father-in-law, that you had not chidden Kristin as you did. If it comfort her to pick at me when she is in ill humour, why — it boots not to speak to her, and she will not suffer any to say a word about her children — "

"And you," asked Lavrans, "mean you to suffer your sons to grow up so uncorrected. Where are they got to, the maids that should keep the children by them and see to them — ?"

"To your serving-men's quarters, I warrant," said Erlend, laughing and stretching himself. "But I dare not say a word to Kristin about her serving-maids — she grows worth in earnest then, and casts it in my teeth that she and I have scarce been a pattern for others."

The day after, as Kristin went plucking strawberries along a meadow south of the manor, her father called to her from the smithy-door, and bade her come to him.

Kristin went, somewhat against the grain. 'Twas Naakkve again, likely — this morning he had left a gate open and the home cattle had gotten into the barley-field.

Her father took a red-hot iron from the forge and laid it on the anvil. The daughter sat and waited, and for long there was no sound but the strokes of the hammer, as they beat the sparking iron into a pot-hook, and the answering clang of the anvil. At last Kristin asked what he would with her.

The iron was cold now. Lavrans put from him tongs and sledge-hammer and came to her. With the soot on his face and hair, his clothes and hands black, and the great leathern apron in front of him, he looked sterner than was his wont.

"I called you to me, my daughter, to say to you this: Here in my house you must show your wedded husband such reverence as beseems a wife. I will not hear my daughter speak to her master as you answered and spoke to Erlend yesterday."

" 'Tis somewhat new, father, that you should deem Erlend to be a man to whom folks should show reverence."

"He is *your* man," said Lavrans. "I put not force upon you either to bring about your match with him. That you should bear in mind."

"You two are such warm friends," answered Kristin. "Had you known him then as you know him now, doubtless you would have been fain to put force on me."

Her father looked down at her, gravely and sadly:

"Now do you speak over-hastily, Kristin, and say that which you know is untrue. I tried not to put force upon you, even when you were bent on throwing off your lawful betrothed husband, though you know that I loved Simon heartily — "

"No — but since Simon too would have none of *me* — "

"Nay. He was too high-minded to stand stiffly on his rights, seeing you were unwilling. But I know not if in his heart 'twould have gone so much against him had I done as Andres Darre would have had me — paid no heed to the wilfulness of you two young folks. And I could well-nigh doubt whether Sir Andres was not right — now when I see that you cannot live in seemly wise with the husband you set all at naught to win — "

Kristin laughed aloud, an ugly laugh:

"Simon! Never would you have forced Simon to take

for wife the woman whom he had found with another man
in such a house — "

Lavrans caught his breath. "House?" he gasped out.

"Aye — a house such as you men call a bordel. She that
owned it — she had been Munan's paramour — she warned
me herself that I should not go thither. I said I went to
meet my kinsman — I knew not that he was *her* kinsman — " She laughed again, a wild, cruel laugh.

"Be silent!" said the father.

He stood still a moment. A shiver passed over his face
— a smile that seemed to blanch it. There came to her the
thought of a wooded hillside — how it whitens when a
storm-gust turns over all its leaves in a wave of palely
glittering light:

"He learns much that questions not — "

Kristin huddled down where she sat on the bench,
leaning on one elbow, and hiding her eyes with the other
hand. For the first time in her life she was in fear of her
father — in deadly fear.

He turned him from her, went and took up the sledge,
and put it in its place amongst the other hammers. Then
he gathered together files and other small tools, and set
himself to placing them in order on the cross-beam between the walls. He stood with his back to her; his hands
shook violently.

"Have you never thought, Kristin — that Erlend kept
silence about this?" He stood before her, looking down
into her white frighted face. "I gave him no for an
answer, curtly enough, when he came to me at Tunsberg
with his rich kinsmen and made suit for you — I knew
not then that I should have been but too thankful that he
was willing to restore my daughter's honour. — Many men
would have let me know it then. . . .

"*He* came again and made suit for you in full honour.
Not all men would have been so steadfast in striving to
win to wife one who was already — was already — what
you then were."

"*That* I trow no man had dared to tell to you — "

"'Tis not of cold steel that Erlend has ever been afraid
— " Lavrans' face had grown unspeakably weary; his voice
had gone hollow and dead. But soon he spoke again
quietly and firmly:

"Ill as all this has been, Kristin — methinks 'tis worst of all that you should tell it, now he is your husband and the father of your sons —

"If it be as you say, you knew the worst of him before you braved all things that you might wed him. And he was willing to buy you at as great a price as though you had been an honest maid. Much freedom to rule your life and manage your affairs has he given you — therefore should you make amends for your sin by managing with understanding, and making up for what Erlend lacks in prudence — so much do you owe to God and your children.

"I have said myself — and others have said the same — that Erlend seemed not to be fit for aught else than to beguile women. You are answerable in part for such things being said — to that you have now yourself borne witness. Since then, he has shown that after all he *was* fit for somewhat else — your husband has won himself a good name for a bold and swift leader in war. 'Tis no small gain for your sons that their father has won fame for boldness and skill in arms. That he was — unwise — *you* should have known best of all of us. Best may you make amends for your shame by honouring and helping the husband yourself have chosen — "

Kristin had bent forward over her lap, her head in her hands. Now she looked up, wildly and despairingly:

"Cruel was it of me to tell you this. Oh — Simon begged me — 'twas the only thing he begged of me — that I should spare you the knowledge of the worst — "

"Simon bade you spare me — ?" She heard the suffering in his voice. And she knew this, too, was cruel of her, to tell him that a stranger had seen she needed to be bidden spare her father.

Then Lavrans sat down by her, took one of her hands between both of his, and laid them on his knee.

"Cruel it was, my Kristin," he said gently and sadly. "Kind are you to all, my child, my treasure, but — I have seen it before this too — you can be cruel to them you love too dear. For Jesu sake, Kristin, spare me from the need of going in such fear for you — fear that this wild heart of yours will yet bring more sorrow over you and yours. You tug and strain like a young horse when 'tis first tied up to the stake, wherever you are tied by your heartstrings."

Sobbing, she sank against him, and her father drew her into his arms and held her close and firmly. They sat thus a long while, but Lavrans said no more. At last he lifted up her head.

"Now you are all black," he said with a little smile. "There is a cloth in yonder corner — but 'tis like it would but make you blacker. You must go home and make you clean again — anyone can see you have been sitting in the smith's lap — "

He pushed her gently out of the door, closed it behind her, and stood still a space. Then he staggered the few steps across to the bench, sank down on it and sat with head thrown back against the log wall and upturned, distorted face. With all his force he pressed one hand against his heart.

Well that it never lasted long. The breathlessness, the black dizziness, the pain shooting out through the limbs from the heart that struggled and shook, gave one or two heavy thuds, and then stood quivering still again. The blood hammering in the neck-veins.

'Twould pass in a little while. It passed always when he had sat still a little. But it came again, more and more often.

Erlend had trysted his ship's crew to meet at Veöy * on the eve of St. James' Mass, but he tarried at Jörundgaard a little longer to go with Simon and hunt down a big bear that had been harrying the sæter cattle. When he came home from the hunt, he found word awaiting him that his men had fought with the townsmen and he must hasten thither to bail them out. Lavrans had an errand in those parts, so he rode along with his son-in-law.

It was nigh the end of the Olav's Mass feast when they came out to the island. Erling Vidkunssön's ship lay there, and at evensong in Peter's Church they met the High Steward. He came with them to the monastery where Lavrans had taken lodging, ate supper with them there, and sent his men down to the ship to fetch some rarely good French wine he had gotten at Nidaros.

* See Note 13.

But the talk went but haltingly over the wine. Erlend sat wrapped in his own thoughts, his eyes bright and eager, as ever when a new venture was before him, but unheedful of the others' speech. Lavrans but sipped the wine, and Sir Erling was quiet and said little.

"You look weary, kinsman," said Erlend to him.

Aye, they had had heavy weather crossing Husastadviken the night before; he had been up —

"And you must ride hard if you would be at Tunsberg by Lawrence Mass day. And much rest or solace you will scarce find there either. Is Master Paal with the King now — ?"

"Aye. Do you touch at Tunsberg on your way?"

"To ask if the King would have me bear his loving and duteous greetings to his mother?" Erlend laughed. "Or if Bishop Aulfinn would send word by me to the Lady — ?"

"Many wonder that you should be journeying to Denmark, now that the chief men of the kingdom are gathering to the meeting at Tunsberg," said Sir Erling.

"Aye, is it not strange that folk must ever be wondering at me? Surely 'tis no marvel if I have a mind to see a little of such manners and breeding as I have not seen since I was last in Denmark — to ride in a tourney once again — and since our kinswoman has bidden us to come. You know well that none other of her kindred in this land will avow her now, save only Munan and I."

"Munan — " Erling frowned, and then laughed. "Is there so much life left in the old boar, I had well-nigh asked, that he can still move his brawn about? — So Duke Knut is to hold a tourney. Doubtless, then, Munan is to ride in the jousting?"

"Aye — 'tis pity of you, Erling, that you cannot bear us company and see that sight." Erling, too, laughed. "I mark well that you fear Lady Ingebjörg has bidden us to this christening-ale so that we may brew the ale for another feast and bid her to it. Nay, you should know best that I am too heavy-handed and too light at heart to be of use in hatching plots. And you have drawn every tooth in Munan's head — "

"Oh, no, we are not so fearful either of plots from that quarter. Methinks it must be clear as day to Ingebjörg Haakonsdatter now that she made forfeit of all rights in her own land when she wedded with the Porse. 'Twill be a hard matter for her to get a foot inside our doors, now she has laid her hand in that man's whose least finger we will never suffer in our affairs."

"Aye, 'twas wisely done of you, indeed, to part the boy from his mother," said Erlend darkly. "He is but a child yet — and already have we Norsemen cause to hold our heads high when we think of the King we have sworn fealty to — "

"Be still!" said Erling Vidkunssön, low and vehemently. "It — for sure it is untrue — "

Erlend said:

" 'Tis whispered in every manor and in every cabin in our north country that Christ's Church burned because our King is unworthy to sit in St. Olav's seat — "

"In God's name, Erlend — I say 'tis unsure whether it be true. And if it were, a child, like King Magnus, we must sure believe is without sin in God's eyes — he can redeem himself — Say you that *we* have parted him from his mother? I say, God punish that mother who betrays her son as Ingebjörg betrayed hers — and put not your trust in such an one, Erlend — bear in mind that they are faithless folk you journey now to meet!"

"Methinks they have kept faith with each other fairly enough. — But, as for you, you talk as if letters from Heaven dropped every day into your lap — 'tis that, maybe, makes you so bold to fight with the Lords of the Church."

"Nay, let us have an end of this, Erlend. Talk of things you have wit to understand, boy, and else be silent." — Sir Erling had risen, and he and Erlend stood face to face, red with anger.

"Oh, take heed of your tongue, Erlend," he went on in a little. "Think twice before you speak, where you are going. And think, and think again twenty times, before you do aught — "

"If 'tis so that *you* do, you who rule the roost here, then I marvel not that all things move but haltingly. But you need not be afraid," he yawned. "I — shall do naught, I trow. But 'tis grown a rare land to live in now, this of ours —

"Aye — you are setting forth early to-morrow. And my father-in-law is weary — "

The two others sat on in silence, after he had bidden them good-night. — Erlend was sleeping aboard his ship. — Erling Vidkunssön sat turning his goblet between his fingers.

"You are coughing?" he said, that he might say somewhat.

"Old men grow full of rheum so easily. We have many plagues, you see, my lord, that you young folk know naught of," said Lavrans smiling.

Again they were silent, till Erling Vidkunssön said, as though half to himself:

"Aye — so think all men — that this land is ill guided. Six years ago at Oslo I deemed I had seen clearly that there was a firm intent to uphold the kingly power — among the men of the houses to whom that duty falls by birth. I — built upon that."

"I believe you saw rightly, my lord. But you said yourself — we are used to rally round our King — and now our King is a child — and half the time in a foreign land — "

"Aye. There be times when I think — nothing is so ill but it is in some ways good. In the old times, when our Kings bore them like wild stallions — there was store of brave foals to rear; our folk needed but to choose the one that was the best fighter — "

Lavrans laughed a little: "Oh, aye — "

"We talked together three years ago, Lavrans Lagmandssön, when you came back from pilgrimage to Skövde and had seen your kinsfolk in Gautland — "

"I remember, my lord, you did me the honour to seek me out."

"Nay, nay, Lavrans, no need for so much *kurteisi*." He struck out a little impatiently with his hand. " 'Tis even as I said then," he went on gloomily. "None now can bring together the nobles of this land. *They* push to the front that are most fain to fill their bellies — there is yet some meat left in the trough. But they that might aspire to win might and riches by such service as was held in honour in our father's days, they come not forth!"

"So would it seem. True it is that honour follows the banner of the chief."

"Then men must deem, I trow, that with my banner there follows little honour," said Erling dryly. "You, too, have held aloof from all that might have made you a name, Lavrans Lagmandssön."

"So have I done ever since I have been a wedded man, my lord. I was wedded early — my wife was sickly, her health could not bear much going about in company. And it seems as if our stock doth not thrive here in Norway. My sons died early, and only one of my brother's sons lived to a man's age."

They sat again in silence awhile.

"Such men as Erlend," said the High Steward low — "they are the most dangerous. They that have thoughts that go a little farther than their own concerns — but not far enough. Aye, is he not like an idle child, this Erlend — ?" He shoved his winecup round about on the table in his vexation. "He is no dullard! And he has birth and valour. But never doth he trouble to hear so much of any matter as to understand it through and through. — And should he ever care to hear a man out, he had forgot the beginning, like enough, before one got to the end — "

Lavrans looked across at the other. Sir Erling had aged much since last he had seen him. He looked worn and weary —seemed not to fill his seat so fully as before. He had fine, clear-cut features, but they were a little too small, and his hue was, as it were, a little faded — had been so always. Lavrans felt that this man — though he was an upright, knightly man, prudent, and willing to serve faithfully and unflinchingly — yet, however measured, was somewhat too small to be the first man in the realm. Had he been a head taller, 'twas like he would more easily have found full following.

Lavrans said low:

"Sir Knut, too, be sure, is wise enough to see this — if they are hatching aught down there — that in any secret counsels he would find Erlend of small use — "

"You like this son-in-law of yours in a fashion, Lavrans," said the other, almost testily. "Though, truth to tell, cause to love him you have not — "

Lavrans sat drawing with his finger on the table in some spilled wine. Sir Erling marked how loosely the rings sat on his fingers now.

"Have *you* cause?" Lavrans looked up with his faint smile. "And yet I trow that you like him too."

"Oh, aye. God knows —

"But you may make your oaths, Lavrans, many things are running in Sir Knut's head now — he is father of a son that is King Haakon's grandson."

"Aye, but even Erlend must sure understand that that child's father has all too broad a back for the little lordling ever to be able to get around it. And the mother has our whole country against her because of this marriage."

A little while after, Erling Vidkunssön stood up and buckled on his sword; Lavrans had courteously taken his guest's cloak from the hook where it hung, and stood with it in his hands — then of a sudden he swayed and would have fallen to the ground, had not Sir Erling caught him in his arms. With much pains and labour he managed to bear the other, a big, tall man, over to the bed. A stroke it was not — but Lavrans lay there, his lips blue and white, with slack, strengthless limbs. Sir Erling ran across the yard and waked up the hospice-father.

Lavrans seemed much abashed when he came to himself. Aye, it was a weakness that took hold on him now and then — 'twas from an elk-hunt he had been on two winters back — he had lost his way in a snowstorm. Maybe something like this was ever needed to teach a man his youth is gone from him — he smiled in excuse.

Sir Erling tarried till the monk had bled the sick man, though Lavrans prayed him not to trouble himself, seeing that he was to set forth at daybreak. . . .

The moon shone bright, riding high above the hills of the mainland, and the water lay black in their shadow, but out on the fjord the light floated in flakes of silver. No smoke from any vent-hole — the grass on the house-roofs glittered with dew in the moonlight. Not a human soul in the one short street of the little town, as Sir Erling walked swiftly the short space, but a few steps, to the King's mansion, where he slept. He looked strangely slim and small in the moonlight — with the black cloak gathered closely about him — shivering a little. A pair of sleepy serving-men, who had sat up for him, came tumbling out into the courtyard with a lanthorn. The High Steward took the lanthorn and sent the men to bed — he shivered a little again as he climbed the staircase to the storehouse-loft, where he was to sleep.

7

A LITTLE after Bartholomew Mass, Kristin set out on her homeward journey, with her great company of children and serving-folk, and all their baggage. Lavrans rode with her to Hjerdkinn.

They walked up and down the courtyard in talk, he and his daughter, the morning he was to set forth down the Dale again. The hill country round them lay in sparkling sunshine — the mosses were red already, and the knolls yellow as gold with the birch copses; out on the upland wastes, tarns glittered and grew dark again, as the shadows of the great shining fair-weather clouds floated across them. They kept rolling up unceasingly, and sank down over the distant clefts and glens, amidst all the grey-stone peaks, and blue mountains with combs of new snow, and old snow-fields, that lay around as far as the eye could reach. The little grey-green patches of corn belonging to the rest-house stood out strangely against the autumn hues of this shining mountain world.

The wind blew fresh and sharp — Lavrans drew up the hood of Kristin's cloak, which had blown back upon her shoulders, and smoothed with his fingers the strip of linen coif that showed beneath it.

"Methinks you have grown pale and thin-cheeked in my house," said he. "Have we not taken good care of you at home, Kristin?"

"That you have. 'Tis not that — "

"Truly, too, this is a toilsome journey for you, with all these children," said her father.

"Oh, aye. Though 'tis not because of these five that my cheeks are pale — " A smile flitted over her face, and when her father looked at her questioningly, in alarm, she nodded and smiled a little again. The father looked away from her, but in a little while he asked:

"Since this is so, then maybe 'twill not be so soon that you can come home to us in the Dale again?"

"At least I hope 'twill not be eight years this time," she said in the same tone. Then she caught a glimpse of the man's face: "Father! Oh — father!"

"Hush, hush, my daughter — " Unwittingly he caught her by the upper arm and stopped her as she would have flung herself upon him. "Nay, Kristin — "

He took her hand firmly in his, and began walking again beside her. They had come away from the houses; they were following now a little path in among the yellow birch thickets — they marked not where they were going. Lavrans jumped over a little rill that crossed the path, turned towards his daughter, and gave her his hand over.

She saw, even in this little movement, he was no longer springy and nimble as of old. She had seen before without marking it — he no longer leapt into the saddle as lightly as he had done; he ran no longer up a loft-stairway; he lifted not a heavy thing easily as he was wont to do. He bore his body more stiffly and carefully — as if he had a slumbering pain within him and went softly so as not to wake it. The blood could be seen beating in his neck-veins when he came in from riding. Sometimes she had seen what seemed swellings or pouches under his eyes — and she remembered that one morning when she came into the hall he lay half dressed in the bed with his naked legs over the bed-foot, and her mother was crouched in front of him, chafing his ankle-joints.

"Should you sorrow for every man that old age strikes down, child, much will be your mourning," he said, evenly and quietly. "You have great sons yourself now, Kristin; sure it cannot come on you unlooked for that you see your father will soon be an old fellow. When we parted in old days while I was yet young — we could not know any more then than we know now, whether 'twas to be our lot to meet again on this earth. I may yet live long — 'twill be as God wills, Kristin."

"Are you sick, father?" she asked, tonelessly.

"Some ailments come with the years," answered her father, lightly.

"You are not old, father. You are but two and fifty years — "

"My father never was so old. Come and sit here by me."

There was a low grass-grown shelf under a rocky wall that leaned over the beck. Lavrans unclasped his cloak, folded it together, and, sitting on it, drew her down beside him. In front of them the beck clucked and rippled over the little stones in its bed, swaying a branch of willow that lay in the water. The father sat with his eyes fixed on the blue and white mountains far off behind the warm-hued autumn uplands.

"You are cold, father," said Kristin; "take my cloak —" She unfastened it; and he drew the skirt of it round his shoulders, so that they both sat in its shelter. Under it he put one arm about her waist.

"You know it well, my Kristin: unwise is he that mourns a man's going hence — let Christ have thee, rather than I — 'tis like you have heard the saying. I trust firmly in God's mercy. 'Tis not so long, the time for which friends are parted. Maybe 'twill seem so to you sometimes, while yet you are young; but you have your children and your husband. When you come to my years you will deem 'tis no time since you saw us who are gone away, and you will wonder, when you reckon up the winters that have passed, that they are so many. . . . To me it seems now 'tis not long since I myself was a boy — and yet it is many years since you were the little light-haired maid that ran about after me, wherever I might stand or go — you followed your father so lovingly — God reward you, my Kristin, for the joy I had of you — "

"Aye — if He reward me as I rewarded you — " She sank on her knees before him, caught his wrists, and kissed the palms of his hands as she hid her weeping face in them. "Oh, father, my dear father — no sooner was I a grown maid than I paid you for your love with the bitterest sorrow — "

"Nay, nay, child; weep not so." He drew his hands loose, and lifted her up beside him, and they sat as before.

"Much joy have I had of you, in these years too, Kristin. Fair and hopeful children have I seen growing up by your knee, a notable and understanding woman are you grown — and I have seen that more and more you have used yourself to seek help where it may best be had, when you were in any trouble. Kristin, my most precious gold, weep not so sorely. You may hurt him you bear under your belt," he whispered. "Nay, sorrow not so."

But he could not check her weeping. Then he lifted his daughter up into his lap and sat her on his knee; now he had her even as when she was a little one — her arms around his neck and her face pressed to his shoulder.

"There is a thing I have never told to any mother's child but to my priest — now will I tell it you. In the days when I was growing up —at home at Skog, and when first I was with the bodyguard — I had a mind to take the vows, as soon as I were old enough. Nay, I made no promise, not even in my own heart. There was much that drew me the other way, too. — But when I lay out fishing in the Botn Fjord and heard the bells ring from the cloister on Hovedö — it seemed that drew me most of all. — Then, when I was sixteen years old, father had made for me that habergeon of mine of Spanish steel plates soldered with silver —Rikard the Englishman at Oslo welded it together; and I got my sword too — the one I use always, and my horse-armour. 'Twas not so peaceful in the land then as in your young years — there was war with the Danes — and I knew I was like soon to have the chance to use my fair weapons. And I could not lay them from me. — I comforted me with the thought that my father would mislike it if his eldest son turned monk, and that I should not cross my parents.

"But 'twas I myself that made choice of the world, and I have striven to think when the world went against me: unmanful would it be to murmur at the lot I had chosen myself. For I have seen it more and more with each year I have lived — no worthier work can there be for a human soul that has found grace to conceive somewhat of God's loving-kindness, than to serve Him and watch and pray for those men whose sight is still darkened by the shadow of the things of the world. Yet must I needs say, my Kristin — hard would it be for me to give up for God's sake the life I have lived on my farms and lands, with cares for earthly things and with worldly cheer — with your mother by my side, and with you my children. Therefore must a man suffer in patience, when he has begotten offspring of his body, that it scorch his heart if he lose them or the world go badly with them. God, who gave them souls, owned them, and not I — "

Kristin's body was still shaken with weeping; and her father began rocking her in his arms like a little child.

"Many things there are that I understood not when I was young. Father held Aasmund dear too, but not so as he loved me. 'Twas for my mother's sake, you see — her he never forgot, though he took Inga because 'twas his father's will. Now would I wish that I could have met my stepmother again in this earthly home and prayed her to forgive that I set no store by her kindness — "

"But you have said often, father, that your stepmother did you neither good nor evil," said Kristin, through her weeping.

"Aye, God help me — 'twas my lack of understanding. Now does it seem a great thing to me that she hated me not, and never gave me an angry word. How would you like it, Kristin, if so it were that you saw a stepson put before your own son, at all times and in all things?"

Kristin was grown somewhat quieter. She lay now with her face turned outwards looking towards the mountain range. A great grey-blue pile of cloud was passing over the sun, darkening the air — some yellow beams stabbed through it, and a sharp glitter was thrown up from the water of the beck.

Then her tears broke out anew:

"Oh, no — father, my father—should I nevermore see you in life — "

"God guard you, Kristin, my child, so that we may find each other again on yonder day, all we who were friends in life — and every human soul. — Christ and Mary Virgin and St. Olav and St. Thomas will keep you all your days." He took her face between his hands and kissed her on the mouth. "God be gracious to you — God give you light in this world's light and in that great light hereafter — "

Some hours later, when Lavrans Björgulfsön rode off from Hjerdkinn, his daughter went with him some way, walking by his horse's side. Lavrans' man had gotten a long way ahead, but he still kept on at a foot's pace. It was grievous to look on her despairing face, all marred by weeping. So had she sat in the guest-shed too, all the time, while he ate and talked to the children, jested with them, and took them in his lap, one by one.

Lavrans said softly:

"Grieve no more for what you have to repent toward *me*, Kristin. But remember it, when your children grow big, and you may deem that they bear them not towards you or towards their father as you might think was right. And remember then, too, what I said to you of my youth. Faithful is your love to them, I know it well; but you are hardest where you love most, and I have marked that in these boys of yours dwells self-will enow," he said with a a little smile.

At last Lavrans bade her turn and go back: "I would not have you go alone any farther from the houses." They were come into a hollow between little hills with birch trees round their foot and stone screes higher up their sides.

Kristin pressed herself against her father's foot in its stirrup. She groped with her fingers over his clothes and his hand and the saddle and the horse's neck and quarters, rocked her head from side to side, and wept with such a deep lamentable sobbing that her father thought his heart must break to see her plunged in such great sorrow.

He sprang from his horse and took his daughter in his arms, holding her in his embrace for the last time. Again and again he made the sign of the cross over her and commended her to the keeping of God and the holy saints. At last he said that now she must let him go.

So they parted. But when he was gone a little way, Kristin saw that her father slackened his horse's pace, and she knew that he was weeping as he rode away from her.

She ran into the birch-wood, hastened through it, and began climbing up over the golden lichen-covered stones of the scree on the nearest knoll. But the stones were big, and hard to climb over, and the little hill was higher than she had thought. At last she was at the top, but by that time he was gone from sight among the low hills. She laid her down in the moss and bear-berry heath that grew on the top of the knoll, and there lay long weeping, with her face hidden in her arms.

* * *

Lavrans Björgulfsön came home to Jörundgaard of an evening late. A cheering little warmth passed over him when he saw folks were still up in the hearth-room house — there was a faint flicker of firelight behind the tiny pane of glass in the pent-house wall. He had always felt that in this house was most of home.

Ragnfrid sat alone in the room with a great seam of work before her on the table — a tallow candle on a brass holder stood by her. She rose to her feet at once, bade him good-evening, put more wood on the hearth-fire, and went herself to fetch food and drink. No — she had sent the maids to bed long ago — they had had a hard day; but now, at any rate, they had barley-bread ready baked to last till Yule. Paal and Gunstein had gone to the hills to gather moss. Speaking of moss — would Lavrans have his winter clothes made from the litmus-dyed web of cloth or from the heather-green? Orm of Moar had been there that morning asking to buy some leather rope. She had taken out the ropes that hung next the door in the shed, and said he could have them as a gift. Aye — his daughter was going on a little better now — the wound on her leg was healing up well. . . .

Lavrans answered or nodded, as he and his man ate and drank. But the master was soon done eating. He stood up, dried his knife on the back of his thigh, and took up a bobbin that lay by Ragnfrid's place. The thread was wound on a pin that was carven into a bird at each end — one of these had had a piece of its tail broken off. Lavrans rounded off the break, and carved a little on it, making the bird dock-tailed. At one time, long ago, he had made a great many such bobbin-pins for his wife.

"Must you mend these yourself?" he asked, looking at her work. It was a pair of his leather hose; Ragnfrid was sewing patches on the inside of the thighs, where they had been worn by the saddle. " 'Tis stiff work for your fingers, Ragnfrid."

"Oh — " His wife laid the pieces of leather edge on edge and bored holes in them with her awl.

The servant bade good-night and went out. Man and wife were alone. He stood by the hearth warming himself, with one foot on the hearth-rim and a hand on the pole of the smoke-vent. Ragnfrid looked across at him. And she grew aware that the little ring with the rubies — his mother's betrothal-ring — was gone from his hand. He saw that she had marked it.

"Aye — I gave it to Kristin," said he. "It has ever been meant for her — methought 'twas as well she should have it now."

At times thereafter one of them would say to the other — maybe 'twas time they went to bed. But he stood up where he was, and she sat on at her work. They spoke some words about Kristin's journey; about work that was towards on the farm, about Ramborg and Simon. Then they said somewhat again about its being, perhaps, well to go to rest — but neither of them moved.

Then Lavrans took the gold ring with the blue and white stone from off his right hand, and went across with it to his wife. Shyly and awkwardly he took her hand and slipped the ring on to it — he had to change it once or twice before he found the finger it fitted. It came to rest on the middle finger, above her wedding-ring.

"I would have you take this now," he said, low, without looking at her.

Ragnfrid sat still as a stone — her cheeks blood-red.

"Why do you this?" she whispered at last. "Think you I grudge our daughter her ring?"

Lavrans shook his head and smiled a little:

"Oh, methinks you know why I do it."

"You said before, this ring you would take with you to the grave," she said in the same whisper. "None was to bear it after you were gone — "

"Therefore must you never take it from off your hand, Ragnfrid — promise me that. I would not have any bear it after you — "

"Why do you this?" she asked again, holding her breath.

The man looked down into her face.

"Last spring 'twas four and thirty years since we were wedded to each other. I was not yet a full-grown man — and all through my manhood you were at my side, both when sorrow came and when things went well with me. God help us — all too little did I understand how heavy was the burden you bore, while we lived together. But methinks now that it ever seemed good to me that you were there. . . .

"I know not whether 'tis so that you have deemed I held Kristin dearer than you. True it is that she was my greatest joy and that she brought me my worst sorrow. — But you were mother to them all. It seems now to me that the worst of all will be to leave you, when I go hence. . . .

"Therefore must you never give my ring to any — not to either of our daughters even — but say that they must leave it on your hand.

"Maybe you deem, wife, you have had more sorrow with me than joy — and in a way things went wrong between us; but yet methinks through all we have been faithful friends. And I have thought that hereafter we shall find each other again in such wise that the wrong will not part us any more, but the love that was between us God will build up again, better than before — "

The wife lifted her pale, furrowed face — her great sunken eyes burned as she looked up at her husband. He held her hand still; she looked at it as it lay in his, lifted a little. The three rings gleamed, one above the other — lowest down the betrothal-ring, then the wedding-ring, and above it this one.

It came over her so strangely. She remembered when he had put the first on her hand — by the hearth in the hall of her Sundbu home, their fathers standing by them. He was white and red, round-cheeked, scarce out of his boyhood — a little bashful as he stepped forward from Sir Björgulf's side.

The other he had set on her finger before the churchdoor at Gerdarud, in God's triune name, under the priest's hand.

She felt it — with this last ring he had wedded her again. When, in a little while, she sat over his lifeless body, he willed she should know that with this ring he had espoused to her the strong and living force that had dwelt in that dust and ashes.

She felt as though her heart was cloven in her breast, and bled and bled, wildly as in youth — for sorrow for the warm and living love she still secretly bemoaned that she had missed, for fearful joy in this pale, shining love that drew her with it towards the uttermost bounds of the earthly life. Through the pitchy darkness that was coming she saw the glimmer of another, milder sun, she smelt the scent of the herbs in the garden at the world's end. . . .

Lavrans laid his wife's hand back in her lap, and sat down on the bench, a little way from her, with his back to the board, and one arm upon it. He looked not at her, but gazed into the hearth-fire.

When she spoke again, her voice was calm and quiet:

"I had not thought, my husband, that I had been so dear to you."

"Aye, but you were"; he spoke as evenly as she.

They sat silent awhile. Ragnfrid moved her sewing from her lap to the bench beside her. In a while she asked in a low voice:

"The thing I told you that night — have you forgotten it?"

"Such things a man cannot forget in this earthly home. And so it is, I have felt myself, that things grew not better between us after I had come to know it. Though God knows, Ragnfrid, I strove hard that you might never mark I thought so much of it — "

"I knew not that you thought so much of it."

He turned sharply towards his wife and looked at her. Then said Ragnfrid:

" 'Twas my fault that things grew worse between us, Lavrans. Methought that since you could be to me in all ways as you were before — after that night — then must you have cared about me even less than I had believed. Had you grown to be a hard husband to me — had you struck me, if only once and when you were drünken — then could I have better borne my sorrow and my remorse. But that you took it so lightly — !"

"Did you deem that I took it lightly?"

The faint tremble in his voice made her wild with longing. She longed to plunge herself within him, to sound the unquiet depths from which his voice came forth strained and labouring. She flamed up:

"Aye, had you taken me in your arms one single time, not for that I was your Christian wedded wife that they had laid at your side, but that I was the wife you had longed for and fought to win — never then could you have been to me as though those words had been unspoken."

Lavrans thought a space:

"No. It — may be I could not. No."

"Had you joyed in the betrothed that was given you, as Simon joyed in our Kristin — "

Lavrans made no answer. In a little he said softly, as though against his will and in fear:

"Why named you — *Simon?*"

"Nay, it could not come to my mind to liken you with the other," answered the wife, herself somewhat confused and fearful, but trying to smile: "You and he are too unlike."

Lavrans rose to his feet and walked a few steps, restlessly — then he said yet lower:

"God will not forsake Simon."

"Seemed it never to you," asked his wife, "as though God had forsaken *you?*"

"No."

"What were your thoughts on that night we sat there in the barn — when you learned in *one* hour that we whom you had held dearest and loved most faithfully, we had both been false to you as we could be — "

"I thought not much, I trow," said the man.

"And since," his wife went on, "when you thought upon it always — as you say you did — "

Lavrans turned away from her. She saw a flush spread over his sun-burned neck.

"I thought on all the time I had been false to Christ," he said very low.

Ragnfrid rose — she stood still a moment before she ventured to go to her husband and lay her hands on his shoulders. When he put his arms around her, she bowed her forehead against his breast; he felt that she was weeping. He drew her close in to him and pressed his face down on her head.

"Now, Ragnfrid, we will go to rest," he said in a little.

Together they went over to the crucifix, bowed before it, and crossed themselves. Lavrans said over the evening prayers; he spoke, low and clearly, in the Church's tongue, and his wife spoke the words after him.

They took off their clothes. Ragnfrid lay down on the inner side of the bed — the pillows were made up now much lower than of old, because her husband in these latter days had been often troubled with dizziness. Lavrans bolted and barred the room-door, scraped ashes over the fire on the hearth, blew out the light, and lay down beside her. They lay in the darkness, their arms touching. In a little the fingers of their hands twined together.

Ragnfrid Ivarsdatter thought — 'twas like a new bridal night, and a strange bridal night. Happiness and unhappiness flowed together and lifted her up on waves so mighty that she felt within her now the first loosening of the roots of her soul — now had death's hand given her, too, a wrench — the first time.

"Speak to me, Lavrans," she prayed him softly. "I am so weary — "

The man whispered:

"*Venite ad me, omnes qui laboratis et onerati estis. Ego reficiam vos,** hath the Lord said."

He passed an arm around her shoulders and drew her in to his side. They lay a little, cheek against cheek. Then she said softly:

"Now have I prayed God's Mother to make for me this prayer, that I may not outlive you, my husband, many days."

His lips and eyelashes touched her cheek in the darkness as lightly as the touch of butterflies' wings.

"My Ragnfrid, my Ragnfrid — "

8

KRISTIN stayed at home at Husaby this autumn and winter and would go nowhither — she gave for a reason that she was ailing. But she was only tired. So tired she had never been before in her life — tired of being merry and tired of sorrowing, and tired, most of all, of brooding.

It would be better when this new child had come, she thought — she longed much for it; it was as though it was to be the saving of her. If it were a son, and her father died before its birth, it should bear his name. And she thought how she would love this child, and nurse it at

---

* Matthew xi. 28.

her own breast — it was so long since she had nursed a child that she could weep with longing when she thought that now she should soon have a suckling in her arms again.

She gathered her sons about her knee again as she had been used to do in early days, and strove to make their upbringing somewhat more orderly and mannerly. She felt that in this she was obeying her father's wish, and this brought some peace into her soul. Sira Eiliv had begun now to teach Naakkve and Björgulf their letters and the Latin tongue, and Kristin often sat over in the priest's house when the children were there at their book. But as scholars they showed not much thirst for knowledge, and all the children were wild and unruly, save Gaute alone, so that he still was his mother's poppet, as Erlend called it.

Erlend came back from Denmark about All Saints' tide, in high feather. He had been entertained with the greatest honour by the Duke and his kinswoman Lady Ingebjörg; they had thanked him right heartily for his gifts of silver and furs; he had ridden in the jousts, and chased the hart and the hind; and when they parted, Sir Knut had bestowed on him a coal-black Spanish stallion, while the Lady had sent loving greetings and two silver-grey greyhounds to his wife. Kristin deemed that these outlandish hounds looked faithless and treacherous, and she was afraid they might do her children hurt. And the folk all through the country-side talked much of the Castilian steed. 'Twas true, Erlend looked well on the long-legged, light-built horse; but such beasts suited not this northern land, and God only knew how the stallion would get about on the hill-paths. Howbeit, Erlend bought up now, wherever he went about in his charge, the bravest black mares, till he had made him a stud that was fair to look on at the least. In other days Erlend Nikulaussön had used to give his riding-horses fine outlandish names: Belkolor and Bayard and suchlike; but this horse, he said, was so rare a beast that he needed no such adornment — plain Soten * should be his name.

Erlend chafed much that his wife would not go about with him anywhither. Sick he could not mark that she

---

* Soten = Soot.

was — she neither swooned away nor vomited this time —
no sign was to be seen upon her yet — and most like her
paleness and weariness were from evermore sitting indoors
brooding and pondering over his misdeeds. Yule-tide came
round — and hot quarrels arose between them. And now
Erlend no longer came afterward to beg forgiveness for
his hot temper, as he had ever used to do before. Till now
he had thought always, when there was strife between
them, that 'twas he was at fault. Kristin was good; she
was ever in the right; and if he wearied and was ill at ease
in his home, it was but that his nature was such that he
grew weary of what was good and right when he had too
much of it. But last summer he had marked more than
once that his father-in-law held with him, and seemed to
deem that Kristin was lacking in wifely gentleness and
forbearance. And then it came into his thought that she
took things to heart in a petty way and was hardly brought
to forgive him small misdeeds that had not been so ill
meant on his part. Always had he prayed her for forgiveness when he had had time to think — and she had
said that she forgave, but afterward he had been made to
see that 'twas hidden away, but not forgotten.

So he was much from home, and now he often took his
daughter Margret with him. The maid's upbringing had
ever been one of the things that set them at odds. Kristin
had never said aught to it, but Erlend knew well what
she — and others — thought. He had dealt with Margret
in all ways as she had been his true-born child, and folk
made her welcome as though she were such when she went
about with her father and stepmother. At Ramborg's
wedding she had been one of the bridesmaids, and had
borne a golden garland on her flowing hair. Many of the
women liked this but ill, but Lavrans had talked them
round, and Simon, too, had said that none must say aught
against it to Erlend or speak of it to the maid herself —
'twas not the fair child's fault that she was so luckless in
her birth. But Kristin saw that Erlend had planned to
wed Margret to an esquire bearing arms, and that he
believed, with the standing he now had, he would be able
to bring this about, although the maid was begotten in
adultery, and 'twould be a hard matter to win for her a

safe, firm footing amongst good folk. It might, perchance,
have been done if folk had felt any right faith that Erlend had it in him to keep up and increase the might and
riches in his house. But though Erlend was liked and
honoured after a fashion, 'twas as though no one had full
faith that the fair fortunes of Husaby would endure. Thus
Kristin feared he would be hard put to it to bring his
plans for Margret to a good end. And though she liked not
Margret overmuch, yet she pitied the maid, and dreaded
the day when, maybe, her pride must be broken — if she
must be content with a much humbler match than her
father had taught her to look for, and quite another way
of life than the one he had nurtured her in.

Thus things were, when just after Candlemas three men
from Formo came to Husaby; they had come across the
hills on ski in haste, bringing Erlend letters from Simon
Andressön. Simon wrote that their father-in-law had now
fallen so sick that 'twas not to be looked for that he could
live long; and that Lavrans had bade him pray that Erlend
would come to Sil if 'twere possible for him; he was fain
to have speech with his two sons-in-law of how all things
should be ordered when he was gone.

Erlend went about casting stolen glances at his wife.
She was far gone with child now, pale and thin-faced —
and she looked so sorrowful — every moment the tears
would be welling up. He began to repent his behaviour to
her in the winter — her father's sickness had not come on
her unlooked for, and if so it was that she had had to bear
about this secret sorrow, he could better forgive her un-
reasonable ways.

Alone he could have made the journey to Sil and back
swiftly enough, going on ski across the mountains. But if
he took his wife with him, 'twould be a slow and toilsome
business. And then he would need to tarry till after the
wapinschaws * in Lent, and set trysts there with his sheriffs; and there were some meetings, too, that he must
attend himself. Before they could set forth, 'twould be
perilously near the time she looked to be delivered — and
Kristin, too, who could nowise endure the sea even when

---

* Wapinschaw = Arms-muster.

she was well! But he was loath to think of her not seeing
her father before he died. In the evening, when they had
lain down, he asked her if she dared make the journey.

It seemed to him he was well repaid when she threw
herself weeping in his arms, full of thankfulness, and of
penitence for her unfriendliness towards him in the winter. Erlend grew soft and tender, as he ever was when he
had brought grief to a woman and had to see her sorrow
it out before his eyes; and he bore well enough afterward
with Kristin's fantasies. He had said from the first, the
children he would not have with them. But the mother
would have it that Naakkve was so great a boy now,
'twould be well for him to see his grandfather's going
hence. Erlend said: No. Then she was sure that Ivar and
Skule were too small to be left behind in the serving-
women's change. No, said their father. And then Lavrans
had been so fond of Gaute. No, said Erlend — 'twas hard
enough as it was — things being as they were with her
— for Ragnfrid to have a childbirth in the house, while
her husband lay on his sick-bed — and for themselves on
the journey homeward with a new-born babe. Either must
she give the child out to nurse on one of Lavrans' farms,
or she must tarry at Jörundgaard till 'twas summer — but
then he must come home before her. He had to put all
this to her, over and over again; but he strove to speak
calmly and reasonably.

Then he bethought him that he should take from Nidaros one thing and another that his mother-in-law might
need for the grave-ale — wine and wax, wheat flour and
millet and the like. But at length they were ready to set
forth, and they came to Jörundgaard the day before
Gertrude's Mass.

But Kristin found 'twas far otherwise than she had
thought to be at home at this time.

She should have rejoiced with all her heart that it had
been given her to see her father once again. And when
she remembered his joy at her coming and how he had
thanked Erlend for it, she was indeed glad. But she felt
that she was shut out now from so much that passed, and
this hurt her.

It was but a short month before her time; and Lavrans

utterly forbade that she should have the least hand in
nursing or tending him; they would not let her watch by
him at night in turn with the others, nor would her mother
suffer her to move a finger towards helping in all the
press of work. She sat by her father all day, but 'twas
seldom that they were alone at any time. Almost daily
there came guests to the house — friends come to see
Lavrans Björgulfsön once again in life. These visits pleased
her father, though they made him exceeding weary. He
talked heartily and cheerfully with all, men and women,
rich and poor, young and old — thanked them for their
friendship, asked their prayers for his soul — and God
grant that we may meet on the day of bliss! At night,
when only his own folk were with him, Kristin lay above
in the upper hall, staring into the dark, and could not
sleep for thinking on her father's going and on her own
heart's folly and wickedness.

Lavrans' end was swiftly drawing nigh. He had kept
himself up and about till Ramborg had borne her child,
and there was no more need for Ragnfrid to be so much
at Formo; he had had himself driven down there one day
and seen his daughter and granddaughter; Ulvhild the
little maid had been called. But after that he took to bed,
and it was not likely he would ever rise again.

He lay in the great room under the upper hall. They
had made a kind of bed for him there on the bench of
the high seat, for he could not endure to have his head
pillowed high — when it was so, he grew dizzy at once
and had swooning-fits and heart-spasms. They dared not
let him blood any more; they had had to do it so often
all through the autumn and winter that now he was quite
drained of blood; and he could eat and drink but little.

Her father's fine and comely features were sharp now,
and the brownness was faded from his face that had
weathered before to so fresh a hue; it was yellow now as
bone, and the lips and corners of the eyes pale and bloodless. The thick hair, flaxen but powdered with white, lay
unclipped, withered and strengthless over the blue-pat-
terned pillow-cover, but what changed him most was the
coarse grey stubble growing now on the lower face and
on the long wide throat, where the sinews stood out like
strong cords. Lavrans had always been so nice in shaving

himself before each holy day. His body was wasted till it was little more than a skeleton. But he said that he was easy so long as he lay stretched out and moved but little. And he was cheerful and glad at all times.

They slaughtered and brewed and baked for the grave-ale; had out bed-gear for all the beds and went through it with care — all that could be got out of hand now was done, so that all might be still and quiet when the last struggle came. It cheered Lavrans greatly to hear of all that was making ready — his last feast would not be the least of all the festivals that had been held at Jörundgaard; in honour and worship would he depart from the governance of his lands and his people. One day he had a mind to see the two cows that were to follow in his funeral train, as gifts for Sira Eirik and Sira Solmund; and they were led into the hall. They had been given double feed all the winter, and they were, in sooth, as fine and fat as sæter cattle at Olav's Mass, though it was now the midst of the spring dearth. He was the one that laughed most when one of them dropped somewhat on the hall-floor. — But he was fearful that his wife would be altogether worn out. Kristin had thought she herself was a notable housewife — she had the name of one at home in Skaun; but it seemed to her now that beside her mother she was naught. None could understand how Ragnfrid was able to compass all she did — and yet she seemed never to be long away from her husband; she took her share of watching every night.

"Never trouble about me, husband," she said, laying her hand in his. "When you are dead, you know that I shall rest altogether from all such cares."

Lavrans Björgulfsön had bought him some years back a resting-place in the church of the Preaching Friars at Hamar, and Ragnfrid Ivarsdatter was set on going thither with his corpse and dwelling there near by it; she was to be a commoner in a hospice the monks owned in the town. But first the coffin was to be taken into the Olav's Church here at home, with great gifts to the Church and the priests; his stallion was to be led after with his armour and weapons, and these were to be redeemed by Erlend for five-and-forty marks of silver. One of his and Kristin's sons would most like be given these things — for choice the child she was soon to bear, if 'twere a son — perhaps he might some day be Lavrans of Jörundgaard, said the sick man with a smile. On the way down through Gudbrandsdal, too, the body was to be taken into certain churches over-night — and these were all remembered in Lavrans' testament with money-gifts and wax tapers.

Kristin felt sick, but it was with sorrow and disquiet of soul. For she could not hide from herself — it hurt her the more, the longer she was at home. Such was her heart that it hurt her to see that, now her father was drawing nigh to death, 'twas his wife that was nearest to him of all.

Ever had she heard her parents' life together held up as a pattern of seemly and worthy wedlock in unity, troth and loving-kindness. But she had felt, though without thinking upon it, that none the less there was somewhat that stood between them — an uncertain shadow, but it dulled the life in their house, though they lived together in peace and kindness. Now was there no shadow any more between her parents. They talked evenly and quietly together, mostly of the little things of every day; but Kristin felt that there was something new in their eyes and in the tone of their voices. She saw that her father missed his wife always when she was not by his side. When he had himself talked her over into going to seek a little rest, he would lie as if waiting somewhat restlessly; and when she came in, 'twas as though peace and gladness came with her to the sick man. One day she heard them speaking of their dead children; yet did they look happily. When Sira Eirik came over and read to Lavrans, Ragnfrid sat ever beside them, and then he would often take his wife's hand, and lie playing with her fingers and turning the rings on them.

She knew that her father loved her not less than before. But she had not seen clearly before now that he loved her mother. And she understood how unlike must be the man's love for his wife, who had lived with him a long life in evil days and good, to his love for the child who had but shared his joys and taken to herself his inmost heart's tenderness. And she wept and prayed God and St. Olav for help — for she remembered that tearful and tender farewell last autumn on the hills, but it *could* not be true that she wished now that farewell had been the last!

On the day that begins the summer half-year * Kristin bore her sixth son, and already, the fifth day after, she was up and had gone across to the hall to sit with her father. Lavrans misliked this — it had never been the use in his house for a lying-in woman to come into the open air before the day she went to be churched. At least, he said, she must never cross the courtyard except when the sun was in the heavens. Ragnfrid was listening when he spoke of this.

"I was thinking but now, husband," she said, "that from us, your womenkind, you have never had great obedience, but we have most often done as we ourselves would."

"Knew you not that before?" asked her husband, laughing. "'Tis not your brother Trond's fault, then — mind you not that he would rail at me always for an old woman because I let you women have the upper hand?"

Sira Eirik came over daily to the dying man. The old parish priest's sight was nearly gone, but the story of the creation in the Norse tongue, and the evangels and psalter in Latin, he read as plainly and flowingly as ever, for he knew the books so well. But Lavrans had gotten a great book some years before down at Saastad, and he was most fain to be read to from it — but Sira Eirik could not manage to read in it, by reason of his bad eyes. So her father prayed Kristin to try if she could read it. And when she was grown a little used to the book, she found, sure enough, that she could read from it fairly and well, and it was a great joy to her that now there was somewhat that she could do for her father.

In this book were such things as debates between Fear and Courage, between Faith and Doubt, Body and Soul. There were likewise some sagas of holy men, and more than one account of men who, while yet alive, had been rapt away in the spirit and had seen the pains of the place of torment, the tribulations of purgatory fire, and the bliss of the heavenly kingdom. Lavrans spoke much now

* Summer day = 14th April.

of purgatory fire, which he looked soon to enter; but he was quite without fear. He hoped for great solacement from his friends' and the priests' intercessions, and trusted firmly that St. Olav and St. Thomas would give him strength in this his last trial, as he had so often felt that they had strengthened him in this life. He had ever heard that he that was firm in the faith would never for a single moment lose from before his eyes the bliss to which the soul was going through the scorching fires. Kristin deemed that her father thought with gladness of what was coming, as of a trial of his manhood. She remembered dimly from her childhood that time when the King's sworn men from the Dale set out for the war against Duke Eirik — it seemed to her that now her father looked forward to his death as he had looked forward then to adventures and battles.

One day she said to him she deemed he had had so many trials in this life that 'twas like he would come off lightly from those of the life to come. Lavrans answered: it seemed not so to him now; he had been a rich man; he had been born of a noble house; friends had he had and good advancement in the world. "My heaviest sorrows were that I never saw my mother's face, and that I lost my children — but soon these will be sorrows no longer. And so it is with other things that have weighed on me while I lived — they are no longer sorrows."

Her mother was often with them while Kristin read; strangers, too; and Erlend now was glad to sit and listen. All these folks had delight of the reading, but she herself was shaken and made hopeless by it — she thought on her own heart that knew so well what good and right were, and yet was ever intent on unrighteousness. And she feared for her little child — scarce dared sleep at night for fear it might die a heathen. She had two women ever to sit up and watch, and yet was she afraid to fall asleep herself. Her other children had all been baptised before they were three days old; but they had put off this one's christening, since it was a big, strong child, and they would fain name it after Lavrans — and in the Dale here folk held stiffly to the custom that children must not be named after living men.

One day when she sat by her father and had the child

in her lap, he prayed her to unwrap its swaddling-clothes; for as yet he had seen no more than the little lad's face. She did so and laid the boy on her father's arm. Lavrans stroked the little rounded breast and took one of the small tight-clenched hands in his:

"Strange it is, kinsman, that you are to bear my breast-plate — now would you fill no more room in it than a worm in a hollow nut, and this hand has a great way to grow before it can grasp my sword-hilt round. When a man sees such things as this little knave, he well-nigh comes to understand that God's will with us was not that we should bear arms. But not much greater shall you be, you little one, before you long to take them up. 'Tis but the fewest men born of women that bear so great love towards God that they will forswear the bearing of arms. I had not such love."

He lay a little, looking at the babe.

"You bear your children under a loving heart, my Kristin — the boy is great and fat, but you are pale and thin as a wand, and so, your mother says, it was with them all when you were delivered of them. Ramborg's daughter was little and thin," he said, laughing, "but Ramborg blooms like a rose."

"Yet seems it to me strange that she would not suckle her child herself," said Kristin.

"Simon would not have it either — he says he will not repay her for the gift by letting her wear herself away. Bear in mind, Ramborg was not full sixteen years — she had scarce worn out her own childish footgear when she had this daughter — and never had she felt an hour's sick-ness before — 'twas no marvel that her patience was short. You were a grown woman when you were wedded, my Kristin."

Of a sudden Kristin fell into a wild weeping — she scarce knew herself what she wept for so. But it was so true — she had loved her children from the first hour that she knew she bore them in her womb; she had loved them while they plagued her with unrest, weighed her down and made her uncomely. She had loved their little faces from the first moment she saw them, and had loved them every hour as they grew and changed. But none had rightly loved them with her, and joyed with her in them

— 'twas not Erlend's way — he was fond of them enough; but Naakkve he deemed had come too early, and as each of the others' came, he had ever thought him one too many. She dimly remembered what she had thought of the fruit of sin the first winter she was at Husaby — she knew that she had been forced to taste its bitterness, though in other wise than she had feared. Something had gone awry between her and Erlend in those first days, and 'twas like it could never be made straight again.

"What it is now, Kristin?" asked her father quietly, in a while. She could not tell him all this. So she said, as soon as she could speak for weeping:

"Should not I sorrow, father, when you lie here — ?"

At length, when Lavrans pressed her, she spoke of her fear for the unchristened child. On this he gave order straightway that the child should be brought to church the next holy-day — he said he believed not that this would slay him before God's good time.

"Besides, I have lain here long enough now," he said, laughing. "Sad work there is over our coming and our going, Kristin — in sickness are we born and in sickness do we die, he that dies not on a sudden. To me when I was young the best death seemed to be slain on the battle-field. But a sinful man may well have need of the sick-bed — though now I cannot feel that my soul is like to grow stronger through my lying on here — "

So the boy was christened the next Sunday, and was given his mother's father's name. Kristin and Erlend were much blamed for this in the country-side, though Lavrans Björgulfsön said to all who came to see him that 'twas done at his desire: he would not have a heathen in his house when death came to the door.

Lavrans began now to grow fearful lest his death should fall in the midst of the spring sowing, thereby putting to great hardship the many folk who would be fain to honour his funeral by following in its train. But one morning, fourteen days after the christening, Erlend came to Kristin in the old weaving-house, where she had lain since the boy's birth. It was well on in the morning, past the dinnertime; but she was still abed, for the boy had been restless. Erlend was much moved; he said to her, quietly and lovingly, that she must get up now and come to her father.

Lavrans had had some fearful heart-spasms at daybreak, and since had lain long swooning. Sira Eirik was with him now and had just heard his confession.

It was the fifth day after Halvard's Mass. Rain was falling gently and steadily. When Kristin came out into the courtyard, there came to her on the soft breath of air from the south the smell of fields new-ploughed and dressed. The country-side lay brown under the spring rain, the air was blue between the high mountains, and the mists drifted along halfway up the hill-sides. A tinkling of little bells came from the thickets along the brimming grey river — the flocks of goats had been let loose, and were nibbling at the blossoming twigs. 'Twas the weather that had ever rejoiced her father's heart, the end of winter and cold for folk and for cattle, the beasts all set free from narrow dark byres and scanty forage.

She saw at once in her father's face that now death was very near. About his nostrils the skin was snow-white, his lips and the circles round his great eyes were bluish, the hair had fallen apart, and lay in damp strands over the broad dewy forehead. But he was in his full senses now, and spoke clearly, though slowly and in a weak voice.

The house-folk went forward to his bed, one by one, and Lavrans took each of them by the hand, thanked them for their service, bade them farewell, and prayed them to forgive him if he had ever wronged them in any wise; he prayed them, too, to think of him with a prayer for his soul. Then he said farewell to his kindred. He bade his daughters bend down so that he might kiss them, and he called down the blessing of God and of all the saints upon them. Both wept bitterly, and young Ramborg threw herself into her sister's arms; then, with arms twined round each other, Lavrans' two daughters went to their place at their father's bed-foot, the younger one still weeping on Kristin's breast.

Erlend's face quivered, and tears ran down over his cheeks when he lifted Lavran's hand and kissed it, while he prayed his wife's father in a low voice to forgive him all his sins against him at all times. Lavrans said he did so with all his heart, and he prayed that God might be with him all his days. There was a strange pale light over

Erlend's comely face when he came softly round and stood at his wife's side, hand in hand with her.

Simon Darre wept not, but he knelt down when he took his father-in-law's hand to kiss, and he stayed kneeling a little while holding it fast. "Warm and good is your hand, son-in-law," said Lavrans with a faint smile. Ramborg turned her to her husband when he came to her side, and Simon threw his arm about her slender girlish shoulders.

Last of all, Lavrans bade farewell to his wife. They whispered some words to each other that none could hear, and exchanged a kiss in the sight of all, as was fitting and seemly, since death was in the room. After this, Ragnfrid kneeled down in front of her husband's couch, with her face turned towards his; she was white and calm and still.

Sira Eirik tarried on after he had anointed the dying man with the sacred oil and given him the viaticum. He sat by the bed-head, saying over prayers; Ragnfrid was sitting now on the bed's edge. Some hours went by. Lavrans lay with half-shut eyes. Now and again he moved his head restlessly on the pillow, groped a little with his hands on the coverlid, and breathed heavily and moaningly once or twice. They deemed that he had lost power of speech, but there were no death-throes.

It grew dusk early, and the priest lit a candle. The folk sat still, looking at the dying man and listening to the dripping and trickling of the rain without the house. Then an unrest seemed to come upon the sick man, his body shook, a blue shade came upon his face, and he seemed to struggle for breath. Sira Eirik passed his arm behind his shoulders and lifted him up to a sitting posture, while he stayed his head on his own breast and held the cross up before his face.

Lavrans opened his eyes, fixed them on the crucifix in the priest's hand, and spoke softly, but so clearly that most in the room heard the words:

"Exsurrexi, et adhuc sum tecum." *

Once more some tremors passed over the body, and his hands groped on the coverlid. Sira Eirik went on holding

---

* Psalm cxxxix. 13.

him close for a little while. Then he warely laid his friend's body back on the pillows, kissed the forehead and smoothed the hair about it, before he pressed the eyelids and nostrils shut, then rose to his feet and began to pray.

Kristin was given leave to take her turn in watching the body at night. They had laid out Lavrans on straw in the upper hall; for there was most room there, and they looked to have a great gathering of folk for the wake.

Her father seemed to her unspeakably beautiful as he lay there in the tapers' light with his pale-golden visage bared. They had turned down the napkin from his face so that it might not be soiled by the hands of the many folk that came to see the corpse. Sira Eirik and the parish priest from Kvam chanted over him — the Kvam priest had come up in the evening to bid Lavrans a last farewell, but had been too late.

But already next day the guests began to come riding to the manor, and now it behoved Kristin, for seemliness' sake, to betake her to bed again, since she had not yet been to church. Now it was her turn to have her bed decked out with silken coverings and the finest cushions the house could furnish. The Gjesling cradle was brought back from Formo on loan; Lavrans the younger was laid in it, and every day people came in and out to see her and the child.

Her father's body kept fresh and sweet, she heard — it was but grown somewhat more yellow. And none had seen before so many candles brought to set about a dead man's bier.

On the fifth day began the grave-ale — and 'twas stately beyond measure in every wise — there were more than a hundred strange horses on the manor and at Laugarbru, and, besides, some guests had to lie at Formo. On the seventh day the heirs divided the lands and goods in all friendliness and concord — Lavrans had taken order for all things himself before he died, and all followed his wishes faithfully.

The next day the body, which lay now in the Olavs' church, was to be brought forth to begin the journey to Hamar.

The evening before — rather 'twas far on in the night

— Ragnfrid came into the hearth-room house, where her daughter lay with her child. The mistress of the house was exceeding weary, but her face was clear and calm. She bade the serving-women go out:

"Every house on the place is full, but I trow you will find a corner somewhere; I have a mind to watch over my daughter myself, on this my last night in my home."

She took the child from Kristin's arms, bore it across to the hearth, and made it ready for the night.

"Strange must it be for you, mother, to flit away from this place where you have lived with my father all these years," said Kristin. "I scarce understand how you can bear it."

"Much less could I bear it, methinks," said Ragnfrid, rocking little Lavrans in her lap, "to live here and not see your father going about among the houses.

"You have never heard how it came about that we flitted hither to the Dale and made our home here," she began again in a while. "The time word came that they looked my father, Ivar, should soon breathe his last, I was unfit to take the road; Lavrans had to journey north alone. I mind well 'twas such fair weather the evening he set forth — already in those days he had come to like riding late, in the cool; so he was to ride to Oslo that night; 'twas just before midsummer. I went with him to where the road from the manor cuts the church road — mind you? there are some great bare rocks there and barren soil round about — the worst lands on all Skog, they ever feel the drought first — but that year the corn grew well and fairly on those fields, and we spoke of it. Lavrans walked, leading his horse; and I had you by the hand — you were four winters old —

"When we came to where the roads joined, I told you to run home to the houses. You were loath to go, but then your father said that you should see if you could find five white stones and set them in a cross in the beck below the spring — 'twas to guard him from the trolls of the Mjörsa Wood when he sailed by there. Then you set off running — "

"Is that a thing that folks there tell?" asked Kristin.

"I have never heard it, either before or since. Your

father must have made it up, methinks, there and then.
Mind you not, he made up so many tales when he played
with you?"

"Yes, I remember."

"I went with him through the wood, all the way to the
dwarf-stone. Then he bade me turn back, and he, too,
went back with me to the cross-roads again — he laughed
and said I might have known he would not let me walk
alone through the wood, and when the sun was gone
down, too. As we stood there at the cross-roads, I put
my arms round his neck; I was so cast down because I
could not come back home — I never could thrive rightly
at Skog, and I longed ever to be back north in the Dale.
Lavrans comforted me, and at last he said: 'When I
come back, if I find you with my son in your arms, then
you may ask me for what you will, and if 'tis in the power
of man to give it, you shall not have asked in vain.' I answered: then would I pray that we might flit up hither
and dwell on the lands of my heritage. Your father liked
this but little, and he said: 'Could you not have found a
greater thing to pray for?' — he laughed a little, and I
thought: he will never do this — and it seemed to me,
too, but reason that he should not. Afterward you know
how things went with me — Sigurd, your youngest
brother, lived not an hour — Halvdan christened him and
he died straightway after. . . .

"Your father came home one morning early— he had
heard at Oslo, the night before, how things stood at home,
and had ridden straight on without tarrying. I was still
in bed; I was so sorrowful I had not the heart to rise — it
seemed to me I would liefest never have risen again. God
forgive me, when they brought you in to me, I turned
my face to the wall, and would not look on you, my little
child. But then said Lavrans, as he sat on my bedside, with
his cloak and sword still on him, that now must we try
whether things would go better for us if we lived here
at Jörundgaard — and 'twas thus we came to make our
flitting from Skog. But since it was so, you may think
that I have no mind to dwell here, now Lavrans is gone."

Ragnfrid came over with the child and laid it to its
mother's breast. She took the silk coverlid that had been

spread over Kristin's bed in the day-time, folded it up, and laid it aside. Then she stood a while looking at her daughter, and touched the thick yellow-brown plait that lay between her white breasts:

"Your father asked me so often if your hair were as thick and fair as ever. 'Twas a great joy to him that you lost not your loveliness through bearing so many children. He rejoiced much over you in these last years, that you had grown to be so notable a woman and still stood fresh and fair with all your fair little sons about you."

Kristin gulped down her tears once or twice.

"To me, mother, he would often speak of how you had been the best of wives — he said I should tell you — " She stopped abashed, and Ragnfrid laughed softly.

"Lavrans might have known that he needed not to have any bear me word of his loving-kindness towards me." She stroked the child's head and her daughter's hand, which was round the little one. "But maybe he would have — It is not so, my Kristin, that on any day I have envied you your father's love. 'Twas but right and reason that you have loved him more than me. You were so sweet and lovely a little maid — I was not grateful enough that God had let me keep you. But I ever thought more on what I had lost than on what I possessed."

Ragnfrid sat down on the bedside:

"They had other ways at Skog than our ways at home here. I cannot call to mind that my father ever kissed me — he kissed my mother when she was laid in the dead-straw. Mother kissed Gudrun at mass, for she stood nearest her, and then sister kissed me — but else we never used to kiss. . . .

"At Skog there was a custom that when we came from church after having taken the sacrament, and we alighted from our horses in the courtyard, Sir Björgulf kissed his sons and me on the cheek, and we kissed his hand. Afterward all married pairs kissed each other, and then we shook hands with all the serving-folk that had been at the service, and wished each other all good of the sacred food. And it was much their use, Lavrans' and Aasmund's, to kiss their father's hand when he made them gifts and at suchlike times. When he or Inga came in, the sons rose always to their feet and stood till they were bidden to sit down. At first these all seemed to me foolish, outlandish ways. . . .

"Afterward, in the years I lived with your father, when we lost our sons, and through all those years when we suffered such great fear and sorrow for our Ulvhild — it was well for me then that Lavrans had been nurtured so — to follow gentler and more loving ways."

In a while Kristin said, low:

"Father never saw Sigurd, then?"

"No," said Ragnfrid in the same low tones. "I saw him not either while he lived."

Kristin lay silent awhile; then she said:

"None the less, mother, so it seems to me, you have yet had much good in your life — "

Tears began to drip down over Ragnfrid Ivarsdatter's white face:

"Aye, God help me. To me, too, it seems so now."

Soon after, she took the babe, which was sleeping now, from its mother's breast, and laid it in the cradle. She fastened up Kristin's shift with its little brooch, stroked her daughter's cheek, and bade her sleep now. Kristin lifted a hand.

"Mother — " she said beseechingly.

Ragnfrid bent down, drew her daughter close in her arms, and kissed her many times. She had not done this before in all the years since Ulvhild died.

Next day 'twas the fairest spring weather, as Kristin stood behind the corner of the hall-house and looked over at the hill-sides beyond the river. The smell of growth was everywhere, and the song of becks set free; there was a tinge of green on all the woods and meadows. Where the road ran along the mountain-side above Laugarbru, a patch of winter rye shone out fresh and bright — Jon had burnt the undergrowth there last year and sowed rye in the burned plot.

She would see the funeral train best when it passed by there. . . .

And there it was, moving slowly along, below the hill-side scree, and above the fresh new rye-field.

She could make out all the priests riding ahead of all;

there were acolytes, too, in the first troop, bearing crosses and tapers. She could not see the flames in the bright daylight, but she saw the tapers themselves as slender white streaks. Then came two horses bearing between them her father's coffin on a litter; and then she could pick out Erlend on his black horse, her mother, Simon and Ramborg, and many of her kinsfolk and friends in the long funeral train.

For a while she could clearly hear the priests' chant above the roar of the Laagen, but then the sound of the hymn died away, lost in the noise of the river and the humming of the becks on the hill-sides. Kristin stood still, gazing, long after the last pack-horse with baggage was lost from sight in the wayside woods.

THE MISTRESS OF HUSABY

PART THREE

ERLEND NIKULAUSSÖN

I

RAGNFRID IVARSDATTER lived not full two years after her husband's death; she died early in the winter of 1332. It is far from Hamar to Skaun, so that they heard not aught of her death at Husaby till she had lain in her grave more than a month. But at Whitsuntide the next year Simon Andressön came to them; there was one thing and another to be talked over between the kinsfolk concerning Ragnfrid's inheritance. Kristin Lavransdatter owned Jörundgaard now, and it was settled that Simon should hold the charge of her lands and goods, and draw her farmers to account; he had managed his mother-in-law's estates in the Dale while she had dwelled at Hamar.

Just at this time Erlend had much trouble and vexation over certain cases that had come up in his Wardenship. The autumn before, Huntjov, the farmer of Forbregd in Updal, had slain a neighbour of his for calling his wife a troll-woman. The parish folk brought the slayer bound to the Warden, and Erlend had him shut up in a loft. But when the cold grew fierce in the winter-time, he let the man go about loose among his followers. Huntjov had been with Erlend in *Margygren* in the north, and had done manful service there. So when Erlend sent in letters touching Huntjov's case, praying that he be given grace * to abide at home till his case was judged, he set the man in the fairest light; and, as Ulf Haldorssön went surety that Huntjov would present him in due time at the Orkedal Thing, Erlend let the man go home for the Yule holydays. But from home he and his wife set out to visit the rest-house keeper in Drivdal — he was a kinsman of theirs — and on that journey they disappeared. Erlend believed that they had lost their lives in the great storm that had

---

* See Note 14.

raged about that time; but many folk said they had run away — the Warden's people might whistle for them now. And then new matters were brought up against the run-aways — that Huntjov had killed a man some years before away in the hills and buried his body in a scree — a man that he thought had slashed his mare on the rump. And it came out clearly that the wife had dealt in witchcraft.

Next the Updal priest and the Archbishop's commissary set to work to search and sift all these rumours of witchcraft. And this led to sorry things coming out about the way folks held by their Christian faith in many parts of the Orkdöla County. 'Twas most in the outlying parishes, like Rennabu and Updalsskog; but the case of one old man from Budvik, too, was brought before the Archbishop's Court at Nidaros. And in this Erlend showed so little zeal that there was much talk of it. It was the old carl Aan, who had lived down by the lake below Husaby, and must well-nigh be reckoned as one of Erlend's house-folk. He had dealt in runes and spells, and there was talk of some images in his hut, which folk said he had used to sacrifice to. But naught of this kind was found after his death. Erlend himself and Ulf Haldorssön had been with him, 'twas known, when he died — and doubtless they had made away with both one thing and another before the priest came, people said. Aye, and when folk came to think of it, had not Erlend's own mother's sister been charged with witchcraft, whoredom and husband-murder? — though Lady Aashild Gautesdatter was too crafty and slippery, and no doubt besides had had too many mighty friends, for aught to be proved against her. And at the same time people called to mind that Erlend in his youth had lived in sadly unchristian fashion, and had set at naught the ban of the Church.

The end of all this was that the Archbishop summoned Erlend Nikulaussön to come and confer with him at Nidaros. Simon went with his brother-in-law into the city; he was to fetch his sister's son from Ranheim, for it was meant that the boy should go back with him to the Dale, and be with his mother for a while.

It was but a week from the time set for the Frostathing, and the city was full of people. When the brothers-in-law came to the Archbishop's palace and were shown into the

hall of audience, a number of Crossed Friars were there,
and some laymen of standing — among them, the Lagmand of the Frostathing, Harald Nikulaussön; Olav Hermanssön, Lagmand of Nidaros; Sir Guttorm Helgessön,
Warden of Jemtland; and also Arne Gjavvaldssön, who
at once came up to Simon Darre and greeted him heartily.
Arne drew Simon apart with him into a window-nook and
they sat down there.

Simon was somewhat ill at ease. He had not met the
other since he had been at Ranheim ten years before, and,
though the Ranheim folk had then welcomed him most
fairly, his visit there on such an errand had left a sore
spot in his mind.

Whilst Arne was bragging of young Gjavvald, Simon
sat watching his brother-in-law. Erlend stood talking with
the Treasurer,* whose name was Sir Baard Peterssön, but
who was not of kin to the Hestnæs house. One could not
have said that Erlend's bearing lacked due courtesy; yet
he was exceeding free and unabashed as he stood there
talking with the old nobleman — swaying a little back
and forth, with his hands laid together behind his back.
As was mostly his use, he was clad in dark colours, but
most richly: violet *kothardi* sitting close to his body and
slashed up the sides, black tippet, with hood thrown back
to show the grey silk lining, silver-mounted belt, and long
red boots that were laced tight round the calves and set
off the man's slender, shapely legs and feet.

In the sharp light from the glass windows of the hall,
'twas plain enough to see that the hair at Erlend Nikulaussön's temples was not a little sprinkled with grey. Round
his mouth and beneath the eyes the fine sunburned skin
was scratched, as it were, with fine wrinkles, and cross-
furrows had appeared in the long, fairly arched neck.
Yet he seemed full young amongst the others there —
though he was in no wise the youngest man in the room.
'Twas that he was slender and lithe as ever, bore his body
in the same supple, somewhat careless fashion as in his
youth, and walked no less lightly and springily, as now,
after the Treasurer had left him, he began to stroll up

* See Note 15.

and down the room, still with his hands clasped behind him. All the other men were seated; they talked a little among themselves in low, dry voices. Erlend's light step and the jingle of his small silver spurs were too clearly heard.

At length one of the younger men testily bade him sit down, "and be a little quieter, man!"

Erlend stopped short and knit his brows, then turned to the man who had spoken.

"Where were you drinking yestereven, kinsman Jon, that your head is so sore to-day?" he said with a laugh, sitting down. When Harald Lagmand came across to him, he rose, indeed, and stood till the other had sat down, but then he dropped down by the Lagmand's side, crossed one leg over the other, and sat with his hands clasped over his knee, while Harald was speaking.

Erlend had told Simon frankly of all the trouble he had fallen into by reason of the manslayer and the witch-woman having slipped through his fingers. But no man could have seemed more care-free than Erlend, as he sat talking the matter over with the Lagmand.

Now the Archbishop entered. He was led to his high seat by two men, who propped him up with pillows. Simon had never before seen Lord Eiliv Kortin. He looked old and feeble, and seemed to be cold, though he was clad in a fur cloak and wore a fur-lined cap. When their turn came, Erlend led his brother-in-law up to him, and Simon knelt on one knee while he kissed Lord Eiliv's ring. Erlend, too, kissed the ring reverently.

He bore him most seemly and reverently, too, when at last he stood forth before the Archbishop, after the churchman had spoken a good while with the others on divers matters. But he answered somewhat lightly the questions one of the Canons put to him, and his mien was that of a man confident in his innocence.

Yes, he had heard the common talk of witchcraft for many years. But, so long as no one had come to him for guidance, surely he could in no wise be bound to search out the truth of all the talk that went on among the womenfolk in the parish. Surely 'twas the priest's affair to make inquiry, if there were grounds for making out a case against any.

Then he was asked of the old man who had dwelt at Husaby, and who folks said was a wizard.

Erlend smiled a little; yes, Aan had bragged of it himself, but no proof of his mystery had Erlend ever seen. From his childhood up he had heard Aan talk of some women he called Hæn and Skögul and Snotra, but he had never taken all this to be aught else but toys and nursery-tales. "My brother Gunnulf and our priest, Sira Eiliv, cross-questioned him once or twice, I know, but I trow they cannot have found aught against him, since they did nothing. The man came to the church each mass-day, and knew his Christian prayers." Great faith in Aan's sorceries he had never had, and since he had seen somewhat in the north of Lapp magic and spells, he had seen full well that the magic Aan dealt in was but foolery.

Then the priest asked if 'twas true that Erlend himself had once been given a thing by Aan — something that was to bring him fortune in *amor?*

Aye, answered Erlend quickly and clearly, with a smile. 'Twas when he was about fifteen, he thought — eight-and-twenty years ago or so. A skin pouch with a little white stone in it, and some dried-up things — bits of some beast, he believed. But he had not had much faith in such things in those days either — he had given it away the next year, the first year he was at the palace. It was in a bath-house up in the town — in a rash, jesting moment he had shown the charm to some other young lads — and afterwards one of the gentlemen of the guard had come to him wanting to buy it. Erlend had given it him in barter for a razor of fine steel.

It was asked who this gentleman might be.

At first Erlend would not come out with it. But the Archbishop himself bade him speak. Then Erlend looked up with a gleam of mischief in his blue eyes. " 'Twas Sir Ivar Ogmundssön."

There came a somewhat strained look into the men's faces. Strange snorting sounds came from old Sir Guttorm Helgessön. Lord Eiliv himself had some ado not to smile. Then Erlend, growing venturesome, went on, with eyes cast down and biting his under lip a little:

"My Lord, I trust you will not trouble the good knight with this ancient matter. As I have said to you, I believe

not much in the thing myself, and I have never marked
that it made any odds to either of us, my giving him this
treasure — "

Sir Guttorm doubled up in a roar, and the other men
had to give in, one after the other, and laugh aloud. The
Archbishop tittered a little, coughing and shaking his
head. It was well known that Sir Ivar's will had ever been
better than his fortune in certain matters.

In a while, however, one of the Crossed Friars grew
sober again and reminded the company they were come
together to speak of grave matters. Erlend asked a little
sharply if a charge had been laid against him from any
quarter, and if he were on his trial — he had not supposed
aught else than that he had been sent for to a friendly
conference. The talk then went on as before, but some
disorder was caused by Guttorm Helgessön bursting out
every now and then into little snickers of laughter.

The day after, when the brothers-in-law rode home
from Ranheim, Simon brought up the matter of this
meeting. It seemed to Simon that Erlend took the thing
over-lightly — he thought he had marked clearly that
more than one of the great folk there would be fain to do
him an ill turn if they could.

Erlend said he knew well enough they would, if they
had the power. For here in the north most men leaned to
the Chancellor's party — not the Archbishop, though: in
him Erlend had a trusty friend. But Erlend's dealings in
all things were conformable to the law — he took counsel
in all cases with his clerk, Klöng Aressön, who was most
skilful in such matters. Erlend spoke gravely now, but he
smiled slightly as he said that he deemed none had looked
that he should be as well skilled in the matters of his
charge as he was — neither his dear friends in this country-
side nor the lords of the Council. For the rest, he was not
sure that he cared to keep the Wardenship if 'twas to be
on other terms than those he had while Erling Vidkunssön
was at the helm. His affairs were now in such a posture —
the more so since his wife's parents' death — that he had
no need to bargain for the favour of the men that had
come to power when the King was declared of age.

Aye, that rotten boy they might just as well call of age
now as later; 'twas unlike he would grow more of a man

by keeping. One would come to know all the sooner what he was planning — he or the Swedish lords that pulled the strings. Folk would soon own that Erling had been clear-sighted after all. It would cost us dear if King Magnus tried to bring Skaane under the Swedish crown, and 'twould mean war with the Danes the moment *one* man, be he Danish or German, came to power in Denmark. And the peace in the north that was to last ten years — half the time was gone by now, and 'twas unsure whether the Russians would hold to the pact even for the five years left. Erlend had little faith in them, nor had Erling either, for that matter. Aye, Chancellor Paal was doubtless a learned man, long-headed too, in many ways, perhaps. But these gentry of the Council who had taken him for their leader — Soten here had more wit than the whole of them put together. Well, now they had got quit of Erling — for the time. And for the time, Erlend, too, had just as lief step aside. But Erling and his friends would doubtless rather that Erlend kept a hold on his powers and fortunes in the north here — so he had not made up his mind.

"Methinks you have learned now to sing Sir Erling's tune," Simon Darre could not help saying.

Erlend answered: aye, it was so. He had dwelt in Sir Erling's house last summer, when he was at Björgvin, and he had learned to understand the man better now. So it was that Sir Erling wished above all things to uphold the King's peace in the land. But he wished, too, that Norway's realm should have the lion's peace — that none should have leave to break a tooth or clip a claw of their kinsman King Haakon's lion — and that it should not be turned, either, into a trained hunting-dog for the people of another land. For the rest, Erling had it much at heart now to bring to an end the old quarrels between Norsemen and Lady Ingebjörg. Now that she had been left a widow by Sir Knut, one could not but wish that she should get some power over her son again. True it was that she bore such exceeding great love to the children she had borne to Knut Porse that it seemed she had in some measure forgotten her eldest son, but doubtless all this would be changed when she came to meet with him again.

Simon deemed that all this sounded as though Erlend were well-informed of what was afoot. But he wondered at Erling Vidkunssön — did the fallen High Steward believe that Erlend Nikulaussön had the wit to form a judgment in such things, or was it that Erling was catching now at any straw within reach? Like enough, the Knight of Giske was loath to loose his grasp on power. None could ever have said of him that he used it for his own profit, but then, with his riches and his standing, he had no need to do so. And all said that as the years of his Stewardship went on he had grown more and more self-willed and wise in his own conceit, and as the other lords of the Council began more and more to withstand him, he had at last grown so masterful that he would scarce deign to listen to a word from any man.

It was like Erlend that he had now, so to speak, gotten aboard Erling Vidkunssön's ship with both feet — just as it had met with head winds and it seemed most unsure whether his throwing himself with all his heart on his rich kinsman's side would profit either Sir Erling or Erlend himself. Yet Simon could not but confess to himself that, rashly as Erlend talked both of people and of affairs, there seemed to be a kernel of good sense in what he said.

But that night he was in a wild and reckless mood. He was dwelling now in Sir Nikulaus' mansion, which his brother had given to Erlend when he took the cowl. Kristin was with him, with three of their children, the two eldest and the youngest, and his daughter Margret.

Late in the evening many folk looked in on them, amongst them some of the men who had been at the meeting at the Archbishop's the morning before. As they sat drinking at the board after supper, Erlend overflowed with noise and laughter. He had taken an apple from a dish upon the table, and he cut scrolls and scratches on it with his knife — and then rolled it across the board into the lap of Lady Sunniva Olavsdatter, who sat over against him.

The lady who sat by Sunniva's side wanted to see the apple, and snatched at it; the other would not give it up, and the two women pushed and struggled with each other, with laughter and little shrieks. But Erlend shouted out that Lady Eyvor should have an apple all to herself. Before long he had thrown apples to all the women in the company, and there were love-runes carved on them all, he said.

"You'll be ruined, lad, should you redeem all these pledges!" cried out a man.

"Then will I let them go unredeemed — 'tis not the first time I have had to," Erlend answered back; and again there was much laughter.

But Klöng, the Icelander, had looked at one of the apples and he cried out that these were not runes, but only meaningless scrolls. He would show them, he said, how runes should rightly be cut. But Erlend cried out he must do no such thing:

"For then 'tis like they would bid me lay you by the heels, Klöng — and I cannot get on without you."

In the midst of all this turmoil, Erlend and Kristin's youngest son had come toddling into the hall. Lavrans Erlendssön was a little over two years old now, and was as comely a child as one could see, fair and fat, with silky-fine yellow curly hair. And so all the women on the outer bench were at once set on getting hold of the child — they passed him from lap to lap, and caressed him, not too gently, for they were now all heated and in wild mirth. Kristin, who sat with her husband in the high-seat against the wall, begged to have the child brought round to her, and the little one whimpered and tried to come at her, but 'twas of no avail.

Of a sudden, Erlend leapt across the table and took the child, which was shrieking now, because Lady Sunniva and Lady Eyvor were dragging at him and struggling over him. The father took the boy up in his arms, coaxing him, and as the little one still went on crying, he began hushing and lulling him, walking up and down with him out in the hall in the half-darkness. It seemed now as though Erlend had quite forgotten his guests. The child's little bright head lay on its father's shoulder under the man's black hair, and now and then Erlend would caress, with lips half opened, the little hand that rested on his breast in front. So he walked up and down till the maid came in that should have looked to the child and put him to bed long before.

Some of the guests called out now that Erlend should sing for them to dance to — he had such a fine strong voice. At first he was unwilling, but then he went over to where his young daughter sat on the women's bench. He put his arm around Margret and drew her out on to the floor.

"You must come along, then, my Margret, and dance with your father."

A young man came forward and took the maid's hand — "Margret has promised to dance with me to-night" — but Erlend lifted his daughter in his arms and set her down on his other side:

"Dance you with your wife, Haakon — never did I dance with others when I was as newly wed as you are."

"Ingebjörg says she cannot dance to-night — and I have promised Haakon to dance with him, father," said Margret.

Simon Darre had no mind to dance. He stood awhile with an old lady, looking on — now and then his glance rested a moment on Kristin. While her serving-maids were clearing the board and wiping it dry and bearing in more drink and dishes of walnuts, she stood up at the end of the table. After, she sat down by the fire-place and talked with a priest who was among the guests. In a while Simon sat down beside the two.

When they had danced one or two dances, Erlend came over to his wife.

"Come and dance with us, Kristin," he said beseechingly, holding out his hand.

"I am tired, " she said, looking up for an instant.

"Do you ask her, Simon — she cannot deny you a dance."

Simon half rose from his seat and reached out his hand, but Kristin shook her head: "Ask me not, Simon — I am so weary — "

Erlend stood there a little; he looked as though he were sorely vexed at this. Then he went back to Lady Sunniva and took her hand in the chain of dancers, while he called out that now Margret should sing for them.

"Who is he that is dancing next your stepdaughter?" asked Simon. He thought in his mind that he liked the man's looks but little — though he was a fine manful-looking young fellow with a fresh brown skin, good teeth and shining eyes — but the eyes were set close in to the bridge of the nose, and, though he had a big strong mouth and chin, his forehead and upper head were narrow. Kristin said 'twas Haakon Eindridessön of Gimsar, grandson of Tore Eindridessön, the Warden of Gauldöla County. Haakon had but just been wedded to the comely little woman that sat there in Olav Lagmand's lap — Olav was her godfather. Simon had marked this woman, for she somewhat favoured his first wife, though she was not so fair. As he found out now that there was a distant kinship, too, he went over and greeted Ingebjörg and sat down and talked with her.

The ring of dancers broke up in a while. The elders betook them to the drinking-board; but the young folks went on singing and disporting themselves out in the hall. Erlend came over to the fire-place along with some of the older men, but he still held Lady Sunniva's hand and led her with him, as if without thinking. The men sat down near the fire; there was no seat for the lady, but she stood before Erlend, eating walnuts, which he cracked for her with his fingers.

"An uncourteous man you are, Erlend, for sure," she said, suddenly. "There you sit, and I have to stand in front of you."

"Nay, do you sit too," said Erlend, laughing, and pulled her down into his lap. She struggled, laughing, and called out to the mistress of the house, asking if she saw how her husband was behaving to her.

"'Tis but the kindness of Erlend's heart," answered Kristin, laughing. "Never does my cat rub herself against his legs but he must needs take her up and lay her in his lap."

Erlend and the lady sat on as before, making no sign, but both had grown very red. He held one arm loosely about her, as if he hardly marked that she sat there, whilst he and the other men talked of the feud between Erling Vidkunssön and Chancellor Paal, which was so much in folks' thoughts just then.

"What are you thinking on now, Kristin?" Simon asked in a while — she was sitting quite still and straight, with her hands folded in her lap. She answered:

"I was thinking now of Margret."

Later in the night, when Erlend and Simon had an errand out in the courtyard, they frighted away from each other a couple that was standing behind the house-corner. The night was clear as day, and Simon knew them for Haakon of Gimsar and Margret Erlendsdatter. Erlend looked after them — he was sober enough — and Simon saw that he misliked this; but he said, as though in excuse, that those two had known each other from childhood and were for ever teasing and jesting with each other. Simon thought that even if there were no other harm, 'twas pity of the young wife, Ingebjörg.

But the day after, when young Haakon came to the house on some errand and asked after Margit, Erlend flamed out at him:

"My daughter is not *Margit* to you. And if so be you left your talk unfinished yesterday, you had best keep awhile what you have to say to her —"

Haakon shrugged his shoulders and, when he left, begged them to greet *Margareta* from him.

The Husaby folk stayed in Nidaros till the Thing was over, but Simon felt none too happy or at home among them. Erlend was apt to fall into fretful moods when in his town-house, because Gunnulf had given the hospital, which lay on the other side of the orchard, the right to use some of the houses that opened from the orchard, and also some rights in the garden. Erlend had set his mind on buying out these rights; he liked not to see the sick folk in the garden and in the courtyard — many of them, indeed, were an ill-favoured sight — and he was fearful lest his children might take some sickness. But he could not come to agreement with the monks who managed the hospital.

Then there was Margret Erlendsdatter. Simon understood that there was much talk about her, and that Kristin was disquieted by it; but the girl's father seemed not to care — it seemed that he felt sure he could guard his own maid and that there was naught to dread. Yet he named one day to Simon that he thought Klöng Aressön had a mind to wed his daughter, and he knew not rightly what to do in the matter. He had naught else against the Icelander than that he was the son of a priest — he was loath it should be said of Margret's children that there was a stain on both their parents' birth. Else was Klöng a man that all liked, cheerful, keen-witted and most learned. His father, Sira Are, had brought him up and taught him himself; he had meant his son to be a priest, and 'twas said he had even taken steps to get a dispensation for him, but Klöng had drawn back and would not take the frock. It seemed as though Erlend was minded to let the matter rest awhile — if no better match offered, he could always give the maid to Klöng Aressön.

Yet was it known that Erlend had already had such a good offer for his daughter that folk had had much to say of his pride and folly in letting the bargain slip. It was from a grandson of Baron Sigvat of Leirhole — Sigmund Finssön was the man's name; he was not rich, for Finn Sigvatssön had had eleven children who all lived; and he could not be called young — he was about Erlend's age — but he was a man in good esteem and of a good understanding. And with the lands that Erlend had given his daughter when he wedded Kristin Lavransdatter, and with all the jewellery and costly gifts he had given the child from time to time, and with the dowry he had agreed with Sigmund to give with her, Margret would have been more than well-to-do. Erlend, indeed, had been glad enough to find such a suitor for his base-born daughter. But when he came home to his daughter with the bridegroom, the maid took a whimsy that she would not have him because Sigmund had some warts on the edge of one of his eyelids, and she said this gave her such a loathing of him. Erlend gave in to the girl; and when Sigmund grew angry and talked of breach of troth, Erlend, too, grew hot, and said that the other must surely understand that all betrothal-pacts were made on condition that the maid was willing — his daughter should not be forced into her bridebed. Kristin thought with her husband in so far as that he should not force the girl, but 'twas known that she had deemed Erlend should have spoken in sober sadness to his daughter and made her understand that Sigmund Finssön was so good a match as, seeing how things stood with her birth, she could not hope to find a better. But Erlend had been wroth with his wife for venturing to speak thus, though she said it to him only. These things Simon had heard at Ranheim. The folk there foretold that all this must needs end badly; 'twas true Erlend was a man of weight now, and the maid was a passing fair maid, but yet 'twas impossible it could be for her good that her father had spoiled her all these years and fed the flames of her pride and self-will.

After the Frosta Thing, Erlend went home to Husaby with his wife and children, and with them went Simon Darre.

Now, when the eldest sons were big enough to ride abroad with Erlend, he had begun to take more heed of the boys. Simon marked that Kristin was not wholly glad of this — she deemed that 'twas not only good they got by going amongst their father's men. And it was about the children that unfriendly words most often passed between this wedded pair — even if they did not quarrel outright, they often came more nigh to it than Simon deemed fitting. And it seemed to him that Kristin was the most at fault. Erlend was quick-tempered and hasty, but she often spoke as though from a deep-hidden grudge. So it was one day when she made some complaint about Naakkve. Erlend answered that he would speak strictly to the boy, but, on his wife's saying something more, he broke out testily, saying he could not well thrash a big boy like him, because of the house-folk.

"No, 'tis too late now; had you done it while he was younger, he might have hearkened to you now. But in those days you never so much as looked his way."

"Oh, but I did. Though it was but reason surely that I should let him go about with you when he was little — and 'tis no work for a man, I trow, to beat little breechless brats."

"You thought not so last week," said Kristin, scornfully and bitterly.

Erlend made no answer, but rose to his feet and went out. And to Simon this seemed an ill speech of Erlend's wife. She was recalling a thing that had befallen the week

before; as Erlend and Simon came riding into the courtyard, little Lavrans had come running towards them with
a wooden sword, and as he ran by his father's horse, he
struck it, in mischievous play, on the leg with his sword.
The horse reared — and next moment the boy had fallen
to the ground beneath its feet. Erlend jerked back and
flung the horse to one side, then leapt down, throwing
the rein to Simon; his face was white with fear when he
lifted the boy up in his arms. But when he saw that the
child was quite unhurt, he laid him over his left arm, took
the wooden sword, and thrashed him with it on the bare
bottom — the boy had not been breeched yet. In his
flurry he knew not how hard he struck, and Lavrans was
still going about black and blue. But since then Erlend had
tried all the time to make friends again with the boy —
while the little man sulked, held to his mother's skirts,
and threatened and slapped at his father. And when Lavrans had been put to rest in the evening in his parents'
bed, where he slept (for he was still nursed by his mother
at night), Erlend sat over on the bedside the whole evening,
looking down on the sleeping child and touching him. He
said himself, to Simon, that this boy was the one of his
sons that he loved best.

When Erlend set forth for the summer meetings in his
charge, Simon took the road for home. He galloped south
through Gauldal, so that the sparks flew from the stones
under his horse's hoofs. Once, when they rode a little
slower up a steep hill, his men asked, laughing, whether
they were to ride three days' march in two. Simon laughed
back, and said he had more than a mind they should —
"for now am I fain to be back at Formo."

He ever longed to be back when he had been away from
his manor awhile — he was a home-loving man, and rejoiced always when he turned his horse into the homeward
road. But it seemed to him that so much as this time he had
never before yearned to get back to the Dale, and his
manor, and his little daughters — aye, and now he longed
for Ramborg too. It seemed to him that he had no good
reason for this great eagerness, but the life at Husaby had
so weighed upon his mood that he deemed he knew now
from himself how the cattle feel when a storm is gathering.

2

ALL the summer through, Kristin thought of little else
than what Simon had told her of her mother's death.

Ragnfrid Ivarsdatter had died all alone — none had been
near when she drew her last breath, saving a serving-
woman, who slept. 'Twas not much comfort, what Simon
said — that, though death came so suddenly, she was yet
well prepared. It seemed like a special providence of God
that, a few days before, she had felt in her such a hunger
for her Redeemer's Body that she had confessed and taken
the sacrament from the priest in the cloister who was her
director. 'Twas certain she had made a good death —
Simon had seen her body, and said it had seemed to him a
marvellous sight. In death she had been so fair; she was,
one knew, a woman nigh threescore years of age, and for
many years her face had been much wrinkled and furrowed; but this was changed altogether: her face was
grown young and smooth, so that she looked like naught
else but a young woman fallen asleep. Now had she been
laid to rest by her husband's side; thither, too, had they
brought Ulvhild Lavransdatter's bones a short time after
her father's death. Over the graves was laid a great stone
slab, divided in two by a fairly carven cross, and on a
winding scroll was written a long Latin verse that the
Prior of the cloister had made; but Simon could not remember it rightly, for he knew but little of that tongue.

Ragnfrid had had a house to herself in the yard up in the
town where the commoners of the cloister lived — a single
room and above it a fair loft-room. There she dwelt alone
with a poor peasant woman who had been taken in by the
friars for small payment, in return for her helping one or
other of the richer women-commoners. But for the last
half-year at least it had rather been Ragnfrid who helped
the other, for the widow — Torgunna was her name —
had been ailing, and Ragnfrid tended her with great kindness and care.

The last evening of her life she had been at evensong in
the cloister-church, and went afterwards into the kitchen
of the commoners' yard. There she cooked a good bowl
of soup with strengthening herbs in it, and said to the

other women who were there that she would give this to
Torgunna, and she hoped the woman would be well
enough in the morning to come with her to matins. This
was the last time any saw the Jörundgaard widow alive.
They came not to matins — neither she nor the peasant
woman — nor yet to prime. When some of the monks in
the choir marked that Ragnfrid was not in the church for
the day-mass either, they began to wonder — she had
never before missed three services in a day. They sent
word up to the town to ask if Lavrans Björgulfsön's widow
were sick. When the folk came into the loft, they found
the bowl of soup standing untouched on the board; in
the bed Torgunna was sleeping sweetly by the wall, but
Ragnfrid Ivarsdatter lay on the outside of the bed with
her hands crossed on her breast, dead, and well-nigh cold
already. Simon and Ramborg had come down to her burial,
and it was a most fair one.

Now that the household at Husaby was grown so great
and Kristin had six sons, she could no longer take a hand
herself in all parts of the housekeeping. She was obliged
to have a housekeeper under her, and so it came about that
most of the time the mistress of the house sat in the hall
sewing; there was ever someone wanting clothes — Erlend, Margret or the boys.

The last she had seen of her mother was riding after
her husband's bier — that bright spring day when she had
stood in the meadow at Jörundgaard and seen her father's
funeral train pass the green patch of winter rye beneath
the scree.

Kristin's needle flew and flew, and she thought on her
parents and their home at Jörundgaard. Now, when all
was memory, she seemed to herself to grow ware of
much that she had not seen when she lived in the midst of
it, and took as things of course her father's tender
guardianship and her silent, sad-faced mother's quiet, constant work and care. She thought on her own children —
they were dearer to her than her own heart's blood; they
were not out of her mind one hour of her waking life.
Yet was there much in her mind that she pondered over
more — she loved her children without brooding on the
matter. She had never thought aught else, when she was
at home, but that her parents' whole life and all their

doings and strivings were for herself and her sisters. Now she seemed to see that betwixt those two, who in their youth had been brought together by their fathers, well-nigh unasked, there had run strong swift currents both of sorrow and of joy — yet *she* knew naught of it save that they had passed now, hand in hand, out of her life. Now she understood that this man's and woman's lives had held much beside their love for their children — and yet that love had been strong and wide and unfathomably deep, while the love she gave them back had been weak and thoughtless and self-seeking, even when, in her childhood, those two had been her whole world. She seemed to see herself standing far, far away — so small, so small beyond that great stretch of time and distance; she stood in the beam of sunlight that streamed down through the smoke-vent in the old hearth-room house at home, the winter-house of her childhood. Her parents stood a little back, in the shadow — they bulked as great as they had seemed to her sight when she was small, and they smiled to her — the smile that she knew now comes to one's face when a little child comes and thrusts aside heavy and troublous thoughts.

"I thought, Kristin, when you had borne a child yourself, you would surely understand better."

She remembered when her mother had said these words. Sorrowfully she thought — it was not true, she feared, even now, that she understood her mother. But she began to understand how much there was she did not understand.

This autumn Archbishop Eiliv died. And about the same time King Magnus changed the terms of service of many of the Wardens, but not Erlend Nikulaussön's. When Erlend was in Björgvin the last summer of the King's nonage, he had been given letters granting him the fourth part of all grace-payments,* fines and forfeitures in his Wardenship — the thing had made much talk, that he should have been given such a grant towards the close of a Regency. Since Erlend now owned much land in the country, and most often lived on his own farms when he moved round his Wardenship, and as he let his farmers

---

* See Note 14.

redeem their land-dues, his incomings were large. True, this meant that the incomings from land-dues in kind were small; and he kept a great and costly train — besides his own manor-folk he had never fewer than twelve men-at-arms with him at Husaby; these were bravely mounted and exceeding well armed; and when he moved about his charge, his men lived like lords.

There was some talk of this one day when Lagmand Harald and Tore the old Warden of Gauldöla County were at Husaby. Erlend made answer that many of these men had been with him when he kept the marches in the north; "and there we shared alike in such cheer as was to be had — dried fish and sour small beer. Now the men I give food and clothes to know that I grudge them not white bread and strong ale; and if now and then, in a rage, I bid them to go to hell, they understand well enough that I mean not they should set forth before I lead the way myself."

Ulf Haldorssön, who was the headman of Erlend's guard now, said afterwards to his mistress that 'twas even so. Erlend's men loved him, and he had them wholly in his hand.

"You know yourself, Kristin, none should take much account of what Erlend says; 'tis what he does one must judge him by."

Another matter that made much talk was that, besides his house-carls, Erlend had men all about the country-side — and not in Orkdöla County only — that he had sworn to his service on his sword-hilt. Some time back he had received royal letters about this matter, but he had answered that these men had made up his ship's crew, and that he had taken oath from them the first spring when he was to sail for the north. Upon this it was enjoined upon him that he should loose these men from their oath at the next Thing he held to publish the judgments and decrees of the Lagthing, and that, to that end, he should summon thither the men from outside the county, bearing himself the costs of their journeys. And in truth he had sent for some of his old sailors from Möre to the Orkedal Thing; but no one heard aught of his having loosed them, or any other man who had ever been his follower, from their oaths. Howbeit the matter was not again opened; and so, as the autumn passed, the talk about it died down.

Late in the autumn Erlend journeyed south, and stayed over Yule-tide at King Magnus' Court, which was at Oslo that year. He was vexed that he could not bring his wife to go with him; but Kristin shrank from the toilsome winter journey, and stayed on at home at Husaby.

Erlend came back three weeks after Yule, bringing fair gifts to his wife and all the children. He had given Kristin a silver bell to ring for her maids; but to Margret he gave a clasp of pure gold, for she had naught of the sort before, though she had many ornaments of every kind of silver and silver gilt. But while the women were putting away these costly gifts in their jewel-chests, something in Margret's chest caught on her sleeve and hung from it. The girl covered it up swiftly with her hand, saying to her step-mother:

" 'Twas my mother left me this, so father would not that I should show it to you."

Kristin had flushed much redder than the maid. Her heart beat hard with fear, but it seemed to her that she *must* speak a word to the young girl to warn her.

In a little she said in a low, faltering voice:

" 'Tis like the gold buckle that Lady Helga of Gimsar used to wear at festivals."

"Aye — many gold things are alike," answered the girl, shortly.

Kristin locked her chest and stood still with her hands resting on it, so that Margret might not see how they shook.

"My Margret," she said softly and gently — she had to stop — but she gathered all her strength and went on:

"My Margret, bitterly have I repented — never could I joy fully in any gladness, though my father forgave me with all his heart for all that I had sinned against him — you know that I sinned against my parents for your father's sake. But the longer I live and the more I come to understand, the heavier it grows for me to remember that I repaid their goodness towards me by bringing them sorrow. My Margret, your father has been good to you all the days of your life — "

"You need not be afraid, mother," answered the girl. "I am not your own daughter; you need not be afraid that I shall ever wear out your dirty shift or stand in your shoes — "

Kristin turned a face flaming with wrath on her step-daughter. Then she clutched the cross she wore about her neck tightly in her hand, and forced back the words that were on her lips.

She went to Sira Eiliv with this the same evening after Vespers, and she looked in vain in the priest's face for a sign — had the worst befallen already, and did he know it? She remembered her own wildered youth, and she remembered Sira Eirik's visage that betrayed nothing, while he lived day by day with her and her trusting parents, with her sinful secret locked in his bosom — and herself hard and dumb under his harsh threats and warnings. And she remembered the time after she had been lawfully be-trothed to Erlend, when she herself showed her mother the gifts he had given her at Oslo. The mother's mien had been immovable in its calmness while she took things in her hand, one by one, looked at them, praised them, and laid them away.

She was in deadly, hopeless fear, and kept as wary a watch as she could on Margret. Erlend marked that there was something amiss with his wife, and one evening, when they had gone to rest, he asked if it was that she was with child again.

Kristin lay silent for a little before she answered that she believed it was so. And when her husband, on this, took her lovingly in his arms and asked no more, she could not bring herself to say that 'twas somewhat else that was weighing on her. But when Erlend whispered to her that this time she must do her devoir and give him a daughter, she had no power to answer, but lay there stiff with dread, thinking that Erlend might come to know all too soon what kind of joy a man has of his daughters.

Some nights after this the folk at Husaby had gone to bed somewhat in drink and heavy with much eating, for it was in the last days before the Fast began — and thus all slept heavily. But well on in the night little Lavrans woke in his parents' bed and, still half asleep, began to whimper and cry for his mother's breast. But the time had now come for him to be weaned. Erlend woke up, grunted angrily, but took the boy and give him milk from a cup that stood on the bed-step, and laid him down then at his other side.

Kristin had sunk again in deep drowsiness, when, of a sudden, she felt Erlend sit upright in the bed. Half awake, she asked what it was — he bade her hush, in a voice she did not know. Without a sound he slipped out of the bed; she marked that he was putting on some pieces of clothing, but when she raised herself on her elbow, he pressed her down again on the cushions with one hand while he leant in above her and took his sword, that hung above the bed-head.

He moved as silently as a lynx; but she felt that he had gone off to the ladder that led up to Margret's bower above the outer room.

For a moment she lay palsied with dread; then she sat up, found her shift and skirt, and groped in the dark for her shoes on the floor by the bed.

At the same moment a woman's shriek rang out from the loft-room — it must have been heard over the whole manor. Erlend's voice shouted a word or two — then she heard the ring of clashing swords and the trampling of feet up above — then the noise of a weapon falling on the floor, and a shriek of terror from Margret.

Kristin knelt crouching by the hearth — raked away the hot ashes with her bare hands and blew on the embers. When she had gotten the fir-root torch alight and held it up in her shaking hands, she saw Erlend high up in the darkness — he leapt down without heeding the ladder, bearing his naked sword in his hands, and ran out of the outer door.

From every side, in the darkness, the boys' heads peeped out. She went to the northern bed, where the three eldest slept, bade them lie down, and shut the bed-door. Ivar and Skule, who sat up on the bench where their beds had been made, blinking in fear and bewilderment at the light, she made creep into her own bed and shut them, too, in. Then she lighted a candle and went out into the courtyard.

It was raining — for one moment, while the light of her candle was mirrored in the wet-shining ice-crust, she saw a crowd of folk outside the door of the nearest house — the servants' quarter where Erlend's house-carls slept. Then her light was blown out — for a moment 'twas pitch-black night — but then Ulf Haldorssön came from the servants' quarter, bearing a lantern.

He bent down over a dark body that lay in a huddled heap on the wet lumpy ice. Kristin knelt down and felt the man's body with her hands — 'twas young Haakon of Gimsar — and he was swooning or dead. Straightway her hands were covered with blood. Helped by Ulf, she turned and straightened out the body. The blood was gushing from the right arm, from which the hand had been cut off.

Unawares she cast a glance upward to where the shutter of the window-hole of Margret's bower was clapping in the wind. She could not see any face up there — but 'twas exceeding dark.

While she knelt in the puddles pressing Haakon's wrist with all her might to stop the gush of blood, she was dimly aware of Erlend's men standing around, half clad. Then she saw Erlend's grey, writhen face — with the skirt of his mantle he was wiping his bloody sword — he was naked beneath and his feet were bare.

"One of you, find me a band," she said, "and you, Björn, go up and wake Sira Eiliv — we must bear him up to the priest's house."

She took the leather strap that someone reached her and wound it tight around the stump of the arm. Of a sudden Erlend said, in a wild, hard voice:

"Let none touch him! Let the man lie where himself has laid him — "

"You know well, husband," said Kristin calmly, though her heart was beating till 'twas like to choke her, "that that cannot be."

Erlend thrust the sword-point hard against the ground.

"Aye — your flesh and blood it is not — that have I been made to feel each day in all these years."

Kristin rose and spoke softly, close in to him:

"Yet would I be fain for her sake that this should be hid — if hid it can be. You men" — she turned to the men that stood about — "are true enough to your master, I trow, not to speak of this till he has told you all of how this strife between Haakon and him came about?"

All the men answered, yes. One ventured forward — they had been wakened, he said, by hearing a woman shriek as though one were ravishing her — then straightway someone had leapt down on the roof of their house, but he must have slipped on the ice-crust, for they heard something sliding down and then a heavy fall in the courtyard. But Kristin bade the man be silent. Now came Sira Eiliv running.

When Erlend turned and went in, his wife ran after him, trying to thrust past him. When he made for the loftladder, she got before him and caught him around the arms.

"Erlend — what would you do with the child?" she gasped out into his grey, wild face.

He made no answer — he tried to fling her aside, but she held fast to him.

"Stay, Erlend, stay — your child! You know not — the man was fully clad," she cried despairingly.

He gave a loud hoarse cry before answering — and she grew deathly white with horror – his words were so gross and his voice so changed by his wild agony.

Again she wrestled dumbly with the raging man that growled and gnashed his teeth together. At last she caught his eyes in the half-light:

"Erlend — let me go to her first. I have not forgotten the day when I was no better than Margret — "

Then he loosed her and, staggering back against the closet-wall, stood there quivering like a dying beast. Kristin went and lit a candle, then came back and went past him up to Margret in her bower.

The first thing the light fell on was a sword that lay on the floor not far from the bed, and, close by, a man's severed hand. Kristin tore off her head-dress, which, hardly knowing it, she had flung loosely round her flowing hair before she went out to the men. Now she threw it over those things lying on the floor.

Margret sat huddled together on the pillows of the bedhead gazing at Kristin's light with great wide-open eyes.

She held the bed-clothes up about her, but her naked shoulders shone white through the golden locks of her hair. There was much blood all about the room.

The strain of Kristin's spirit burst in a vehement fit of weeping — 'twas so miserable a sight to see the fair child amidst all this horror. Then Margret shrieked aloud:

"Mother — what will father do with me?"

Kristin could not help it — in the midst of her deep pity for the girl, her heart seemed to grow small and hard in her breast. Margret asked not what her father had done with Haakon. In a flash it came before her — Erlend lying on the ground, her father standing over him with a bloody sword, and she herself — But Margret had not moved from the spot. She could not hinder the old scornful dislike for Eline's daughter from coming up in her again, as Margret clung to her, shaking, well-nigh crazy with fear, and she sat down on the bed's edge and strove to quiet the child a little.

So they sat when Erlend came up through the trapdoor. He was fully clad now. Margret shrieked again, and hid in her stepmother's arms — Kristin looked up at her husband for a moment — he was calm now, but pale and strange of face. For the first time he looked as old as he was.

But when he said quietly: "You must go down, Kristin — I would speak with my daughter alone," she obeyed. She laid the girl down in the bed carefully, covered her up to the chin with the clothes, and then went down the ladder.

As Erlend had done, she dressed herself fully — 'twas certain none at Husaby would sleep any more that night — and set herself to quiet the frightened children and serving-women.

The next morning, in a driving snowstorm, Margret's maid went weeping off the place with all her worldly goods in a sack on her back. Her master had driven her out with the direst words, and threats that she should be flayed alive because she had sold her mistress thus.

Then he put the rest of the serving-folk to the question — had not the maids suspected mischief when Ingeleiv in the autumn and winter had begun sleeping with them, instead of in Margret's bower? And how came it that the dogs had been locked inside their house? But they denied all, as 'twas like they would.

Last of all, he took his wife to task, they two alone. Heart-sick and weary, Kristin listened to him and strove to turn aside his injustice with soft answers. She denied not that she had been fearful, and she refrained her from saying that she had not spoken to him of her fear, because she had never reaped aught but unthankfulness from him when she had tried to counsel either him or Margret for the maid's good. But she swore by God and Mary Virgin that she had never known, nor could have thought, such a thing as that this man came to Margret in the loft at night.

"You!" said Erlend scornfully. "You say yourself you mind the time when you were no better than Margret — and the Lord God in heaven knows that you have let me mark, every day of the years we have lived together, that you remembered the wrong I did you — though your will was as good as mine, and your father and not I caused much of the trouble by denying to let you wed me — *I* was willing enough from the first hour to make amends for the sin. When you saw the Gimsar gold" — he gripped her hand tightly and held it up so that the two rings she had had of him at Gerdarud glittered in the light — "knew you not what it meant? You have worn every day in these years the rings I gave you when you gave me your honour — "

Kristin was ready to sink down with weariness and sorrow; she answered in a low voice:

"I marvel, Erlend, if you still remember that time you overcame my honour — "

He buried his head in his arms, and flung himself down on the bench, tossing and writhing. Kristin sat down a little way off — she wished she could help her husband. She understood that this calamity fell yet more hardly upon him because he himself had sinned against others in the same sort as now he had been sinned against. And he, who had never been willing to look his fault in the face in any trouble he himself had caused, could never bear to take the blame for this — and there was none else but she on whom he could fasten it. But she was not so much

angered as sorrowful, and fearful of what now might come
to pass.

Now and again she was above with Margret. The girl
lay white and unmoving, staring before her. She had not
yet asked what had befallen Haakon — Kristin knew not
whether 'twas that she dared not, or that she was quite
dulled by her own misery.

Well on in the afternoon Kristin saw Erlend and Klöng
the Icelander going together through the thick-falling
snow to the armoury. But a short time passed, and Erlend came back alone. Kristin looked up a moment as he
came into the light and passed her by — afterward she did
not dare to turn her eyes toward the corner of the hall
where he hid himself away. She had seen that he was
quite broken.

Soon after, when she had an errand over at the storehouse, Ivar and Skule came running and told their mother
that Klöng the Icelander was going away that evening —
the boys were sad about it, for the clerk was a good friend
of theirs. He was packing his things now, and was to go
down to Birgsi to-night —

She had guessed already what must have befallen. Erlend had offered his daughter to the clerk, and he had refused to have a fallen maid. But what that parley must
have meant to Erlend — she grew dizzy and sick and could
not bear to think the thought out.

The day after, word came from the priest's house.
Haakon Eindridessön prayed that he might have speech
with Erlend. Erlend sent back in answer that he had
naught more to say to Haakon. Sira Eiliv said to Kristin
that if Haakon lived he would be crippled wholly — besides that he had lost his right hand, he had hurt his back
and hips badly in falling from the roof of the servants'
house. But he was set on coming home, even as he was,
and the priest had promised to get him a sleigh. He repented his sin now with all his heart — he said that Margret's father had been within his rights, however the law
might stand; but he was most fain that all should do their
best to hush up the matter, so that his misdeed and
Margret's shame should be hidden as much as might be.
In the afternoon he was borne out to the sleigh, which Sira

Eiliv had borrowed from Repstad, and the priest himself rode with him to Gauldal.

Thus the next day, which was Ash Wednesday, the Husaby folk had to go down to Vinjar to the parish Church. But at the time of vespers Kristin had the acolyte let her into the chapel at home.

She could feel the ashes still on her forehead when she knelt down by her stepson's grave and said over the paternosters for his soul.

Not much but bones would be left of the boy now, down under his stone. Bones and the hair, and some shreds of the clothes they had been laid in. She had seen the bones of her little sister when they took her up that they might bring her to her father at Hamar. Dust and ashes — she thought of her father's comely visage, of her mother with the great eyes in the furrowed face, and the form that still kept so strangely young and slim and light, though her face grew old so early. There they lay under a stone, falling in sunder, as houses fall to ruin when the folks that lived there are gone. Pictures flitted and faded — the burned ruins of the church at home; a farm in Silsaadal that they used to ride by when they went to Vaage — the houses stood empty and were falling in pieces, the folk that tilled the lands dared not go nigh after the sun was down. She thought upon her beloved dead — their looks and their voices and smiles and ways and bearing — now that they themselves were gone away to yonder other country, to think on their shapes was sore; 'twas like remembering one's home when one knew that it stood desolate, and the rotting timbers were sinking into the soil.

She sat on the wall-bench in the empty church, and the smell of cold stale incense held her thoughts fast bound to pictures of death and the decay of all earthly things. And she was powerless to lift up her soul to see a glimpse of the land where her beloved were, whither all goodness and love and truth at last were taken away and there treasured up. Every day when she prayed for the peace of their souls, it seemed to herself strange and unmeet that she should pray for them whose souls already on this earth had possessed a peace far deeper than she had ever known since she grew to be a woman. Sira Eiliv, indeed, said that prayer for the dead was good always — good for oneself, even if those others were already inheritors of God's peace.

But it helped not her. It seemed to her that when her weary body at last was rotting under a tombstone, her restless spirit would still be doomed to wander about somewhere near by, as an unhappy ghost wanders lamenting round the tumble-down houses of a ruined farm. For in her soul sin still had its being, as the root-tissue of the weeds is inwoven in the soil. It flowered and flamed and scented the air no longer, but 'twas still there in the soil, bleached, but strong and full of life. In despite of all the tenderness that welled up in her heart when she saw her husband's despair, she had not will or strength to stifle the voice of her that cried out, in bitterness and anger: Can you speak thus to *me*? Have you forgotten the time when I gave you my troth and my honour? Have you forgotten the time when I was your dearest love? And yet she knew that as long as this voice questioned thus within her, so long would she speak to him as though *she* had forgotten.

She flung herself in her thoughts before St. Olav's shrine; caught at the mouldering bones of Brother Edvin's hand far off in the church at Vatsfjeld; clenched her hands about the reliquaries with the shreds of a dead woman's shroud, and the splinters of the bones of an unknown blood-witness — caught for a safeguard at the small remains that through death and nothingness had kept a little of the virtue of the departed soul — like the magic power that clings about the rust-eaten swords dug up from ancient warriors' barrows.

The day after, Erlend rode into the city, taking with him only Ulf and one other man. All through the fast-time he came not home to Husaby, but Ulf came to fetch his body-guard and took them to met him at the mid-fast Thing in Orkedal.

In talk with Kristin alone, Ulf told her that Erlend had agreed with Tiedeken Paus, the German goldsmith at Nidaros, that Margret should wed Tiedeken's son Gerlak as soon as Easter was past.

Erlend came home at Easter. He was quiet and calm now, but Kristin thought she could see that he would not be able to shake this off as he had shaken off so much else — whether 'twas because he was no longer so young, or because nothing before had ever humbled him so deeply. Margret seemed quite heedless how her father was ordering things for her.

But one evening when man and wife were alone, Erlend said:

"Had she been my true-born child — or her mother an unwed woman — never would I have given her to a stranger while things are so with her; I could have sheltered and guarded both her and hers. This is an ugly way out, but, seeing what her birth is, a wedded husband can best safeguard her."

While Kristin was making all ready for her stepdaughter's going, Erlend said one day, curtly:

"Belike you are scarce well enough to go with us to the city?"

"If you wish it, you know that I will go," said Kristin.

"Why should I wish it? Since before you have never stood in a mother's stead to her, there is no need you should do so now — and a joyous wedding 'twill scarce be. Lady Gunna of Raasvold and her son's wife have promised to come for our kinship's sake."

So Kristin stayed at Husaby, while at Nidaros Erlend gave his daughter to Gerlak Tiedekenssön.

3

THAT summer, just before St. John's Mass, Gunnulf Niku-laussön came back to his cloister. Erlend was in the city then for the Frosta Thing; he sent word to his wife asking her if she deemed she was able to come in thither to meet her brother-in-law. Kristin was none too well, but yet she came. When she met Erlend, he told her that his brother's health seemed to him quite broken down. They had made but little speed with their undertaking in the north, the Friars of Munkefjord. The church they had built they could never get consecrated, for the Archbishop could not journey so far north in these unquiet times; they had

had to say mass the whole time at their travelling-altar. At length they came to lack both bread and wine and candles and oil for the services; and when Brother Gunnulf and Brother Aslak set sail for Vargöy to fetch these things, the Lapps had cast a spell on them, so that they capsized and had to sit for three days and nights on a rocky islet — after this they both fell sick, and Brother Aslak died some time after. They had suffered much from scurvy in the Long Fast, for they lacked both meal and herbs to eat with the dry fish. Therefore had Bishop Haakon of Björgvin and Master Arne (who were at the head of the Cathedral Chapter at Nidaros while the new Archbishop, Sir Paal, was gone to the Curia to be consecrated) ordered the monks who yet lived to come back home, and that the priests of Vargöy should tend the flock at Munkefjord till further order.

But though she was thus not unprepared, yet was Kristin dismayed when she saw Gunnulf Nikulaussön once more. She went with Erlend to the cloister the next day, and they were led into the parlour. The monk came in — his form was bent and crooked, the ring of hair was grown quite grey, under the sunken eyes the skin was wrinkled and dark-brown, but on the smooth white skin of the face were lead-coloured spots, and his hand showed like patches when he drew it out of the sleeve of his gown and held it out towards her. He smiled — and she saw that many of his teeth were gone.

They sat down and talked awhile, but it was as though Gunnulf had forgotten how to speak. He said as much himself before the others left him.

"But you, Erlend, are still the same — you seem not to have grown older," he said with a little smile.

Kristin knew well enough that she herself looked wretchedly just now. And Erlend was a comely sight, as he stood there tall and slender and dark and richly clad. And yet Kristin thought that he, too, had changed much — 'twas strange that Gunnulf saw it not — he had used to be so sharp-sighted.

One day of late summer Kristin was in the clothes-loft, and Lady Gunna of Raasvold was with her — she had come to Husaby to help Kristin, now her lying-in was at hand. Standing there, they heard Naakkve and Björgulf singing out in the courtyard, while they sharpened their knives, a coarse ribald song that they were bawling at the top of their voices.

Their mother was beside herself with anger — she went down to the boys and chid them with the harshest words. And then she said she must know whom they learnt such things from — most like 'twas in the servants' house, but which of the men was it that taught children such things? The boys would not answer. Then Skule came out from under the loft-stairway, and said mother had best be still, for they had learnt the song from hearing father sing it.

Lady Gunna spoke up then; had they so little fear of God that they could sing such things — and now, when they could not know, any night when they lay down to rest, that they might not be motherless before cock-crow? Kristin said no more, but went quietly into the house.

After, when she had lain down for a little on her bed, Naakkve came in and went over to her. He took his mother's hand, but said nothing, and then he began to cry quite quietly. She spoke to him then mildly and jestingly, telling him not to weep or wail; she had won through this trouble six times, and surely she would win through the seventh time too. But the boy wept more and more. At last she had to let him creep in between her and the wall, and there he lay weeping with his arms around her neck and his head against her breast; but she could not make him say for what he sorrowed so, though he lay there by her until the serving-woman bore in the supper.

Naakkve was now in his twelfth year; he was a great boy for his age, and was most fain to bear himself grown-up and manly; but he had a soft heart, and the mother could see sometimes that he was still most childlike. He was old enough to have been able to understand his half-sister's mischance; the mother wondered if he understood, too, how much since then his father was changed.

Erlend had always been a man who could say the worst of things when he was enraged — but before this he had never given hard words to any save in wrath; and he had been quick to make all good when he himself was cool again. But now he could say hard and ugly things in cold blood. He had been a terrible man for cursing and swearing; yet had he, in some measure, left off this evil habit, because he saw it hurt his wife and gave offence to Sira Eiliv, for whom he had grown, little by little, to feel much respect. But never had he been foul-mouthed or unseemly in his talk — in that matter he had been much more modest than many a man who had led a purer life. Sorely as it hurt Kristin to hear such words on her young sons' boyish lips, most of all in the state she was in, and to hear that they had learnt them of their father, there was yet another thing which left the bitterest taste of all in her mouth: she saw that Erlend was still childish enough to deem he could brave out the shame of his daughter's fall by taking impure and unseemly words on his lips.

Fru Gunna had told her that Margret had had a still-born son a while before Olav's Mass. The lady had come to know, she said too, that Margret was already not so ill-content — she agreed well with Gerlak and he was kind to her. Erlend went to see his daughter when he was in the city, and Gerlak made a great to-do of his wife's father, though Erlend was none too forward to own the other as kinsman. Erlend himself had not named his daughter's name since she left Husaby.

Kristin bore yet another son; and he was christened Munan after Erlend's father's father. In all the time she lay in the little hall, Naakkve came daily in to his mother with berries and nuts he had plucked in the woods, or wreaths of healing herbs that he had plaited. Erlend came home when the new son was three weeks old; he sat much with his wife, and strove to be kind and loving — and this time he made no complaint because the new-born child was not a little maid, or because it was weakly and throve but ill. But Kristin made not much answer to his kind words; she was quiet and sadly brooding — and this time her strength came back to her exceeding slowly.

All through the winter Kristin was ailing, and the child seemed little like to live and thrive. Thus its mother had little thought to spare for aught else than the poor little being, and she heard with but half an ear all the talk of the great tidings that were stirring this winter. King Magnus had fallen into the greatest straits for money by reason of his endeavors to win the lordship over Skaane,

and he had called for succours from Norway. Some of the
lords of the Council were willing to stand by him in this
matter. But when his messengers came to Tunsberg, the
Treasurer had gone away, and Stig Haakonssön, who was
Governor in Tunsberghus, shut the gates of the castle
against the King's men, and made ready to hold the place
by force of arms. He had but few folk with him, but
Erling Vidkunssön, who was his uncle by marriage, and
was then at his manor in Aker, sent forty of his men-at-
arms to strengthen the fortress, while he himself sailed
westwards. Much about the same time the King's cousins,
Jon and Sigurd Haftorssön, rose against the King, on account of a judgment that had been passed against some of
their men. Erlend laughed at this, and said the Haftorssöns
had shown themselves raw and foolish. There was great
discontent now with King Magnus throughout the land.
The nobles demanded that a High Steward should be put
at the head of the affairs of the realm and the great seal
placed in the hands of a Norseman, since the King, for the
sake of his affairs in Skaane, seemed minded to spend most
of his time in Sweden. The townsmen and the clergy in
the cities had been frighted by the rumours of the King's
borrowings from the German cities. The haughtiness of
the Germans and their flouting of the laws and customs of
the land were already greater than could be borne, and
now 'twas said that the King had promised them yet
greater rights and franchises in Norwegian cities, so that
things would become quite unbearable for the Norse trad-
ers, who were already hard put to it. Among the commons
the rumours concerning King Magnus' secret sin were
still widespread, and many of the parish priests and of the
wandering monks were at one in this, if in nothing else,
that they believed this was the cause why the Olav's
Church in Trondheim had burned down. And so the farmers, too, sought in this the reasons for the many mis-
chances that in these last years had visited now one and
now another country-side — plagues among the cattle,
blight in the corn, causing sickness and disease to man and
beast, and bad harvests of corn and hay. So Erlend said if
only the Haftorssöns had had wit enough to hold still
awhile yet and win themselves a name for open-handed-

ness and chieftainly dealings, for sure folk would soon
have called to mind that they were King Haakon's grandsons too.

These disorders quieted down, but their upshot was that
the King made Ivar Ogmundssön High Steward in Norway. Erling Vidkunssön, Stig Haakonssön, the Haftorss-
söns and all their following were threatened with attainder
of high treason. On this they yielded, came in, and made
their peace with the King. There was a powerful man of
the Uplands named Ulf Saksessön, who had joined in the
Haftorssöns' rising; and he did not go with the others to
make his peace, but came to Nidaros after Yule. He was
much with Erlend in the city, and from him the folk north
of Dovre had accounts of all these affairs, in the light he
saw them in. Kristin greatly mistrusted this man; she
knew him not, but she knew his sister, Helga Saksesdatter,
who was wed with Gyrd Darre of Dyfrin. She was fair,
but exceeding proud and haughty, and Simon liked her
not, though Ramborg agreed well with her. Soon after the
Fast was begun, letters came to the Wardens ordering that
Ulf Saksessön be proclaimed an outlaw at the Things, but
by that time he had left the land, having sailed in the
depth of winter.

That spring Erlend and Kristin were at their town-
house for Easter, and they had their youngest child,
Munan, with them, for there was a sister at Bakke Cloister
who was so skilled in leech-craft that all the sick children
who were put in her hands got well, if so be it were not
the will of God that they should die.

One day just after the holy-day, Kristin came home
from the cloister with the little one. The serving-man
and the maid who had been with her came with her into
the hall. Erlend was alone there, lying on one of the
benches. After the man was gone out and the woman had
laid by their cloaks — Kristin had sat down by the fire
with the child, and the maid was warming some oil they
had gotten from the nun — Erlend began asking, from
where he lay, what Sister Ragnhild had said of the child.
Kristin gave short answers, as she sat unwrapping the
child's swaddling-clothes; and at last she made no answer
at all.

"Are things so ill with the child, Kristin, that you have no mind to speak of it?" he asked with a little impatience in his voice.

"You have asked this before, Erlend," answered his wife coldly, "and I have told you all there is to tell many times. But since you care not enough for the boy to remember it from one day to another — "

"It has chanced to me, too, Kristin," said Erlend, rising and coming over to her, "that I have had to answer you two and three times about things you yourself had asked of me, because you cared not to remember what I said."

"They were not things of such import as the children's health, I trow," she said in the same tone.

"They were not trifles, either — this last winter; I, at least, had them much at heart."

" 'Tis not true, Erlend. 'Tis many a long day since you talked to me of the things you have most at heart."

"Go out, Signe," said Erlend to the maid. He had flushed to the forehead — now he turned to his wife. "I understand what you would speak of. Of that matter I would not speak to you in your serving-maid's hearing — even if you are such good friends with her that you count it for nothing she should be by when you pick a quarrel with your husband, and tell me I speak untrue — "

" 'Tis the last thing a man sees, the beam in his own eye," said Kristin shortly.

"I understand not well what you mean. Never have I spoken ungently to you in strangers' hearing, or forgot to show you all honour and worship before our serving-folk."

Kristin burst into a strange, heart-sick, unsteady laugh.

"You are good at forgetting, Erlend! Through all these years has Ulf Haldorssön lived with us. Mind you when you sent him and Haftor to bring me to you in the sleeping-loft of Brynhild's house at Oslo?"

Erlend sank down on the bench, gazing at his wife, with parted lips. But she went on:

"Not much has befallen at Husaby — or elsewhere — of unseemly or dishonourable that you have taken thought to hide from your serving-folk — whether 'twere yourself or your wife it put to shame — "

Erlend sat still, looking at her in dismay.

"Mind you the first winter we were wed? I was with child of Naakkve, things were so that 'twas hard enough for me to win obedience and honour from my household. Mind you how you helped and stayed me? Mind you when your foster-father came to be our guest with strange ladies and maids and men, and our own folk sat at the board with us — mind you that Munan dragged off from me every rag I might have hid me with, and you sat mute and dared not stop his mouth — ?"

"Jesus! Have you stored this up against me for fifteen year!" He looked up at her — his eyes, in that glance, seemed strangely light-blue, and his voice was weak and helpless. "Yet, my Kristin — methinks 'tis worse than this that we two should say unfriendly and bitter words to each other — "

"Aye," said Kristin, "worse indeed did it cut into my heart that time at our Yule-tide feast when you chid and rated me because I had thrown my cloak over Margret — and ladies from three counties were standing by and listening — "

Erlend made no answer.

"And now you blame me because things went with Margret as they did — when each time I tried to correct her with a word she would run to you, and you would bid me, in unfriendly words, to let the maid be — for she was yours and not mine — "

"Blamed you — I have not!" answered Erlend in a laboured voice, striving hard to speak calmly. "Had one of our children been a daughter, it had mayhap been easier for you to understand how such things as this that befell my daughter — how they pierce a father to the marrow — "

"I deemed I had shown you last year that I understood," said his wife in a low voice. "I needed but to think of my own father — "

"For all that," said Erlend, speaking quietly as before, "this was a worse thing. I was an unwed man. This man — was — wedded. I was not bound — I was not so bound," he corrected himself, "that I could never be set free — "

"And yet you did not free yourself," said Kristin. "Mind you how it came to pass that you were set free — ?"

Erlend sprang up and struck her in the face. Afterward

he stood gazing, aghast — a red mark came out on her white cheek. But she sat stiff and silent, with hard eyes. The child had begun to cry with fear — she rocked it a little in her lap and hushed it.

" 'Twas — 'twas cruelly spoken, Kristin," said the man, in a shaken voice.

"Last time you struck me," she said softly, "I bore your child beneath my heart. Now have you struck me whilst I sat with your son upon my lap — "

"Aye, these children — we are never without them — " he cried impatiently.

They fell silent. Erlend began walking swiftly up and down the hall. She bore the child into the closet and laid it on the bed; when she came out through the closet-door, he stopped in front of her:

"I — I should not have struck you, my Kristin. I wish with my heart I had not done it — I shall repent it, I trow, for as long as I repented the last time. But you — you have taunted me because you deem I forget too lightly. But you forget naught — no single wrong that I ever did you. Yet I have tried — I have tried to be a good husband to you; but that, I trow, you deem not worth remembrance. You — you are fair, Kristin — " He looked after her as she went past him.

Aye, the housewife's still and stately bearing was as beautiful as had been the young maid's supple loveliness; her bosom and hips were grown broader, but she was taller, too; she held herself upright, and the neck bore up the little round head proudly and graciously as ever. The pale, close-shut face with its great dark-grey eyes stirred and kindled him even as the round, rosy child-face had stirred and kindled his restless soul by its mysterious calm. He went over and took her hand:

"For me, Kristin, you are and ever will be the fairest of all women, and the dearest — "

She let him hold her hand, but gave not back the pressure of his. Then he flung it from him, as his bitterness overcame him again.

"Forgotten, say you I have? I trow 'tis not ever the worst of sins — to forget. I have never set up to be a pious man; but I remember what I learned of Sira Jon when a child, and God's ministers have reminded me of it since.

'Tis sin to brood and call back to mind the sins we have confessed to the priest and done penance for before God, and been granted His forgiveness for through the priest's hand and mouth. And 'tis not from holiness, Kristin, that you are ever tearing open these old sins of ours, but 'tis to have a weapon against me each time I go against you in aught — "

He walked away from her, and then came back.

"Greedy to rule — God knows that I love you, Kristin — yet do I see that you are greedy to rule, and never have you forgiven me that I did you wrong and tempted you to wrong. Much have I borne from you, Kristin, but I will no longer bear never to be left in peace for these old mis-chances, nor to have you speak to me as though I were your thrall — "

Kristin was shaking with passion as she answered:

"Never have I spoken to you as though you were a thrall. Have you *once* heard me speak harshly or angrily to any human being that could be counted as lesser than I — if it were the worst and most worthless of our servant-folk? I know myself free before God from the sin of having offended His poor in word or deed. But you should be my *master;* you should I obey and honour, bow myself before and stay myself on, next to God — according to God's law, Erlend. And if so be I have lost patience, and have spoken to you in such wise as it befits not a wife to speak to her husband — I trow it has been because you have many a time made it hard for me to bow my sim-plicity before your better understanding, to honour and obey my husband and lord so much as I fain would have done — and maybe I looked that you — maybe I deemed I might spur you on to show that you were a man, and I but a poor simple woman —

"But take comfort, Erlend. I shall not offend you with my words any more, for after this day never shall I forget to speak to you as gently as though, in truth, you were born of thralls — "

Erlend's face flushed darkly red — he lifted his clenched fist — then turned sharp about on his heel, seized his cloak and sword from the bench by the door, and rushed out.

* * *

Without there was sunshine and a sharp wind — it was cold, but the bright sparkles that besprinkled him from house-eaves and from wind-swept trees were drops of water thawed out and frozen again in the air. The snow on the house-tops shone like silver, and behind the dark-green tree-clad hills around the town the mountains glittered cold blue and shining white in the bitter, bright, wintry spring day.

Erlend passed through streets and lanes — swiftly, at haphazard. He was boiling inwardly — *she* had been wrong, 'twas as clear as day she had been wrong from the first, and he had been right; and he had played the fool and struck her and made him seem less right — but the wrong *was* with her. What he should do with himself now, he knew not. He had no mind to go to any acquaintance' house, and home he would not go.

There was some hurry and bustle in the city. A big trader from Iceland — the first of the year — had come in to the wharves in the morning. Erlend wandered westward through the lanes, came out by St. Martin's Church, and went down toward the waterside alleys. Though 'twas early in the afternoon, already there was noise and yelling in the ale-houses and from the taverns. In his youth he had been able to go into such houses himself, with his friends and fellows. But now all the folks would stare the eyes out of their heads, and talk themselves hoarse afterwards, if the Warden of Orkdöla County, with his great house in the city, with ale and mead and wine at home to his heart's desire, should come into an ale-house and ask for a drink of their bad small beer. Yet truly this was what he had a mind to do — to sit and drink with small farmers in town for the day and serving-men and seamen. There was no to-do when those fellows caught their woman a buffet on the ear; and after that all was well again — hell and furies! how should a man rule a woman when he cannot thrash her soundly, by reason of her birth and his own honour? — at bandying words the devil himself could be no match for them. Troll she was — and so fair, too — if only he could beat her till she grew good again. . . .

* * *

The bells began ringing from all the city churches to call folks together to Vespers — the spring wind mingled all the notes together in the unquiet air above his head. 'Twas like she was going to Christ's Church now, the holy troll — to bemoan her to God and Mary Virgin and holy Olav that her husband had hit her a buffet on the ear. Erlend sent up towards his wife's guardian saints a greeting of sinful thoughts, while the bells clashed and clanged and resounded. He made his way towards St. Gregory's Church.

His father and mother's graves were before St. Anne's altar in the northern aisle. Whilst he said over his prayers, he caught sight of Lady Sunniva Olavsdatter and her maid coming in at the church-door. When he had done praying, he went across and greeted her.

It had been the way of these two, ever since he had come to know the lady, that whenever they met they fell to somewhat free toying and jesting. And this evening as they sat on the wall-bench waiting for evensong to begin, he was so forward that she had more times than one to remind him that they were in church and that folk kept coming by them.

"Aye, aye," said Erlend, "but you are so fair to-night, Sunniva! 'Tis so good to jest with a lady that has such gentle eyes — "

"Little do you deserve, Erlend Nikulaussön, that I should look at you with gentle eyes," said she, laughing.

"Then will I come and jest with you when 'tis dark," said Erlend, laughing too. "When evensong is over, I will go home with you — "

Then the priests came into the choir, and Erlend went across to the southern aisle to take his place among the men.

When the service was at an end, he went out at the great door. He saw Lady Sunniva and her maid a little way down the street, and thought he had best not go with her, but go home straightway. At that moment a band of Icelanders from the trader came up the street; they staggered along, holding on to each other, and seemed as they would block the two women's passage. Erlend ran after the lady. As soon as the sailors saw a gentleman with a sword at his belt come toward them, they swerved aside and made room for the women to pass.

"I trow I had best go home with you, after all," said Erlend. "The city is none too quiet to-night."

"Can you believe it, Erlend? — old a woman as I am, maybe I like it not ill that some men think I am yet so fair, 'tis worth while blocking my way — "

There was but one answer that a courteous man could make to this.

He came home to his own house the next morning in the grey of dawn, and stood a little outside the locked door of the hall, frozen, dead-weary, heart-sore and sickened. Rouse the household with his knocking; go in and creep into bed beside Kristin lying with the child at her breast — no! He had on him the key of the eastern storehouse loft; there were some goods stored there that he was answerable for. He let himself in, pulled off his boots, and got together some webs of wadmal and some empty sacks and spread them on straw in the bedstead. He wrapped his cloak about him, crept under the sacks, and, tired out and harassed as he was, was able at last to forget everything in sleep.

Kristin was pale and weary with waking when she sat down to the morning meal with her house-folk. One of the men told her that he had prayed the master to come to breakfast — he was sleeping up in the east storehouse loft, he said — but Erlend had bidden him go to the devil.

Erlend had a tryst out at Elgesæter after the day-mass; he had to witness some dealings in land. But he managed to slip away from the feast that followed in the refectory, and from Árne Gjavvaldssön, who, like himself, could not stay and drink with the Brothers, but was set on having Erlend go with him to Ranheim.

Afterwards he repented that he had parted company with the others — dismay came over him as he went back alone to the town — now must he needs think over what he had done. For a moment he had a mind to go straightway to St. Gregory's Church — he had leave to confess to one of the priests there, when he was in Nidaros. But if he did this again, after he had confessed, the sin would be much greater. 'Twere better to wait awhile —

She must think now, Sunniva, that he was a chicken she had caught with her bare hands. But devil take him if he had ever thought a woman had been able to teach him so much that was new — here was he going about yawning still from the adventure. He had flattered himself he was not unskilled in *ars amoris*, or whatever the learned men called it. Had he been young and green, like enough he had been proud of himself and deemed it fine and brave. But he liked not the woman — the mad creature — he was sick of her; he was sick of *all* women save his wife — and he was sick of her too! By the Cross itself — he had been so wedded to her that he had grown most holy himself — for he had believed in her holiness — but 'twas a fair reward he had had from his holy wife for his faithfulness and love — troll that she was! He remembered her scorching venomous words of the day before — so she deemed he bore himself as though he were born of thralls! — And the other, Sunniva, she thought, doubtless, he was naught but a raw weakling, since he had let himself be taken by surprise and had shown some dismay at her way of love-making. He would show her now that he was no more of a holy man than she was a woman. He had promised her to come down to Thorolf's townhouse that night, and he might e'en as well go; the sin was sinned: why not enjoy any disport it might bring with it?

Since he had broken his troth to Kristin already — and she herself had brought it about, by her hateful and unjust ways towards him —

He went home, and wandered about the stables and out-houses seeking for somewhat to find fault with; he had words with the priest's serving-maid from the hospital because she had brought malt into his drying-house, though he knew well that his house-folk would have no use for the house this time while they were in the town. He wished he had had his boys here — they would have been some company — he wished he could set forth back to Husaby at once. But he must needs wait in the city for letters from the south — 'twas too venturesome to have such things come to hand at one's home in the country.

The mistress of the house came not in to supper — she

was lying down on the bed in the closet, said Signe, her maid, looking at her master reproachfully. Erlend answered harshly that he had not asked after her mistress. When the house-folk had left the hall, he went into the closet. It was pitch-dark in there. Erlend bent over the bed.

"Are you weeping?" he asked very low, her breath came so strangely. But she answered in a thick, husky voice that she was not weeping.

"Are you weary? Aye — I will go to rest too, now," he said in the same low voice.

Kristin's voice quivered as she said:

"Then I had liefer, Erlend, that you should go and lie to-night where you lay last night."

Erlend made no answer. He went out and fetched the candle from the hall into the closet, and opened his chest of clothing. He was well clad enough already to go wherever he listed, for he still bore the violet *kothardi* he had worn at Elgesæter in the morning. But now he changed his clothes slowly and deliberately — put on a red silk shirt and a mouse-grey knee-long velvet coat with little silver bells on the sleeve-points, brushed his hair, and washed his hands. Time and again he looked over towards his wife — she lay silent and motionless. Then he went out without bidding good night. Next day he came home openly at the breakfast-hour.

So things went on for a week. Then Erlend came home one evening from Hangrar where he had been on an errand and learnt that Kristin had ridden off that morning home to Husaby.

It had grown clear to him already that never had any man had less joy from a sin than he from these dealings of his with Sunniva Olavsdatter. In his heart he was so deadly weary of the crazy creature — sick of her even while he caressed and toyed with her. 'Twas a mad and reckless thing, too — like enough it was all over the town and the country-side by now, that he had his nightly resort in Thorolf's house — and to smirch his name for Sunniva's sake! Now and then he had thought, too, that the thing might raise some trouble — the woman had a husband such as he was, old and sickly — 'twas pity of Thorolf that he should be wed to such a wild and witless woman — most like *he* was not the first that had made free

with the husband's honour. And Haftor — he had clean
forgotten, when he had to do with Sunniva, that she was
Haftor's sister — he remembered it only when it was too
late. All was as bad as it could be — and now he could see
that Kristin knew it.

She surely could not take it in her head to bring suit
against him before the Archbishop — crave leave to depart
from him. She had Jörundgaard to take refuge in — but
it was impossible to travel thither over the mountains at
this time of year, quite impossible if she would take the
small children with her — and Kristin would not leave
them. No — he thought, to comfort himself — and she
could not go by sea either, with Munan and Lavrans, so
early in the spring. Oh, but 'twould be unlike Kristin to
crave the Archbishop's help against him — though she had
good reason for it — but he would keep away from her
bed of his own accord — till she saw that he repented
from his heart. Kristin could never wish to have this thing
publicly brought to question. But he knew in his heart that
'twas long since he had rightly known what his wife
could or could not do.

He lay at night in his own bed, letting his thoughts go
hither and thither. It dawned upon him that he had behaved yet more witlessly than he had understood at first, to
let himself be tangled in this wretched adventure, now,
when he was in the thick of the greatest plans of State.

He cursed himself that he was still such a fool about his
wife that she had been able to drive him to this. He cursed
both Kristin and Sunniva. In the devil's name, sure he was
no fonder on women than other men — rather he had had
to do with fewer of them than most others that he knew
of. But 'twas as though the foul fiend himself had the ordering of things for him — he could not come near a
woman without finding himself up to the neck in a
bog. . . .

There should be an end of it now. God be thanked and
praised that he had other things on hand. Soon, soon, for
sure, would he have Lady Ingebjörg's letters. Aye, in that
matter, too, he had to reckon with women's whims; belike
that was God's punishment for the sins of his youth. Erlend laughed to himself in the dark. The lady must see that
things were as they had so clearly set them forth to her.

The question was whether it should be one of her sons or
the sons of her base-born sister that the Norsemen set up
against King Magnus. And she loved her children by Knut
Porse as she had never loved her other children.

Soon, soon, 'twould be the sharp wind and the salt sea
breakers that should fill his embrace. God in heaven! it
would be good to be drenched by the waves once again
and have the wind blow freshness to his very marrow —
be quit of all womankind for a long, delicious time.

Sunniva — she might think what she would. Thither he
would go no more. And Kristin might fare to Jörundgaard
if she would, for him. 'Twould mayhap be best and safest
for her and the children if they were well out of the way
in Gudbrandsdal this summer. Afterward there was no
fear but he could make friends with her again. . . .

Next morning he rode up into Skaun. Say what he
would, he could not rest till he had made sure what his
wife meant to do.

She met him with gentle, cold courtesy when he reached
Husaby late in the day. She spoke not a word to him of
her own accord, nor any unfriendly word; and she said
naught against it when, in the evening, as if feeling his
way, he came over and lay down in their bed. When
they had lain awhile, he tried, falteringly, to lay a hand
on her shoulder.

Kristin's voice shook, but Erlend could not tell whether
it were in sorrow or from anger, as she whispered:

"So base a man I trow you are not, Erlend, that you
would make this worse for me than need be. I cannot
strive with you, with our children sleeping around us.
And since I have seven sons by you, I would be loath that
our house-folk should see that I know I am a wronged
wife — "

Erlend lay long and silent before he ventured to answer:

"Aye. God have mercy on me, Kristin, I have wronged
you. I had not — had not done it if I could have taken
more lightly the cruel words you said to me that day at
Nidaros. It is not so that I am come home to beg you for
forgiveness; for I know well that that would be a great
thing to ask you now — "

"I see that Munan Baardssön spoke true," answered his
wife; "the day will never come when you will stand up

and take the blame on yourself for what you have done amiss. 'Twere best you turned you to God and sought to make your peace with Him — you have less need to ask my forgiveness than His — "

"Aye, so much I can see," said Erlend bitterly. After this they spoke no more. And the next morning he rode back to Nidaros.

He had been in the city some days when Lady Sunniva's woman came to him one evening in St. Gregory's Church. It seemed to Erlend that after all it were well he should speak with the lady one last time; and he bade the girl keep watch that night — he would come by the same way as before.

He had had to creep and clamber about like a poultry-thief to come up into the loft where they had their meetings. It made him sick with shame now to think he had been such a fool — a man of his age and his place. But at first he had deemed it sport to play such youthful pranks.

The lady was in bed when he came in.

"Come you at long last and so late?" she laughed, and yawned. "Quick now, love, and come to bed; and we can talk afterward about where you have been so long — "

Erlend knew not rightly what to do, or how to tell her what he had at heart. Without thinking, he began to loose the fastenings of his dress.

"Foolhardy is it, this that we have done, Sunniva — I trow it were not well that I should stay here to-night. Thorolf must be looked for home ere long?" he said.

"Are you frighted at my husband?" asked Sunniva teasingly. "You saw yourself Thorolf never so much as pricked up an ear when we toyed and jested before his very eyes. Should he hear that you have been coming about the house, I warrant I make him believe that it is but the old foolery. He trusts me all too well — "

"Aye, it seems indeed that he trusts you all too well," laughed Erlend, burying his fingers in the bright hair on her firm white shoulders.

"Say you so?" She caught him round the wrist. "Yet you trust your wife too. And *I* was yet modest and shamefast when Baard wedded me — "

"*My* wife we will leave outside of this matter," said Erlend sharply, letting go his hold.

"Why so — ? Think you 'tis more unseemly that we should speak of Kristin Lavransdatter than of Sir Thorolf, my husband?"

Erlend set his teeth hard and made no answer.

"Methinks you are one of those men, Erlend," said Sunniva mockingly, "that think you are so winning and fair that it can scarce be reckoned a fault in a woman that her virtue was as frail as glass against you — she may be staunch as steel against all others."

"Of you I have never thought so," answered Erlend coarsely.

Sunniva's eyes gleamed:

"What would you with me, then, Erlend — since you were so happily wedded?"

"I have said, you shall not name my wife — "

"Your wife or my husband — "

" 'Twas ever you that began to speak of Thorolf; and 'twas you that scoffed at him worst," said Erlend bitterly. "And if you had not flouted him in words — 'twas plain enough how dear you held his honour, when you took another man to you in your husband's place. *She* — is not brought low by my misdoing."

"Is this what you would say to me — that you love Kristin, though you like me well enough to play with me — "

"I know not how well I like you — you showed that you liked me — "

"And Kristin sets not your love at its true worth?" she scoffed. "I have seen well enough how gently she is used to look on you, Erlend."

"Hold your tongue, now!" shouted the man. "Maybe she knew what I deserved!" he said harshly and savagely. "You and I may well be each other's like — "

"Is it so," asked Sunniva threateningly, "that I was to be but a whip for you to lash your wife with?"

Erlend stood breathing hard:

"Call it so, if you will. But you laid yourself ready to my hand."

"Beware," said Sunniva, "that that whip smite not yourself — "

She sat up in the bed and waited. But Erlend offered not to gainsay her or to seek to make up the quarrel. He re-dressed fully and left her without a word.

He was not greatly pleased with himself, or with the fashion in which he had parted from Sunniva. There was small honour in it for him. But it could not be helped — and at least he was quit of her now.

4

THIS spring and summer not much was seen of the master at Husaby. At such times as he was at the manor, he and his wife met each other with courtesy and friendliness. Erlend in no way tried to break down the wall she now built up between them, though he would often look after her searchingly. For the rest, he seemed to have much to think on outside his home. Touching the management of the estate he never asked a single word.

It was this matter of the estates his wife brought forward when, just after the spring Holy Cross day, he would have had her go with him to Raumsdal. He had business in the Uplands — would she not take the children with her, stay awhile at Jörundgaard, and see her kinsfolk and friends in the Dale? But Kristin would on no account agree to this.

He was in Nidaros at the time of the Lagthing, and afterward out in Orkedal; then he came home to Husaby, but at once busied himself making ready for a journey to Björgvin. *Margygren* was lying out at Nidarholm, and he but waited for Haftor Graut, who was to sail along with him.

Three days before Margaret's Mass * they began the hay-harvest at Husaby. It was the fairest weather, and when the hay-makers went back to the meadows after the mid-day rest, Olav, the foreman, got leave for the children to go with them.

Kristin was in the clothing storehouse that was in the second story of the armoury building. The house was so built that an outer stairway led up to this room, which

* 10th June.

had a balcony before it; the third story — the armoury itself — stood out above the balcony, and it could be reached only by a loose ladder from the clothing loft leading to a trap-door on the floor. The trap-door was open, for Erlend was up in the armoury.

Kristin bore out the fur cloak that Erlend was to take with him for the sea-voyage, and shook it out in the balcony. Then she was ware of the noise of a great company of horsemen, and at the same moment she saw folks come riding out of the woods on the Gauldal road. The next moment Erlend stood by her side.

"Was it so, did you say, Kristin, that the fire in the kitchen was put out this morning?"

"Aye — Gudrid upset the broth-cauldron. We must borrow a light from Sira Eiliv — "

Erlend looked across at the priest's house.

"No; he must not be mixed in this. Gaute," he called softly to the boy, who was loitering under the balcony, lifting one rake after another, unwilling to set out to the hay-making. "Come up hither, up the stairs — no farther, or they might see you."

Kristin gazed at her husband. Like this she had never seen him before — the strained, alert calm in his voice, in his face, as he spied out southward along the road — in the whole of his tall supple form as he ran into the loft and came back at once with a flat packet, sewn up in linen cloth. He gave it to the boy:

"Hide this in your breast — and mark well what I say to you. You must save these letters — more is at stake than you can understand, my Gaute. Put your rake over your shoulder and go quietly down across the fields till you come to the alder-thickets. Keep well among the bushes till you get down to the wood — you know all the paths there, I know — and creep through the thickest brush all the way across to Skjoldvirkstad. When you get there, make sure first that all is quiet on the farm. Should you see signs of aught amiss or of strangers about, then hide you. But should you be sure all is safe, go down to the farm and give this to Ulf, if he be at home. But if you cannot give the letters into his hands while you are sure that none is near, burn them the moment you can come by the wherewithal to do it. But be sure that both writing and

seals are altogether burnt up, and that they come not into
any man's hands but Ulf's. God help us, my son — these
be great matters to put in the hands of a boy of ten winters — the lives and welfare of many good men — understand you that much is at stake, Gaute?"

"Aye, father. I have understood all that you have said
to me." Gaute looked up from the stairway, his little fair
face full of earnestness.

"Say to Isak, if Ulf is not at home, that he must ride
straight out to Havne and on all night — and tell them he
wots of, that a head wind has sprung up, and I fear me
evil spells have spoiled my journey. Do you understand?"

"Aye, father. I mind well all you have said to me."

"Go, then. God keep you, my son."

Erlend ran up into the armoury, and would have closed
the trap-door, but Kristin was already half-way through
the opening. He waited till she was up, then shut the trap,
ran over to a chest, and took out some written parchments. He tore the seals off and trampled them to pieces
on the floor, tore the parchments into rags and wrapped
them together round the key of the chest and dropped
the little bundle from the window-hole into the midst of
the nettles that grew high behind the storehouse. With his
hands on the window-frame, he stood gazing after the
little boy walking along the edge of the cornfield down
towards the meadow, where the mowers were moving
forward in a line, plying their scythes and rakes. When
Gaute disappeared into the little copse between the cornfield and the meadow, he closed the shutter. The noise
of hoofs came loudly now and from near by.

Erlend turned towards his wife:

"Can you have what I threw out but now made away
with? — send Skule, he has his wits about him — tell him
to fling it down into the pit behind the byre. Like enough
they will keep an eye on you, and mayhap on the big boys
too. But they will scarce search you — " He put the
fragments of the seals down inside the bosom of her dress.
"None could make them out, methinks; but yet — "

"Are you in peril, Erlend?" she asked quietly. When he
had looked into her face, he drew her outspread arms
around him. For a moment he pressed her to him:

"I know not, Kristin. We shall see soon, I trow. Tore Eindridessön rides at the head of the men, and Sir Baard is with them, if I saw aright. I can scarce deem that Tore comes hither for any good — "

The horsemen were in the courtyard now. Erlend stood still a moment. Then he kissed his wife vehemently, opened the trap-door, and ran down. When Kristin came out into the balcony, she saw Erlend in the courtyard helping the Treasurer, who was an old man and a heavy, to dismount from his horse. There were thirty men-at-arms, at least, with Sir Baard and the Warden of Gauldöla County. As Kristin crossed the yard, she heard the Warden say:

"I bring you greetings from your cousins, Erlend. Borgar and Guttorm Trondssöns are the King's honoured guests at Veöy, and I trow Haftor Toressön will have paid a friendly visit by this time to Ivar and the young lad at their home at Sundbu. Sir Baard took the Graut into keeping in the city yestermorn."

"And now I see you are come hither to bid me to this same meeting of the guardsmen," said Erlend, smiling.

"So it is, Erlend."

"And doubtless you would search my manor here too? Oh, I have done my part in such affairs so often that I ought to know the way 'tis done."

"Such great affairs as a high treason cause you have scarce had in your hands," said Tore.

"No, not before now," said Erlend. "And it looks as though I were playing with the black men, Tore, and you had mated me — is it not so, kinsman?"

"We must find forth the letters you have had from Lady Ingebjörg Haakonsdatter," said Tore Eindridessön.

"They are in the chest covered with red leather up in the armoury — but there is not much in them save such greetings as dear kinsfolk use to send to each other — and they are all old. Stein here can take you up."

The stranger horsemen had dismounted now, and the house-folk were coming crowding into the courtyard.

"There was more in the one we took from Bogar Trondssön," said Tore. Erlend whistled softly.

"We had best go into the hall," he said; "it begins to grow crowded here."

Kristin followed the men into the hall. At a sign from Tore, two of the stranger men-at-arms came with them.

"You must give up your sword, Erlend," said Tore of Gimsar, when they stood within, "for a sign that you are our prisoner."

Erlend smote his thighs to show that he wore no other weapon than the dagger in his belt. But Tore said again:

"You must reach us your sword for a token —"

"Aye, aye, if all is to be done to such a nicety —" said Erlend, laughing a little. He went over and took his sword from its peg, and holding the scabbard, held out the hilt to Tore Eindridessön with a little bow.

The old man of Gimsar loosened his fastenings, drew the blade right out, and ran his fingers along its groove.

"Was it this sword, Erlend, that you — ?"

Erlend's blue eyes glittered like steel, his mouth grew narrow and straight-lipped:

"Aye. 'Twas with this sword I chastised your grandson when I found him with my daughter."

Tore stood holding the sword; he looked down at it and spoke threateningly:

"You that should uphold the law, Erlend — you must sure have known that that time you went a little farther than the law would follow you —"

Erlend threw back his head, flushing, and said hotly:

"There is a law, Tore, that cannot be set aside by Kings or Thingmen — his women's honour a man may guard with the sword —"

"Well for you, Erlend Nikulaussön, that no man has put that law to use against you," answered Tore of Gimsar malignantly. "Else had you need of as many lives as a cat —"

Erlend said with stinging slowness of speech:

"Think you not this matter is so grave that 'tis untimely to mix up with it old stories from my youth?"

"I know not if Thorolf of Lensvik deems those matters are so old."

Erlend flamed up and would have answered; but Tore shouted:

"You should try first, Erlend, whether your mistresses have skill to read writing, before you run about to nightly trysts with secret letters in your waist-band! Ask you of

Baard there who it was that warned us that you were hatching traitorous counsels against your King that you have sworn troth to, and hold your place in fee from — "

Unwittingly Erlend raised a hand to his breast — he glanced for a moment at his wife, and his face flushed darkly. Then Kristin ran forward and threw her arms about his neck. Erlend looked down into her face — he saw naught in it but love:

"Erlend — husband."

The Treasurer had scarce spoken a word hitherto. Now he went over to the pair and said softly:

"Dear lady — mayhap it would be best that you take the children and your serving-women with you into the women's house, and stay there as long as we are on the manor."

Erlend loosed his wife, with a last pressure of his arm about her shoulders.

"It is best so, my own Kristin — do you as Sir Baard counsels."

Kristin lifted herself on tiptoe and offered him her mouth to kiss. Then she went out into the courtyard. And out of the confused throng of folks she gathered together her children and serving-women, and carried them with her into the little hall — other women's house there was not at Husaby.

For some hours they sat there, and the mistress' calm and steadfastness kept the frightened little company's terror somewhat in check. Then Erlend came in, disarmed and clad as for a journey. Two stranger men-at-arms took post down by the door.

He took his eldest sons by the hand, and then lifted the little ones in his arms, while he asked where Gaute was — "but you must greet him from me, Naakkve. He is run off to the woods with his bow, I warrant, as his wont is. Tell him he can have my English long-bow after all — I denied it him when he asked for it on Sunday."

Kristin crushed him to her without a word.

"When will you come back, Erlend dearest?" she whispered pleadingly.

"That must be as God will, my wife."

She drew back from him, struggling that she might not break down. He was never used to speak to her but by her

christened name, and these last words of his shook her to
the very heart. It was as if only now she understood to the
full what it was that had befallen.

At the sunset hour Kristin sat up on the hill north of the
houses.

She had never seen the sky so red and golden before.
Above the hill right over against her lay a great cloud; it
was shaped like a bird's wing, it glowed within like iron in
the forge, it shone clear as amber. Little golden wisps like
feathers loosed themselves from it and floated out into the
sky. And deep below on the lake in the valley-bottom
the sky was mirrored with the cloud and the hill-side
above it — it seemed as though it were from down there
in the depths that the burning glow streamed up to tinge
all that lay before her.

The grass in the meadows was seeding, and its silken
spikes shone darkling red in the red light from the sky;
the barley was in ear, and caught the radiance on its young
silky-bright beards. The turf house-roofs of the manor
were thick with sorrel and buttercups, and the sunlight
lay in broad rays across them; the blackish shingles of the
church-roof glowed darkly, and the light stones of the
walls were softly gilded.

The sun broke forth beneath the cloud, rested on the
mountain-crest and sent his light out over range behind
range of wooded hills. The evening was so clear — the
light showed up to sight little clearings among the pine-
covered hill-sides; she could see sæters and little farms in
among the woods that she had never before known one
could see from Husaby. Great hill masses, deep violet
in hue, rose in the south, in toward Dovre, where else
there were wont to be clouds or haze.

The least of the bells in the chapel below began to ring,
and the church-bell at Vinjar answered. Kristin sat bowed
over her folded hands till the last of the three triple
strokes died away on the air.

Now the sun was below the mountain-top, the golden
radiance grew paler and the red more rosy and soft. After
the bells had fallen silent, the soughing of the woods
seemed to grow again and spread abroad; the noise of the
little beck that ran through the leaf-woods down in the

valley sounded louder on the ear. From the close near by came the well-known clinking from the bells of the home cattle; a flying beetle hummed half-way round about her, and was gone.

She sent a last sigh after her prayers — a prayer for forgiveness because her thoughts had been elsewhere while she prayed.

The great goodly manor lay below her on the hill-side, like a jewel on the hill's broad bosom. She looked out over all the lands that she had owned along with her husband. Thoughts of this estate, cares for it, had filled her mind to the brim. She had worked and striven — never till to-night had she known herself how she had striven to set this manor on its feet and keep it safe — nor all she had found strength to do and how much she had compassed.

She had taken it as her lot, to be borne patiently and unflinchingly, that all this rested on her shoulders. Even so she had striven to be patient and to hold her head high under the burden her life laid on her, each time she knew she had again a child to bear under her heart — again and again. With each son added to the flock, she had felt more strongly the duty of upholding the welfare and safety of the house — she saw to-night, too, that her power to overlook the whole, her watchfulness, had grown with each new child she had to watch and strive for. Never had she seen so clearly as this evening what fate had craved of her and what it had granted her, in giving her these seven sons. Over again and over again had joy in them quickened the beating of her heart, fear for them pierced it — they were her children, these great lads with their lean angular boys' bodies, as they had been when they were so small and plump they could scarce hurt themselves when they tumbled in their journeys between the bench and her knee. They were hers, even as they had been when, as she would lift one of them from the cradle up to her breast for milk, she had to hold up its head, because it nodded on the slender neck as a bluebell nods on its stalk. Wherever they might wander out in the world, whithersoever they might fare, forgetful of their mother, she felt as though for her their life must still be an action of her life, they must still be as one with herself as they had been when she alone in all the world knew of the

new life which lay hidden within and drank of her blood
and made her cheeks pale. Over again and over again had
she proved the sickening sweating terror when she felt:
now her time was come again, now again was she to be
dragged under in the breakers of travail—till she was
borne up again with a new child in her arms; how much
richer and stronger and braver with each child, never till
to-night had she understood.

And yet she saw to-night that she was still the Kristin
of Jörundgaard, who had never learned to endure an
ungentle word, because she had been shielded all her
days by so strong and tender a love. In Erlend's hands she
was still the same. . . .

Aye. Aye. Aye. 'Twas true that she had gone on storing
up, year in, year out, the memory of every wound he had
dealt her—though she had known always that he had
wounded her, not from ill will as a grown man wounds another, but as a child strikes his playfellow in their play.
She had tended the memory of each time when he had
offended her, as one tends a festering sore. And every
abasement he had brought upon himself by following his
own every whim struck her like a whip-stroke on the
flesh and left a running weal. 'Twas not so that she willingly and of purpose stored up grudges against her husband; she knew that towards others she was not petty-
minded, but when he was concerned she grew so straightway. When Erlend was in question, she could forget nothing, and every least scratch on her soul went on smarting
and bleeding and swelling and throbbing, when 'twas he
that had dealt it her.

Towards him she never grew wiser, never stronger. She
might strive to seem, in her life with him too, capable and
brave and strong and pious, but 'twas not true that she was
so. Ever, ever had longing gnawed within her—the
longing to be again his Kristin of the woods of Gerdarud.

In those days she had been willing to do all that she
knew was evil and sinful rather than lose him. To bind
Erlend to her, she had given him all that was hers: her
love and her body, her honour and her part in the salvation of her Lord. And she had given him what she could
find to give that was not hers: her father's honour and his
trust in his child; all that wise and prudent grown men

had built up to safeguard a little maid in her nonage she had overturned; against their plans for the welfare and advancement of their race, against their hopes that their work would bear fruit when they themselves lay under the mould, she had set her love. Much more than her own life had she staked in the game, wherein the sole prize was Erlend Nikulaussön's love.

And she had won. She had known, from the time when he kissed her for the first time in the garden by Hofvin, even till to-day when he kissed her in the little hall before he was led away a prisoner from his home — Erlend loved her as his own life. And if he had not guided her life well, yet had she known well-nigh from the first hour she met him how he had guided his own. If he had not ever dealt well by her, yet had he dealt better by her than with himself.

Jesus, how she had won him! She confessed to herself this evening — she herself had driven him to breach of his marriage-vows by her coldness and by her venomous words. She confessed to herself now — even in these years when she had looked on, time and again, at his unseemly dalliance with this woman Sunniva, and had been angered by it, she had felt in the midst of her wrath a haughty and defiant joy — none knew of any open stain on Sunniva Olavsdatter's honour, yet Erlend talked and jested with her as a serving-man might with an ale-house wench. And of her he had known that she could lie and betray those that trusted her best, that she willingly let herself be lured to the most shameful places — yet had he trusted her, yet had he honoured her so far as in him lay. Easy as it was for him to forget all fear of sin, easy as it had been for him to break his promise made to God at the altar, yet had he sorrowed over his sins against her, and had struggled for years that he might be able to keep his promises to her.

She had chosen him herself. She had chosen him in a frenzy of love, and she had chosen anew each day of those hard years at home at Jörundgaard — chosen his wild reckless passion before her father's love that would not suffer the wind to blow ungently upon her. She had thrown away that lot her father had shaped for her, when he would have given her to the arms of a man who would surely have led her by the safest ways, and would have stooped down, to boot, to take away each little stone that she might have dashed her foot against. She had chosen to follow the other, who she knew was straying in perilous paths. Monks and priests had pointed the way of repentance and atonement to lead her home to peace — she had chosen turmoil rather than let slip her darling sin.

So there was but one way for her — not to murmur or cry out, whatever should befall her at this man's side. Dizzily far behind her it seemed now, the time when she had left her father. But she saw his beloved face, remembered his words that day in the smithy when she dealt his heart the last stab, remembered their talk together, up in the mountains in the hour when she saw that the door of death stood ajar waiting for him. Unworthy is it to murmur at the lot one has chosen for oneself. — Holy Olav, help me that now I may not show me altogether unworthy of my father's love!

Erlend, Erlend — When she met him in her youth, life had become for her a swift river rushing over rocks and rapids. In these years at Husaby, life had spread out, lying wide and ample like a lake, mirroring all that surrounded her. She remembered, at home, when the Laagen overflowed in spring-time, and lay grey and mighty in the valley-bottom, bearing on its bosom the driftage that came floating, while the tops of the growing leaf-trees in its course swayed on the waters. Out in its midst showed little dark threatening eddies, where the current ran swift and wild and perilous under the shining surface. Now she knew that even so had her love for Erlend run like a swift and perilous current beneath the surface of her life through all these years. Now 'twas bearing forward — she knew not to what.

Erlend, beloved — !

Once more Kristin breathed an Ave out into the evening glow. Hail Mary, full of grace! I dare pray thee but for one mercy, that see I now: Save Erlend, save my husband's life — !

She looked down at Husaby and thought of her sons. Now when the manor lay there in the evening light like a dream vision that might melt away — now that fear for her children's doubtful fate shook her heart, it came to her mind: Never had she thanked God fully for the rich fruits her toil had borne in these years; and never had she thanked Him as she ought that seven times He had granted her a son.

Out of the dome of the evening heaven, from all the country-side beneath her eyes, came the murmured words of the mass that she had heard thousands of times, in her father's voice that had set forth the words for her when she was a little child at his knee: Thus sings Sira Eirik in the *Præfatio,* when he turns him to the altar, and thus it says in the Norse tongue:

It is truly meet and just, right and available to salvation, that we should always, and in all places, give thanks to Thee, O holy Lord, Father Almighty, eternal God.

When she lifted up her face from her hands, she saw Gaute coming up the hill. Kristin sat still and waited till the boy stood before her, then she stretched out her hand and took his. There was grassy sward a good way round the stone where she was sitting, with no place where any might hide.

"How did you do your father's errand, my son?" she asked softly.

"As he bade me, mother. I came to the farm so that none saw me. Ulf was not at home, so I burned what father had given me on the hearth-fire in the hall. I took it out of the cloth." He hesitated a little. "Mother — there were nine seals to it — "

"My Gaute." The mother moved her hands up till they rested on the boy's shoulders, and looked into his face. "Your father has had to lay great matters in your hands. If you deem that you cannot but speak of them to someone, then tell your mother what is weighing on you. But best of all would it like me if you could altogether be silent, my son."

The light-hued face under the smooth flaxen hair, the great eyes, the full, firm red mouth — how like he was to her father now! Gaute nodded. Then he laid an arm on his mother's shoulder.

Sweetly and sorrowfully it came on Kristin that now she could lean her head against the boy's spare little breast; he was so tall now that when he stood and she sat, her head reached to just above his heart. For the first time it was she that leaned against her child.

Gaute said:

"Isak was alone at home. I showed him not what I was bearing, but said only that I had somewhat I must burn. So he made up a big fire on the hearth before he went out to saddle his horse."

His mother nodded. Then he let go of her, turned towards her, and said, with childish awe and wonder in his voice:

"Mother, know you what they say? They say that father — would have been *King* — "

"It sounds not over-likely, boy — " she answered with a smile.

"But he is of kingly birth, my mother," said the boy earnestly and proudly. "And methinks father would be fitter to be a King than most men."

"Hush." She took his hand again. "My Gaute — you must understand, now that father has shown such trust in you — you and we all must say nor think nothing, but must watch our tongues well till we have learnt somewhat, so that we can judge whether and how we should speak. I ride to Nidaros to-morrow, and if I come to speech with your father alone at any time, I shall surely tell him that you have done his errand well — "

"Take me with you, mother!" begged the boy vehemently.

"We must not let any think, Gaute, that you are aught else than a thoughtless child. You must try, little son, to play about here at home and be as joyful as you can — so will you serve your father best."

Naakkve and Björgulf came slowly up the hill. They came up to their mother and stood there, their young faces strained with feeling. Kristin saw they were still so far children that they took refuge with their mother in this disquietude — and yet were come so far towards manhood that they would fain have comforted her and heartened her, if they could but find the way. She held out a hand to each of the boys. But not much was said between them.

Soon after, they all went down, Kristin with a hand on a shoulder of each of her two eldest sons.

"Why are you looking so at me, Naakkve?" But the boy grew red, turned away his head, and made no answer. He had never before thought about his mother's looks. It was many a long day since he had begun to compare his father with other men — his father was the most comely of them all, and the most like a chieftain. His mother was the mother that had new children, who, as they grew from out the hands of the women, joined the little troop of brothers, to share its life in fellowship, its friendship and its strife; mother had open hands through which flowed all that they needed; mother knew what to do for well-nigh all ills; mother was like the fire on the hearth, she bore the life of the home as the lands round about Husaby bore the crops year by year; life and warmth streamed from her as from the cattle in the byre or the horses in the stalls. The boy had never thought of likening her to other women. . . .

This evening he saw all at once: she was a proud and fair lady. With the broad, white forehead under the linen coif, the steel-grey eyes' straight gaze under the calmly arched brows; with the heavy bosom and the long, shapely limbs. She bore her tall body straight as a lance. But he could not speak of this; he walked on flushed and silent with her hand upon his neck.

Gaute walked behind Björgulf and held by his mother's belt. The elder brother began grumbling because Gaute was treading on his heels — the two set to pushing and scuffling a little. The mother stopped the quarrel and bade them be quiet, and, so doing, her grave face softened into a smile. After all they were but children, her sons.

Kristin lay awake at night — Munan was sleeping at her breast, and Lavrans between her and the wall.

She tried to come to some judgment of her husband's case. She could not believe the peril could be so great. Erling Vidkunssön and the King's cousins at Sudrheim had been charged with disloyalty and treason; yet were they back at their homes now as safe and as rich as ever, though they stood not so high as before in the King's favour.

'Twas like that Erlend had engaged in some unlawful courses to serve Lady Ingebjörg. In all these years he had kept up the friendship with his noble kinswoman; she knew that he had given her some sort of unlawful aid that had to be kept secret, five winters ago when he was her guest in Denmark. Now that Erling Vidkunssön had taken up the lady's cause, and would have put her in possession of the estates she owned in Norway, it might well be that Erling had counselled her to go to Erlend, or that she had turned to her father's kinsman of her own accord, after Erling had made his peace with the King. And that Erlend had dealt foolhardily with the matter ....

But she could not well understand how her kinsfolk at Sundbu could be mixed up in this.

Yet 'twas impossible that the end of the matter could be aught else than that Erlend came to full atonement with the King, if all his offence was to have been too zealous in the King's mother's service.

High treason. She had heard of Audon Hestakorn's downfall — and his death on the gallows at Nordnes — it was in her father's youth that it befell. But frightful misdeeds had been charged against Sir Audun. No; she would not think on such things. 'Twas so little likely that Erlend's cause should have a worse outcome than — than Erling Vidkunssön's and the Haftorssön's, for example.

— Nikulaus Erlendssön of Husaby. Ah, now it seemed so to her too — Husaby was the fairest manor in Norway's land.

She would go to Sir Baard and find out all that could be known. The Treasurer had always been her friend. Olav Lagmand too — in former days. But Erlend had taken it in such bad part when the Lagmand's decree went against him in that suit about the townhouse. And Olav had taken so much to heart the mischance that had befallen his god-daughter's husband.

Near kinsfolk they had none, either Erlend nor she — widespread as their kindred was. Munan Baardssön scarce counted any more. He stood condemned for unlawful dealings when he held the Wardenship of Ringerike; he sought too eagerly to get his many children on in the world — four he had had in wedlock and five outside it. And 'twas said he had fallen away much since Lady Katrin's death. Inge of Ryfylke; Julitta and her husband; Ragnrid, who was married in Sweden, knew but little of Erlend — these were the other children left by Lady Aashild. Between the Hestnes folk and Erlend there had been no friendship since Sir Baard Peterssön died; Tormod of Raasvold was in his second childhood; his and Lady Gunna's children were dead, and their grandchildren still in their nonage.

She herself had no other kin on her father's side in Norway than Ketil Aasmundssön of Skog, and Sigurd Kyrning, who was wedded to Aasmund's eldest daughter. The other daughter was a widow, and the third a nun. The men of Sundbu seemed to be all four mixed in the case.

The sick monk in the Preacher's Cloister was Erlend's only near kinsman. And the man who stood nearest to her in the world was Simon Darre, since he was wed with her only sister.

Munan awoke, whimpering. Kristin turned in her bed and laid the child to her breast on the other side. She could not take him with her to Nidaros, uncertain as all things were. Maybe this would be the last time this little one would drink from his own mother's breast. Maybe this would be the last time in this world that she should lie thus, holding a little child close to her, so blissfully, so blessedly. If Erlend's life were forfeit — ! Blessed Mary, Mother of God, had she on any day or in any hour murmured by reason of the children God had granted her — ? Would this be the last kiss she should ever have from such a little milky mouth as this — ?

5

KRISTIN went to the palace the next evening as soon as she was come into the city. Where in all this great mansion have they put Erlend? she thought, as she looked round at the many stone houses. It seemed to her she thought more of how Erlend might be faring than of what she might have to learn. But she was told, on asking, that the Treasurer had left the town.

Her eyes smarted after the long boat-journey in the glittering sunlight, and her breasts, overfilled with milk, troubled her. When the serving-folk who lay in the room were gone to sleep, she got up and walked the floor all through the night.

Next day she sent Haldor, her own man, to the palace. — He came home, terrified and unhappy — Ulf Haldors-sön, his father's brother, had been taken prisoner on the fjord, trying to come over to the cloister at Holm. The Treasurer was not yet come back.

This tidings put Kristin, too, in the greatest fear. Ulf had not dwelt at Husaby this last year, but had been work-ing as one of the Warden's sheriffs, for the most part at Skjoldvirkstad, of which he now owned the greatest part. What kind of cause could this be, in which so many men were entangled? She could no longer keep at bay the worst fears, sick and worn with waking as she was now.

On the morning of the third day Sir Baard was still not come home. And a message that Kristin had tried to send to her husband did not reach him. She thought of seeking out Gunnulf in the cloister, but felt she could not. She walked and walked up and down the hall at home, with half-shut burning eyes. Sometimes 'twas as though she were walking half in her sleep; but as soon as she lay down, the fear and the pains came over her again so strongly that she was forced to rise again, broad awake, and walk, so as to endure them.

Just after nones Gunnulf Nikulaussön came in to her. Kristin went swiftly to meet the monk.

"Have you seen Erlend — Gunnulf, what is it they charge against him — ?"

"There are heavy tidings, Kristin. No, they will not let any come near Erlend — least of all us cloister-folk. They believe that Abbot Olav has been privy to his plans. 'Tis true he borrowed money there, but the Brothers all swear that they knew naught of what he meant to do with it, when they put the Convent's seal to the deed. And the Abbot will give no account of his doings — "

"Aye. But what is it? — Is it the Duchess that has lured Erlend on to this — ?"

Gunnulf answered:

"It seems rather as though they had to press her hard

before she would join in their plans. The letter that —
someone — has seen a draft of, which Erlend and his
friends sent her last spring, they can scarce lay hands on,
I trow, except they can force the lady to give it up to
them. And they have found no draft. But by the answering
letter and Sir Aage Laurissen's letter, which they took
from Borgar Trondssön at Veöy, it seems sure enough
that she has had such a writing from Erlend and the men
who had bound them to stand by him in this plot. It
seems to be clear that she was long afraid to send Prince
Haakon to Norway — but that they urged upon her that,
whatever the outcome of the matter might be, 'twas not
possible that King Magnus should harm the child, his own
brother. Should Haakon Knutssön not win the crown of
Norway, he would not be much worse off than before —
but these men were willing to venture their lives and all
their goods to set him upon the throne."

For a long time Kristin sat quite still.

"I understand. These are greater matters than those that
were between Sir Erling or the Haftorssöns and the
King?"

"Aye," said Gunnulf in a low voice. "'Twas to be given
out that Haftor Graut and Erlend sailed for Björgvin.
But it was Kalundborg they were bound for, and they
were to bring Prince Haakon back with them to Norway,
while King Magnus was yet abroad about his wooing — "

A little after, the monk said, still low:

"'Tis more than — 'tis well-nigh a hundred years now
since any Norwegian noble has dared the like of this; tried
to overthrow him who was King by inheritance, and set
up a rival King — "

Kristin sat gazing before her. Gunnulf could not see her
face.

"Aye," she said in a while, thoughtfully. "The last men
who ventured on that game were your forbears and Erlend's — and that time, too, my dead and gone kinsmen of
the Gjesling house stood by King Skule."

She met Gunnulf's questioning look, and burst out,
hotly and vehemently:

"I am but a simple woman, Gunnulf — little heed did I
pay when my husband spoke with other men of such
things — and unwilling was I to listen when he would

have spoken with me of them — God help me, such
weighty matters were beyond my understanding. But,
simple woman as I am, unskilled in aught but my household work and the nurture of my children — even I know
that right and justice had too far to travel before any
man's cause could win through to the King and back
again whence it sprung; and I have understood, too, that
the common folk of this land have less prosperity and
harder times now than when I was a child and King
Haakon of blessed memory was our overlord. My husband" — she breathed quickly and tremulously once
or twice — "my husband, I see it now, had taken up a
cause so great that none of the other chiefs of this land
dared set a hand to it — "

"That had he." The monk clenched his hands tightly
together; his voice sank to a whisper. "So great a cause
that many will deem it an ill thing that he should himself
have brought about its downfall — and in such wise — "

Kristin cried out and started up. The sudden violent
motion made the pain in her breast and arms so sharp that
her whole body was bathed in sweat. Wildly and fever-
ishly she turned on the monk and cried loudly:

"'Twas not Erlend did it — it was doomed so to be — it
was his evil chance — "

She flung herself forward on her knees, with her hands
pressed against the bench, and lifted her flushed, despairing visage towards the monk:

"You and I, Gunnulf — you, his bother, and I, his wife
for thirteen years — we should not throw blame on
Erlend, now he is a poor prisoner, in peril of his life
maybe — "

Gunnulf's face quivered. He looked down at the kneeling woman.

"God requite it to you, Kristin, that you can take this
matter so." Again he wrung his wasted hands together.
"God — God grant Erlend life, and the power to repay
you for your faithfulness. God turn away this evil from
you and from your children, Kristin — "

"Speak not so!" She drew herself upright on her knees
and looked up into the man's face. "No good has come
of it, Gunnulf, when you have stepped in, in Erlend's

affairs and mine. None has judged him so hardly as you —
his brother and God's servant!"

"Never has it been my will to judge Erlend more
harshly than — than I must." His white face had grown
yet whiter. "None upon earth have I held so dear as my
brother. It may well be therefore that it wrung my soul
as it had been my own sin that I must myself atone for,
when Erlend offended against you. And then there is
Husaby — 'twas for Erlend alone to carry onward the
race that is mine as well as his. The greatest part of my
heritage I gave into his hands. Your sons are the men that
are nearest to me in blood — "

"Erlend has *not* offended against me! I was no better
than he! Why speak you so to me, Gunnulf? — you have
never been my confessor. Sira Eirik did not blame my
husband to me — he corrected *me* for my sins when I laid
bare my troubles to him. He was a better priest than you
— and he it is that God has set over me, that I should
hearken to him — and he has never said that I suffered
wrong. I will hearken to him!"

Gunnulf had risen when she stood up. He muttered,
his face pale and troubled:

"You speak the truth. 'Tis Sira Eirik you must hearken
to — "

He turned to go, but she caught his hand impetuously:

"Nay, go not from me so! I remember, Gunnulf — I
remember when I was your guest here in this house —
'twas yours then; and you were good to me. I mind the
first time I met you — I was plunged in fear and pain — I
remember you spoke to me in Erlend's excuse — you
could not know — You prayed and prayed for my life
and my child's life. I know that you wished us well, you
held Erlend dear —

"Oh, speak not hardly of Erlend, Gunnulf — which of
us is clean before God? My father grew to be fond of
him; our children love their father. Remember, he found
me weak and easy to lure astray, and he set me in a good
and honourable place. Oh, aye, 'tis fair at Husaby — the
last evening before I left home it was so fair, the sunset
that evening was so beautiful. We have lived many a good

day there together, Erlend and I. — However it may go, however it may go, yet is he my husband, my husband whom I love — "

"Kristin — trust not to the sunset glow and to the — love — you remember now that you fear for his life.

" — I remember a thing when I was young — a sub-deacon only. Gudbjörg, that was after wedded with Alf of Uvaasen, was a servant at Siheim then; she was charged with stealing a golden ring. It came out that she was guilt-less; but the shame and fear had so shaken her soul that the enemy won power over her; she went down to the lake and would have thrown herself in. And she often bore witness for us afterwards that, as she went in, the world seemed to her so red and golden and fair, and the water shone bright and felt warm and comforting, but when she stood in it to her middle she was moved to name Jesu name and cross herself — and then did the whole world grow grey, and the waters cold, and she saw whither she had been bound — "

"Then will I not name it," Kristin spoke softly — she stood stiffly upright — "if I could believe that then I would be tempted to forsake my lord in his need. But methinks 'twould not be Christ's name, but rather the enemy's, that could do the like — "

"I meant not that; I meant — God strengthen you, Kris-tin, to bear your husband's faults with a loving spirit — "

"You see that I do so," said the wife in the same tone.

Gunnulf turned from her, white and trembling. He passed his hand over his face:

"I will go home. I can more easily — at home I can more easily gather my thoughts — that I may do all that lies in my power for Erlend and you. God — God and all holy men preserve my brother's life and freedom. Oh, Kristin — never believe that my brother is not dear to me — "

But after he was gone, Kristin deemed that all things were grown worse. She would not have the serving-folk in with her, but walked and walked, wringing her hands and moaning softly. The evening had grown late, when there came a noise of people riding into the courtyard. A moment after, the door was opened, and a tall, stout man in a riding-cloak, first dimly seen in the twilight, came quickly towards her with jingling spurs and trailing sword.

When she knew him for Simon Andressön, she burst out into loud sobbing and ran towards him with outstretched arms, but she cried out in pain when he drew her to him.

Simon loosed his hold. She stood with her hands on his shoulders and her forehead leaned upon his breast, sobbing helplessly. He put his arm lightly around her:

"In God's name, Kristin!" — It seemed as though there was rescue in the very tones of his dry, warm voice, in the living smell of man that came from him — mixed of sweat, dust of the highroad, horses and leather garments. "In God's name — 'tis all too soon to lose hope and courage yet. — There must be some way out, be sure — "

In a little she had grown calm enough to beg him for pardon. She was quite sick and wretched, she said, for she had had to take her youngest child from the breast so suddenly.

Simon learned how she had fared these three days and nights. He called her serving-maid and asked angrily if there was not a single woman in the house that had wit enough to know what ailed their mistress. But the woman was a raw young maid, and the bailiff of Erlend's townhouse was a widower with two unmarried daughters. Simon sent a man out into the town to fetch a leech-woman, and bade Kristin go to bed. When she had grown a little easier, he would come in and speak with her.

While they waited for the leech-woman, he and his man were served with food in the hall; and over his meal he talked with her as she undressed in the closet. Yes, he had ridden north as soon as he had heard what had befallen at Sundbu — he had come hither, and Ramborg gone thither to be with Ivar's and Borgar's wives. Ivar they had taken to Mjös Castle, but Haavard they had left at large, yet had he been made to promise not to leave the parish. 'Twas said that Borgar and Guttorm had been lucky enough to get clear away — Jon of Laugarbru had ridden out to Raumsdal for tidings, and was to send word hither. Simon had been at Husaby at midday, but had not tarried long. All was well with the boys, but Naakkve and Björgulf had begged and begged him to take them with him.

Kristin had got back her courage and calm when, late in the evening, Simon sat by her on the bed's edge. She lay, in the grateful weariness that comes after racking pains

are gone, and looked at her brother-in-law's heavy sun-
burned face and small strong eyes. It stayed and comforted
her that he had come. Simon grew most grave, indeed,
when he heard more fully how the matter stood, but yet
he spoke hearteningly.

Kristin lay looking at the elkskin belt round his bulky
waist. The great flat buckle of copper thinly coated with
silver, without other adornment than a pierced A and M,
betokening Ave Maria; the long dagger with the silver-
gilt mountings and great rock-crystals set in the hilt; the
poor little table-knife with the handle of cracked horn
mended with brass bands — all this she had known ever
since she was a child as part of her father's everyday gear.
She remembered when Simon had got these things — just
before her father died, he had been minded to give Simon
his gilded best belt, and silver to make plates enough to
lengthen it to fit his son-in-law. But Simon begged he
might be given this one — and when Lavrans said he was
cheating himself, Simon would have it that the dagger, at
least, was a costly piece. "Aye, and then the knife," said
Ragnfrid with a little smile; and then the men laughed
and said: "Aye, the knife, to be sure." For about this knife
her father and mother had had so much debate. It was a
daily and hourly vexation to Ragnfrid to see such an
ugly, paltry thing at her husband's belt. But Lavrans
swore she should never gain her end and part him from it.
"Never have I drawn it against you, Ragnfrid; and 'tis as
fine a knife as any in Norway's land to cut butter with —
when 'tis hot enough."

She begged Simon to let her see the knife, and lay
awhile with it in her hands.

"I could wish that I owned this knife," she said softly
and beseechingly.

"Aye — that I can well believe — glad am I that I own
it — I would not sell it for twenty marks." He caught her
wrists laughingly, and took the knife back from her.
Simon's small plump hands were always so good to touch,
so warm and dry.

A little after, he bade her good-night, took the candle,
and went into the hall. She heard him kneel before the
crucifix in there, stand up again, throw off his boots on the
floor, then in a little lie heavily down on the bed by the

north wall. Then Kristin sank into a fathomless, sweet sleep.

She did not wake till far on in the next day. Simon Andressön had gone out long before, and the house-folk had orders to pray her from him that she stay quietly at home in the house.

It was well-nigh the time of nones before he came back; but he said at once:

"I bear you greetings from Erlend, Kristin — I came to speech with him."

He saw how young her face grew, how soft and anxiously tender. So he took her hand in his while he told his story. It was not much that he and Erlend had been able to say to each other, for the man that had brought Simon to the prisoner had stayed by them all the time. Olav Lagmand had got Simon leave for this visit, for the sake of the kinship that had been between them while Halfrid lived. — Erlend sent loving greeting to her and the children; he had asked much about them all, but most of all about Gaute. Simon thought that in some days Kristin might get leave to see her husband. Erlend had seemed calm and in good heart.

"Had I gone out with you to-day, I might have seen him too," said the wife softly.

Simon said no; he had got in because he came alone. "In many a wise, Kristin, it may be easier for you to make way when a man goes ahead of you."

Erlend was kept in a room in the East Tower, out towards the river — one of the gentlefolks' rooms, though a small one. Ulf, they said, was in the dungeon. Haftor in another cell.

Warily feeling his way, to make sure how much she could bear, Simon told her what he had heard in the town. When he saw that she already understood the case to the full, he hid not from her that he, too, thought it a perilous matter. But all those he had talked to said that 'twas not possible Erlend could have dared to plan such an undertaking, and to carry it forward as far as he had done, except he was sure that he had a great part of the knights and the nobles at his back. And since the malcontent great folk were so strong in numbers, it was not like that the King would dare to deal too hardly with their leader; but rather that he must let Erlend make his peace with him in some fashion.

Kristin asked in a low voice:

"Where does Erling Vidkunssön stand in this matter?"

"*That* I can see many a man would give something to know," said Simon.

Though he said it not to Kristin, and had not said it to the men with whom he had talked of this, it seemed to him little likely that Erlend should have at his back any powerful band of men who had bound themselves to risk life and goods in such a perilous affair — had it been so, they would scarce have chosen him their chief, for that Erlend was rash and unstable all his fellows must know well. It was true that he was kinsman to Lady Ingebjörg and the young pretender; he had enjoyed much power and esteem in these latter years, he was not quite so unpractised in war as were most men of an age with him — was known as one that his men liked and followed — and though he had so often borne him witlessly, yet could he, when he would, speak well and to the point, so that it might well be thought that he had now at long last learnt prudence from his mischances. Simon thought 'twas most like there were some who had known of Erlend's undertaking and had pushed him forward; but he could scarce believe that they had bound them so strongly that they could not draw back now and leave Erlend to bear the brunt alone.

Simon deemed he had seen that Erlend himself looked for naught else, and that his mind seemed made up to having to pay dearly for his desperate venture. "When kine lie mired, 'tis for the owners to hang on to their tails," he said, laughing a little. But, to be sure, Erlend had not been able to say much, with a third man listening.

Simon marvelled that this meeting with his brother-in-law had moved his so much. But the narrow little turret-chamber, where Erlend had prayed him to take a seat on the bed — it went from wall to wall and filled half the room — Erlend's straight, slender form, as he stood by the little slit in the wall whence the light came — Erlend quite unafraid, clear-eyed, untroubled by either fear or hope — he was a fresh, cool, manful fellow, now all the clogging cobwebs of love-dalliance and foolery with women had been blown from off him. True, it was women and the commerce of love that had brought him hither, with all his daring plans that were ended ere yet he had brought them out into the light. But on that Erlend seemed not to think. He stood there like a man who had dared a desperate throw, had lost it, and knew how to suffer defeat well and manfully.

And his wondering and joyous thankfulness when he saw his brother-in-law sat well on him. Simon had said, when he saw it:

"Mind you not, kinsman, that night we watched to-gether by our father-in-law? We gave one another our hands, and Lavrans laid his hand on ours — and we promised him and each other that we would stand together like brothers all our days."

"Aye." Erlend's face lit up with a smile. "Aye, and I trow Lavrans thought not that you would ever need *my* help."

"Nay," said Simon, unmoved, " 'tis most like he deemed you, as you were placed, might well prove a stay to me, and not that you were like to need help from me."

Erlend smiled again:

"Lavrans was a wise man, Simon. And, strange as it may sound — I know that he liked me well."

Simon thought, aye, strange it was, God knew — yet even he himself — despite of all he knew of Erlend, and of all the other had done to him — even he could not help now feeling somewhat of a brother's tenderness towards Kristin's husband. Then Erlend asked of her.

Simon told him of how he had found her, sick and full of fear for her husband. Olav Hermanssön had promised to do his best to have her let in to see him, as soon as Sir Baard came home.

"Not before she is well!" said Erlend quickly. A strange flush, like a young girl's, passed over his brown unshaven face. " 'Tis the one thing I dread, Simon — that I should not have strength to bear it well when I see her."

But in a little he said, as calmly as before:

"I know that you will stand by her faithfully, if she should be left a widow this year. Penniless they will not be, she and the children, having her heritage from Lavrans. And she will have you near by, should she dwell at Jörundgaard."

The day after the Nativity of Mary,* the High Steward, Sir Ivar Ogmundssön, came to Nidaros. Twelve of the King's sworn liegemen from north of Dovre were named now as a Court to try Erlend Nikulaussön's cause. Sir Finn Ogmundssön, the High Steward's brother, had been chosen to set forth the charge against him.

Some time before this, in the summer, Haftor Olavssön of Godöy had slain himself with the little knife that each prisoner had been allowed to keep to cut his food with. Folk said that prison had so told on Haftor that he had not been in his right mind. Erlend said to Simon, when he heard of it, that now he need have no fear of Haftor's tongue. But yet he was much moved.

As time went on, it happened now and then that the guards would make themselves an errand without when Simon or Kristin was with Erlend. Both of these two saw — and spoke of it to each other — that Erlend's first and last thought was to come through this business without the names of those with him in the plot being discovered. To Simon one day he said so straight out. He had promised all who had joined in his counsels that he would hold the rope so that if it came to the worst, the blow should fall on his hands only, "and never yet have I betrayed any that put their trust in me." Simon looked at the man — Erlend's eyes were blue and clear; 'twas plain that he said this of himself in all good faith.

Nor had the King's agents yet been able to track down any other who had taken part in Erlend's treason, save the brothers Greip and Torvard Toressöns of Möre; and these would not confess that they had known the intent of Erlend's plan to be aught else than that he and other men had moved the Duchess to let Prince Haakon Knutssön be brought up in Norway. Afterwards it was meant that the chiefs should make prayer to King Magnus that 'twould be for the good of both his kingdoms if he gave his half-brother the name of King in Norway.

---

* 8th September.

Borgar and Guttorm Trondssön had been lucky enough to escape from the palace at Veöy — none could say how, but folk guessed that Borgar had got help from some woman — he was a comely youth and something of a light liver. Ivar was still in prison in Mjös Castle; young Haavard his brothers seemed to have kept outside their counsels.

While the meeting of King's-men sat in the castle, the Archbishop held a councilium in his palace. Simon was a man with many friends and acquaintances; he could thus tell Kristin what was going forward. All deemed it likely that Erlend would be outlawed and banished and his lands and goods be forfeit to the King. Erlend, too, said 'twas like it would be so; he was in good heart — he meant to seek refuge in Denmark. As things stood in that land, the road to advancement was ever open to a man of mettle and skill in arms, and Lady Ingebjörg would surely welcome his wife as her kinswoman and keep her with her in all seemly honour. Simon would have to take the children, save the two eldest sons, whom Erlend was minded to have with him in Denmark.

Kristin had not been outside the city for a day in all this time, and had not seen her children, save Naakkve and Björgulf; they had come riding into the courtyard one evening alone. Their mother kept them with her for some days; but then she sent them to Raasvold, where Lady Gunna had taken the little ones to be with her. Erlend wished it should be so. And Kristin was afraid of the thoughts that might arise in her if she were to see her sons about her, listen to their questions, and try to make things clear to them. She strove to thrust away from her all thoughts of her wedded years at Husaby. So rich had been those years that they seemed to her now to have been one great calm — even as there seems a sort of calm on a billowy sea when one stands high enough above it on a great cliff. The waves that chase each other seem everlasting and unchanging — even so had life billowed through her soul in those spacious years.

Now was it with her again as in her youth, when she had pitted her will to win Erlend against all things and all men. Now again was her life but a waiting from hour to hour, between the hours when she saw her husband, sat by his side on the bed in the turret-chamber of the castle, talked with him calmly and evenly — till by some chance they would be left alone for a moment, and would fall into each other's arms, with endless passionate kisses and wild embraces.

At other times she sat in Christ's Church, hours at a time. She kneeled, gazing up at St. Olav's golden shrine behind the grated lattice of the choir. Lord, I am his wife. Lord, I held fast to him when I was his in sin and unrighteousness. Through God's mercy were we two, all unworthy, joined together in holy wedlock. Seared with the brands of sin, weighed down with sin's burden, we came together to the threshold of God's house, and together received the body of our Redeemer from the priest's hands. Should I now murmur if God puts my faithfulness to the proof? Should I think of aught else than that I am his wife and he is my husband as long as we both live — ?

The Thursday before Michaelmas, the meeting of the King's-men's court was held and judgment given on Erlend Nikulaussön of Husaby. He was found guilty of having plotted to despoil King Magnus of land and lieges by treachery, to raise revolt against the King within the land, and to lead into Norway forces hired from without. After having made search into all such cases in former times, the judges decreed that Erlend Nikulaussön had forfeited his life and all his goods into King Magnus' hands.

Arne Gjavvaldssön came down to Simon Darre and Kristin Lavransdatter in the Nikulaus town-house. He had been at the meeting.

Erlend had not tried to deny what he had done. Clearly and firmly he had acknowledged his purpose, by these measures to force King Magnus to give his young half-brother, Prince Haakon Knutssön Porse, the kingship of Norway. Arne had deemed that Erlend spoke exceeding well. He had pointed to the great hardships and troubles suffered by the people of the land, by reason that the King, in these later years, had scarce set foot on Norwegian soil, and had ever shown unwillingness to appoint Stadtholders who could do justice and wield the kingly powers. By reason of the King's undertakings in Skaane,

and of the wastefulness and the unwisdom in money matters that were shown by the men he most hearkened to,
the folk suffered oppression and impoverishment, and
could never feel safe from new demands for help and new-
fangled taxes. Since the Norwegian knights and esquires
bearing arms had far fewer rights and liberties than the
Swedish knighthood, it was hard for them to contend with
the Swedes on equal terms, and it was but reason and nature that a man like Sir Magnus Eirikssön, young and
unskilled in affairs, should hearken more to his Swedish
lords and love them more, since they had greater riches
and therefore more power to support him with well-
armed and well-trained warriors.

He and his confederate friends had deemed they had
such sure knowledge of the minds of the greatest part of
the folk, both nobles, peasants and townsmen, in the north
and west of Norway, that they had doubted not at all they
would find full following there, if they could bring forward a Princ' as nearly akin to their dear lord, King
Haakon of blessed memory, as he they had now. He had
looked that then the folk of the land would agree together
that we should move King Magnus to let his brother
mount the throne here; while Prince Haakon should swear
to maintain peace and brotherhood with King Magnus, to
guard the realm of Norway in accordance with the
ancient boundaries, to uphold the rights of God's Church,
the laws and customs of the land as handed down from of
old, and the rights and liberties of both country folk and
townsmen; and to put a stop to foreigners' forcing their
way into the kingdom. This plan it had been his and his
friends' intent to put before King Magnus in friendly
wise. Yet had it ever from of old been the right of the
Norwegian farmers and chieftains to set aside a King
who tried to rule unlawfully.

Of Ulf Saksesön's doings in England and Scotland, he
said that Ulf's intent was but to win favour there for
Prince Haakon, if so be God would grant that he became
our lord. With him in this undertaking there had been no
Norseman, saving Haftor Olavssön of Godöy (to whose
soul God be merciful), his kinsmen Trond Gjesling's three
sons of Sundbu, and Greip and Torvard Toressöns of the
Hatteberg kindred.

Erlend's words had moved his hearers strongly, said
Arne Gjavvaldssön. But at the end, when he spoke of the
support they had looked for from the Church's men, he
had recalled those old rumours of the time when King
Magnus was not yet grown up, and that, Arne deemed,
had been unwise. The Archbishop's officer had taken him
sharply to task — Archbishop Paal Baardssön, as they
knew, both while he was Chancellor and since, bore great
love towards King Magnus, by reason of the King's godly
turn of mind; and folk were fain now to forget that such
rumours had ever been spread about their King; besides,
he was even now about to wed a lady, the Count's daughter of Namur — had there ever been any truth in the
matter, it must be deemed that now Magnus Eirikssön
had altogether turned him from all such things.

— Arne Gjavvaldssön had shown Simon Andressön the
greatest friendliness while Simon had been in Nidaros. It
was Arne, too, who now reminded Simon that it must be
open to Erlend to appeal from this judgment, as unlawfully come to. According to the words of the law, the
charge against Erlend should have been brought forward
by one of his peers; but Sir Finn of Hestbö was a knight,
and Erlend but an esquire. It might well be, thought Arne,
that a new court would find that a harder punishment
than outlawry could not be awarded Erlend.

As for what Erlend had set forth, concerning the kind
of kingly rule he deemed this land would best be served
with, it had sounded fair and fine, truly. And all men
knew where the man was to be found who would have
been glad enough to take the helm and steer this course
while the new King was in his nonage — Arne scratched
the grey stubble on his chin, and glanced across at Simon.

"None has heard from him, or of him, this summer?"
asked Simon in a low voice.

"No. He says, I did hear, that he is out of favour with
the King and stands outside all such matters. 'Tis many a
long day since he has been content to sit so long at home
and listen to Lady Elin's talk. His daughters are as fair
and as dull as their mother, folks say."

Erlend had heard the doom of the court with steadfast calm, and he had saluted the members of the court
as mannerly, freely and fairly when he was led out as

when he had come in. He was calm and cheerful when Kristin and Simon were granted speech with him next day. Arne Gjavvaldssön was with them, and Erlend said that he would follow Arne's counsel.

"Never could I bring Kristin here to come with me to Denmark in former days," he said, putting an arm about his wife's waist. "And I had ever such a mind to go forth into the world with her."

A kind of quiver passed over his face, and of a sudden he kissed her pale cheek vehemently, heedless of the two lookers-on.

Simon Andressön rode out to Husaby to take order for the moving of Kristin's goods to Jörundgaard. He counselled her to send the children, too, to Gudbrandsdal, at the same time. Kristin said:

"My sons shall not depart from their father's house till they are driven out."

"I would not wait for that, if I were in your place," said Simon. "They are so young, they can scarce rightly understand this matter. 'Twere better you let them leave Husaby believing that they are going but to visit their mother's sister, and to see to their mother's heritage in the Dale."

Erlend held with Simon in this. But the upshot was that only Ivar and Skule went south with their aunt's husband. Kristin could not send the two little ones so far away from her. When Lavrans and Munan were brought to her in the town-house, and she saw that her youngest son knew her not again, she quite broke down. Simon had not seen her shed a tear since the first evening he came to Nidaros; but now she wept and wept over Munan as he sprawled and struggled, close pressed in his mother's arms, striving to come to his foster-mother; and she wept over young Lavrans, who crept up into her lap and caught her round the neck, weeping because she wept. So she kept the two with her, and Gaute too — he was unwilling to go with Simon, and it seemed to her unwise to let this child, who was bearing a load all too heavy for his years, out of her sight.

Sira Eiliv had brought the children to the city. He had prayed the Archbishop to let him leave his church awhile and visit his brother at Tautra; and this was readily granted to Erlend Nikulaussön's house priest. And since it seemed to him that Kristin could scarce take care of so many children while staying alone in the city, he proffered to take Naakkve and Björgulf out with him to the cloister.

The last night before the priest and the boys were to set forth — Simon had left already with the twins — Kristin made confession to the holy and pure-hearted men who had been her spiritual father all these years. They sat together many hours, and Sira Eiliv was instant with her to be humble and obedient towards God, and patient, faithful and loving towards her wedded husband. She knelt by the bench where he sat; then Sira Eiliv rose up and knelt by her side, still bearing the red stole, the token of the yoke of Christ's love, and prayed long and fervently without words. But she knew that he prayed for the father and mother and the children and all the household, whose souls' health he had so faithfully striven to further all these years.

The day after, she stood on the shore at Bratören and watched the lay brothers from Tautra setting sail on the boat that was to bear away the priest and her two eldest sons. On the way home she went into the Minorites' Church, and tarried there till she deemed she was strong enough to venture back to her own house. And in the evening, when the two little ones were gone to sleep, she sat with her spinning and told Gaute stories till it was bedtime for him too.

6

ERLEND was held prisoner in the castle till nigh upon Clement's Mass.* Then there came word and letters ordering that he should be taken south under safe-conduct,† to be brought before King Magnus. The King purposed to hold the Yule-tide feast at Baagahus ‡ that year.

---

* 23rd November.
† See Note 16.
‡ See Note 17.

Kristin was thrown into deadly fear. With unspeakable struggles she had used herself to keep calm, with Erlend a prisoner under doom of death. Now was he to be taken far away to a doubtful fate; folk said all manner of things about the King, and in the band of men who were about him Erlend had no friends. Ivar Ogmundssön, who now was Governor of the castle at Baagahus, had spoken of Erlend's treason in the harshest words. And 'twas said he had been set against Erlend the more by being told of some malapert speech of Erlend's about him in former days.

But Erlend was glad of the tidings. Kristin saw, indeed, that he took not the parting now at hand lightly. But this long imprisonment had begun to wear so upon him that he grasped eagerly at the thought of the long sea journey, and seemed careless well-nigh of all else.

In three days all was ready, and Erlend set sail in Sir Finn's ship. — Simon had promised to come back to Nidaros before Advent, when he had cast about him a little and ordered his affairs at home; but if before that there were any new tidings, he had prayed Kristin to send him word, and he would come at once. Now it came to her that she would journey south to him, and from thence would she go on to where the King was, and would fall at his feet and pray for mercy for her husband — gladly would she offer all her possessions to redeem his life.

Erlend had sold or pledged his mansion in Nidaros to divers people; the Nidarholm cloister owned the hall-house now, but Abbot Olav had written lovingly to Kristin, praying her to use it as long as she had need. She was there alone with one serving-maid, Ulf Haldorssön (who had been set free, since they had not enough proof against him), and his nephew Haldor, Kristin's own man.

She took counsel with Ulf, and at first he showed himself somewhat doubtful — he deemed it would be too hard a journey for her across the Dovrefjeld; much snow had fallen in the hills already. But when he saw the woman's anguish of soul, he turned round and counselled that she should go. Lady Gunna took the two little children out to Raasvold; but Gaute would not be parted from his mother, and she felt, too, that she dared not well let the boy stay north of Dovre and out of her sight.

They met such hard weather when they came south on to the high mountains, that by Ulf's counsel they borrowed ski at Drivstuen and left their horses behind there, lest they should be forced to pass the next night in the open. Kristin had not had ski on her feet since she was a little maid, and it was hard for her to make headway on them, though the men upheld and helped her to the utmost. They could come no farther that day than midway in the hills between Drivstuen and Hjerdkinn; and when it grew dark, they had to seek shelter in a birchwood and dig themselves down into the snow. At Toftar they were able to hire horses again; here they plunged into mists, and when they came a little down into the Dale, they found rainy weather. When, some hours after dark had fallen, they rode into the Formo courtyard, the wind howled about the house-corners, the river roared, and a rushing, soughing sound came from the hill-side woods. The courtyard was like a swamp, and deadened the sound of the horses' hoofs — in the Saturday evening holiday from work there was no sign of life on the great manor, and neither the folk nor the dogs seemed to be ware of their coming.

Ulf thundered on the door of the hall-house with his spear; a serving-man came and opened. A moment after, Simon himself stood in the outer-room door, broad and dark against the light behind, with a child on his arm; he drove the barking dogs behind him. He gave a cry when he saw his wife's sister, set down the child, and drew her and Gaute in, taking off their soaking outer garments himself.

It was goodly and warm in the hall, but the air was very thick, for it was a fire-place room with flat ceiling under the upper hall. And 'twas full of folk, and children and dogs seemed swarming out of every corner. Then Kristin made out her two little sons' faces, red, warm and joyous, in behind the table where a candle stood burning. They came forward now and greeted their mother and brother a little shyly. Kristin saw that she had broken in here into the midst of these good folks' comfort and cheer. For the rest, the whole room was in a litter, and at each step she took she trod on crunching nutshells — they were scattered all over the floor.

Simon sent the serving-men and women out on divers errands, and the most part of the dogs and children followed them, as well as the grown folks — these were neighbours with their following. — While he questioned and listened to her, he fastened up his shirt and coat, which had stood open, showing his naked, hairy chest. The children had made him in this plight, he said in excuse. He was indeed in sad disarray: his belt was twisted awry, his hands and clothes dirty, his face sooty, and his hair full of dust and straw.

Soon after came two serving-women and brought Kristin and Gaute over to Ramborg's ladies' hall. A fire had been lit there, and busy serving-maids lighted candles, made up the bed, and helped her and the boy into dry clothes, while others set the board with meat and drink. A half-grown maid with silk-bound plaits brought Kristin a foaming mug of ale. The girl was Simon's eldest daughter, Arngjerd.

Then he himself came in; he had made himself trim and was more as Kristin was used to see him — well and richly clad. He led his little daughter by the hand, and Ivar and Skule came with him.

Kristin asked after her sister, and Simon said Ramborg had gone with the ladies from Sundbu down to Ringheim; Jostein had come to fetch his daughter Helga, and he had wished to take Dagny and Ramborg with him too; he was a cheerful kindly old man, and he had promised to take the best care of the three young wives. So maybe Ramborg would stay there through the winter. She looked to have another child at St. Matthew's Mass or thereabout — and then Simon had thought that 'twas like he might have to be away from home this winter; so she would be better off with her young kinswomen. Oh, no — for the housekeeping here at Formo, it made no odds whether she were at home or away, laughed Simon — for he had never craved of a young child like Ramborg that she should wear herself out with the drudgery of a great household.

On hearing Kristin's plans, Simon said at once that he would go with her south. He had so many kinsfolk there, and so many old friends of his father's and his own, that he hoped he could be of more service to her there than in

Trondheim. Whether it would be wise for her to seek the King herself, he could better judge when they came thither. He would be ready for the road in three or four days.

They went together to mass the next day, being Sunday, and afterward went to see Sira Eirik at Romungaard. The priest was old now; he welcomed Kristin lovingly, and seemed most sorrowful at her mischance. Then they went round by Jörundgaard.

The houses were the same, and in the rooms were the same beds and benches and tables. This was now her own manor, and it seemed most like that 'twas here her sons would grow up, and that here she herself would one day lay her down and close her eyes for ever. But never had she felt so clearly as in this hour that it was on her father and mother all the life of this home had rested. Whatever hidden troubles they might have had to struggle with, warmth and help, peace and safety had flowed out from them to all that lived about them.

Restless and heavy of heart as she was, it wearied her a little when Simon talked of his own affairs, his estates and the children. She saw herself that there was no reason in it; he was ready to help her with all his might; she saw that 'twas most good of him to be willing to leave his home at Yule-tide, and to be parted from his wife at a time like this — he surely was thinking much on whether he would soon have a son now — for he had only the one child by Ramborg as yet, though they would soon have been wedded six years. She could not look that he should take her mischance and Erlend's so much to heart as quite to forget all joy in his own happy lot; but it was strange to go about with him here, and see him so joyful and warm and secure in his own home.

Unwittingly Kristin had thought that Ulvhild Simonsdatter would be like her own little sister, whom she had been named after — would be fair-haired and slender and clear-skinned. But Simon's little daughter was round and fat, with cheeks like apples and a mouth like a red berry, quick grey eyes like her father's in his youth, and with his goodly brown curling hair. Simon loved the bonny, lively child much, and was proud of her quick-witted prattle.

"Though yet this little girl is so ugly and loathly and ill-favoured," he said, putting his hands on each side of her chest and twirling her while he lifted her into the air, "I deem 'tis a changeling that the trolls up in the fell here have brought for her mother and me and put into the cradle, such a grim and grum little thing is it"; then he set her down suddenly, and hastily made the sign of the cross over her three times, as though frighted by his own rash words.

His base-born daughter, Arngjerd, was not fair, but she looked good and understanding, and her father took her about with him as often as 'twas possible. He was full of praises of her handiness. — Kristin was made to look into Arngjerd's chest, and see all the things she had spun and woven and sewed already for her dowry.

"The day I lay the hand of this daughter of mine in a good true-hearted bridegroom's hand," said Simon, looking long after the girl, "will be one of the gladdest days I have known."

To save the cost and to get the journey over quickly, Kristin would not take with her any maid, nor any man other than Ulf Haldorssön. Fourteen days before Yule, then, she and he set forth from Formo, in company with Simon and his two stout young serving-men.

When they came to Oslo, Simon soon learnt that the King would not come to Norway — he was to hold the Yule-tide feast at Stockholm, it seemed. Erlend was in the castle at Akersnes; the Governor of the castle was away, so that in the meantime 'twas not possible for any of them to see the prisoner. But the Under-Treasurer, Olav Kyrning, promised to let Erlend know they were in the city. Olav showed much friendliness toward Simon and Kristin, for his brother was wedded to Ramborg Aasmundsdatter of Skog, so that he counted him a far-off kinsman of Lavrans' daughters.

Ketil of Skog came into the city and bade them out to Skog to drink Yule-tide with him; but Kristin would not keep the holy-days with feasting while things stood thus with Erlend. And Simon would not go alone, though she prayed him much to do so. Simon and Ketil knew somewhat of each other, but Kristin had seen her uncle's son but once since he was grown up.

Kristin and Simon took lodging in the same mansion where she had once been his parents' guest when they two were betrothed; but they dwelt in another house. There were two beds in the room; she slept in one, and Simon and Ulf in the other; the men lay in the stable.

On Christmas Eve, Kristin wished to go to the midnight mass in Nonneseter church — she said it was because the sisters sang so sweetly. So they all five went together. The night was clear starlight, mild and fair, and it had snowed a little in the evening, so that 'twas somewhat light. When the bells began ringing from the churches, folk streamed out of all the houses, and Simon had to lead Kristin by the hand. Now and then he stole a glance at her. She was grown greatly thinner this last autumn, but it was as though her tall, straight form had got back somewhat of the young maid's supple and tranquil grace. Over her pale face there was come again the look she had had in youth of calm and gentleness covering a deep and hidden, listening expectancy. She had taken on a strange ghostly likeness to the young Kristin of that Yule-tide long ago. — Simon pressed her hand, and knew not that he had done so till he felt an answering pressure. He looked up — she smiled and nodded, and he understood that she had taken his hand-clasp as a warning to her to be brave — and now she was striving to show him that indeed she was brave.

Toward the end of the holy-days, Sir Munan came to her — he had only now heard she was in the city, he said. He greeted her heartily, likewise Simon Andressön and Ulf, whom he spoke to at every second word as "kinsman" and "dear friend." It might be hard for them to gain sight of Erlend, he said; he was most strictly guarded — *he* had not been able to win in to see his cousin. But Ulf said, with a laugh, when the knight was gone, that he deemed not Munan had pressed so exceeding hard to gain entrance — he was in such deadly fear of being tangled in the affair that he could scarcely bear to hear it named. Munan had grown exceeding old, exceeding bald, and wasted in flesh; the skin hung lose on his bulky frame. He dwelt out at Skogheim, and had with him one of his base-born daughters who was a widow. The father would gladly have been quit of her, for none of his other children, neither those born in wedlock nor the others, would come near him so long as this half-sister ruled his house; she was an overbearing, greedy and shrewish woman. But Munan dared not bid her begone.

At length, some time on in the new year, Olav Kyrning got leave for Erlend's wife and Simon to see him. And now again it fell to Simon's lot to bear the sorrowful wife company at these heart-breaking meetings. Much stricter watch was kept here than at Nidaros to see that Erlend spoke with none, except the Governor's folk were by.

Erlend was calm as before, but Simon saw that this waiting was beginning to wear upon the man. He made no complaint at any time, and said that he suffered no ill-usage and that all was done for him that could be done; but he owned that he was much plagued with the cold — there was no fire-place in the cell. And 'twas not in his power to indulge himself overmuch in cleanliness — though, he laughed, had he not had the lice to fight with, like enough the time would have seemed yet longer out here.

Kristin, too, was calm — so calm that Simon waited in breathless fear for the day when she should break down altogether.

King Magnus made his royal progress in Sweden, and there seemed no likelihood that he would cross the boundary soon, or that any change in Erlend's state was at hand.

On the day of Gregory's Mass,* Kristin and Ulf Haldorssön had been at church at Nonneseter. When, on their way back, they had crossed the bridge over the Nonnebeck, she took not the way down towards her lodging, which lay near the Bishop's palace, but turned eastward towards the open place by Clement's Church, and into the narrow lanes between the church and the river.

The day was grey and thick — there had been soft weather for a time — and their footgear and the skirts of their cloaks quickly grew wet and heavy with the yellow clay of the riverside. They came out on the open lands towards the high bank of the river. Once their eyes met.

---

* 13th February.

Ulf laughed noiselessly, and his mouth twisted into a sort of grimace, but his eyes were sorrowful; Kristin smiled — a strange, sick smile.

Soon after they stood at the edge of the high ground; there had once been a landslip in the clay bank here, and Fluga's house lay right under it, so close up against the dirty yellow slope, where a few black stunted weeds were growing, that the stench of the pigsty, which they looked down on, came rankly up to them — two fat sows were snuffling about in the black mud. The riverbank was but a narrow strip here; the muddy grey river-current, with the jostling ice-flakes on it, came right up to the tumble-down houses with their bleached grey shingle roofs.

Whilst they stood there, a man and a woman came up to the fencing of the sty and looked at the pigs — the man leaned over and began scratching one of the sows with the shaft-end of the silver-mounted light axe he was using as a staff. It was Munan Baardssön himself, and the woman was Brynhild. He looked up and was ware of them — he stood gaping up at them, and Kristin called out a cheerful greeting.

Sir Munan fell a-laughing loudly.

"Come down and have a drink of warm ale to keep out this filthy weather," he called up.

As they went down to the gate of the houses, Ulf told Kristin that Brynhild Jonsdatter kept neither lodging-house nor ale-tap any more. She had been in trouble many times, and at last had been threatened with flogging, but Munan had got her out and gone surety for her that she would altogether cease her unlawful traffickings. Her sons, too, had now got so far on in the world that for their sake their mother was forced to think of bettering her ill repute. After his wife's death, Munan Baardssön had taken up with her again, and he was often to be found in her house.

He met them at the gate.

"Here we are — kinsfolk, all four of us, in a fashion," he snickered — he was a little in drink, but not much. "You are a good woman, Kristin Lavransdatter, pious but not proud. — Brynhild, too, is an honourable, worshipful woman now — and I was not yet a wedded man when I got the two sons I have by her — and they are much the best of all my children. — I have told you as much every day in all these years, Brynhild. Inge and Gudleik are dearest to me of all my children — "

Brynhild was comely still, but her skin was a pale yellow, and looked as if it must feel clammy, Kristin thought, as when one has stood all day over the fat-cauldrons. But her house was well kept, the food and drink that she put on the board were of the best, and the vessels fair and clean.

"Aye, I look in here when I have an errand in Oslo," said Munan. "You understand, the mother is fain to hear tidings of her sons. Inge writes to me from time to time, for he is a learned man, Inge; a Bishop's commissary must be so, you know — and I got him well wedded too, with Tora Bjarnesdatter of Grjote; think you many men have got such a wife for their bastard? So we sit here and talk about this, and Brynhild bears in the meat and ale to me, just as she used in days gone by, when she bore the keys of my house at Skogheim. 'Tis heavy work sitting out there now thinking of my wife that is gone. — So I ride in hither to find a little comfort — when it chances that Brynhild is in such mood that she grudges me not a little friendliness and comfort."

Ulf Haldorssön sat with his chin resting on his hand, looking at the lady of Husaby. Kristin sat and listened and answered quietly and gently and mannerly — as calm and courtly as if she had been a guest at one of the great folk's manors at home in Trondheim.

"Aye, you, Kristin Lavransdatter, you won a wife's name and came to honour," said Brynhild Fluga, "though you came willingly enough to meet Erlend in my loft. I have been called slut and loose woman all my days — my stepmother sold me into his hands, there — I bit and scratched, and left the marks of every one of my nails on his face before he had his will of me — "

"Must you speak of this ever?" grumbled Munan. "Be sure — I have told it you so often too — I had let you go in peace had you borne you like a human creature and bidden me spare you — but you flew at my face like a wildcat before I was well inside the door — "

Ulf Haldorssön laughed softly to himself.

"And I dealt well by you evermore thereafter," said Munan. "You had but to point to a thing and I gave it you — and our children — aye, for sure they are far better off and safer this day than Kristin's poor sons — God guard the poor young lads, so as Erlend has guided things for his children! Methinks that should mean more than the name of wife to a mother's heart — and you know that I wished often your birth had been such that I could have wedded you lawfully — no woman have I liked as well as you — though you were but seldom kind or good to me — and the wife I did get, God reward her — ! I have set up an altar for my Katrin and myself out in our church at home, Kristin — I have thanked God and Our Lady every day for my wedded life — no man has had a better — " He whimpered and sniffed.

Soon after, Ulf Haldorssön said that they must go. He and Kristin exchanged not a word on the way back. But outside the door she reached out her hand to the man:

"Ulf — my kinsman and my friend."

"If it could help aught," he said in a low voice, "I would gladly go to the gallows in Erlend's stead — for his sake and for yours!"

That evening, a little before bedtime, Kristin sat alone in the hall with Simon. Of a sudden she began telling him where she had been that day. She told of the meeting at Brynhild Fluga's.

Simon sat on a stool not far from her. Leaning forward a little, with his arms resting on his thighs, and hands hanging down, he sat looking up at her with a strange, searching look in his small, sharp eyes. He spoke not a word, and not a muscle moved in his big, heavy face.

Then she let fall that she had told all to her father, and what he had said to her.

Simon sat as before, immovable. But in a while he said calmly:

" 'Twas the one thing I ever prayed of you, in all the years we have known each other — if I mind aright — that you would not — but if you could not keep silence to spare Lavrans, why — "

Kristin's whole body trembled:

"Aye! But — oh, Erlend, Erlend, Erlend — "

At the wild cry the man leaped up — Kristin had thrown herself forward, with her head between her arms, and was rocking her body from side to side, still calling on Erlend, between quivering, moaning sobs that seemed to tear their way out of her body.

"Kristin — in Jesu name!"

When he seized her upper arm and tried to stay her sobbing, she flung herself on him with all the weight of her body, and caught him around the neck, while through her weeping she went on calling her husband's name.

"Kristin — be still — " He held her tight in his arms and saw that she marked it not — she was weeping so that she could not stand upright. Then he lifted her up in his arms — crushed her to him a moment, and then bore her over and laid her on the bed.

"Be still," he prayed her again, in a choking, almost a threatening voice — he laid his hands over her face, and she caught his wrists and arms and clung close to him.

"Simon — Simon — oh! he must be saved — "

"I do what I can, Kristin — but now you *shall* be still." He turned sharply, walked over to the door and out into the yard. He shouted, till the echoes rang back from the house-walls, for the maid whom Kristin had hired in Oslo. The girl came running, and Simon bade her go in to her mistress. In a moment she came out again — her mistress would be alone, she said affrightedly to Simon, who still stood on the same spot.

He nodded and walked over to the stables; and stayed there till Gunnar, his man, and Ulf Haldorssön came to give the beasts their evening feed. Simon talked with them awhile, and then went with Ulf back again to the hall.

Kristin saw not much of her brother-in-law the next day. But after nones, as she sat sewing on a garment she meant to take out to her husband, he came running in, said naught to her and looked not at her, but flung open the lid of his travelling-chest, filled his silver goblet with wine, and rushed out again. Kristin stood up and went after him. Before the door of the hall stood a strange man, still holding his horse — Simon drew a gold ring from his finger, dropped it into the goblet, and drank to the newcomer.

Kristin guessed what this must be, and cried out joyfully:

"You have a son, Simon!"

"Aye." He slapped the messenger on the shoulder, as the man, thanking him, put up the goblet and the ring safely inside his belt. Then Simon caught his wife's sister round the waist and whirled her round and round. He looked so glad that Kristin could not but put her two hands on his shoulders — and then he kissed her full on the mouth and laughed aloud.

"Then 'twill still be the Darre stock that will hold Formo when you are gone, Simon," she said joyfully.

"Aye, so will it be — if God will. — No, to-night I would go alone," he said, when Kristin asked if they should go together to evensong.

That night he said to Kristin that he had heard Sir Erling Vidkunssön was now at his manor of Aker near Tunsberg. And that morning he had hired him passage in a ship going down the fjord — he was minded to speak with Sir Erling of Erlend's case.

Kristin said not much in answer. They had barely touched on it before — had kept them from going much into the question — whether Sir Erling had been privy to Erlend's plans or not. Simon said now it would be well that he should ask Erling Vidkunssön's counsel as to Kristin's plan that Simon should go with her to Lavrans' powerful kinsfolk in Sweden to claim kinship and crave their help.

Then Kristin spoke her thought:

"But, now you have had these great tidings, brother-in-law, methinks that it were but reason you should put off this journey to Aker — and first ride up to Ringheim and see to Ramborg and your son."

He was forced to turn away, so overcome was he. He had waited so for this — whether Kristin would make any sign that she understood how he longed to see his son. But when he had mastered himself somewhat, he said, with some shyness in his voice:

"I have been thinking, Kristin — mayhap God will

vouchsafe that the boy thrive and prosper the better if I can be patient and hold in check my longing to see him till I have managed to help you and Erlend a little forward in this matter."

The day after, he went out and bought rich and costly gifts for his wife and the boy — and for all the women, too, who had been with Ramborg when she bore the child. Kristin took out a fair silver spoon she had had from her mother's heritage — it was to be for Andres Simonssön — but to her sister she sent the heavy silver-gilt chain that Lavrans had given her in her childhood along with the reliquary cross. The cross she fastened now to the chain Erlend had given her as a betrothal-gift. The next day at midday Simon sailed.

In the evening the ship lay to under an island in the fjord. Simon stayed on board; he lay in a sleeping-bag of hide with some pieces of wadmal over him, looking up at the starry skies, where the constellations seemed to climb and dive again as the boat pitched on the sleepily gliding swell. The water splashed and the ice-flakes scraped and thumped softly against the vessel's sides. 'Twas almost comforting to feel the cold creeping farther and farther through his body. It deadened —

Yet now he was sure: so ill as things had been with him they could never be again. Now that he had a son. It was not that he thought he could be fonder of the boy than he was of his daughters. It was somewhat else. For all the heart's gladness that his little maids could give him, when they sought their father with their games and laughter and prattle — sweet as it was to hold them in his lap and feel their soft hair against his chin — yet in this wise a man could never take his place in the succession of the men of his house, if his lands and goods and the memory of his doings in the world must pass with a daughter's hand over into a strange kindred. But now, when he might hope, if God would but grant that this little son should grow to manhood, that a Formo son should come after father — Andres Gudmundssön, Simon Andressön, Andres Simonssön — now it must surely follow of itself that he must stand before Andres as his father had stood before himself, an honest man in his secret thoughts no less than in his open acts.

— Sometimes things had been so that he understood not how he could bear it any longer. Had he seen but *one* token that she understood aught! But she was to him as though they had been brother and sister by blood — careful for his well-being, kind and loving and gentle — And he knew not how long this would last — how long they should live together in this wise in one house. Did the thought never come to her that he could not forget — that even though he was wedded with her sister, he could never quite forget that they two had been meant once to live together in wedlock?

But now he had this son. He had ever been ashamed to add in his own words aught of his wishes or his thanks when he said over his prayers. But he deemed that Christ and Mary Virgin knew well what it had meant that he had said double number of Paternosters and Aves each day in these last days. And he would keep on with this so long as he was from home. And in other wise, too, he would show his thankfulness in fitting and open-handed fashion. And thus maybe he would win help on this present journey too.

Though, indeed, he deemed himself there was little reason to look for much from this visit. Sir Erling was quite estranged now from the King. And however powerful and secure the former Regent of the realm might be, and however little he needed to fear the young King, who was much more ticklishly placed than was he — the richest and most high-born man in Norway — yet it could not be looked for that he should be willing to anger King Magnus yet more against himself by pleading Erlend Nikulaussön's cause, and bringing suspicion on himself of having been privy to Erlend's treason. Even if he had had a part in it — aye, even if he had been at the bottom of the whole plan, ready to step in and have himself placed at the head of affairs the moment a minor King was once more upon the throne — he would scarce feel himself bound to venture aught to help the man who had brought the whole plan to ruin for the sake of a shameful love-adventure. 'Twas as though Simon half forgot it when he was with Erlend and Kristin — for they, too, seemed scarce to remember it any more. But so it was that Erlend had himself wrought the mischief — 'twas his doing that naught else had come of the whole undertaking than ruin for himself — and for the good men who had been betrayed by his wantonness and folly.

But he must try all shifts to help her and her husband. And now he began to hope; for mayhap God and Mary Virgin, or some of the saints whom he had used to honour with offerings and almsgivings, would vouchsafe their aid in this as well.

He came to Aker somewhat late the next evening. A steward met him, and sent off men, some with the horses, some with Simon's man to the serving-men's house; while Simon himself went to the loft-room where the knight was sitting drinking. Sir Erling himself came out into the balcony straightway, and stood there while Simon mounted the stair; then he greeted his guest courteously enough, and led him into the hall, where was Stig Haakonssön of Mandvik with Erling's only son, Bjarne Erlingssön, a quite young man.

He was welcomed fairly enough — the serving-folk took from him his outer garments and bore in meat and drink. But he guessed that the men guessed — at least Sir Erling and Stig — what he was come for, and he felt that they were holding back. So when Stig began saying how rarely he was to be seen in this part of the country now — how seldom he darkened the doors of his former kinsfolk — and asked if he had ever been further south than Dyfrin since Halfrid died — Simon answered: No, not before this winter. But now he had been in Oslo some months with his wife's sister, Kristin Lavransdatter, who was wed with Erlend Nikulaussön.

On this there was a short silence. Then Sir Erling asked courteously after Kristin and Simon's wife and his brothers and sisters; and Simon asked after Lady Elin and Erling's daughters, and how things were with Stig, and how his old neighbours at Mandvik were, and what were the tidings from there.

Stig Haakonssön was a stoutly made, dark-haired man, some years older than Simon, son of Halfrid Erlingsdatter's half-brother, Sir Haakon Toressön, and brother's son of Erling Vidkunssön's lady, Elin Toresdatter. He had lost the Wardenship of Skidu and the Governorship of the

Tunsberg castles two winters back, when he fell out with the King; but still was well enough off with his Mandvik estates. But he was a widower and childless. Simon knew him well, and had been good friends with him, as with all his first wife's kin — even if the friendship had not been over-hot. He knew exceeding well what they had all thought of Halfrid's second marriage — Sir Andres Gudmundssön's younger son was doubtless a man of substance and of good birth, but an even match for Halfrid Erlingsdatter he was not — besides that he was ten years younger; they could not understand why she had set her heart on this young man — but they must e'en let her do as she would, since she had suffered such unbearable misery with her first husband.

Erling Vidkunssön Simon had met but few times before; and then it had ever been in Lady Elin's company; and at such times no sound ever came from him — none needed say more than yes or no where she was in presence. Sir Erling had aged not a little since that time — he was grown somewhat stouter, but his form was still comely and noble, for he bore himself exceeding fairly, and it suited him well that his pale, reddish-yellow hair was now turned a shining silver-grey.

The young Bjarne Erlingssön Simon had not seen before. He had been brought up near Björgvin in the house of a cleric, a friend of Erling's — 'twas said among the kindred that this was because his father would not have him live out at Giske in the midst of a pack of foolish women. Erling himself was there no more than he was forced to be, and he dared not take the boy with him on his constant journeys, for Bjarne had been weakly and ailing as a growing lad, and Erling had lost two other sons in their childhood.

The boy looked exceeding comely as he sat with the light behind him, showing his side-face. Black, tightly curling hair rolled forward over his forehead, his great eyes seemed black, his nose was large and strongly curved, the mouth full and firm and fine and the chin well formed. Withal he was tall, slender and broad-shouldered. But when Simon had to sit down to the table to eat, the serving-man moved the candle, and now he saw that the skin of Bjarne's neck was quite eaten up with the scars of scrofula — they stretched on both sides right up under the ears and forward beneath the chin, dead, dull-white patches and bluish-red stripes and swollen knots. And then Bjarne had a trick of time and again pulling up the hood of the round fur-edged velvet cape that he wore even here in the room — pulling it up half-way over his head. When, soon after, it grew too hot for him, he would turn it down, and then draw it up again — he seemed not to know that he did it. Simon felt his hands grow quite restless at length with but looking on at this — though he tried to keep from looking.

Sir Erling scarcely took his eyes from his son — but he, too, seemed not to know that he was gazing so intently at the boy. Erling Vidkunssön's face was set and unchanging and his pale-blue eyes showed his feeling but little — but beneath the somewhat vague and watery glance there seemed to lie the cares and thought and love of endless years.

So the three elder men exchanged mannerly, sluggish talk, while Simon ate, and the youth sat fiddle-faddling with his hood. Afterwards all four sat drinking for a fitting space of time, and then Sir Erling asked if Simon were not weary with his journey, and Stig asked if he would be pleased to sleep with him. Simon was glad to be able to put off speaking of his errand. This first evening at Aker had left him not a little cast down.

The next day, when he broached the matter, Sir Erling's answer was much what Simon had looked for. He said that King Magnus had never hearkened to him willingly, and he had seen, from the time Magnus Eiriksön was old enough to have a will of his own, that his will had been that Erling Vidkunssön should have naught to say in his affairs when once he was of age. And since the quarrel between him and his friends on the one side, and the King on the other, had been made up, he had heard, and tried to hear, naught of the King or the King's friends. If he pleaded Erlend's cause with King Magnus, it would scarce avail the man much. He knew well enough that many in this land believed that he had been in some wise at the bottom of Erlend's undertaking. But, whether Simon believed him or not, neither he nor his friends had known aught of what was hatching. Had this matter come to light

in another fashion, or had these venturesome young dare-
devils risked their throw and failed, then would he have
stepped in and striven to make their peace. But as things
had gone, he deemed not that any could justly crave of
him that he should come forward and thereby strengthen
all men's suspicion that he had played a double game.

But he counselled Simon to have recourse to the Haftorssöns. They were the King's cousins, and, when they
chanced not to be at feud with him, they kept up between
them a friendship of a kind. And, so far as Erling understood the matter, the men whom Erlend was shielding
were rather to be found amongst the Haftorssöns' party,
and among the youngest of the nobles.

Now, as Simon knew, the King was to hold his wedding
in Norway this summer. And there might then be a fitting
occasion for Magnus to show mildness and clemency to
his enemies. And the King's mother and Lady Isabel would
doubtless come to the wedding-feast. Since Simon's mother
had been Queen Isabel's maid of honour in her youth,
Simon might turn him to Lady Isabel, or Erlend's wife
might throw herself before the King's bride and Lady
Ingebjörg Haakonsdatter with prayers for their intercession.

Simon thought that the last shift of all to try would be
for Kristin to kneel to Lady Ingebjörg. Had the Duchess
understood what honour was, she had sure long since
come forward and rescued Erlend from his straits. But
when he had named this once to Erlend, he had but
laughed and said, the lady had always so many ticklish
matters of her own to see to; and doubtless she was angry,
since it now seemed but little like that her dearest child
should ever bear the name of King.

7

SPRING was come when Simon Andressön journeyed north
to Toten, to fetch his wife and his little son and take them
home to Formo. Then he stayed there awhile to see a little
to his own affairs.

Kristin would not remove from Oslo. And she dared not
yield to her hungry, burning longing for her three sons

who were up there in the Dale. That she might still hold out and endure the life she was now living from day to day, she must not think on her children. She held out, she seemed calm and brave; she spoke with strangers and listened to strangers, and bore with their counsels and comfort; but to do this she must hold fast to the thought of Erlend, of Erlend alone! In the stray moments when she failed to hold fast her thoughts in the grasp of her will, pictures and thoughts flashed through her mind: Ivar stood in the wood-shed at Formo with Simon, watching his uncle intently, as he searched out a piece of wood to helve the boy's hatchet, bending and testing the sticks with his hands. Gaute's fair boyish face set manfully as, bending forward, he struggled against the snowstorm that grey winter day in the mountains — his ski slipped back, he sank backward some way down the slope, and landed deep in a snow-drift — and for a moment his manful mien was all but gone, and he was an over-wearied, helpless child. Her thoughts would turn to the two little ones; 'twas like Munan could both walk and talk a little now — was he as lovely as the others had been at that age? Lavrans had perhaps forgotten her. And the two big ones out in the cloister at Tautra — Naakkve, Naakkve, her first-born — How much did the two big ones understand, and what were their thoughts — how did Naakkve, child as he was, endure the thought that nothing in life now was like to be for him as she and he himself and all men had deemed that it would be?

Sira Eiliv had sent her a letter, and she had told Erlend what was written in it of their sons. Else they never spoke of their children. They spoke not any more either of the past or of the future. Kristin brought him a piece of clothing or a dish of food; he asked her how things had gone with her since he saw her last; they sat hand in hand on his bed. Then sometimes it might chance that they were left alone a moment in the small, cold, dirty, stinking room — and they clung together with dumb, burning caresses, hearing, without marking it, Kristin's woman laughing with the watchmen outside on the stair.

Time enough, when he had either been taken from her or given back to her, to face the thought of their troop of children and the change in their lot — of all else in her life save this man beside her. She could not bear to lose an hour of the time that was left them together, and she dared not think of the meeting again with the four children she had left in the north — so she was fain to assent when Simon Andressön proffered to go alone to Trondheim and, along with Arne Gjavvaldssön, watch over her interest in the settlement of the forfeited estates. Much richer King Magnus was not like to be for Erlend's possessions — the man was more heavily in debt than he himself had had any knowledge of, and he had raised moneys that had been sent off to Denmark and Scotland and England. Erlend shrugged his shoulders and said with a half-smile that he looked not now to reap any return from *them*.

Thus Erlend's case stood much as before when Simon came back to Oslo about Holy Cross day in the autumn.* But he was dismayed to see how worn out they looked, both Kristin and his brother-in-law, and he felt a strange, sinking qualm at his heart when they both had yet enough self-mastery to thank him for coming hither at this time of year, when he could least well be spared from his estates at home. But now were all folks' faces set towards Tunsberg, where King Magnus was come to await his bride.

A little on in the month, Simon managed to hire passage in a ship bound thither, with some merchants who were to sail in eight days time. Then one morning a strange serving-man came to pray Simon Andressön to be at the pains of coming at once to St. Halvard's Church — Olav Kyrning waited for him there.

The Under-Treasurer was vehemently stirred. He was holding charge at the castle, whilst the Treasurer was at Tunsberg. And, the evening before, there had come a company of gentlemen who showed him a letter under King Magnus' seal, signifying that they were to inquire into Erlend Nikulaussön's case; and he had had the prisoner brought in to them. Three of them were foreigners, Frenchmen doubtless — Olav had not understood their speech, but the chaplain had spoken with them in Latin this morning, and " 'tis said they are kinsmen of the lady who is to be our Queen — a fair beginning!" They had put

---

* 14th September.

Erlend to the question by torture — they had with them a kind of ladder and some fellows used to work such things. To-day he had denied to bring Erlend out of his chamber, and had set a strong watch — for so much he was ready to answer, for these were lawless doings, such as never were heard tell of in Norway before!

Simon borrowed a horse from one of the priests of the church and rode with Olav straightway out to Akersnes.

Olav Kyrning looked a little fearfully at the other's grimly set face, over which stormy waves of red were beating. Now and again Simon made a wild, violent movement, as if knowing not what he did — and the strange horse leapt aside, reared and balked under his rider.

"One can see on you, Simon, that you are angered," said Olav Kyrning.

Simon scarce knew himself what was uppermost in his mind. He was so stirred to the depths that at times he felt qualms of sickness. The blind and wild feeling that struggled in him and goaded him to utmost fury was a kind of shame — a helpless man, without weapon or defence, forced to suffer strangers' fists in his clothes, strangers manhandling his body — it was like hearing of the outraging of women; he grew dizzy with thirst for revenge, with longing to see blood shed for it. No — such things had never been the use and wont of this land — would they accustom Norwegian nobles to suffer such things — ? That should never be!

He was sick for horror of what he was to see — fear of the shame he must bring on another man by seeing him in such a pass over-powered him above all other feelings, when Olav Kyrning opened to him the door of Erlend's cell.

Erlend lay on the floor, stretched out aslant from one corner of the room to the other; he was so tall that only thus could he find room to stretch out at full length. Some straw and clothes had been laid under him on the thick layer of filth that covered the floor, and his body was covered over with his dark-blue fur-lined cloak right up to the chin, so that the soft grey-brown marten fur mingled with the curly, tufted black beard Erlend had grown while in his prison.

His mouth showed white through the beard; his face

was snow-white. The great, straight-lined triangle of the nose stood out monstrously high above the sunken cheeks, the grizzled hair lay in clammy, separate wisps back from the high, narrow forehead — on each of the hollow temples was a great bluish-red mark, as though something had pressed on or gripped him there.

Slowly, with labour, he opened his great sea-blue eyes; essayed a sort of smile when he recognized the man; his voice sounded a little veiled and like a stranger's.

"Sit down, brother-in-law —" He moved his head slightly towards the empty bed. "Aye — now have I learned somewhat new — since we last met —"

Olav Kyrning bent over Erlend and asked if there was aught he would. There was no answer — doubtless because Erlend could not speak — and he took away the cloak from over him. Erlend had on him naught but a pair of linen drawers and a rag of shirt — and the sight of the swollen and discoloured limbs shook and maddened Simon like some loathsome horror. He wondered whether Erlend had a like feeling — a shade of red came over his face as Olav passed a wet cloth, which he dipped into a vessel of water, down over his arms and legs. And when he laid the cloak over again, Erlend pushed it into place with some small movements of his limbs and by drawing the hood up with his chin, till he was quite covered up.

"Aye," said Erlend — he was a little more like himself in voice now, and the smile on his pale mouth was a little plainer; "next time — will be worse! But I am not afraid — none need be afraid — they will break naught out of me — in this way —"

Simon felt within himself that the man spoke the truth. Torture would not force a word out of Erlend Nikulaussön. There was naught that he might not do, might not reveal, in anger or in recklessness — he would never be moved a hair's breadth by force. And Simon felt that the shame and insult that he himself suffered on another man's behalf, Erlend scarce felt at all — he was filled with an obstinate joy in defying his torturers and a contented trust in his hardihood. He, who ever broke down so pitifully when he came up against a firm will, who might himself doubtless have been cruel in a moment of fear, rose above himself now that in this cruelty he scented an opposer weaker than himself.

But Simon's answer, growled through his teeth, was:

"Next time — I trow there will be none! What say you, Olav?"

Olav shook his head, but Erlend said, with a shadow of his old reckless flippancy in his voice:

"Aye, if I could but — believe it — as firmly as you! But these gentry will scarce — be content with this — " He grew ware of the working of Simon's heavy, sinewy face: "Nay, Simon — kinsman!" — Erlend would have raised himself on his elbow; the pain forced from him a strange muffled groan, and he sank back in a swoon.

Olav and Simon ministered to him clumsily. When the swoon was over, Erlend lay a little with open eyes; he spoke then, more gravely:

"See you not — it means — much — for Magnus — to get on the track of — what men he would better not trust — farther than he can see them? So much unrest — and discontent — as there has been — "

"Aye — if he deems that this will quench the discontent — " said Olav Kyrning threateningly. Then said Erlend, in a low and clear voice:

"I have dealt so in this matter — that few will deem — it matters much how it goes with me — I know that myself — "

The other two men reddened. Simon had thought that Erlend saw not this himself — and never before had Lady Sunniva been so much as hinted at between them. Now Simon broke out desperately:

"How could you have borne yourself so recklessly — so madly?"

"Nay, I understand it not either — now," said Erlend simply. "But — in hell's name! — how could I have thought she could read writing? She seemed — most unlearned — "

His eyelids drooped and closed, he was nigh swooning once more. Olav Kyrning muttered about fetching something, and went out. Simon bent over Erlend, who now again lay with eyes half opened.

"Brother-in-law — was — was Erling Vidkunssön with you in this?"

Erlend shook his head a little, with a slow smile:

"By Jesus, no! We thought — either he would not be bold enough to join with us — or else he had kept all things in his own hands. But ask not, Simon — I will say naught — to any — so only am I sure not to let aught slip out — "

On a sudden Erlend whispered his wife's name. Simon bent over him again — he looked the other should ask him to bring Kristin to him now. But he said quickly, as if in a flicker of fever:

"She must not hear of this, Simon. Say order has come from the King that none is to come near me. Take her out to Munan — to Skogheim — hear you? — these French — or Moorish — new friends — of our King's — will not give up yet! Get her out of the city before it is noised about! Simon?"

"Aye." How he was to bring this about he knew not.

Erlend lay a little with his eyes shut. Then he said, with a kind of smile:

"I thought last night — of the time she bore our eldest son — she was in no better case than I — if a man may judge by her crying. And if she has been able to bear it — seven times — for the sake of our joys — I trow that I can — "

Simon was silent. The fearful shrinking he felt — from looking into life's deepest secrets of torment and of joy — Erlend seemed yet to have no touch of. He played with the worst and the loveliest things as simply as a guileless boy, whose friends have brought him with them to a bordel, drunken and curious. . . .

Erlend shook his head impatiently:

"These flies — are the worst — Methinks they are the foul fiend himself — "

Simon took his cap, and smote high and low at the thick clusters of blue-black flies, so that they flew up in the air in buzzing noisy clouds — and trampled furiously into the mud of the floor those that fell stunned. It could not avail much, for the window-hole in the wall stood open — the winter before, it had been closed with a wooden shutter with a bladder-covered port-hole, but that had made the room too dark.

But he was still at this when Olav Kyrning came back

with a priest bearing a drinking-cup. The priest lifted Erlend's head, and stayed it while he drank. A great deal of the liquor ran out into his beard and down his neck, and he lay quiet and untroubled as a child when the priest afterwards wiped it away with a cloth.

Simon felt his whole being in a ferment; the blood thumped and thumped in his neck below the ears, and his heart beat strangely and unsteadily. He stood a moment gazing back from the doorway at the long body outstretched beneath the cloak. The flush of fever came and went now in waves over Erlend's face; he lay with half-open glittering eyes, but he smiled to his brother-in-law, the shadow of his strangely boyish smile.

The next day, as Stig Haakonssön of Mandvik sat at the breakfast-board with his guests, Sir Erling Vidkunssön and his son Bjarne, the hoof-beats of a single horse were heard in the courtyard. The next moment the door of the hall was flung open, and Simon Andressön came swiftly towards them. He wiped his face with his sleeve as he came — he was splashed to the neck with mud from his ride.

The three men rose to meet the comer, with little outbursts, half of greeting, half of wonder. Simon answered not their greetings — he stood leaning on his sword, both hands upon the hilt, and said:

"Would you hear strange tidings? They have taken Erlend Nikulaussön and stretched him on a rack — some foreigners the King sent to put him to the question — "

With a cry the men gathered round Simon Andressön. Stig smote one hand into the other:

"What has he said — ?"

At the same moment both he and Bjarne Erlingssön turned, as though unwittingly, towards Sir Erling. Simon burst out into laughter — he laughed and laughed.

He sank down on the chair that Bjarne Erlingssön had drawn forward for him, took the ale-bowl that the youth proffered him, and drank greedily.

"Why laugh you?" asked Sir Erling sternly.

"I laughed at Stig." He sat a little bent forward, with his hands resting on his muddy thighs — yet once or twice again little bursts of laughter came from him. "I thought

— we are sons of nobles, all of us here — I had looked that
you would be so wroth that one of our fellows should be
so dealt with, that you would have asked first how such
things can be. . . .

"I cannot say that I know to a hair how the law stands
in such matters. Since my lord King Haakon died, it hath
been enough for me that I owed him that came to his
throne my service when he listed to call for it, in war
and in peace — else have I dwelt in quiet upon my estates.
But I cannot see aught else than that in this case against
Erlend Nikulaussön there have been unlawful doings. His
fellows had sifted his case and given judgment in it — with
how much right they doomed him to death, I know not
— then was he offered reprieve and safe-conduct till he
could be brought to a meeting with the King, his kinsman
— that perchance he might grant Erlend grace to make his
peace with him. . . . Since has the man lain in the tower
of Akers Castle nigh on a year, and the King has been
abroad well-nigh all that time — some letters have passed
to and fro — naught has come of it. Then he sends hither
some varlets — Norsemen they are not, nor of the King's
guard — and dares to put Erlend to the question in such
wise as none ever heard that a Norseman with a Guardsman's rights was dealt with before — this while there is
peace in the land, and Erlend's fellows and his kinsmen
are gathering at Tunsberg to honour the King's wedding. . . .

"What think you of this, Sir Erling?"

"I think — " Erling sat down on the bench over against
him. "I deem that you have set the matter forth clearly
and plainly, Simon Darre, as it stands. I see not that the
King can do aught but one of three things: Either must he
let Erlend pay the penalty according to the doom given
at Nidaros — or he must choose out a new court of
Guardsmen and have the case against Erlend set forth
by a man who bears not the knightly name, and they
must doom Erlend to outlawry with such respite as the
law allows for him to remove himself from King Magnus'
realm — or he must grant Erlend grace to make his peace
with him. And that would be the wisest thing that he
could do.

"This matter seems to me now so plain that whomso-

ever you will lay it before at Tunsberg will join with you
and take up your cause. Jon Haftorssön and his brother
are there. Erlend is their kinsman no less than the King's.
And the Ogmundssön brothers must see that this is injustice and folly. 'Twere best you went first to the Lord
Marshal — move him and Sir Paal Eirikssön to call a
meeting of the sworn King's-men that are in the town
and that seem fittest to take the matter in hand — "

"Will not you and your kinsmen go with me, my lord?"

"We mean not to go to the wedding-feast," said Erling
shortly.

"The Haftorssöns are young— and Sir Paal is old and
ailing — and the others — You know best yourself, my
lord, — they doubtless have some small power, through
the King's favour and such-like, but — Erling Vidkunssön,
what are they all beside you? You, sir, you have such
power in this land as no other chief has had since — I know
not when. Behind you, sir, are the old houses that folk
in this land have known man by man, so far back as
record goes of evil times and good times in these our
country-sides. On the father's side — what is the birth of
Magnus Eirikssön or Haftor of Sudrheim's sons beside
yours — are their riches worth naming beside yours?
These counsels you give me — all this will take time, and
these Frenchmen are in Oslo, and you may stake your
soul that they will not give over. . . . Olav Kyrning has
sent letters, and all gentlemen he could find to join with
them, the Bishop promised he would write — but all this
unrest and strife, Erling Vidkunssön, *you* could end it, in
the same hour you stepped forth before King Magnus.
You stand foremost among the heirs of those that ruled
this land in the old age — the King knows that you would
have us all at your back." . . .

"I can scarce say that I marked as much some time
back," said Erling bitterly. "You speak out warmly for
your brother-in-law, Simon — but can you not understand? *Now* I cannot move. It would be said: the very
moment they put such duress on Erlend that a man might
fear he would not be able to keep his tongue between his
teeth — that moment I came forward!"

There was silence for a while. Then Stig asked again:
"Has — Erlend spoken?"

"No," answered Simon impatiently. "He has held his tongue. And I trow he will go on holding it. Erling Vidkunssön," he said beseechingly, "he is your kinsman — you were friends — "

Erling breathed short and heavily once or twice.

"Aye. — Simon Andressön, have you fully understood *what* Erlend Nikulaussön had undertaken? To put an end to this sharing of our King with the Swedes — this way of rule that never has been tried before — that seems to bring more and more hardships and troubles on this land with each year that passes; to bring us back to the kingship that we knew of old, and that we know brings welfare and good fortune. See you not that this was a wise and a bold counsel — and see you not that this counsel can now hardly be taken up by others after him? He has ruined the cause of Knut Porse's sons — and other men of the kingly house there are none for the folk to rally round. You will say, mayhap, had Erlend carried through his intent and brought Prince Haakon to Norway, then had he played into *my* hands. Much further than the boy's landing had these — young boys — scarce been able to carry forward their plans without having need of prudent men to come forward and work out what remained to do. So it is — I dare avow it. Yet God knows that I gained nothing — rather had I to set aside the care of my own affairs — in those ten years when I lived in disquiet and toil and strife and troubles without end — some few men in this land have understood so much, and with that I must e'en be content." He struck his hand hard against the table. "See you not, Simon, that the man who had taken on his shoulders such mighty plans that none knows whether the welfare of us all in this land and of our children for long ages to come was not the stake in them — and flung it all from him with his breeches on a harlot's bedside! — God's blood! — he would be full well served with the measure that was dealt to Audun Hestakorn?"

He went on in a little more quietly:

"Yet would I be fain that Erlend should be saved; and believe not that I, too, am not angered at the tidings you have brought us. And I deem that, should you follow my counsel, you will find men and enough to join with you in this matter. But I believe not that my company would be of such great help to you that for the sake of it I should do well to come uncalled before the King."

Simon rose, stiffly and heavily. His face was streaked and grey with weariness. Stig Haakonssön went over and took him by the shoulders — now he would have some food brought; he had but bided till they had had their talk out before having serving-folk in. But now Simon must strengthen himself with meat and drink, and must sleep on top of it. Simon thanked him — he must ride on in a little while, if Stig could lend him a fresh horse. And would he give shelter to his man, Jon Daalk, to-night? — Simon had had to leave the man behind on the road, for his horse could not keep up with Digerbein.* Aye, he had ridden the most of the night — he had thought, for sure, that he knew the road out hither well enough — but yet he had gone astray more than once.

Stig bade him stay till to-morrow, and he would ride with him himself — at the least a part of the way — aye, and he might as well go on to Tunsberg with him too. . . .

Simon said:

"Here is naught more for me to stay for. I would but go over to the Church — seeing that I am here once more, I would yet fain say a prayer where Halfrid lies — "

The blood rushed and tingled in his weary body; his heart beat deafeningly. It was as though he must drop down headlong; he was as one but half awake. But he heard his own voice say, calmly and evenly:

"Will not you bear me company, Sir Erling? I know she held you dearest of all her kinsmen."

He looked not at the other, but felt him stiffen. In a little he heard, through the rushing and singing of his own blood, Erling Vidkunssön's clear and courteous voice:

"That will I, willingly, Simon Darre. — It is rough weather," he said, as he buckled on his sword and threw a thick cloak about his shoulders. Simon stood still as a stone till the other was ready. Then they went out.

Without, the autumn rain poured down, and the mist drove in from the sea so thickly that they could scarce see more than a couple of horses' lengths over the fields

---

* Digerbein = Big-legs.

and the yellow tree-clumps that bordered the path on each side. It was no long way to the church. Simon fetched the key from the chaplain's house near by — he was glad when he saw they were new folk, come since his time, since so he was spared much talk.

It was a little stone church with a single altar. Unheedingly Simon saw again the same pictures and ornaments he had seen so many hundred times, while he knelt by the white marble tomb a little way from Erling Vidkunssön, saying over his prayers, and crossing himself where 'twas fit — without knowing what he did.

He understood not himself that he had been able to try this. But now he was in the midst of it. Of what he should say, he had no guess — but, sick with horror and shame at himself as he was, he knew that he would make the trial at all costs.

He remembered the ageing woman's white, suffering face deep in the half-darkness of the bed, her lovely gentle voice — that afternoon when he sat on her bedside and she told him. It was a month before the child came — and she herself looked that it should cost her life — and she was willing and glad to buy their son so dear. The poor little soul that lay here beneath the great stone in a little coffin by his mother's side — No, no man could do what he had meant to do. . . .

But Kristin's white face. She knew what had befallen, when he came home from Akersnes that day. Pale and calm she was as she spoke of it and questioned him — but he had seen her eyes in one short glimpse, and he had not dared to meet them after. Where she was now, or what she had done, he knew not — whether she had stayed in her lodging or was with her husband, or whether they had prevailed on her to go to Skogheim; he had left it in Olav Kyrning's and Sira Ingolf's hands — he could do no more, and he deemed that he must lose no time. . . .

Simon knew not that he had hidden his face in his hands. Halfrid — there is naught in it of shame or of sin, my Halfrid. — And yet — what she had said to him, her husband — of her sorrow, and of her love that had made her stay on under the old devil's roof. Once already had he killed his child under its mother's heart — and she had stayed on with him because she would not tempt her dearest love. . . .

Erling Vidkunssön knelt, his colourless, clear-cut face showing no sign. His hands he held close in to his breast, with the palms pressed together; from time to time he crossed himself with a quiet, supple, gracious gesture, then brought his finger-tips again together.

No, thought Simon. This was so hateful a thing that no man could do it. Not even for Kristin's sake could he do it. — They rose together, made obeisance to the altar, and went down the nave; Simon's spurs jingled a little at each step he took on the stone pavement. As yet they had spoken no word together since they had left the manor, and Simon knew not at all what would now come.

He locked the church-door; and Erling Vidkunssön walked ahead through the graveyard. Under the little roof of the lich-gate he stopped. Simon came up; they stood a little in the shelter before going out into the pouring rain.

Erling Vidkunssön spoke quietly and evenly, but Simon felt the dull, measureless rage muttering deep within the other — he dared not look up.

"In the devil's name, Simon Andressön, what mean you by — devising — this?"

Simon could not answer a word.

"Think you that you can threaten me — force me to do your will — because, maybe, you have heard some lying rumours of things that befell when you were scarce yet weaned from the breast — ?" His rage growled nearer the surface now.

Simon shook his head:

"I thought, my lord, when you called to mind her who was better than the purest gold — mayhap you might take pity on Erlend's wife and children."

Sir Erling looked at him — he made no answer, but began stripping moss and lichen from the stones of the churchyard wall. Simon swallowed, and wet his lips with his tongue:

"I scarce know what I thought, Erling Vidkunssön — maybe that when you remembered her that suffered all those evil years — without other comfort or help than God alone — that then you would help these many unhappy beings — for you can! — since you could not help her. If you have repented at any time that you rode away from Mandvik yonder day and let Halfrid remain behind in Sir Finn's power — "

"But I have not!" Erling's voice was piercing now. "For I know that *she* never did — but this I trow *you* could not understand. For had you ever understood for one hour how proud she was, the lady you won to wife" — he laughed in his wrath — "then had you not done this. I know not how much you know — but you may as lief know this. They sent me — for Haakon lay sick then — to fetch her home to her kindred. Elin and she had grown up together as sisters — they were well-nigh of an age, though Elin was her father's sister; — we had — things had come about so that, had she come home from Mandvik, we had been forced to meet daily and hourly. We sat and talked, a whole night through, in the balcony of the dragon-house — every word that was spoken both she and I can answer for to God on the day of doom. And then let *Him* answer *us*, why it should have been so —

"Though, indeed, God rewarded her holiness in the end. Gave her a good husband to comfort her for the one she had had before — a whelp of a boy like you — who lay with her serving-women in her own house — and had her bring up your bastards — " He flung away the ball of moss he had kneaded together.

Simon stood motionless and dumb. Erling peeled off a flake of moss again and flung it away:

"I did what *she* bade me. Have you heard enough? There was no other way. Wherever else in the world we might have met, we had — we had — 'Adultery' is no fair word. 'Incest' — is yet uglier — "

Simon moved his head in a stiff little nod.

He felt it himself — it would be laughable to say what he thought. Erling Vidkunssön had been a man in the twenties, courtly and gallant. Halfrid had loved him so that she would fain have kissed his footprints in the dewy grass of the courtyard that morning in spring. *He* was an ageing hulking, ugly farmer — and Kristin? Never, for sure, would the thought come in her head that there would be peril to the soul of either, should they live under

one roof for twenty years. Surely he had learnt to understand that well enough. . . .

So he said in a low voice, almost humbly:

"She had not the heart to suffer the innocent child,
even though 'twas her woman's child by her husband, to
fare ill in the world. *She* it was that prayed me to do it
right and justice so far as lay in my power. Oh, Erling
Vidkunssön — for Erlend's poor innocent wife's sake —
She will grieve to death. Methought I could not leave
any stone unturned in seeking help for her and all her
children — "

Erling Vidkunssön stood leaning against the gate-post.
His face was calm as it was wont to be, and his voice cool
and courteous, when he spoke again:

"I liked her well, Kristin Lavransdatter, the little I have
seen of her — a fair and stately woman she is — and I have
told you already, Simon Andressön, I deem full surely
that you will find help if you will follow my counsel. But
I understand not rightly what you mean by this — strange
device. You surely cannot think that because I had to
suffer my father's brother to rule the matter of my wedding, being then in my nonage, and because the maid I
liked best was betrothed elsewhere when we first met —
And so innocent as you say, I trow Erlend's wife is not,
either. Aye, you are wedded to her sister, I know it; but
you and not I have brought about this — strange parley —
and so you must suffer that I name it. I mind me there
was talk enough about it, the time Erlend was wed with
her — 'twas against Lavrans Björgulfssön's counsel and his
will that that bargain came about; but the maid had
thought more of having her own will than of obeying her
father or guarding her honour. Aye, she may be a good
wife none the less — but she *won* Erlend after all, and
they have doubtless had their time of joy and mirth. I
trow that Lavrans had never much joy of that son-in-law
— *he* had chosen another man for his daughter ere she
came to know Erlend — she was promised in marriage, I
know — " He stopped short, looked at Simon a moment,
and turned his head aside in some confusion.

Flushing red with shame, Simon bent his head on his
breast, but he spoke, none the less, low but firmly:

"Aye; she was promised to me."

For a moment they stood, not venturing to look at each other. Then Erling Vidkunssön threw away the last ball of moss he had gathered, turned, and went out into the rain. Simon was left standing alone — but when the other was some way off in the mist, he stopped and beckoned impatiently.

Then they went back together, as silently as they had come. When they had well-nigh reached the manor, Sir Erling said:

"I will do it, Simon Andressön. You must wait until to-morrow; then we can ride in company, all four together."

Simon looked up at the other — with a face all drawn with pain and shame. He would have given thanks, but could not; he had to bite his lip hard, his lower jaw trembled so violently.

As they were passing through the door of the hall, Erling Vidkunssön touched Simon's shoulder, as it were by chance. But each knew that neither of them dared look at the other.

Next day, when they were making them ready for the journey, Stig Haakonssön pressed Simon to let him lend him clothes — Simon had brought no change of garments with him. Simon looked down at himself — his man had brushed and cleaned up his dress, but it had suffered past remedy in his long ride through foul weather. But he slapped himself on the thighs:

"I am too fat, Stig. — And I go not thither to be a guest at the feasting."

Erling Vidkunssön stood with one foot on the bench, while his son buckled the gilded spur on it — it seemed as though Sir Erling tried to keep his serving-folk at a distance to-day as much as might be. The knight laughed in an oddly vexed fashion:

" 'Twill do no hurt, I trow, if it should show on Simon Darre that he has not spared himself in his kinsman's service, but bursts right in from the high-road with his bold and subtle speech. He is no tongue-tied loon, this one-time kinsman of ours, Stig. One thing only I fear — that he may not know himself when he should stop —"

Simon stood there, flushing darkly, but he said no word.

In all that Erling Vidkunssön had said to him since the
day before, he marked a grudging mockery — and a
strange, unwilling kindness — and a firm will to see this
matter through — since, once and for all, he had taken
it up.

So they rode with him north from Mandvik, Sir Erling,
his son and Stig, with, in all, ten fairly clad and well-
armed yeomen. Simon with his single follower thought
now he could have chosen to come to the meeting more
fittingly attended and equipped — Simon Darre of Formo
had no need to ride with his former kinsmen in the guise
of a small franklin that had sought their aid in his helplessness. But he heeded not much. He was so weary and
so broken with what he had gone through the day before,
that almost it seemed to him now he cared not what the
outcome of this journey might be.

Simon had ever averred that he put no faith in the ugly
rumours about King Magnus. He was no such saint but
that he could suffer a gross jest amongst grown-up menfolk. But when people stuck their heads together and
muttered shudderingly of dark and secret sins, he ever
grew ill at ease. And it seemed to him unseemly to
believe or to hearken to aught of the kind about the
King among whose sworn men he was counted.

Yet was he filled with wonder when he stood before
the young King. He had not seen Magnus Eiriksson since
the King was a child, and, in spite of his disbelief, he had
looked to find something womanish, soft or unhealthy
about him — but this was one of the properest young men
Simon had set eyes on — and he looked manly and kingly
too, despite of his youth and slender fineness.

He wore a flowing robe of light blue shot with green,
falling to his feet, and girt about his slim waist with a
gilded belt, and he bore his tall, lean body with exceeding
grace in the heavy dress. King Magnus had light hair,
which lay smoothly on his well-formed head, but was
cunningly curled at the ends, so that it seemed to toss and
wave about his neck's broad, free-standing pillar. His
features were fine and boldly cut, the hue of his skin
fresh, with red cheeks and a yellowish tingle of sunburn;
he had clear eyes and an open look. He bore him fairly

and with winning gentleness as he greeted his liege men. Then in a while he laid his hand on Erling Vidkunssön's sleeve and drew him some steps apart from the others, while he thanked him for his coming.

They talked together awhile, and Erling let fall that there was a special matter wherein he had to crave the King's grace and bounty. The royal ushers then set a chair for the knight in front of the King's high seat, showed the other three men to places somewhat further down the hall, and then went out.

Without effort Simon seemed to have found again the mannerly and courtly bearing he had learned in his youth, and, since he had yielded and taken from Stig the loan of a long brown dress of state, in outward looks, too, he differed in no wise from the other men. But as he sat there he felt as though he were in the midst of a dream — he was and he was not the same as yonder young Simon Darre, the quick-witted and *kurteis* son of a noble knight, who had borne napkin and taper before King Haakon in Oslo Palace an endless tale of winters agone — he was and he was not Simon, the esquire of Formo, who had lived a life of freedom and cheer away north in the Dale through all these years — free from care, after a fashion, though he had known all his days that within him lay this glowing ember — but he turned his thoughts away from it. A dull and threatening humour of revolt rose up in the man — it was no willed sin or fault of his that he knew of, but fate, that had blown the embers into a blaze, so that he must strive and make no sign while roasting over a slow fire.

He stood up when all the others did so — King Magnus had risen:

"Dear kinsman," came his young, fresh voice, "methinks the matter stands thus. The Prince is my brother, but we have never tried to keep court together with a common guard — the same men cannot serve us both. Nor does it seem that Erlend had meant that things should continue in such wise — even though for a time he did hold his Wardenship under my hand, while at the same time he was Haakon's sworn man. But those of my men who would liefer follow my brother Haakon shall have leave from my service and freedom to seek their fortune in his house. Who they may be — that I mean to learn from Erlend's mouth."

"Then must you, Sir King, try if you can come to agreement with Erlend Nikulaussön in this matter. You must keep the promise of safe-conduct that you have given, and grant your kinsman an audience — "

"Aye, he is my kinsman and your kinsman, and Sir Ivar moved me to promise him safe-conduct — but *he* kept not his oaths to me, and *he* remembered not the kinship betwixt us." King Magnus laughed a little and again laid his hand on Erling's arm. "My kinsmen seem to be faithful to the byword we have in this land: None so unkind as kin. Now is it my full will to show my kinsman Erlend of Husaby grace for the sake of God and Mary Virgin, and for my own lady's sake; life and goods and leave here to abide, if he will make his peace with me — lawful respite to remove him from my lands, if he would betake him to his new master, Prince Haakon. — The same grace will I grant to every man who has been leagued with him — but I will know who they are, and which of my men dwelling up and down this land of ours has been a false servant to his lord. What say you, Simon Andressön? — I know that your father was my grandfather's trusty henchman; you yourself served King Haakon with honour — think you not that I have the right to make inquiry in this matter?"

"I think, my lord King" — Simon stepped forward and again made obeisance — "that so long as your grace rules according to this land's law and custom, mercifully, you will surely never learn who the men may be that had planned to have recourse to lawlessness and treason. For as soon as the people of this land see that your Grace will hold fast to the right and justice that your forefathers have set up, of a surety no man in this realm will think of troubling the peace. And those will be silent and will bethink them again, to whom for a time it may have seemed hard to believe that you, my Lord, young as you are, could rule two great kingdoms with wisdom and strength."

"It is so, my lord King," put in Erling Vidkunssön. "No man in this land has thought of denying you obedience in aught that you may command rightfully — "

"Have they not? Then you deem, maybe, that Erlend has not been guilty of disloyalty and treason — when we look more closely into the case?"

For a moment Sir Erling seemed at fault for an answer, and Simon took the word:

"You, Sir, are our lord — to you each man looks to punish lawbreaking by the law. But if you should follow where Erlend Nikulaussön has led the way, it might well befall that the men whose names you now so hotly seek to know should come forth and name them aloud, or other men who may begin to ponder over the rights and wrongs of this matter — for much talk will there come to be of it if your Grace should deal as you have threatened with a man so well-known and so high-born as Erlend Nikulaussön."

"What mean you, Simon Andressön?" said the King sharply — he grew red as he spoke.

"Simon means," Bjarne Erlingssön broke in, "that it might do your Grace an ill service if folk should begin to ask why Erlend must suffer such dishonour as the law warrants all men against, save thieves and nithings.* They might come then to think on King Haakon's other grandsons — "

Erling Vidkunssön turned sharply on his son — he looked angry — but the King only asked dryly:

"Count you not traitors and rebels as nithings?"

"None *call* them so, sir, if their plans speed well," answered Bjarne.

For a moment all stood silent. Then Erling Vidkunssön spoke: "Whatever Erlend should be called, my lord, it beseems not that you should override the law to come at him — "

"Then should the law be mended in this matter," said the King vehemently. "If 'tis so that I have no power to get me by all means the knowledge of how folk mean to keep faith with me — "

"Yet can you not act upon a mended law before it is changed," said Sir Erling doggedly, "without oppressing the folk of the land — and that folk has ever found it hard to use itself to oppression from its Kings."

---

* See Note 18.

"I have my knighthood and my sworn King's-men to back me," answered Magnus Eirikssön with a boyish laugh. "What say you, Simon?"

"I say, my lord — it might well prove that that was no such sure backing — to judge by the measure the knight-hood and the nobles of Denmark and Sweden have dealt their Kings when the commons had no strength to back up the kingly power against them. But if your Grace be set upon such counsels, then would I pray that you will loose me from your service — for then would I liefer be found among the common folk."

Simon had spoken so calmly and soberly that it seemed as though the King at first understood not his meaning. Then he laughed:

"Is this a threat, Simon Andressön? — Is it so that you would throw down your glove to me?"

"That must be as you will, my lord," said Simon as evenly as before; but he took his gloves out of his belt and held them in his hand. Then young Bjarne bent forward and took them:

"These are not seemly wedding-gloves for your Grace to buy!" He held the thick, worn riding-gloves in the air and laughed. "If it should come out, Sir, that you are seeking for such gloves, you might well have proffered you all too many of them — and all too cheap."

Erling Vidkunssön uttered a cry. With a sharp move-ment he seemed to sweep the young King to one side, and the three men to another; and he drove the men down the hall toward the door:

"I must speak with the King alone."

"No, no! I would speak with Bjarne!" cried the King, running after. But Sir Erling pushed his son out with the others.

They loitered about awhile in the castle yard and on the hill outside — none of them spoke a word. Stig Haakonssön looked doubtfully, but held his tongue as he had done throughout; Bjarne Erlingssön went about all the time with a little, hidden smile. In a while Sir Erling's weapon-bearer came and prayed them from his master to wait for him at their lodging— their horses were in the castle yard.

Afterward they sat in the inn. They were shy of speech

about what had just befallen — at last they fell into talk of their horses and hounds and hawks. The end of it was that Stig and Simon sat far on into the evening telling stories about women. Stig Haakonssön had always great store of such tales, but with Simon the worst was that most of those he called to mind Stig straightway began to tell, and 'twas ever so that either the thing had happened to himself, or it had befallen of late at some place near Mandvik — even if Simon remembered having heard the tale in his boyhood from the house-carls at home at Dyfrin.

But he chuckled and laughed as heartily as Stig. From time to time it was as though the bench rocked beneath him as he sat there — he was afraid of something, but dared not think what it was. Bjarne Erlingssön laughed quietly, drank wine and munched apples, fiddled with the hood of his cape, and told now and then a little snatch of a tale — they were the worst of all, but they were so cunningly veiled that Stig did not understand them. Bjarne had heard them from a priest in Björgvin, he said.

At last Erling came. His son went to meet him and take from him his outer garments. Erling turned angrily on the youth:

"You!" He flung his cloak into Bjarne's hands — and there flitted across his face, as though against his will, the shadow of a smile. He turned to Simon:

"Aye — now you must be content, Simon Andressön! I make no doubt that now you may safely hope the day is not far off when you shall sit in peace and comfort together on your neighbouring manors — you and Erlend — and his wife and all their sons."

Simon had grown a shade paler when he stood up and thanked Sir Erling. — He knew now what the fear was that he had not dared to look in the face. But now there was no way out. . . .

About fourteen days after, Erlend Nikulaussön was set free. Simon, with his two men and Ulf Haldorssön, rode out to Akersnes and fetched him.

The trees were almost bare already, for it had blown hard the week before. A black frost had set in now — the earth rang hard under the horses' hoofs, and the fields were wan with rime, as they rode in towards the town. It looked as though snow were coming — the heavens were evenly overspread with cloud and the daylight was sullen and chilly grey.

Simon had seen that Erlend dragged one foot a little, as he came out into the castle yard, and he seemed somewhat stiff and unhandy in mounting. He was very pale, too. He had had his beard taken off and his hair cut and made trim — his upper face was now a dull yellow, and, below, the blue of his shaven beard showed against white cheeks and chin; there were hollows beneath his eyes. But he made a stately figure in his long dark-blue robe and cloak, and, as he said farewell to Olav Kyrning, and made gifts of money to the men who had guarded him and brought him food in prison, he bore him like a chieftain parting from the house-folk at a wedding-feast.

At first, as they rode, he seemed to feel cold; he shivered more than once. Then a little colour came into his cheeks — his face lighted up — it was as though sap and life were welling up in him. Simon thought: sure it was Erlend was no easier to break than a willow wand.

They came to the lodging, and Kristin went to meet her husband in the courtyard. Simon tried not to look thither, but he could not forbear.

They gave each other their hands and exchanged some words, in quiet, clear tones. They managed this meeting in the sight of all the people of the house in fair and seemly wise enough. Only that both flushed red, looked at each other a second, and then both dropped their eyes. Then Erlend proffered his hand again to his wife, and they went together towards the loft-room where they were to dwell whilst they were in the city.

Simon turned towards the room where he and Kristin had lived till now. Then Kristin turned at the lowest step of the stairway and called to him, in a wonderful ringing voice:

"Will you not come, brother-in-law? — get you some food first — and you, Ulf!"

She seemed so young and supple as she stood there, turned a little from the hips, and looked back over her shoulder. As soon as she came to Oslo, she had begun to fasten her head-gear in another fashion. Here in the

south it was only small farmers' wives who wore the
linen head-dress in the old-world way she had used ever
since she was wedded; tight round the face like a nun's
coif, with the falls fastened cross-wise over the shoulders
so that the neck was quite hid, and with many folds on
the sides and over the knot of hair at the back of the
head. In Trondheim it was accounted, so to speak, a
token of piety to set up the coif in this fashion, which
Archbishop Eiliv praised always as the most fitting and
modest way for wedded women. But so as not to be too
much marked out, she had taken up now with the fashion
of these southern parts: the linen cloth laid smoothly
over the crown of the head and hanging straight down
behind, so that the front hair showed, and the neck and
shoulders were free — and then it was the proper thing
that the plaits should be but tied up so that they did not
show under the edge of the coif, while the linen fitted
close above, throwing out the form of the head. True,
Simon had seen this before and deemed that it became
her well — yet he had not seen before how young it made
her look. And her eyes shone like stars.

Farther on in the day many folk appeared to greet
Erlend— Ketil of Skog, Markus Torgeirssön, and later
in the evening Olav Kyrning himself, Sira Ingolf and
Canon Guttorm, a priest of St. Halvard's Church. When
the two priests came it had begun to snow — a slight dry
fine-grained drizzle — and they had missed the path and
come in among some burdocks — their clothes were full
of the burrs. Everyone set to work to pluck burrs off the
priests and their followers — Erlend and Kristin rid Canon
Guttorm of his — from time to time their faces flushed,
and their voices were strangely unsteady, as they jested
with the priest and laughed.

Simon drank much in the first part of the evening, but
he grew not at all light-headed with the drink — only a
little heavy in the body. He heard each word that was
said with unnatural sharpness. The others soon grew free-
spoken — none of them were friends to the King.

He was heartily sick of it all now 'twas all over. Foolish
prate it was that they babbled forth as they sat there —
loud-voiced and heated. Ketil Aasmundssön was somewhat

simple, and his brother-in-law Markus was none too wise either; Olav Kyrning was a right-minded and sensible man, but short-sighted — neither did the two priests seem to Simon too clear-witted. They all sat there and listened to Erlend and chimed in with him, and he grew more and more like himself as he had ever been, wild and reckless. He had taken Kristin's hand now, and laid it over his knee, and sat playing with her fingers — they sat so that their shoulders touched. Now the deep flush showed clear through her skin, she could not take her eyes from him — when he stole an arm about her waist, her mouth trembled, so that she had much ado keeping her lips shut. . . .

Then the door opened and Munan Baardssön stepped in.

"Last came the great bull himself!" shouted Erlend, laughing, and leapt up to meet him.

"Help us God and Mary Virgin — I believe you care not a straw, Erlend," said Munan in vexation.

"Aye, deem you, then, it would help aught to whimper and sorrow now, kinsman?"

"Never have I seen the like of you — all your welfare have you cast away — "

"Aye — for I was never the kind to go unbreeched to hell, to save my breeches from the burning," said Erlend, and Kristin laughed softly and dizzily.

Simon laid him down over the table, his head between his arms. If only they might think he was so drunken already that he had fallen asleep! — he would fain be left in peace.

Nothing was otherwise than he had looked that it would be — should have looked it would be, at the least. Nothing — not even she. Here she sat, the only woman amongst all these men — as gentle and bashful and fearless and secure as ever. Even so had she been yonder other time — when she betrayed him — shameless or innocent, he knew not which. Oh, no, 'twas not so either; she had not been so secure, she had not been shameless — behind the calmness of her bearing she had not been calm. — But that man had bewitched her — for Erlend's sake she would gladly tread over red-hot stones — and she had trodden over him as though she knew not he was aught but a cold stone.

Oh, all this was folly — her mind was set on having her own way, and she heeded naught else. Let them have their joyance, he need not care a jot. What mattered it to him if they had seven sons more, so that there would be fourteen to part betwixt them the half of Lavrans Björgulfssön's estate? It looked not as though *he* would need to be careful and troubled for his children — Ramborg was not so quick at bearing children as her sister — but as though in due time he would leave behind his children and children's children in riches and power. But 'twas all one to him — to-night. He would fain have drunk more — but he knew that to-night God's gifts would not cheer him — and then he would have had to raise his head and perhaps to join in the talk.

"Aye, you think, I trow, *you* were the man for Regent of the realm," said Munan scornfully.

"Nay, surely you must know we had meant that place for you," laughed Erlend.

"In God's name, heed your tongue, man — " The others laughed.

Erlend came over and touched Simon's shoulder:

"Are you asleep, brother-in-law?" Simon looked up. The other stood before him with a goblet in his hand. "Come, drink with me, Simon. You I have to thank most of all that I came off with my life — and, such as it is, 'tis dear to me, lad! You stood by me like a brother — had you not been my brother-in-law, I trow I had lost my head for sure. — And then could you have wed my widow — "

Simon sprang up. A moment the two stood looking at each other — Erlend grew white and sobered, his lips parted in a gasp. — With his clenched fist Simon struck the goblet out of the other's hand — the mead splashed on the floor. Then he turned and went out of the room.

Erlend stood there alone. Without knowing what he did, he dried his hand and wrist with the skirt of his coat, then looked behind him; the others had not marked aught. With his foot he thrust the goblet in under the bench — he stood still a moment — then went quietly out after his brother-in-law.

Simon Darre stood at the foot of the stairway — Jon

Daalk was leading his horses out of the stable. He made no movement when Erlend came down to him:

"Simon! Simon — I knew not — I knew not what I said!"

"You know it now."

Simon's voice was toneless. He stood quite still, not looking at the other.

Erlend looked about him, as at his wit's end. The moon showed dimly, a pale patch, through the veil of cloud; small hard grains of snow showered down on them. Erlend made a shivering motion.

"Where — where are you bound?" he asked dully, looking at the man and the horses.

"To seek me another lodging," said Simon shortly. "Maybe you can understand that *here* I care not to be — "

"Simon!" Erlend burst out. "Oh, I know not what I would not give if it could be unsaid — !"

"Nor I either," answered the other as before.

The door of the upper room opened. Kristin came out on the balcony with a lanthorn in her hand; she bent over the railing and threw the light down on them.

"Why stand you there?" she asked in a clear voice. "What would you without the house?"

"I felt I must go out and see to my horses — as 'tis the courtly fashion to say," answered Simon, laughing up to her.

"But — you have taken your horses out," she said, in laughing wonder.

"Aye — a man will do strange things in his cups," said Simon in the same tone.

"Well, come back now," she broke in, brightly and gladly.

"Aye. In a moment." She went in, and Simon called out to Jon to take back the horses. He turned towards Erlend — the man was standing there, strangely helpless in looks and bearing. "I shall come in a little. We must — try to bear as though this were unspoken, Erlend — for our wives' sake. But so much maybe you too can understand — you were the last man on earth that I would — would have had know of — this. And forget not that I am not so forgetful as you!"

The door above was opened again; the guests came out in a troop; Kristin was with them, and her woman, bearing the lanthorn.

"Aye," tittered Munan Baardssön, "the night is well worn already — and these two good folks would fain to bed, I trow — "

"Erlend — Erlend — Erlend!" Kristin had thrown herself into his arms the moment they stood alone within the loft-room door. She clung close and tightly to him. "Erlend — you look so sorrowful — " she whispered in fear, with her half-open lips close to his mouth. "Erlend — " She pressed the palms of her hands to his temples.

He stood a little with his arms laid loosely around her. Then, with a soft moaning sound in his throat, he crushed her to him.

Simon went across to the stable — he would have said somewhat to Jon, but he forgot it on the way. For a while he stood in the stable-door, looking up at the light-haze of the moon and the tumbling snow — it had begun now to fall in great flakes. Jon and Ulf came out and shut the door behind them, and the three men went together across to the house where they were to sleep.

NOTES

THE MISTRESS OF HUSABY

P. 4.   1. *Husaby (See Sketch Map)*

THE old manor of Husaby, comprising some thirty greater and smaller farms and homesteads, lies about twenty English miles southwest of Trondhjem (Nidaros). The head-quarter buildings are on a broad mountain slope above a little lake, about ten English miles from the nearest point on the Trondhjem Fjord (Birgsi), and between the great valleys of Guldal and Orkedal, which stretch southward up to the Dovrefjeld. The journey from Husaby to Nidaros would ordinarily be made by horse to Birgsi and thence by boat. The path followed by Kristin on her pilgrimage ran across the hills to the estuary of the Gula River, and thence across Bynes, the high promontory which shoots into the fjord between the Gula and Nid estuaries.

P. 5.   2. *Hall at Husaby (see Plan)*

The hall at Husaby was a large, ancient stone building, in the style of the Saga-times, when the chieftain and his house-carls dwelt and slept under one roof — the serving-women being quartered in a separate women's house (the "little hall" of this book).

Two rows of wooden pillars supported the roof. Between the line of pillars and the wall on each side was the sleeping-accommodation — two box-beds with doors at one end of the hall, and two broad fixed benches running the rest of the length of the hall. These benches were divided into sleeping-places for the warriors (originally called "rooms"), and were wide enough to admit of each man's keeping his belongings by him, while his weapons hung on the wall above him. As in the "hearth-room house" of the later mediæval manor, the room was heated from a hearth in the middle of the floor, and the only daylight came from the smoke-vent above this or from the door; but the hearth was much longer, and several fires were ordinarily kept burning on it, when artificial heat was required. On festival occasions long tables were set up — otherwise each man ate

his meals sitting in his "room," with his porridge-bowl or his
meat in his lap; or drank his beer sitting on the floor by the
hearth.

In later times, when a separate house was assigned to the
servingmen, the "rooms" in the hall were left free to be used,
when necessary, as sleeping-places for the family or guests. The
high seat was moved up to the east end of the hall, and tem-
porary tables were put up daily for the chief meal of the day
(supper).

Later, yet further changes were, of course, made. Thus the
plan shows the hall as arranged for the banquet described in
Part I, chap. iii. By this time, it will be seen, a smaller table
has been put up across the hall, before the high seat, with a
loose outer bench for the servants.

Above the "outer-room" and the "closet" at the lower end
of the hall, there was, at Husaby, a loft-room (Margret's
bower) reached by a ladder leading up from the hall.

P. 62.   3. *Gunnulf's Song*

Master Gunnulf must evidently have come upon an early
version of the old English ballad, "The Falcon," and have
adapted it freely, with an eye to edification.

P. 62.   4. *Lady Midwives*

The Church laws, as well as the custom, of mediæval Norway
(and Denmark) made it the duty of all married women to act
as midwives to the women of their neighbourhood. The house-
wives living in the neighbourhood (*grannekoner*) were bound
to come, each accompanied by a serving-woman (*gridkone*).
Professional midwives were unknown, but any woman who
gained a name for skill in midwifery was bound to go wherever
she was sent for within half a day's horseback journey. No fee
could be claimed for this assistance, but custom required that
the father should send the ladies and women away with gifts
when they left his house. The value of the gifts was propor-
tioned to the importance of the event — in the case of a first-
born son the father would be expected to show special gener-
osity.

When an heir to an estate was expected, the father should,
months before, have prayed every neighbouring housewife of
social standing to come to his wife's assistance. This was a
matter of practical importance, as the ladies' evidence might be
required to settle questions of inheritance. For instance, if the
mother and child both died, the ladies could testify whether the
child had been born alive and been christened, and whether it

had died before or after its mother. In the former case the wife's family inherited her share of the joint estate; in the latter the husband inherited from the child. A conclave of *sages femmes* of quality would always be a guarantee as to the identity, etc., of an heir, if doubts of any kind arose.

Thus Erlend's failure to invite the ladies of the country-side to his wife's lying-in was a culpably rash, as well as a scandalous, omission.

P. 106.   *5. Trondhjem Cathedral*

In the steep sand-bank by the River Nid, where King Olav's body had lain buried the first winter after his death at the battle of Stiklestad (A.D. 1030), a spring welled up, the waters of which were credited with healing powers. On this site, first marked by a small chapel, a Bishop's Church, completed about the year 1075, was erected, in honour of Christ and St. Olav, the shrine containing the saint's remains being transferred to it from the wooden church in which his successors had lodged it.

In 1152 Norway was organized as an independent ecclesiastical province, and the Archbishopric of Nidaros created, by Cardinal Nicholas Breakspeare, afterwards Pope Hadrian IV. Shortly after, the building of the great Cathedral was begun, on the site of the existing Bishop's Church. "The man who had the most far-reaching influence on the work," says Professor Nordhagen, "was Eystein Erlandson, the imperious and highly gifted third Archbishop (1161–1188)." The work was begun in the Norman style, but, on returning from a three years' exile in England (1180–1183)—a result of his quarrel with the no less imperious and forceful King Sverre—Eystein brought with him the new ideas in architecture which had shortly before reached England and were even then being put into practice in the building of the choir of Canterbury Cathedral. Later, the work continued to be influenced by the contemporary developments of Gothic in England. Completed, in its first form, in the fourteenth century (about the period of this book), the Cathedral was the wonder of the North, and was sought by thousands of pilgrims, come to visit the shrine of the saint and the wonder-working well. The press was greatest at the time of the great Olav's Festival—the 29th July. *Feginsbrekka* — "The Hill of Joy" — was the hill on the pilgrims' route from the top of which they first caught sight of the city and the Cathedral.

The first of numerous fires from which the Cathedral has suffered was that of 1328, referred to in this book. It was again swept by fire in 1432 and 1531, yet even as it stood half destroyed in 1567 it is described by a Norwegian writer of the

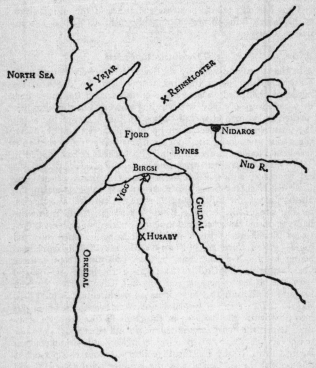

NORTH SEA

+ YRJAR

× REINSKLOSTER

FJORD

BYNES

NIDAROS

NID R.

BIRGSI

VIGG

GULDAL

× HUSABY

ORKEDAL

SKETCH MAP SHOWING HUSABY

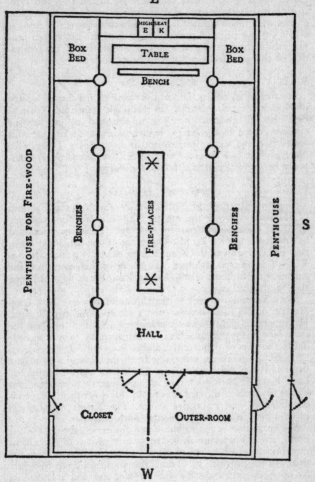

period as "the crown, the flower and ornament of the king-dom." There were further visitations by fire in 1708 and 1719, so that in the eighteenth and nineteenth centuries its original glories were much obscured. Very elaborate schemes of restora-tion were put in hand towards the end of the nineteenth cen-tury, and have in great part been carried out; though the renova-tion of the west front remains to be completed.

P. 116.   6. *Court of Six*

Practically all nobles, and a great part of the landed gentry of Norway, were *hirdmænd* (in this book, translated "Guards-men" or "King's-men"); that is to say, had at one time or another served in the King's household. Suits at law between such *hirdmænd* were referred, not to the ordinary *Things* (see Note 9), but to a special Court of Six (*seksmandsdom*) com-posed of six of the parties' fellow-*hirdmænd*.

P. 116.   7. *Norway, 1319–1335*

Magnus VII. (King of Norway 1319–1343) was the grandson of Haakon V., being the son of Haakon's daughter Ingeborg by Duke Eirik of Sweden. In 1318 Duke Eirik was murdered by his brother, King Birger. The party of the murdered Duke rose and drove away the King, making the baby prince Magnus King of Sweden in his place. The leader of their army was the gallant and handsome Danish knight, Knut Porse, who had been the Duke's loyal vassal.

Since Magnus was heir to the throne of Norway also, King Haakon on his death-bed foresaw a union of the two Crowns. He appointed eight Norwegian lords as a Council of State for the minor King, and made them swear solemnly to rule accord-ing to the laws of the country, and to keep out foreigners from all posts of influence, especially from the strongholds of the Crown. When, on the death of Haakon in 1319, Magnus be-came King of both countries, an agreement was entered into by the Councils of the two realms, providing for their independ-ent relations and respective duties. The King's mother was given a very influential position in both countries.

She very soon began to abuse her power in favour of Knut Porse, and the attachment of the young widow to her cham-pion became much talked of. Knut Porse's plan was to bring the then Danish province of Skaane (Scania in southern Sweden) by conquest under the Swedish Crown, and win a Dukedom for himself, so as to be in position to marry the Lady Inge-borg. When, in 1321, King Magnus'- little sister Eufemia was betrothed to a son of Duke Henry the Lion of Mecklenburg, in the presence of members of both the Swedish and Norwegian governments, a secret compact was made between the Duke and Sir Knut, by which the Duke pledged himself to furnish soldiers in support of the Skaane enterprise. Some Norwegian noblemen, among them Sir Munan Baardssön, who was an intimate friend of Sir Knut and Lady Ingeborg, attached their seals to this compact. And Lady Ingeborg, who had unlawfully carried the Great Seal out of Norway, tried by all means to raise funds for the war.

But in the summer of 1322 the Swedish lords met in parliament at Skara, and by a *coup d'état* took all power out of the hands of the lady; and in February, 1323, the Norwegian nobles, under the leadership of Archbishop Eiliv, gathered at Oslo and followed the Swedish example. A young lord of the highest descent in the land and of great wealth, Sir Erling Vidkunssön, was made Regent of Norway, with the title of *Drotsete* (here translated High Steward), with the Council of nobles to assist him.

As a result of the union of the two Crowns, Norway had been dragged into the wars between Sweden and Russia, and for some years from 1323 onwards the Russians made a series of destructive raids on the coasts of Northern Norway, coming as far south as Haalogaland (the modern Nordland's Amt). Sir Erling took measures to defend the country, and achieved some success, but he wanted money badly, as Lady Ingeborg had left him an empty treasury. He sought assistance from the Bishops, as the Russians were accounted heretics or worse; the Archbishop stood by him, but Bishop Audfinn of Bergen refused help. In the years following, Erling had several quarrels with the Bishop, who was a staunch partisan of Lady Ingeborg. In particular the Bishop defended the lady's rights to some estates in his diocese, when, in 1326, she further enraged her native country by marrying Knut Porse, who, by one of the vicissitudes in the struggle between the Danish King, Christopher II., and his nobles, had now become a Danish Duke. Knut Porse, however, died in 1330, leaving his widow with two little sons, Haakon and Knut.

In 1326 peace was made between Norway and Russia, on terms not unfavourable to Norway.

In 1330 King Magnus, now sixteen years old, was declared to be of age, and Erling Vidkunssön resigned. King Magnus took up his stepfather's plan of winning Skaane, stayed on in Sweden, and made an old antagonist of Sir Erling's, Paal Baardsen, his Chancellor for Norway and bearer of the Great Seal. But when, in 1333, Paal was chosen Archbishop in succession to Archbishop Eiliv, Magnus failed to appoint a new Chancellor; and, as the King still stayed on in Sweden, always in want of money and demanding supplies, and leaving Norway without any lawful government, a party among the Norwegian nobles, headed by Erling Vidkunssön and the King's young cousins, Jon and Sigurd Haftorssön, rose against him. The matter was, however, settled the same year without bloodshed; the leaders of the revolt were forgiven, and kept their position and titles, King Magnus appointed Sir Ivar Ogmundssön *Drotsete* and Chancellor for Norway, the Council's powers were strengthened, and Norway once more had a working government. This seems to have been all that was aimed at by Erling Vidkunssön, whom history represents as an upright, honourable, brave and sensible man, though somewhat lacking in the vigour which achieves great things.

Erlend Nikulaussön's subsequent attempt to separate the Crowns of Norway and Sweden by placing Lady Ingeborg's son Haakon on the Norwegian throne has escaped the notice of history, which, however, records the wedding of King Magnus at Tunsberg in 1335 to the Countess Blanche of Namur.

P. 122.   8. *War Levies*

The word translated "levy" or "war levy" is in the original *leding*. All landholders were bound to pay to the Crown an annual contribution (*leding*) towards the defence of the country. This was due in time of peace as well as in war-time; but they might be called on for additional voluntary assistance in emergencies.

Failure to pay the tax rendered the defaulter liable to fine. Lavrans undertakes to pay any fines that may be imposed on his tenants owing to their refusal to comply with the illegal demands made upon them under colour of the enforcement of the contribution.

P. 136.   9. *Things*

At the period of this book there were three classes of *Things* (popular assemblies):

I. The parish Thing (held ordinarily at regular intervals, but which could also be specially summoned) for the transaction of all sorts of local business. Appeals from its decisions could be taken to:

II. The county Thing (*Herreds,* or *Fylkesthing*), which was held in each *Fylke* (county) twice a year — in the middle of Lent, and three weeks after the return of the county representatives from the *Lagthing* (see below). These county assemblies were known as *sysselmandsthing,* as they were convened by the *sysselmand* (warden) of the county (see Note 12, *Wardens*). Lawsuits which were to be brought before a *Lagthing* must be announced at these county *Things;* and their sentences and decisions (arrived at by a jury, not by the *sysselmand*, who had no judicial functions) were appealable to the *Lagthing*.

III. The *Lagthing*. Norway was divided into four sections called *lagdömmer*, each of which had its annual *Lagthing*. These were known as *Frostathing*, *Gulathing*, *Eidsivathing* and *Borgarthing*, and, at the time of the story, all met on the same day, St. Botolph's day (17th June). The *Frostathing*, which represented the northern section of the country, met at Nidaros.

Each of the four *lagdömmer* was a complete legislative and judicial entity. The *Lagthing* was composed of representatives chosen by the *sysselmænd* from the members of the various county *Things*, and was attended by all the *sysselmænd* of the section, whose duty it was to follow the cases from their several counties, and to report the decisions, and any new legislation, to the county *Things* summoned by them to meet three weeks after their return from the *Lagthing*.

The chairman of the *Lagthing* was known as the *Lagmand*. His functions corresponded roughly with those of the Speaker of the early House of Commons in England.

P. 151. 10. *Fourteenth-Century Rome*

Master Gunnulf had, of course, visited Rome during the so-called "Babylonian exile" of the Popes at Avignon, which began in 1305.

P. 164. 11. *Land Measurements*

The words here translated "half a hide" are in the original *to maanedsmatsbol*, meaning literally "two months'-meat's-area" — *i.e.* twice as much land as will feed one man for one month.

P. 190. 12. *Wardens*

The word translated "Warden" is in the original *sysselmand*. The *sysselmand* was a high official in charge of a district, roughly corresponding to a county in England, his duties being those of a chief administrator, military commander and chief police officer. The appointment was made by the Crown from among gentlemen of distinction — at the period of the story a *sysselmand* would be usually, though not always, a knight. He had to maintain a certain number of armed men and subordinate officials. His remuneration varied in different cases — he might be paid directly by the Crown or remunerated by a share in fines and fiefs.

Among other duties, the *sysselmand* had to hold each year in Lent a "wapinschaw" (*vaabenting*), at which he reviewed the weapons in the possession of the men of his district, to make sure that each man had the weapons he was bound by law to possess, according to his station. He had also to choose the delegates from his district to the annual *Lagthing* of his section of the country (see Note 9, *Things*), and to attend at the *Lagthing* to follow all cases relating to his district. Within three weeks from his return home he had to call a *sysselmands-thing* and there communicate to the people all decisions and sentences of the *Lagthing* affecting his district, all new laws passed, etc. He had to account for the fines and other dues of the Crown to the Treasurer of his section of the country (see Note 15, *Treasurers*).

P. 235.   13. *Veöy*

A little island near Molde, where in the Middle Ages there was a small market town, with a couple of churches and a royal mansion.

Pp. 274, 291.   14. *Grace to Criminals*

When a man had committed any offence punishable with outlawry (such as manslaughter or the abduction of a woman), he might, on making a payment to the Crown, be given leave to remain at his home under the protection of the law till his case was judged.

P. 276.   15. *Treasurers*

For purposes of financial administration, Norway was divided into four Treasury Districts (*fehirdsle*), each under a Treasurer (*fehirde*). The Treasurer of the District which had its head-quarters at Tunsberg was a sort of Minister of Finance, supervising the whole.

P. 354.   16. *Safe-Conduct*

As a *Hirdmand* (Guardsman — see Note 6), Erlend has the right to be tried by a court of his peers within the boundaries of the *lagdömme* (section — see Note 9) where he is domiciled. Having appealed against the finding and sentence of the court on the ground of the illegality of its composition, and claimed a fresh trial, he is granted a safe-conduct in order that he may be taken outside the bounds of the *lagdömme* to be brought before the King. The effect of this is that, if he fails to make his peace with the King, he is entitled to be sent back in safety to his own *lagdömme*, there to be given a fresh trial or to suffer the execution of the sentence of the original court, according as his appeal is or is not admitted.

P. 354. 17. *Baagahus*

Now Baahus in Sweden, near Gothenburg. The boundary between Norway and Sweden at this time was the Göta River,
and Baagahus was the Norwegian frontier town.

P. 392. 18. *Immunities of Freemen*

This passage, translated as literally as possible, would run:
"why Erlend must not enjoy such personal immunities [*mandhelg*] as are the right of all men save thieves and nithings."

By the ancient laws every free man of Norway was guaran-
teed *mandhelg* — *i.e.* immunity from dishonouring bodily pun-
ishments and outrages against his person and honour. This
involved, in Saga-times, his right to avenge himself and his kin.

ABOUT THE AUTHOR

SIGRID UNDSET (born 1882) grew up in Oslo, Norway, the daughter of a Norwegian archaeologist, whose early death left his family in difficult circumstances. From 1899 to 1909, she supported her family as an office clerk, writing at night, and won her first success in 1912 with her third novel, *Jenny*, the story of an urban working girl. Other modern novels followed during her thirteen-year marriage to a painter, during which she also raised six children. But her interest in the Middle Ages gave rise to her masterpieces: the three-volume *Kristin Lavransdatter* (1920–22), and the four-volume *The Master of Hestviken* (1925–27). She was awarded the Nobel Prize for Literature in 1928. Undset was divorced in 1925, and turning her back on the liberal, feminist circles of her youth, converted to Roman Catholicism the same year. Her later novels (she published fourteen during her lifetime) returned to modern settings and were overtly religious in tone. During World War II, she escaped to the United States, having been a vocal and bitter opponent of Nazism. She returned to Norway to die in 1949.